New Carbon Based Materials for Electrochemical Energy Storage Systems: Batteries, Supercapacitors and Fuel Cells

NATO Science Series

ASeries presenting the results of scientific meetings supported under the NATO Science Programme.

The Series is published by IOS Press, Amsterdam, and Springer in conjunction with the NATO Public Diplomacy Division

Sub-Series

I. Life and Behavioural Sciences	IOS Press
II. Mathematics, Physics and Chemistry	Springer
III. Computer and Systems Science	IOS Press
IV. Earth and Environmental Sciences	Springer

The NATO Science Series continues the series of books published formerly as the NATO ASI Series.

The NATO Science Programme offers support for collaboration in civil science between scientists of countries of the Euro-Atlantic Partnership Council. The types of scientific meeting generally supported are "Advanced Study Institutes" and "Advanced Research Workshops", and the NATO Science Series collects together the results of these meetings. The meetings are co-organized bij scientists from NATO countries and scientists from NATO's Partner countries – countries of the CIS and Central and Eastern Europe.

Advanced Study Institutes are high-level tutorial courses offering in-depth study of latest advances in a field.
Advanced Research Workshops are expert meetings aimed at critical assessment of a field, and identification of directions for future action.

As a consequence of the restructuring of the NATO Science Programme in 1999, the NATO Science Series was re-organised to the four sub-series noted above. Please consult the following web sites for information on previous volumes published in the Series.

http://www.nato.int/science
http://www.springer.com
http://www.iospress.nl

Series II: Mathematics, Physics and Chemistry – Vol. 229

New Carbon Based Materials for Electrochemical Energy Storage Systems: Batteries, Supercapacitors and Fuel Cells

edited by

Igor V. Barsukov
Superior Graphite Co., Chicago, IL, U.S.A.

Christopher S. Johnson
Argonne National Laboratory, Argonne, IL, U.S.A.

Joseph E. Doninger
Dontech Global, Inc., Lake Forest, IL, U.S.A.

and

Vyacheslav Z. Barsukov
Kiev National University of Technologies and Design,
Kiev, Ukraine

Springer

Published in cooperation with NATO Public Diplomacy Division

NATO - Carbon Advanced Research Workshop and Conference
New Carbon Based Materials for Electrochemical Energy Storage Systems
Argonne, Illinois, U.S.A.
19–24 October 2003

A C.I.P. Catalogue record for this book is available from the Library of Congress.

ISBN-10 1-4020-4811-4 (PB)
ISBN-13 978-1-4020-4811-1 (PB)
ISBN-10 1-4020-4810-6 (HB)
ISBN-13 978-1-4020-4810-4 (HB)
ISBN-10 1-4020-4812-2 (e-book)
ISBN-13 978-1-4020-4812-8 (e-book)

Published by Springer,
P.O. Box 17, 3300 AA Dordrecht, The Netherlands.

www.springer.com

Printed on acid-free paper

Printed in the Netherlands.

TABLE OF CONTENTS

PREFACE

Carbonaceous materials play a fundamental role in electrochemical energy storage systems. Carbon in the structural form of graphite is widely used as the active material in lithium-ion batteries; it is abundant, and environmentally friendly. Carbon is also used to conduct and distribute charge effectively throughout composite electrodes of supercapacitors, batteries and fuel cells. The electronic conductive pathways are critical to delivering and extracting current out of the device. However, many challenges and the understanding of the role of carbon and its stability and efficiency in charge storage applications still exists. This NATO-ARW volume contains a diverse collection of papers addressing the role of carbon in some key electrochemical systems, both conventional and emerging. These papers discuss the latest issues associated with development, synthesis, characterization and use of new advanced carbonaceous materials for electrochemical energy storage. Such systems include: metal-air primary and rechargeable batteries, fuel cells, supercapacitors, cathodes and anodes of lithium-ion and lithium polymer rechargeable batteries, as well as nanocarbon materials of the future.

In the present volume, these papers originate from most of the lectures given at the NATO-Carbon Advanced Research Workshop and Conference (NATO-CARWC). Forty-one papers and six chapters are presented in this book. The order of papers in each chapter in this volume followed the order of lectures during the conference. This workshop and conference featured papers from leading researchers in the field of supercapacitors and batteries as well as papers from scientists and engineers from the former Soviet Union (Russia and Ukraine) and other Eastern European countries, such as Bulgaria and Poland. The workshop and conference was held at Argonne National Laboratory in Argonne, Illinois from October 20 to 23, 2003. A total number of registrants of approximately 90 were in attendance. Of these, approximately 21 Ukrainian and Russian delegates were present. Roughly 16 other presenters were from NATO and Mediterranean countries, and an additional 10 from the United States of America. The remainder participants came from national laboratories, government agencies, academia, and industry around the United States whom are interested in energy storage research and technologies.

We thank all the presenters who contributed to this book and appreciate their time spent in preparing their presentations for the conference and manuscripts for this volume. We acknowledge the National University of Technologies and Design (Kiev, Ukraine), Center for Research on Divided Materials (CRDM-Orleans, France), and Poznan University of

Technology (Poland) for their participation and assistance in identifying speakers from the Ukraine and Russia. We also thank all other registrants for coming and making the workshop and conference a success.

Finally we acknowledge the NATO Science Committee, and are grateful to the Civilian Research and Development Foundation (CRDF), which provided funding in the form of individual travel grants for the invited scientists and engineers from the Ukraine and Russia. In addition we would like to thank the NATO-Science for Peace Program, Superior Graphite Co., Argonne National Laboratory, the National Renewable Energy Laboratory, Lawrence Berkeley National Laboratory, Illinois Institute of Technology, Gas Technology Institute and Broddarp of Nevada, Inc. for financial sponsorship assistance.

We would like to specifically acknowledge logistical support from the following individuals: Laurie Carbaugh and Joan Brunsvold (Argonne National Laboratory), Maritza Gallego, Gabriela Uribe and Peter R. Carney (Superior Graphite Co.), Dr. Chris De Wispelaere (Science for Peace program, NATO), Dr. Fausto Pedrazzini (ARW Program, NATO), Dr. Michael Onischak and Michael Romanco (Gas Technology Institute), Professor Jai Prakash and Dr. Said Al-Hallaj (Illinois Institute of Technology), Ross Campbell (Next Steps to Market Program, U.S. Civilian Research and Development Foundation), Chris Hilton (Travel Grants Program, U.S. Civilian Research and Development Foundation).

Last but not least, the editors wish to express gratitude to those loved members of our families and friends, who inspired and supported us during preparation of this volume for press.

Igor V. Barsukov
Christopher S. Johnson
Joseph E. Doninger
Vyacheslav Z. Barsukov

ARGONNE NATIONAL LABORATORY

9700 SOUTH CASS AVENUE, ARGONNE, ILLINOIS 60439-4832

OFFICE OF THE DIRECTOR (630) 252-2481 / FAX (630) 252-7923

October 20, 2003

Dear Participants,

On behalf of Argonne National Laboratory, I would like to welcome you to Argonne National Laboratory and the NATO Carbon Advanced Research Workshop and Conference. This event, which brings together innovative scientists and engineers from many organizations and nations, is part of the *NATO Science for Peace* program initiative to encourage fruitful collaborations between former Warsaw pact countries and the United States and other Western nations. International collaborations like these have become especially important in today's climate, and I am pleased that so many leading experts from government, industry, and academia are in attendance.

I invite you to take advantage of the tours of the laboratory. As the nation's first national laboratory, Argonne has a unique mix of history and leading-edge technology throughout its 1,500-acre site. Events will come to a close with a private tour and dinner at the Museum of Science and Industry, one of Chicago's great treasures.

I am confident that this meeting will be a stimulating and productive one. As scientists and engineers dedicated to peaceful collaboration, I trust your vision and efforts will carry electrochemical technology to new heights.

Sincerely,

Hermann A. Grunder

Hermann A. Grunder
Director of Argonne National Laboratory

OPERATED by THE UNIVERSITY of CHICAGO for THE UNITED STATES DEPARTMENT of ENERGY

PHOTOGRAPHS AND COMMENTS

Group Photograph, Argonne National Laboratory, Argonne, IL USA (Courtesy C. Johnson)

Technical discussions during the poster session. Left to right: Professor R. Yazami (CALTEC / CNRS, USA-France), Dr. I. Goncharova (Dniepropetrovsk National Univ. of Railway Transport, Ukraine) and Professor M. Winter (Graz Technical Univ., Austria). Courtesy M. Gallego. →

Eastern European participants of the NATO-ARW during the downtown Chicago city tour. Courtesy Dr. V. Matveev. ↓

A satellite meeting during the NATO-ARW. ↑Planning of what later resulted in a start of a high profile U.S. Department of Energy Initiative for Proliferation Prevention project between the Argonne National Laboratory, Superior Graphite Co. (Chicago, IL, USA) and three ex-weapons institutes, and other organizations in Ukraine (Headquarters SGC, downtown Chicago).

PHOTOGRAPHS AND COMMENTS

Courtesy Prof. Doninger.

← At the renowned Chicago's Museum of Science and Industry, posing in the Space Exhibit are from left: Prof. J. Machnikowski (Wroclaw Univ. of Technology, Poland), M. Gallego and J. Chavez (Superior Graphite, USA), Y. Il'in (Kiev National University of Technology & Design, Ukraine), Prof. F. Beguin (CNRS, France), Dr. E. Frackowiak (Inst. of Chemistry & Technical Electrochemistry, Poland) and Dr. Neal White (Sion Power Research, Great Britain).

← Conducting a technical tour of Superior Graphite's Peter R. Carney Technology Center is one of the inventors of expanded graphite for batteries, Peter Zaleski, Lab Director (center); Listening left to right are: Prof. Y. Maletin (National Technical Univ. of Ukraine "KPI"), Dr. N. White (Sion Power Research, Great Britain), Dr. A. Hull (Argonne National Lab, USA), Mr. M. Walkowiak (CLAiO, Poland), and Mr. V. Khomenko (Kiev National University of Technology & Design, Ukraine).

A honorary NATO-CARWC award for significant contribution to the science of carbon for power sources is being given by Dr. I. Barsukov (workshop co-organizer) to Prof. D. Aurbach (Bar-Ilan Univ., Israel). Sitting are other awardees: S. Dimovski (Drexel Univ., USA (left), and Prof. R. Yazami (Caltech/CNRS – USA/France).

Dr. N. Vlasenko (left) of the Institute of General & Inorganic Chemistry, Ukraine and Dr. V. Kotlyar of Research Institute "Istochnik", Russia observe a high temperature fuel cell, as shown by Dr. R. Remick during a tour to Gas Technology Institute (a leading fuel cell technology organization in greater Chicago area). Courtesy V. Matveev.

ORGANIZATION

(i) USA Director:

Dr. Joseph E. Doninger– President, Dontech Global, Inc.
Dontech Global, Inc.
427 East Deerpath Rd., Lake Forest, IL 60045, USA
(224) 436-4835 (phone); +1 (847) 234-4835 (fax);
JDoninger@dontechglobal.com

(ii) USA Co-director:

Dr. Christopher S. Johnson – Staff Chemist
Electrochemical Technology Program
Chemical Engineering Division
Argonne National Laboratory
9700 S. Cass Ave., Argonne, IL 60565, USA
(630)-252 4787 (phone); (630)-252 4176 (fax);
johnsoncs@cmt.anl.gov

(iii) Ukraine Co-director:

Prof. Dr. Vyacheslav Z. Barsukov – Head of Chemistry Department
National University of Technologies and Design, Department of Chemistry, 2, Nemirovich-Danchenko street, Kiev, 252011, Ukraine
+ 38 (044) 291-2102 (phone); + 38 (044) 290-1603 (fax);
chemi@mail.kar.net

Local Organizing Committee

(i) **Dr. Igor Barsukov** Superior Graphite Co., Peter R. Carney Technology Center, 201 West 36th Street, Chicago, IL, 60632, USA, Phone: (773) 890-4117, Fax: (773) 890-4121, E-mail: IBarsukov@SuperiorGraphite.com
(ii) **Dr. Christopher S. Johnson, Chair** (Argonne National Laboratory, ANL)
(iii) **Dr. Joe Doninger** (Superior Graphite Co.)
(iv) **Mr. Gary Henriksen** (ANL), phone (630) 252-4591
(v) **Dr. James Miller** (ANL), phone (630) 252-4537

LIST OF PARTICIPANTS

Oleksandr Andriiko	Chemical Engineering Faculty, Kiev National Technical University, Polytechnik Institute (NTUU KPI), 7 Peremogy Prospekt, 03056 Kiev, Ukraine, flat@andriiko.kiev.ua
Terrill Atwater	Army Power, U.S. Army CECOM, AMSEL-RD-C2-AP-B, Fort Monmouth, NJ 07703, USA, Terrill.b.atwater@us.army.mil
Doron Aurbach	Department of Chemistry, Bar-Ilan University Remat ban 53900, Israel, aurbach@mail.biu.ac.il
James Barnes	U.S. Department of Energy, EE-2G, 1000 Independence Avenue, SW, Washington, DC 20585, USA, james.barnes@ee.doe.gov
Igor Barsukov	Department of Research & Development, Superior Graphite Company, 4201 West 36th Street, Chicago, IL 60629, USA, ibarsukov@superiorgraphite.com
Vyacheslav Barsukov	Electrochemical Power Engineering & Chemistry Kiev National University of Technologies an Design Nemirovich-Danchenko str. 2 02011 Kiev, Ukraine chemi@mail.kar.net
James Beckwith	Power Sources Center, 3206 Kinross Circle, Herndon, VA 20171, USA, jrbeckwith@cox.net
Francois Béguin	CRMD, CNRS – University of Orléans 1b, rue de la Férollerie, 45071 Orléans, France beguin@cnrs-orleans.fr
Alexey Belyakov	Management Department, ELIT Electrochemical Power Sources Co., 40, Rpospect Leninscogo Komsomola, 305026 Kursk, Russia elit@pub.sovtest.ru
Ira Bloom	Chemical Engineering Division, Argonne National Laboratory, Bldg. 205, 9700 South Cass Avenue, Argonne, IL 60439, USA bloom@cmt.anl.gov
Peter Booth	Technology Department, Superior Graphite Company 10 South Riverside Plaza, Chicago. IL 60606 USA pbooth@superiorgraphite.com
Ralph Brodd	Broddarp of Nevada, Inc., 2161 Fountain Springs Drive, Henderson, NV 89074 dbrodd@ix.netcom.com
Ross Campbell	Next Steps to Market, U.S. Civilian Research & Development Foundation (CRDF), 1530 Wilson Boulevard, Arlington, VA 22209, USA rcampbell@crdf.org

Peter Carney	Executive Department, Superior Graphite Company 10 South Riverside Plaza, Chicago, IL 60606, USA
Alexi Churikov	Chemistry Department, Saratov State University 83, Astrakhanskaya, 410012 Saratov, Russia churikovav@info.sgu.ru
Svetlana Dimovski	Department of Materials Science and Engineering Drexel University, Lewbow 344, 3141 Chestnut Street, Philadelphia, PA 19104, USA sd46@drexel.edu
Joseph Doninger	Dontech Global, Inc., 427 East Deerpath Rd., Lake Forest, IL 60045, USA jdoninger@superiorgraphite.com
Charles Drews	EMS, Inc., 165 King Street, Elk Grove Village, IL 60007, USA, sales@elecmet.com
Harvey Drucker	Energy & Environmental Science & Technology Argonne National Laboratory, Bldg. 202 9700 South Cass Avenue, Argonne, IL 60439, USA drucker@anl.gov
Elzbieta Frackowiak	Institute of Chemistry & Technical Electrochemistry Poznan University of Technology, Piotrowo 3, 60-965 Poznan, Poland, fracko@fct.put.poznan.pl
Arkadiy Furman	EMS, Inc., 165 King Street, Elk Grove Village, IL 60007, USA, arkadiy@elecmet.com
Maritza Gallego	Department of Research & Development, Superior Graphite Company, 4201 West 36th Street, Chicago, IL 60632, USA, mgallego@superiographite.com
Rex Gerald	Chemical Engineering Division, Argonne National Laboratory, Bldg. 205, 9700 South Cass Avenue, Argonne, IL 60439, USA, gerald@cmt.anl.gov
Steve Gerts	R&D Department, Amtek Research International 250 North Hansard Ave., Lebanon, OR 97355, USA sgerts@entek-international.com
Yury Gogotsi	Department of Materials Science & Engineering Drexel University, Lebow 344, 3141 Chestnut Street Philadelphia, PA 19104, USA, Yg36@drexel.edu
Irina Goncharova	Department of Chemistry & Engineering Ecology Dnieproppetrovsk National University of Railway Transport, 2 Academika Lazaryana St., 49010 Dniepropetrovsk, Ukraine Gonch_irina@hotmail.com
Diane Graziano	Chemical Engineering Division, Argonne National Laboratory, Bldg. 205, 9700 South Cass Avenue, Argonne, IL 60439, USA, graziano@cmt.anl.gov
Gary Henriksen	Chemical Engineering Division, Argonne National Laboratory, Bldg. 205, 9700 South Cass Avenue, Argonne, IL 60439, USA, henriksen@cmt.anl.gov

Francois-Xavier Henry Department of Research & Development, Superior Graphite Company, 4201 West 36th Street, Chicago, IL 60632, USA, fhenry@graphitesgc.com

Amy Hull Energy Technology Division, Argonne National Laboratory, Bldg. 212, 9700 South Cass Avenue, Argonne, IL 60439, USA, amyhull@anl.gov

Ilia Illiev Electrochemistry of Biocatalytic & Metal-Air Systems, Central Laboratory of Electrochemical Power Sources, Acad. G. Bonchev Str., bl. 10, 1113 Sofia, Bulgaria, ilia@cleps.bas.bg

Yevgen Ilyin Electrochemical Energetics & Chemistry, Kiev National University of Technologies & Design Nemirovich-Danchenko str., 2, 02001 Kiev, Ukraine Ujin-ilyin@ukr.net

Andrew Jansen Chemical Engineering Division, Argonne National Laboratory, Bldg. 205, 9700 South Cass Avenue, Argonne, IL 60439, USA, jansen@cmt.anl.gov

Christopher Johnson Chemical Engineering Division, Argonne National Laboratory, Bldg. 205, 9700 South Cass Avenue, Argonne, IL 60439, USA, johnsoncs@cmt.anl.gov

Donald Joyce Office of the Laboratory Director, Argonne National Laboratory, Bldg. 201, 9700 South Cass Avenue, Argonne, IL 60439, USA, donjoyce@anl.gov

Arthur Kahaian Chemical Engineering Division, Argonne National Laboratory, Bldg. 205, 9700 South Cass Avenue, Argonne, IL 60439, USA, kahaian@cmt.anl.gov

Anastassia Kaisheva Electrochemistry of Biocatalytic & Metal-Air Systems, Central Laboratory of Electrochemical Power Sources, Acad. G. Bonchev str., bl. 170, 1113 Sofia, Bulgaria, kaisheva@cleps.bas.bg

Sun-Ho Kang Chemical Engineering Division, Argonne National Laboratory, Bldg. 205, 9700 South Cass Avenue, Argonne, IL 60439, USA, kangs@cmt.anl.gov

Volodimir Khomenko Electrochemical Power Engineering & Chemistry Kiev national University of Technologies & Design, Nemirovich-Danchenko Str., 2, 02011 Kiev, Ukraine vkg@svitonline.com

Jeom-Soo Kim Chemical Engineering Division, Argonne National Laboratory, Bldg. 205, 9700 South Cass Avenue, Argonne, IL 60439, USA, kimjs@cmt.anl.gov

Sviatoslav Kirillov Institute for Sorption & Problems of Endoecology, National Academy of Scienes, 13, Naumov St., 03164 Kiev, Ukraine, kir@i.kiev.ua

Arkadiy Klementov JME, Inc., 17210 Parkland Drive, Shaker Heights, OH 44120, USA, klementov@att.net

Svitlana Kochetova	N11-High Temperature Electrochemical Synthesis, National Academy of Sciences, Institute of General & Inorganic Chemistry, 32/34 Palladin Avenue, 03680 Kiev, Ukraine, ktumanova@ionic.kar.net
Robert Kostecki	Environmental Energy Technologies, Lawrence Berkeley National Laboratory, 1 Cyclotron Road, Berkeley, CA 94720, USA, r_kostecki@lbl.gov
Vladimir Kotlyar	Reliability Department, Accumulator Research & Design Institute "Istochnik", 10, Dahl Street, 197376 Saint-Petersburg, Russia, vkotlyar@nwgsm.ru
Igor Kovalenko	Ministry for Emergency Management of Chemistry, Ukrainian Fire Safety Research Institute, 18, Rybarskya Street, 01011 Kiev, Ukraine igorkio@kiev.ldc.net
Alexander Kozhevniko	N20, Acumulator Research & Design Institute "Istochnik", 10, Dahl Street, 197376 Saint-Petersburg, Russia, kojevan@rambler.ru
David Laughton	Sales Department, Superior Graphite Company 10 South Riverside Plaza, Chicago, IL 60606, USA dlaughton@graphitesgc.com
Andrzej Lewandowski	Faculty of Chemical Technology, Poznan University of Technology, ul. Piotrowo 3, 60-965 Poznan, Poland, alew@fct.put.poznan.pl
David Lewis	Chemical Engineering Division, Argonne National Laboratory, Bldg. 205, 9700 South Cass Avenue, Argonne, IL 60439, USA, lewisd@cmt.anl.gov
Jun Liu	Chemical Engineering Division, Argonne National Laboratory, Bldg.205, 9700 South Cass Avenue, Argonne, IL 60439, USA, liuj@cmt.anl.gov
Jacek Machnikowski	Institute of Chemical & Technology of Petroleum and Coal, Wroclaw University of Technology, Gdanska 7/9, 50 344 Wroclaw, Poland, jacek.machnikowski@pwr.wroc.pl
Irina Najteteva	Electrochemical Power Engineering & Chemistry, Kiev National University of Technologies & Design, Nemirovich-Danchenco str., 2, 02011 Kiev, Ukraine, makeeva05@yahoo.com
Yurii Maletin	Department of Physical Chemistry, National Technical University of Ukraine "KPI", 37 Prospect Peremohy, 03056 Kiev, Ukraine, maletin@xtf.ntu-kpi.kiev.ua
Vadym Matveyev	Scientific Research Laboratory of Storage-Battery System, Ukrainian State University of Chemical Engineering, 8, Gagarin Avenue, 49005 Dniepropetrovsk, Ukraine, vavlma@mail.ru

Luidmila Matzui	Department of Biophysics, Kyiv National Shevchenko University Volodymyrska Str., 64, Kyiv, Ukraine, matzui@mail/univ/kiev/ua
Dayal Meshri	Advance Research Chemicals, Inc., 1110 West Keystone Avenue, Catoosa, OK 74015, USA, dr.meshri@fluoridearc.com
James Miller	Chemical Engineering Division, Argonne National Laboratory, Bldg. 205, 9700 South Cass Avenue, Argonne, IL 60439, USA, millerj@cmt.anl.gov
Vadim Mochalin	Chemistry of Molecular Complexes, National Academy of Sciences, L.M. Litvinenko Institute of Physical Organic & Coal Chemistry, R. Luxemburg Str., 70, 83114 Donetsk, Ukraine, mail2mochalin@yahoo.com
Michael Onischak	Department of Hydrogen & Distributed Energy, Gas Technology Institute, 1700 South Mount Prospect Road, Des Plaines, IL 60018, USA, mike.onischak@gastechnology.org
Oksana Pendelyuk	National Academy of Sciences, Institute for Sorption and Problems of Endeocology, 13, Naumov St., 03164 Kiev, Ukraine, khain@ispe.kiev,ua
Iouri Prokofiev	Nuclear Engineering Division, Argonne National Laboratory, Bldg. 206, 9700 South Cass Avenue, Argonne, IL 60439, USA, prokofiev@anl.gov
Encarnacion Raymundo-Pinero	CRMD, CNRS – University of Orléans, 1B rue de la Férollerie, 45071 Orleans, France, raymundo@cnrs-orleans.fr
Michael Romanco	Department of Hydrogen & Distributed Energy, Gas Technology Institute, 1700 South Mount Prospect Road, Des Plaine, IL 60018, USA, michael.romanco@gastechnology.org
Jean-Noel Rouzaud	CRMD, CNRS – University of Orléans, 1B rue de la Férollerie, 45071 Orleans, France, rouzand@cnrs-orleans.fr
Giselle Sandi-Tapia	Chemistry Division, Argonne National Laboratory, Bldg. 200, 9700 South Cass Avenue, Argonne, IL 60439, USA, gsandi@anl.gov
Michael Savoskin	Chemistry of Molecular Complexes, National Academy of Sciences, L.M. Livinenko Institute of Physical Organic & Coal Chemistry, R. Luxemburg Str., 70, 83114 Donetsk, Ukraine, m-savoskin@yandex.ru
Sergiy Yu Sayenko	Institute of Solid State Physics, National Science Center – KIPT, Materials Science and Technologies, 1, Akadamicheskaya, 61108 Kharkov, Ukraine, sayenko@kipt.kharkov.ua

Sergiy Sazhin	Department of Lithium & Rechargeable Technologies, Rayovac Corporation, 630 Forward Drive, Madison, WI 53711-2497, USA, sazhin@rayovac.com
Kathryn Striebel	Environmental Energy Technologies Division, Lawrence Berkeley National Laboratory, MS: 70R0108B, 1 Cyclotron Road, Berkeley, CA 94720, USA, kastreibel@lbl.gov
Jeffrey Swoyer	Research Department, Valence Technology, Inc., 301 Conestoga Way, Henderson, NV 89015, USA, jeff.swoyer@valence.com
Michael Thackeray	Chemical Engineering Division, Argonne National Laboratory, Bldg. 205, 9700 South Cass Avenue, Argonne, IL 60439, USA, thackeray@cmt.anl.gov
Thanh Tran	Sentech, Inc., 4733 Bethesda Avenue, Bethesda, MD 20814, USA, lreese@sentech.org
Gabriela Uribe	Research & Development, Superior Graphite Company, 4201 West 36th Street, Chicago, IL 60632, USA, guribe@graphites.gc.com
Jack Vaughey	Chemical Engineering Division, Argonne National Laboratory, Bldg. 205, 9700 South Cass Avenue, Argonne, IL 60439, USA, vaughey@cmt.anl.gov
Donald Vissers	Chemical Engineering Division, Argonne National Laboratory, Bldg. 205, 9700 South Cass Avenue, Argonne, IL 60439, USA, vissers@cmt.anl.gov
Natasha Vlasenko	Photoelectrochemistry & Chemical Power Sources, National Academy of Sciences, Institute of General & Inorganic Chemistry, 32/34 Palladin Prospect, 03680 Kiev – 142, Ukraine, vlasenko05@yahoo.com
Mariusz Walkowiak	Lithium Batteries Department, Central Laboratory of Batteries & Cells, Ul. Forteczna 12, 61-362 Poznan, Poland, mw@plusnet.pl
Qingzheng Wang	Chemical Engineering Division, Argonne National Laboratory, Bldg. 205, 9700 South Cass Avenue, Argonne, IL 60439, USA, wangq@cmt.anl.gov
Robert Waterhouse	Department of Research & Development, Amtek Research International, 250 North Hansard Avenue, Lebanon,OR, 97355, USA, rwaterhouse@amtek-research.com
Neal White	Sion Power Research, Inc., Sion Power Corporation, Tanfield Lea Industrial Park, Stanley, Durham DH9 9QF, United Kingdom, nealwhite@sionpowerruk.fsbusiness.co.uk
Martin Winter	Institute of Chemical Technology of Inorganic Materials, Graz University of Technology, Stremayrgasse 16, A-8010 Graz, Austria, martin.winter@tugraz.at

Mathis Wissler	Product Management, Superior Graphite Europe, Kungsmatt 8, Ch 5643 Sins, Switzerland, mwissler@superiorgraphite.com
Rachid Yazami	Department of Materials Science, CRNS/Caltech, M/C 183-78, Pasadena, CA 91125, USA, yazami@caltech.edu
Peter Zaleski	Laboratory Department, Superior Graphite Company, 4201 West 36th Street, Chicago, IL 60632, USA, pzaleski@graphitesgc.com

CHAPTER 1:

NEW CARBON MATERIALS FOR SUPERCAPACITORS

Chapter 1: Subject Overview

A classic definition of electrochemical ultracapacitors or super-capacitors summarizes them as devices, which store electrical energy via charge in the electrical double layer, mainly by electrostatic forces, without phase transformation in the electrode materials. Most commercially available capacitors consist of two high surface area carbon electrodes with graphitic or soot-like material as electrical conductivity enhancement additives. Chapter 1 of this volume contains seven papers with overview presentations, and development reports, as related to new carbon materials for this emerging segment of the energy market.

In the first paper by E. Frackowiak *et al.*, a French-Polish team of authors describes properties of new materials produced by KOH activation of microporous carbon precursors, as well as carbons obtained through the so-called template synthesis. The authors also devise a carbon-carbon composite, which has been synthesized via a process of carbonization of carbon nanotubes with a polyacrylonitrile conducting polymer. It is claimed that a synergistic effect has been observed with the above composite, while its individual ingredients show instead negligible capacitance values. An attempt made by the authors to explain the reported synergistic capacitance enhancement effect by, exclusively, nitrogen-based surface groups on carbon, is an interesting, and, perhaps, debatable concept. It may likely stimulate further discussions and investigative research on the subject.

The second paper, by A. Belyakov, comes from the Russian supercapacitor industry. This applied work highlights modern industrial approaches used for boosting the power output characteristics of aqueous commercial supercapacitors. The author discusses two supercapacitor chemistries: carbon-carbon and NiO_x-carbon. A key emphasis is given to the role of conductive additives, which constitute up to 40wt% of electrode composition. Several classes of conductive diluents were investigated, e.g. Ni powder, colloidal carbon black, expanded and boron-doped flake graphite. The author concludes that for the carbon-carbon type of supercapacitors, the most preferable conductive additive should be carbon material, but the exact type of carbon is of a secondary importance. However, in the case of asymmetric NiO_x-carbon capacitor, boron-doped graphite showed the biggest promise, due to it being prone to oxidation.

In the third paper by French and Ukrainian scientists (Khomenko *et al.),* the authors focus on high performance a-MnO_2/carbon nanotube composites as pseudo-capacitor materials. Somewhat surprisingly, this paper teaches to use carbon nanotubes for the role of conductive additives, thus suggesting an alternative to the carbon blacks and graphite materials – low cost, widely accepted conductive diluents, which are typically used in todays' supercapacitors. The electrochemical devices used in the report are full symmetric and optimized asymmetric systems, and are discussed here for the first time. The editors tend to agree with author's conclusions that carbon nanotubes may become quite efficient conductive diluents at low loadings in the electrode matrixes, due to their unique structure, especially after the costs of carbon nanotubes are brought in line with alternative conductive additives.

The fourth contribution (Khomenko *et al.*) is a product of an extensive international study as presented by an alliance of academia from France, Poland and Ukraine. In this interesting work, the authors furnish carbon nanotubes and/or expanded graphite as a backbone, highly conductive electrode network, onto which the following conducting polymers have been deposited: polyaniline, polypyrrole, and poly(3,4-ethylenedioxythiophene). In so doing, these authors propose an interesting approach to solving a major drawback of modern pseudocapacitor technologies, e.g. stability of their performance characteristics over time.

Ukraine's Y. Maletin *et al.* presented a comprehensive overview describing state of the art as well as future development trends in supercapacitors, as the fifth paper in this chapter. The authors establish key performance bars for supecapacitors; upon meeting those, supercapacitors may start to compete with batteries. Also, this paper highlights so-called hybrid applications where supercapacitors complement operation of batteries and/or fuel cells. Optimization of supercapacitor performance through varying electrode thickness is contemplated in length.

In the sixth paper of this chapter, Kierzek *et al.*, mainly focus on modeling of pore formation vs surface area growth phenomena upon activation of coal and pitch-derived carbon precursors. These authors briefly touch on other precursor carbons as well. The properties of newly synthesized materials are being looked at from the point of view of their application as active materials in the supercapacitor electrodes. Editors thought this work by the Institute of Chemistry and Technology of Petroleum and Coal in Poland, could be of genuine interest to the practical developers of carbon materials for the supercapacitor industry.

In the last paper, A. Lewandowski *et al.* of Poland, examines the role of ionic liquids as new electrolytes for carbon-based supercapacitors. Although not directly addressing the role of new carbon materials (the area of major focus of this book), this interesting theoretical work seeks to optimize electrolyte media, which is in contact with carbon electrodes.

We hope you will enjoy reading this chapter.

NOVEL CARBONACEOUS MATERIALS FOR APPLICATION IN THE ELECTROCHEMICAL SUPERCAPACITORS

[1]E. Frackowiak*, [2]J. Machnikowski, [3]F. Béguin

[1]*Poznan University of Technology, 60-965 Poznan, ul. Piotrowo 3, Poland*
[2]*Wroclaw University of Technology, 50-344 Wroclaw, ul. Gdanska 7/9, Poland*
[3]*CRMD, CNRS-University, 45071 Orléans, 1B rue de la Férollerie, France*

Abstract

Three types of novel carbon materials, such as KOH-activated microporous carbons, highly ordered carbons obtained by a template technique, and a special carbon/carbon composite based on carbon nanotubes with carbonized polyacrylonitrile have been used as electrodes for supercapacitors in aqueous and organic electrolytes. The various carbon materials used allow their capacitance properties to be correlated with the total surface area, micro and meso pore volume, elemental composition. Especially, the role of mesopores in charge propagation was strongly underlined. The capacitance values widely range from 100 to 300 F/g depending on the carbon microtexture but also on the heteroatom (nitrogen) content responsible for pseudocapacitance effects. Comparison of these three classes of carbon materials gives a basic knowledge for optimization of supercapacitor electrode materials depending on electrolytic solutions used.

1. INTRODUCTION

Electrochemical capacitors are power storage devices, whose performance is based on the charge accumulation from an electrolytic solution through electrostatic attraction by polarized electrodes. The capacitance of this system is directly proportional to the electrode surface, therefore carbons are very efficient for this application because of various possibilities of their modification and creation of a controlled pore size distribution [1-3]. The electrostatic attraction of ions takes place mainly in micropores, however, the presence of mesopores is necessary for efficient

---

* Corresponding author. E-mail: fracko@fct.put.poznan.pl

I.V. Barsukov et al. (eds.), New Carbon Based Materials for Electrochemical Energy Storage Systems, 5–20.

charge propagation. It is noteworthy that the electrochemically active surface is not the same as the physically described surface area from nitrogen adsorption data, and is determined by the ability of ions to be trapped (function of the size of ions, wetability, conductivity of carbon, etc.). Hence, in general, well-balanced micro/meso porosity is an important criterion for the selection of the carbon material.

In supercapacitors, apart from the electrostatic attraction of ions in the electrode/electrolyte interface, which is strongly affected by the electrochemically available surface area, pseudocapacitance effects connected with faradaic reactions take place. Pseudocapacitance may be realized through carbon modification by conducting polymers [4-7], transition metal oxides [8-10] and special doping via the presence of heteroatoms, e.g. oxygen and/or nitrogen [11, 12].

In this article, we will pay a special attention to three novel types of carbonaceous materials:

1) Carbons activated with KOH, featuring a highly developed surface area;
2) highly ordered carbons prepared from silica templates;
3) carbon/carbon composite based on multi-walled carbon nanotubes (MWNTs) and polyacrylonitrile (PAN).

1.1. Chemical Activation by KOH

The first type of materials, i.e. KOH activated carbons constitute an interesting class of capacitor electrodes due to their highly developed surface area of the order of 3000 m^2/g. Especially, inexpensive natural precursors are well adapted for this process. The activation process is strongly affected by the C:KOH ratio, temperature and time. The optimal ratio seems to be 4:1 and the temperature for activation ca. 800°C. The total activation process is quite complicated and proceeds via different pathways and by-products. The following reactions can be considered:

$$2KOH \rightarrow K_2O + H_2O \quad (1)$$

$$C + H_2O \rightarrow CO + H_2 \quad (2)$$

$$CO + H_2O \rightarrow CO_2 + H_2 \quad (3)$$

$$K_2O + H_2 \rightarrow 2K + H_2O \quad (4)$$

$$K_2O + C \rightarrow 2K + CO \quad (5)$$

Under severe conditions (above 700°C), a potassium vapor is formed. It plays a special role in the activation of carbonaceous materials, easily penetrating in the graphitic domains that form cage-like micropores. The efficient development of micropores, which often gives a few-fold increase of the total specific surface area, is very useful for the application of these materials in supercapacitors [13-14].

Capacitance of this type of material reaches extremely high values over 300 F/g. Our target is to correlate it with the total specific surface area, pore size distribution, particle size and elemental composition of carbon.

1.2. Carbons Produced by the Template Technique

The second class of materials, which we will consider herein are carbons with a highly ordered porosity prepared by a template technique [15-18]. The pores are characterized by a well-defined size determined by the wall thickness of the silica substrate used as substrate for carbon infiltration. They can be also interconnected, that is very useful for the charge diffusion in the electrodes. Figure 1 presents the general principle of the carbon preparation by a template technique, where the silica matrix can be, for example, MCM-48 or SBA-15.

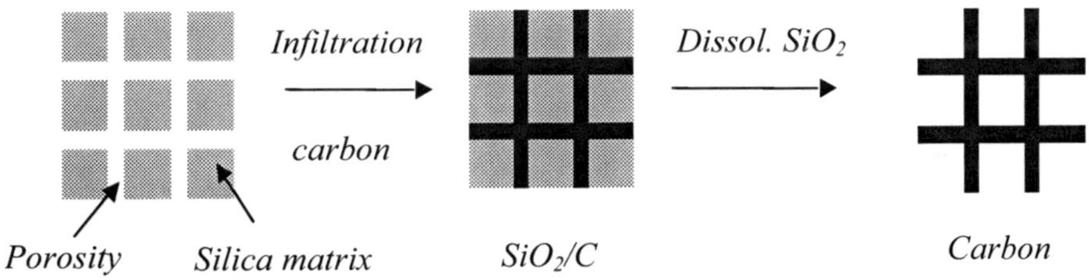

Figure 1. Schematic representation of carbon elaboration by a template technique.

Figure 2. Two examples of silica matrices: (a) MCM-48 and (b) SBA-15.

Schematic representations of organization of the porous network in MCM-48 and SBA-15 are given in Fig. 2 [16]. The mesoporous silica material referred to as MCM-48 displays a cubic structure, which consists of two independent and intricately interwoven networks of mesoporous channels. The structure of the material named SBA-15 is composed of a hexagonal arrangement of 1-D silica rods forming cylindrical pores interconnected through smaller pores, which are randomly located perpendicular to the 1-D channels [15]. Both silica materials exhibit a highly ordered porous structure, a high specific surface area and a narrow pore size distribution centered at 2.6 nm and 5.2 nm for MCM-48 and SBA-15, respectively.

The pores of the silica template can be filled by carbon from a gas or a liquid phase. One may consider an insertion of pyrolytic carbon from the thermal decomposition of propylene or by an aqueous solution of sucrose, which after elimination of water requires a carbonization step at 900°C. The carbon infiltration is followed by the dissolution of silica by HF. The main attribute of template carbons is their well sized pores defined by the wall thickness of the silica matrix. Application of such highly ordered materials allows an exact screening of pores adapted for efficient charging of the electrical double layer. The electrochemical performance of capacitor electrodes prepared from the various template carbons have been determined and are tentatively correlated with their structural and microtextural characteristics.

1.3. Carbon/Carbon Composites

Another interesting type of novel carbons applicable for supercapacitors, consists of a carbon/carbon composite using nanotubes as a perfect backbone for carbonized polyacrylonitrile. Multiwalled carbon nanotubes (MWNTs), due to their entanglement form an interconnected network of open mesopores, which makes them optimal for assuring good mechanical properties of the electrodes while allowing an easy diffusion of ions.

Due to their moderate specific surface area, carbon nanotubes alone demonstrate small capacitance values. However, the presence of heteroatoms can be a source of pseudocapacitance effects. It has been already proven that oxygenated functional groups can significantly enhance the capacitance values through redox reactions [11]. Lately, it was discovered that nitrogen, which is present in carbon affects also the capacitance properties [12].

In this work, MWNTs will be mixed with polyacrylonitrile (PAN) and the MWNTs/PAN composite will be carbonized, giving rise to a new C/C composite with interesting capacitance properties. It is striking that the composite components alone give negligible capacitance values below 20

F/g, whereas the composite in the optimal proportions can supply quite interesting capacitance behavior explained only by a positive effect of nitrogen.

2. EXPERIMENTAL

2.1. Preparation and Characteristics of the Carbon Materials

2.1.1. KOH Activated Carbons

The following natural precursors have been selected for KOH activation: coal (C), coal semi-coke (CS), pitch semi-coke (PS) and pitch mesophase (PM). An industrial activated carbon (AC) was also used. Activation was performed at 800°C in KOH with 4:1 (C:KOH) weight ratio, for 5 hours, followed by a careful washing of the samples with 10% HCl and distilled water. The activation process supplied highly microporous carbons with BET specific surface areas from 1900 to 3150 m^2/g. The BET surface area together with the micro and the total pore volume of the KOH-activated carbons are presented in Table 1. The mean micropore width calculated from the Dubinin equation is designed as L_D.

Table 1. Porosity parameters of the KOH activated carbons (A-C means activated carbon from coal C etc.)

Sample	S_{BET} $m^2\ g^{-1}$	V_{DR} $cm^3\ g^{-1}$	V_{total} $cm^3\ g^{-1}$	L_D nm
A-C	3150	0.951	1.612	1.39
A-CS	3190	0.936	1.446	1.36
A-PM	2660	0.839	1.209	1.37
A-PS	2750	0.859	1.227	1.34
A-AC	1900	0.609	1.051	1.29

2.1.2. Template Carbons

Two types of organized mesoporous silica, MCM-48 and SBA-15, which display different crystallographic structure, size and shape of the pores, were selected as templates for carbons.

Pyrolytic carbon was inserted into the pores of these silica matrices by chemical vapor infiltration (CVI). The silica template was contacted with a flow of propylene Pr, (2.5 vol%) diluted in argon at 750°C during 15 hours. A quite uniform pore filling can be obtained by CVI. At the end, the carbon represents about 50 wt% of the C/SiO_2 material. Since the deposition time was the same for the two silica precursors, the fraction of the pore volume filled with carbon is function of the template. The C/SiO_2 material was then treated with hydrofluoric acid (HF) to remove the silica template and the carbons obtained are designed CPr48 and CPr15.

The second type of template carbon was formed by liquid impregnation of the pores of MCM-48 and SBA-15 silica by a sucrose (S) solution, followed by a carbonization treatment at 900°C under vacuum. After a second impregnation step, the amount of carbon introduced in the silica matrix represents 33wt% which is not far of the theoretical amount expected, taking into account the pore volume of the silica matrix and the concentration of the sucrose solution, i.e. 36wt% of the total mass of the C/SiO_2 composite. The silica template is eliminated by dissolution in HF to recover the carbon material denominated here as CS48 and CS15.

Silica matrices MCM-48 and SBA-15, as well as carbons obtained by replica method are characterized by nitrogen adsorption at 77K and X-ray diffraction (Table 2).

Table 2. Characteristics of the silica templates and the corresponding carbon materials; a: unit cell parameter; S_{BET}: specific surface area; Vp: total pore volume (at P/Po=0.95); Pore size determined according to the BJH method - Maximum value of the BJH pore size distribution peak calculated from the adsorption branch of the N_2 isotherm.

Materials	*a (nm)*	*S_{BET} (m^2/g)*	*V_p (cm^3/g)*	*Pore size (nm)*
MCM-48	8.6	1330	1.07	2.6
CPr48	8.4	850	0.56	-
CS48	7.6	2000	1.10	2.7
SBA-15	11.3	807	0.97	5.2
CPr15	10.5	713	0.60	3.4
CS15	9.4	1470	1.11	3.1

2.1.3. CNTs/PAN Composites

The high purity carbon nanotubes (CNTs) used in this study were obtained by decomposition of acetylene over a powdered $Co_xMg_{1-x}O$ solid solution catalyst [19]. Different proportions of CNTs from 15 to 70% and polyacrylonitrile (PAN, Aldrich) have been mixed in an excess of acetone to obtain a slurry. After evaporation of acetone, precursor electrodes were formed by pressing the CNTs/PAN mixture at 1-2 tons/cm^2. The C/C composites were formed by carbonisation of the pellets at 700-900°C for 30-420 min under nitrogen flow [20]. The optimal capacitance properties of the composite were obtained for a mixture CNTs/PAN (30/70 wt%) treated at 700°C. Such C/C composite remains still quite rich in nitrogen (9 at% of N) demonstrating that PAN is an efficient nitrogen carrier. On the other hand, the noticeable amount of oxygen (4.1 at%) in the composite could be due to oxygen addition on the dangling bonds when the C/C composite is exposed to air after its formation.

During the pyrolysis of the CNTs/PAN mixture, the pellets keep their original shape while becoming rigid, without any noticeable cracks or defects. The carbonised C/C composites are quite compact, with a strong bonding between CNTs and the carbon matrix. Hence, CNTs play a very important role during the C/C composite formation, by creating a stable network which prevents the dimensional changes and shrinkage. However, when the CNTs content is less than 15%, the weight loss increases dramatically and shrinkage appears. Such self-standing composite pellets were used as electrodes for capacitance investigations.

2.2. Electrochemical Measurements

The capacitor electrodes were pellets formed by pressing a mixture of the active carbon material (85wt%) with a PVDF binder (10wt%) and acetylene black (5wt%). In the case of the C/C composite, PVDF and acetylene black were not used because the composite forms self-standing electrodes. The capacitance measurements were performed in 6 mol l^{-1} KOH, 1 mol l^{-1} H_2SO_4 and 1 mol l^{-1} $TEABF_4$ in acetonitrile. The values of capacitance were estimated by voltammetry (scan rate of potential from 2 to 10 mV/s) and galvanostatic charge-discharge cycling (VMP-Biologic, France and ARBIN BT2000, USA). Impedance spectroscopy (AUTOLAB-ECOCHEMIE BV) allows the capacitance (F/g) dependence versus frequency (Hz), series resistance, time constant and charge propagation to be evaluated. For some samples a careful analysis of leakage current, self-discharge and capacitors cycleability was performed.

3. RESULTS AND DISCUSSION

3.1. Capacitance Properties of the KOH Activated Carbons

From three electrochemical techniques described in the previous section, the most reliable capacitance results are obtained from galvanostatic discharge, however each method supplies complementary information. A typical galvanostatic charge-discharge characteristic for carbon A-PM is shown in Figure 3. The curve presents a correct triangular shape without a significant ohmic drop. The values of specific capacitance per mass of carbon material and per surface area estimated by all three electrochemical methods using 1 mol l^{-1} sulfuric acid solution are given in Table 3. All the carbons present a very satisfactory capability of charge accumulation in the electrical double layer with capacitance values ca. 300 F/g, especially for the activated carbons from coal (A-C) and mesophase pitch (A-PM). The highest capacitance value for A-C well correlates with a maximum total pore volume of 1.6 cm^3/g and its BET surface area of 3150 m^2/g. However, a careful comparison of the carbons characteristics (Table 1) with the capacitance values (Table 3) shows that there is not a proportional relation between the surface area or the pore volume and the electrochemical behavior. On the other hand, from the low values of capacitance per surface area (7 to 11 $\mu F/cm^2$) one can assume that not all the micropores take part in the charge accumulation. It is clear that the micropores not adapted to the size of the solvated ions will not take part in the double layer charging. The charging of the double layer is very complex and depends also on other parameters such as the pore size distribution, the affinity to the electrolytic solution, the hydrophobic-hydrophilic character, the particle conductivity and their size.

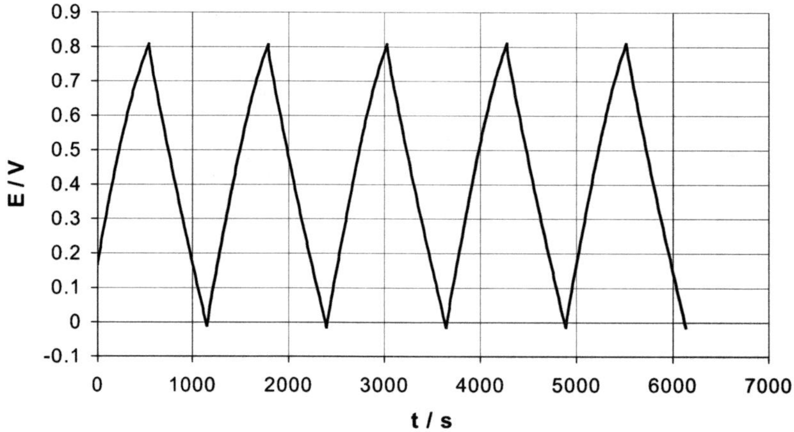

Figure 3. Galvanostatic charge/discharge characteristics of a capacitor built from KOH activated carbon A-PM (mass of electrodes 12.2 mg/12.8 mg) I = 2 mA. Electrolytic solution: 1mol l^{-1} H_2SO_4.

Table 3. Capacitance values (F g^{-1}) of the KOH activated carbons (A-C; A-CS; A-PM; A-PS; A-AC) estimated by galvanostatic discharge, cyclic voltammetry and impedance spectroscopy. Specific capacitance $\mu F\ cm^{-2}$ calculated per surface area of carbon.

Sample	Galvanostatic discharge C / F g^{-1}	Cyclic voltammetry C / F g^{-1}	Impedance spectroscopy C / F g^{-1}	Specific capacitance $\mu F\ cm^{-2}$
A-C	312	317	282	9.9
A-CS	223	235	206	7.0
A-PM	294	299	273	11.0
A-PS	261	241	227	9.5
A-AC	198	198	193	10.4

Even if the KOH activated carbons supply high capacitance values, the practical application of such materials is determined by the supercapacitor cycleability, quick charge propagation at different loads, a low self-discharge. Highly microporous carbons supply always some diffusion limitation. This effect can be observed at quick scan rates during voltammetry experiments and during impedance spectroscopy measurements. Figure 4 shows the impedance characteristic for the carbon A-PM with almost perpendicular dependence of imaginary part to real one that is a proof for capacitive response; however, a small diffusion slope is slightly marked.

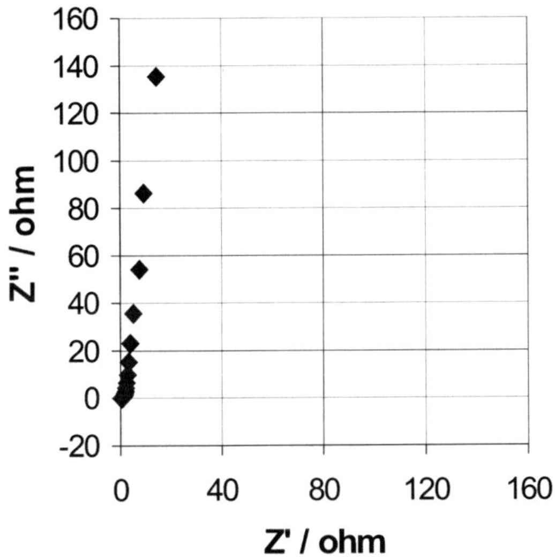

Figure 4. Impedance spectroscopy characteristic for the carbon sample A-PM. Mass of electrodes: 12.2/12.8 mg. $C = 273\ F\ g^{-1}$ (at 1mHz).

It is important to stress that the capacitive behaviour of the microporous carbons could be further improved by enhancing the mesopore volume. The presence of mesopores plays a crucial role for the ion transportation to the active surface. Hence, a development of mesopores in these materials, and a strict control of the micropore-mesopore volume ratio is necessary.

3.2. Template Carbons as Capacitor Electrodes

The total surface area of the template carbons prepared by sucrose impregnation is significantly higher than the surface area of the corresponding silica template (Table 2), that confirms the formation of micropores during the carbonization. Just an opposite tendency is observed when propylene is used as carbon precursor. In the latter case, the pore volume of the template carbon is consistent with an uniform pore filling, and the values of surface area are easily justified by different pores and walls diameters in the pristine template. The higher pore volume and surface area of sucrose derived template carbons is attributed to the release of water and gases during the thermal treatment.

The capacitance values determined by the three techniques are reported in Table 4 for the different types of carbons (CS48, CS15, CPr48, CPr15) depending on the electrolytic medium (acidic, basic and organic). For each kind of capacitor, it can be seen that a satisfactory correlation is found between the three techniques. The various electrode materials give rise to different values of capacitance due to their different physico-chemical characteristics as seen in Table 2. The voltammograms of CS15 and CS48 for the acidic and organic media have a box-like shape which is characteristic for an ideal capacitor with quick charge propagation. As seen in Table 4, the two materials exhibit high values of capacitance both in aqueous and organic medium; the capacitance values of CS48 reach 200F/g and 110 F/g in acidic and organic medium, respectively. These values are promising compared to other reported in the literature for various activated carbons with comparable area. Hence, the additional porosity created during the carbonization step is profitable for the performance of these template carbons. Comparing the performance of the electrodes made from CS48 and CS15, the specific capacitance per surface unit in acidic medium is roughly the same for the two materials, ca.10 $\mu F/cm^2$ and 12 $\mu F/cm^2$, respectively. This would lead to the conclusion that the main parameter controlling the electrochemical performance in these materials is the total surface area (Table 2). However, taking into account that the surface area determined by nitrogen adsorption and the electrochemically active surface area are different, other factors might explain the different capacitance values observed for CS48 and CS15, e.g. the different pore arrangement which affects the diffusion of the solvated ions, the presence of a more marked secondary microporosity (1.5 nm $< D <1.8$ nm) for CS48 and consequently, smaller pore size compared to CS15.

The electrochemical performance of CS15 looks to be similar in basic and acidic medium, while smaller values are observed for CS48 in basic medium (Table 4). Such a difference might be related to the different surface functionality of the two materials, the latter controlling the wettability of the materials. In the case of CS48, surface oxygenated groups which dissociate differently depending on the electrolyte pH, could be responsible of different adsorption of the ions on the carbon surface in both media. On the other hand, CS15 behaves rather as a material, which has a poor surface functionality. A careful analysis of the surface properties of these carbons is necessary to support these hypotheses.

Table 4. Capacitance values in F/g (per gram of carbon material) for the carbons CS48, CS15, CPr48, CPr15 in acidic, basic and organic media[†].

Material: CS48	**C (F/g)** 1M H_2SO_4	**C (F/g)** 6M KOH	**C (F/g)** $TEABF_4$ in CH_3CN
G (2mA)	206	184	115
G (5mA)	188	170	116
P (2mV/s)	193	181	118
P (10mV/s)	170	159	114
I (1mHz)	210	177	96

Material: CS15	**C (F/g)** 1M H_2SO_4	**C (F/g)** 6M KOH	**C (F/g)** $TEABF_4$ in CH_3CN
G (2mA)	169	174	93
G (5mA)	158	161	90
P (2mV/s)	162	170	98
P (10mV/s)	144	156	94
I (1mHz)	171	172	78

Material: CPr48	**C (F/g)** 1M H_2SO_4	**C (F/g)** 6M KOH	**C (F/g)** $TEABF_4$ in CH_3CN
G (2mA)	101	74	65
G (5mA)	99	87	62
P (2mV/s)	113	82	64
P (10mV/s)	99	75	62
I (1mHz)	113	81	57

Material: CPr15	**C (F/g)** 1M H_2SO_4	**C (F/g)** 6M KOH	**C (F/g)** $TEABF_4$ in CH_3CN
G (2mA)	62	58	48
G (5mA)	61	55	46
P (2mV/s)	62	56	50
P (10mV/s)	58	51	49
I (1mHz)	78	59	42

Figure 5 presents the capacitance-frequency dependence from impedance spectroscopy measurements for CS48 and CS15 in acidic and organic medium. In the low frequency region (from 1mHz to 100mHz) nearly a complete penetration of the ions into the pores is allowed and the quite stable values indicate the domination of the capacitive behavior at the electrolyte/carbon interface. All the curves show a typical drop of

[†] In the table, G, P, I stand for measurements by galvanostatic, potentiodynamic and impedance spectroscopy, respectively.

capacitance with frequency, which becomes quite significant at frequencies higher than 1Hz. In the case of the sample CS15, the decrease of capacitance is shifted to slightly higher frequency compared to CS48, both in aqueous and organic medium. The ions access to the active surface, although the pore diameter is larger than the size of the solvated ions in both cases.

It means that, for a given frequency, the ion diffusion to the active surface is more efficient in the case of CS15. Comparing CS15 and CS48, this better frequency behavior of CS15 could be due to the presence of slightly larger mesopores (Table 2), or mesopores of appropriate shape, which favor a better efficiency. However, one must be careful with the interpretation of adsorption data, because the pores of the template carbons, which are the walls replica, are no longer cylindrical, and the determination of the pore size by the BJH method (which assumes cylindrical pores) might be slightly misleading.

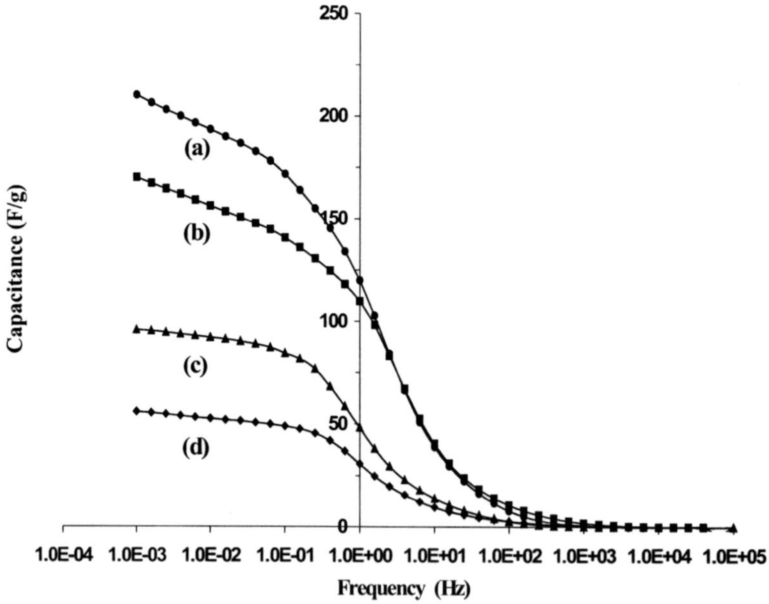

Figure 5. Capacitance vs frequency for carbons from CS48 (a,c) and CS15 (b,d) in 1M H_2SO_4 (a,b) and 1M $TEABF_4$ in acetonitrile (c,d).

The capacitance values of the carbons from propylene, i.e. CPr in the three electrolytic media are lower than for the materials from sucrose CS (Table 4), and they decrease with the total surface area of the carbon materials, i.e. with the filling rate of carbon in the silica porosity [18]. These results are not surprising since the carbon filling is quite uniform during the CVI process, and consequently, the fraction of micropores formed is much less important than for CS. The same as for the CS materials, the carbon made from the MCM-48 template gives higher capacitance values than the carbons resulting from the SBA-15 template. Since the wall thickness of MCM-48 (1 to 1.5 nm) is smaller than that of SBA-15 (4 nm) templates, the template carbon CPr48 possess a more important amount of secondary micropores than CPr15, that is profitable for charging the double layer. The presence of a well-balanced micro/mesoporosity in CPr48 could explain the relatively high capacitance per surface unit, e.g. 12 $\mu F/cm^2$, against only 8 $\mu F/cm^2$ for CPr15.

In conclusion, although the porous texture of these materials is of limited interest for getting high capacitance values, it allows to clearly demonstrate the beneficial effect of mesopores on the capacitor performance.

3.3. C/C Composite for Supercapacitor Electrodes

The composites based on multiwalled carbon nanotubes (CNTs) and N-enriched carbons are novel materials, which may be interesting for supercapacitor application. The example of the composite material prepared with the nitrogenated carbon from polyacrylonitrile (PAN) is analyzed in this paper. The strongly entangled network of the nanotubes reinforces the composite, preventing its shrinkage during the carbonization process. The observations by Scanning Electrone Microscopy technique (Figure 6) show that the C/C composites are formed of a nanotube backbone and individual or interconnected aggregates of pyrolysed PAN.

The capacitance of cells built from the C/C composites were compared with the values given by the single components of the composite. The pristine CNTs gave 19 F/g and the value for carbonized PAN was negligible. The formation of pores, due to PAN decomposition and conversion into microporous carbon matrix, as well as the nitrogen presence could be the reasons of the higher capacitance values. Table 5 summarizes the values of capacitance measured on supercapacitors based on the C/C composite electrodes prepared in different conditions, without oxidative stabilization of PAN. By comparison with the single components, the most striking fact for the C/C composites, was a noticeable increase of capacitance ranging from 50 to 100 F/g. The highest capacitance of 100 F/g was obtained for the composite with 30wt% of CNTs and 70 wt% of PAN, carbonized at 700°C in inert atmosphere for 180 min. Comparable values of atomic percentages of nitrogen (7 and 9 at%) were found from XPS and elemental analysis, respectively on this C/C composite. It demonstrates that nitrogen is well distributed on the surface as well as in the bulk of electrode. On the other hand the BET surface area of this composite is only 210 m^2/g.

These extraordinary properties of the C/C composite prepared with the CNTs/PAN (30wt%/70wt%) mixture are the consequence of several effects provided by the combination of the two components. The PAN matrix is auto-activated by the gas evolution during the thermal decomposition, giving rise to the creation of pores with the help of the CNTs backbones which prevent the matrix from shrinkage. However, taking into account the relatively high value of capacitance compared with the BET surface area, the most important contribution to capacitance in this material seems to be the pseudocapacitance effects provided by the nitrogen functionality.

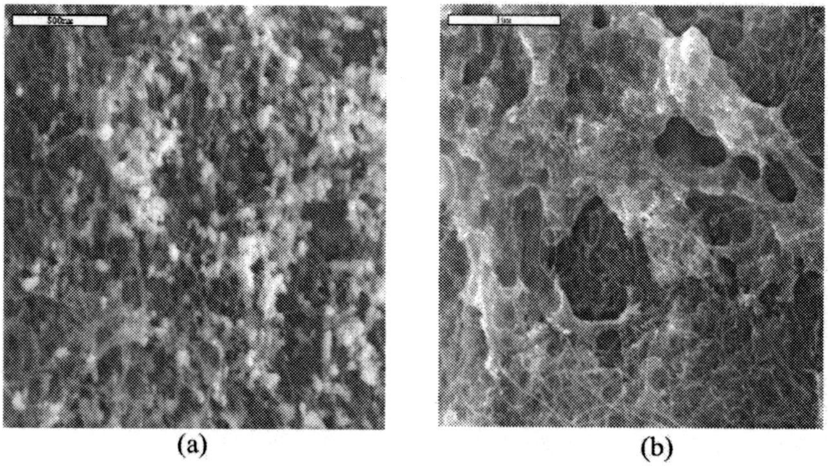

Figure 6. SEM images of the C/C composite showing individual (a) or interconnected (b) aggregates of PAN.

Table 5. Relation between the C/C composites composition, HTT, time of carbonisation and the capacitance values (F/g).

NTs (%)	PAN (%)	HTT (^{0}C)	Time (min.)	C (F/g)
70	30	900	30	48
70	30	700	30	53
30	70	900	30	55
30	70	900	90	48
30	70	700	30	60
30	70	700	90	76
30	70	700	180	100
30	70	700	420	60
15	85	700	180	58
50	50	700	180	57

4. CONCLUSION

The KOH-activated carbons give excellent capacitance values reaching ca. 300 F g^{-1} despite a rather moderate specific capacitance per surface area, being in the range of 7-11 μF cm^{-2}. Apparently, this can be considered as due to a restricted accessibility to the micropores active surface in the case of the KOH activated materials with extremely developed porosity. It seems that the carbon A-PM from mesophase pitch, because of a good precursor organization, gives pores of more adapted size as a consequence of intercalation phenomena during KOH activation. Hence, A-PM is a definitively better candidate for capacitor electrode than the other investigated carbons. A detailed analysis proved that wide micropores and narrow mesopores play a crucial role for ion transportation during charging of the electrical double layer using carbons with extremely developed surface area. Additionally, the low cost of the carbons based on natural precursors makes them extremely attractive for capacitor application. However, for a long-term capacitor performance, the carbons with a low leakage current and limited self-discharge should be selected to fulfil the practical demand.

Interesting carbon materials were prepared by a templating procedure from MCM-48 or SBA-15 highly ordered mesoporous silica. The perfectly controlled texture of the template carbons allows a better understanding of the electrochemical performance in supercapacitors. Capacitance values as high as 200F/g and 110F/g in aqueous and organic media are reached with the carbons CS from sucrose presenting secondary micropores. Taking into account these high values, it is obvious that the presence of interconnected mesopores and secondary micropores makes the active surface more available for charging the electrical double layer than in a strictly microporous material. An improved capacitance-frequency dependence has been found for the essentially mesoporous carbons. To our knowledge, it is the first time that the beneficial effect of mesopores on the diffusion of ions to the active surface is so clearly demonstrated. The template carbons, with a well-tailored pore size and pore distribution are of prime interest for designing high performance supercapacitors.

The N-doped carbons with a nanotube backbone combine a moderate presence of micropores with the extraordinary effect of nitrogen that gives pseudocapacitance phenomena. The capacitance of the PAN/CNts composite (ca. 100 F/g) definitively exceeds the capacitance of the single components (5-20 F/g). The nitrogen functionalities, with electron donor properties, incorporated into the graphene rings have a great importance in the exceptional capacitance behavior.

ACKNOWLEDGMENTS

This research was supported by the NATO Science for Peace project number 973849. Thanks are due to C. Vix from ICSI, Mulhouse, for providing some of the materials investigated in this paper.

REFERENCES

1. Conway B.E. Electrochemical Supercapacitors. Kluwer Academic/Plenum Publishers, New York, 1999.
2. Frackowiak E., Béguin F. Carbon materials for the electrochemical storage of energy in capacitors. Carbon 2001; 39: 937-50.
3. Mayer S.T., Pekala R.W., Kaschmitter J.L. The aerocapacitor: an electrochemical double-layer energy-storage device. J Electrochem Soc 1993; 140:446-51.
4. Laforgue A., Simon P., Sarrazin Ch., Fauvarque J-F. Polythiophene-based supercapacitors. J Power Sourc 1999; 80:142-8.
5. Frackowiak E., Jurewicz K., Delpeux S., Béguin F. Nanotubular materials for supercapacitors. J Power Sourc 2001; 97-98:822-5.
6. Jurewicz K., Delpeux S., Bertagna V., Béguin F., Frackowiak E. Supercapacitors from nanotubes/polypyrrole composites. Chem Phys Lett 2001; 347: 36-40.
7. Arbizzani C., Mastragostino M., Soavi F. New trends in electrochemical supercapacitors. J Power Sourc 2001; 100:164-170.
8. Miller J.M., Dunn B., Tran T.D., Pekala R.W. Morphology and electrochemistry of ruthenium/carbon/aerogel nanostructures. Langmuir 1999; 15:799-806.
9. Toupin M., Brousse T., Bélanger D. Influence of microtexture on the charge storage properties of chemically synthesized manganese dioxide, Chem Mater 2002; 14:3946-52.
10. Wu N.L. Nanocrystalline oxide supercapacitors. Mater Chem Phys 2002; 75:6-11.
11. Frackowiak E., Méténier K., Bertagna V., Béguin F. Supercapacitor electrodes from multiwalled carbon nanotubes. Appl Phys Lett 2000; 77:2421-3.
12. Jurewicz K, Babel K., Ziolkowski A., Wachowska H. Ammoxidation of active carbons for improvement of supercapacitor characteristics. Electrochim Acta 2003; 48:1491-8.
13. Kierzek K., Frackowiak E., Lota G., Gryglewicz G., Machnikowski J. Electrochemical capacitors based on highly porous carbons prepared by KOH activation Electrochim Acta 2004; 49:515-23.
14. Frackowiak E., Delpeux S., Jurewicz K., Szostak K., Cazorla-Amoros D., Béguin F. Enhanced capacitance of carbon nanotubes through chemical activation. Chem Phys Lett 2002; 361: 35-41.
15. Galarneau A., Cambon H., Di Renzo F., Fajula F. True microporosity and surface area of mesoporous SBA-15 silicas as a function of synthesis temperature. Langmuir 2001; 17:8328.
16. Ryoo R., Joo S.H., Kruk M., Jaroniec M. Ordered mesoporous carbons Adv Mater 2001; 13:677-81.
17. Fuertes A.B. Template synthesis of mesoporous carbons with a controlled particle size, J. Mater. Chem. 2003; 13:3085-8.
18. Jurewicz K., Vix C., Frackowiak E., Saadallach S., Reda M., Parmentier J., Patarin J., Béguin F. Capacitance properties of ordered porous carbon materials prepared by a templating procedure. J Phys Chem Solids (2004) in press.
19. Delpeux S., Szostak K., Frackowiak E., Bonnamy S., Beguin F. High yield carbon nanotubes from the catalytic decomposition of acetylene on in-situ formed Co nanoparticles. J Nanosci Nanotech 2002; 2: 481-4.
20. Szostak K., Frackowiak E., Delpeux S., Béguin F. CD-ROM G-1 Carbon'03 Oviedo Spain (2003).

EFFECT OF CARBONACEOUS MATERIALS ON PERFORMANCE OF CARBON-CARBON AND CARBON-Ni OXIDE TYPES OF ELECTROCHEMICAL CAPACITORS WITH ALKALINE ELECTROLYTE

Alexey I. Belyakov*

ELIT JSC (Electrochemical Power Sources)
40, Prospekt Leninskogo Komsomola, Kursk, 305026, Russian Federation

1. INTRODUCTION

Carbonaceous materials play a key role in achieving the necessary performance parameters of electrochemical capacitors (EC). In fact, various forms of carbon constitute more than 95% of electrode composition [1]. Double layer capacity and energy storage capacity of the capacitor is directly proportional to the accessible electrode surface, which is defined as surface that is wetted with electrolyte and participating in the electrochemical process.

Different types of activated carbon are among the most suitable materials for this purpose. For this reason specialists, involved in development of active materials for EC try to increase carbon's specific surface as much as possible and to optimize the internal structure of the carbon porous structure.

Activated carbons possess sufficient volumetric conductivity for electrolyte/collector current interchange. However, contact resistance between carbon particles in the electrode limits charge/discharge currents of the porous volumetric system and therefore EC's power capability.

It should be also mentioned that an increase of activated carbon's specific surface (over activation) always leads to increase of its specific resistance. Different methods of obtaining activated carbon with an optimized volumetric structure were developed [2, 3], but they have not yet found industrial application.

At this point of time, key improvements of volumetric conductivity are being achieved through introduction of highly conductive, finely

* E-mail: elit@pub.sovtest.ru

I.V. Barsukov et al. (eds.), New Carbon Based Materials for Electrochemical Energy Storage Systems, 21–32.

dispersed additives, such as carbon black, graphite, metal powders, carbon fibers, other powdered materials into activated carbon's active mass. Depending on the power demand of a particular EC, percentage of the required conductive additive in the electrode varies in the range of 5 – 40wt%.

Finely dispersed carbon substances introduced into the active mass of pseudocapacitive electrodes of "hybrid" (asymmetric) capacitors play the most important role in operation of an efficient volumetric collector.

A classic case is an EC of a faradic type in which an electrode is comprised of: $Ni(OH)_2$, MnOOH, etc. active materials. Since in these chemistries the conductivity depends on electrode state-of-charge charge level, they require presence of additional stable conductive skeletons in their structure. Noteworthy mentioning that besides traditional forms of carbon or other conductors that may form such a skeleton, the latest progressive investigations demonstrate the possibility of application of different nanostructured forms of carbon, such as single-wall and multi-wall carbon nanotubes [4, 5]. Yet, for the industrial application, highly conductive carbon powders, fibers and metal powders dominate at present.

Summarizing the above, it may be stated that activated carbons and pseudocapacitive materials in EC electrode structure are responsible for the energy storage parameters (specific energy), while non-active highly conductive carbon additives are responsible for the electrode internal resistance (EC specific power).

The present work describes the results of investigations of carbon materials used by us for creation of conductive skeletons in the EC electrode body. Investigations were carried out in full-size Ni Oxide and carbon-carbon supercapacitor systems with aqueous solution of KOH.

2. EXPERIMENTAL

2.1. Conductivity of Feed Materials and Their Dry Mixtures

Before measuring conductivity, carbon materials and their mixtures were dried to residual moisture content of less than 5wt%. A sample of the investigated material was put into a measuring cell, representing a dielectric cylinder closed at both ends by the measuring electrodes.

Electrode pressure on the sample material approximately corresponded to the pressure used in a practical EC in its operational mode and was equal to 8 $kgf{\cdot}cm^{-2}$ (this pressure was applied for purposes of minimization of particle contact resistance).

Conductivity measurements were carried out by an alternating current method using 1 kHz, and measurements were taken at a current density in the range of 2,2-270 $mA{\cdot}cm^{-2}$. Six samples of each formula were analyzed in order to ensure repeatability of data points. In order to reduce the influence of the current density on resistance, as well as for comparability of results with different materials, a diagram for extrapolation of "Zero" current density (J=0) was developed.

2.2. Testing in the Carbon-Carbon type of Electrochemical Capacitors

In our tests, we used pasted mixtures of carbon-carbon electrode components with KOH solution having a density of 1,26 $g \cdot cm^{-3}$. Positive and negative electrodes were pasted onto the conductive polymer film, separated by ionoconductive separator, made out of special paper, pressed between external collectors of nickel-plated copper with pressure of 8 $kgf \cdot cm^{-2}$.

Each cell was sealed via welding all around its edges. The welding seal has been formed on the part of the cell, where conductive polymer makes a boundary with the layer of nonconductive polymer. Electrode thickness did not exceed 150μm; separator thickness was approximately 50μm; active electrode surface dimensions were: 128 x 148 mm. Electric capacity was measured within the voltage range of 0,8v-0,4v by a constant current method. ESR (electrode specific resistance) was measured at a frequency of 1 kHz at measuring current value of 1A. Before the test, all the cells had been kept during 24 hours under working voltage and additionally had been subjected to 10 charge-discharge cycles. Test temperature was kept at 22-24°C.

While choosing an optimum conductive additive, dozens of various carbon and metal powders were researched. In this report, the characteristics of three main representatives are given. They are: Nickel (produced in Russian Federation by means of carbonyl method), colloidal graphite, refined graphite and refined Boron-doped graphite. Superior Graphite Co., Chicago, IL, USA supplied all grades, of high purity graphite mentioned in this paper.

2.3. Construction of NiOx-Carbon Ultra Capacitors

2.3.1. NiOx Electrode

For electrode manufacturing, powder of $Ni(OH)_2$ was used, produced by H.C.Starck, USA. The product has mean particle size of d_{50}~6μm. It was mixed with conductive additive, such as, for instance, refined graphite powder.

The quantity of conductive component varied from 15wt% to 30wt%. Solution of thermoplastic polymer in organic solvent was prepared separately. Then, a mixture of dry components $Ni(OH)_2$ + graphite was added to that solution and mixed one more time. Content of polymer binder in composition of a dry electrode was 5-10wt%.

Electrode film was formed from a mixture of components in organic solvent by means of casting, which after drying (removing of solvent) was rolled by dry method to the desirable thickness (150-200μm). As a result, stable forms of self-supported electrodes were obtained. Then these electrodes were subjected to formation during two charge-discharge cycles in the solution of electrolyte, consisting of 40wt% KOH and $10 g \cdot l^{-1}$ LiOH.

2.3.2. Carbon Electrode

Activated carbon cloth of 1-1.1mm thickness was used as the initial material for most of experiments. Ribbon, made of carbon cloth, was soaked in the solution with 40wt% KOH and $10\ g \cdot l^{-1}$ LiOH for 72 hours, and afterwards, was cut into electrodes of the appropriate dimensions. Electrolyte surplus was removed by pressing.

2.3.3. Electrochemical Cells

An electrode block, which included a positive NiOx electrode, a separator, made of special paper (60μm), and a negative electrode, were placed between two sheets of conductive plastic of 50-60μm thick.

The upper and the lower plastic sheets were welded along the edges through the insolating layer with an exception of the area which was used preliminary as a ventilation outlet.

The assembled cell was put between two metal collectors and pressed to the level of operating pressure for purposes of reduction of the contact resistance of assembly parts between power covers of the clamp unit.

3. RESULTS AND DISCUSSION

3.1. Conductivity of Electrodes

Following the procedure described in section 2.1, initially, conductivity parameters of dozens of materials were analyzed. Below, we present data for conductivity of typical representatives of families of conductive materials, which we used in our study. Data is given in Table 1.

The table shows, that resistance of dry activated carbon is more than 20 times higher than that of the least resistive graphite. In this table, Boron-doped natural graphite, 2939APH–RG follows next after activated carbon.

More interesting data was found in the mixtures: we have seen that carbon additives tend to reduce specific resistance of the mixture, while active material (activated carbon) increases resistance (54 10^{-3} ohm·m vs. 49 10^{-3} ohm·m). This phenomenon suggests that carbon-carbon component mixtures have better "packing" characteristics.

Table 1. Specific Resistance of Dry Powdered Materials.

TYPE OF CONDUCTIVE MATERIALS	SPECIFIC RESISTANCE, j = 0, ohm·m
Activated Carbon (grade SKT)	$49 \cdot 10^{-3}$
Carbon Black (grade P- 267)	$2{,}1 \cdot 10^{-3}$
Expanded Graphite (grade ABG1005, Superior Graphite)	$1{,}1 \cdot 10^{-3}$
Purified Flake Graphite (grade 2939APH, Superior Graphite)	$2{,}6 \cdot 10^{-3}$
Boron-doped purified flake graphite (grade 2939APH-RG, Superior Graphite)	$0{,}9 \cdot 10^{-3}$
Ni Powder (Carbonylic Method)	$0{,}75 \cdot 10^{-3}$
ABG1005 (8wt%) + Activated Carbon (92wt%)	$35 \cdot 10^{-3}$
Ni Powder (10wt%) + Activated Carbon (90wt%)	$54 \cdot 10^{-3}$
Carbon Black (15wt%) + Activated Carbon (85wt%)	$11 \cdot 10^{-3}$

Notes for the Table: Material particle size was as follows:
- activated carbon: < 60μm (10-20)
- carbon black: < 5μm
- nickel powder: < 5 μm
- graphite powders "Superior Graphite": <20μm.

It is noteworthy mentioning that above experiment allows drawing only a preliminary conclusion, as far as selection of an optimum conductive additive is concerned.

It should be noted, that ESR of electrode, having minimum resistance in dry state may be significantly increased during impregnating it with electrolyte solution (said resistance increase is typical in n-type solid phase composites). It is caused by the expansion effect due to formation of electrolyte films during wetting of the surface of electrode particles. Thus, "dry" conductivity data may serve only as an initial estimation during optimization of volumetric structure of an EC electrode.

3.2 Effect of Carbonaceous Materials on Performance of Carbon-Carbon EC in presence of an Alkaline Electrolyte

In active material of both electrodes of the EC with purely "double electric layer", volume changes do not take place during charge-discharge processes. That's why it is not expedient to add binder into active material.

However, that is observed only within the range of the so-called "safe" voltage window of operation – for alkaline aqueous electrolytes it covers the range of 1,12-1,24v, and depends on material purity and current density. Exceeding this voltage causes gassing, electrode swelling and loss of electric contact between particles in the electrode volume.

Summarized "as measured" and "calculated" data for electrochemical cells and active electrodes with practical dimensions of the electrodes of 128x148mm are shown in Table 2 (please see cell assembly detail in section 2.2).

Table 2. Parameters of capacitors as a function of conductive additive used in electrode composition.

PARAMETERS	**Conductive Additive and its Concentration in the Electrode**				
	Expanded Graphite, ABG1005, 8wt%	Natural Flake Graphite, 2939APH, 8wt%	Boron-Doped Flake Graphite, 2939APH-RG, 8wt%	Carbon Black, P-267, 5wt%	Ni Powder, 10wt%
Capacity, F	90	88	92,4	98	104
ESR, 1 kHz, Ohm	0,92	1,03	0,84	0,75	1,07
Energy Density, $kj\cdot kg^{-1}$	2,12	2,04	2,29	2,5	2,30
RC-time constant, ms (milliseconds)	83	91	78	70	111
Power Density, $W\cdot g^{-1}$	12,8	11,3	14,8	15,2	10,7

Analyzing the data one may notice, that cells with carbon as conductive powders have advantage over cells using metal as conductive powders, possessing 1,5 times higher specific power at comparable specific energy level. Carbon black P-267 and graphite 2939APH-RG have the best performance indexes. Its higher specific surface area in comparison with that of graphite explains the higher carbon black indexes. Higher surface area provides higher specific sorption of electrolyte and improvement of electrode conductivity according to the conductivity of the second kind.

As far as cycle life of the ECs is concerned, with the above conductive additives, and the active mass, consisting of activated carbon, for all groups of EC we built for this work, it exceeded 10,000,000 cycles (at depth of discharge of 30%); that is quite sufficient for the main spheres of application.

3.3 Effect of Carbonaceous Materials on Performance of Carbon-Ni Oxide Type of EC with Alkaline Electrolyte

The character of carbonaceous material's influence on performance of asymmetric EC systems (carbon – NiO_x) is more complicated than that of carbon-carbon system. It is determined by the higher operating potential of NiOx electrode, due to which oxidation and volume changes in electrode active mass upon charge-discharge processes are taking place as the reaction of transformation of $Ni(OH)_2$ into NiOOH is occurring.

1 - Current Collector;
2 - Volume Collector - Chain Structures;
3 - Particles of Active Material;

Rki - Contact Resistance:
Volume Collector / Current Collector;
Rci - Contact Resistance between particles of
Volume Collectors;
Rvi - Bulk Resistance of materials of particles;
Rmi - Contact Resistance:
Volume Collector / Active material;
Rb - Resistance, determined by binder,
increased the all Ri;
AM - Active material.

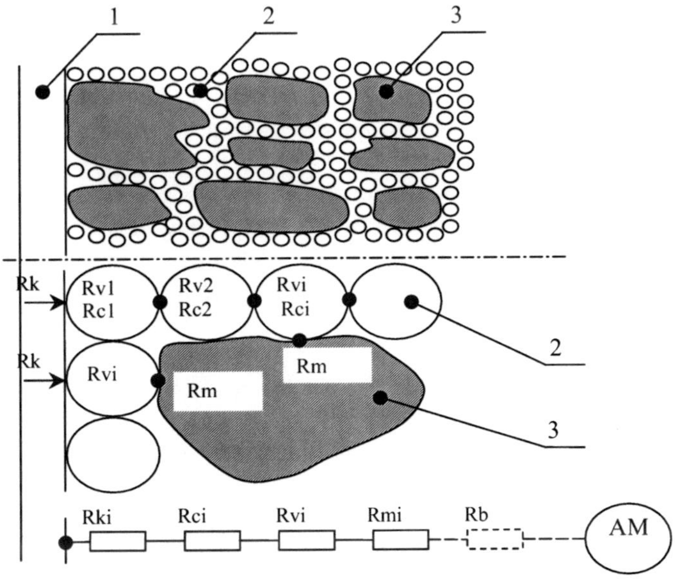

Figure 1. Simplified model of agglomerate electrode.

First of all let us consider the morphological structure of an agglomerate electrode [6] by way of example of the model shown in Figure 1. This schematic represents a multiphase system with no fixed connection between its components. As a rule, the active mass of an electrode is a mixture of Nickel hydroxide (oxyhydroxide) with conductive carbon or a metal, which are well dispersed mechanically in the matrix.

The NiO_x particles, when in operation, are closely packed with conductive additive by means of pressing of the electrode block or addition of a binder into the active mass slurry. The quantity of conductive additive should be chosen pursuing the necessity to create a spatially ramified grid providing current transmission from the active material particles to an external collector. In dependence on reciprocal dispersion of conductive additive and active material, the quantity of conductor in the mass may vary from 10wt% to 70wt% of the total electrode mass.

The conductive particles in the volume of electrode make a so-called "chain" structures. As it is seen from the model, the electrical circuit assumes interconnection of the active mass particles in the solid phase of the electrode, while the external collector consists of a long chain of resistors.

In accordance with this model, the current path is made through the contact resistance: external collector – volume collectors (Rki), then goes through many other resistances, such as: bulk resistance of the conductive particles (Rvi) and the contact resistance between the conductive particles in the volume of the electrode (Rci); at the end, there is contact resistance between the conductive particles and active mass particles (Rmi).

Considering this chain of electrical resistors, we should take into account the influence of the binder (dielectric) effecting on all the above contact resistances in the circuit and increasing their value (Rb). Evidently, the main influence on the ESR of electrode is this circuit because of *a big number of contact resistances distributed in the bulk of the electrode.*

We consider that the practical electrode's efficiency at operation under high current density conditions and during service life is determined by the state of the particles of conductive additive's surface. With reference to this, we can point out two main factors effecting fundamentally reliable operation of the NiO_x electrode.

(a) Electrochemical oxidation of the particles' surface of the conductive additive. Potential of charged NiOx in KOH electrolyte is 0,45v to 0,5v vs SHE (standard hydrogen electrode). This value is sufficient for fast oxidation of carbon materials and appreciable for metal powder. Even insignificant oxidation of particle surfaces causes dramatic ESR increase on electrode, because we have here a long series circuit of contact resistances between the particles.

(b) Electrode swelling. The agglomerate electrode, when it even has binder inside, is not a solid structure. Volumetric changes in morphology (α, β-$Ni(OH)_2$; β, γ-NiOOH) arise thickening of electrode. It is evident even in

tightly squeezed electrode stacks. Thus, a mechanical movement of the volumetric conductor particles from the active mass particles, and from each other takes place. Simultaneously, the contact area between the particles diminishes with time, the growth of the contact resistance increases the total ESR of the electrode.

Combined influence of both of the above factors significantly limits the performance and operation capability of the NiOx electrodes of agglomerate type in the asymmetric C/NiOx capacitors. Good initial performance may change for accelerated failure either in the floating mode or in heavy duty cycling mode of operation.

3.4. Electrochemical Testing

In the experiments discussed hereafter, carbon black was not taken into account as the loss of the performance parameters of EC due to ESR growth appeared soon after several charge–discharge cycles with this type of conductive additive.

Initially, the amount of conductive additives was matched up to the active mass of Ni Oxide electrode, and then its amount was fine-tuned going by the criteria of having a minimum ESR and reaching highest oxidation resistance. Thus, for the colloidal graphite the optimum quantity of conductor became 12wt%, for Ni powder – 60wt%, for the refined graphite – 25wt%. The design of experimental cells are thoroughly described in our previous papers [7], and also detail is given in section 2.3.

Assembled cells were cycled in the most "severe" mode; known as "continuous day cycling". This is an uninterruptible (without pauses) cycling during 8 hours in the voltage window 1,5 – 1,1v.

The upper level of charging voltage, 1,5v is a critical value for the electrochemical system C–NiO_x in aqueous electrolyte. Here, formation of some quantity of atomic oxygen on the NiOx electrode takes place, which oxidizes electrode components more rapidly. Figure 2 shows a typical test cyclogram. Upon such cycling, an electrochemical device goes through approximately 1400 cycles in a day.

Figure 3 shows changes of the ESR and capacity during cycling. High stability of conductive carbon, represented by Boron-doped refined graphite (2939APH–RG) in this critical test conditions was observed.

For comparison, less superior test results of ECs with conductive additives of colloidal graphite and powdered nickel are shown in Figure 4. ESR of EC with refined graphite has increased by 56% after 36,000 cycles, and with nickel powder by 81% already after 1,400 cycles. Resistance of EC colloidal graphite increased by as high as 4,5 times compared to initial values already by the 500th cycle.

Capacity loss for EC with graphite 2939APH–RG was 13% during the 36,000th cycle, while with nickel powder, it reached 43% by approximately the 1,000th cycle; and with colloidal graphite the capacity loss came to be 32% by the 500th cycle.

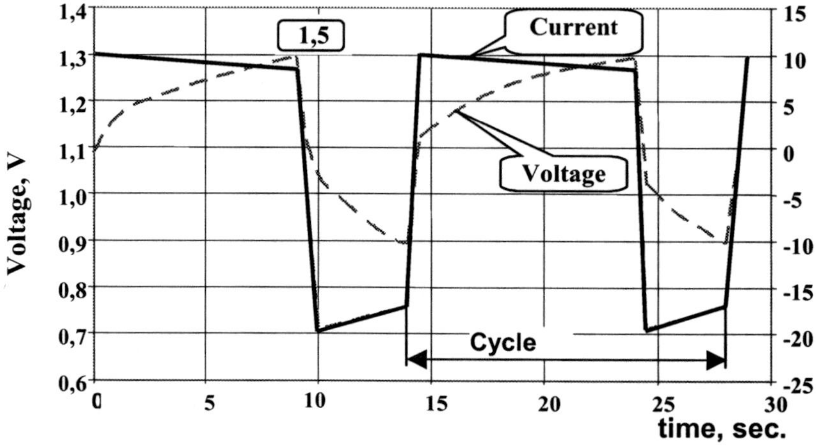

Figure 2. Cycling diagram in the "day-cycling" mode.

Carried out testing confirmed the supposition about the degradation mechanism of electrode in NiOx-carbon ultra capacitors (Figure 1). Obviously, the smaller are the dimensions of particles of conductive additive (colloidal graphite) and the more active its surface is (nickel powder), the faster oxidation processes are taking place on the surface of these particles.

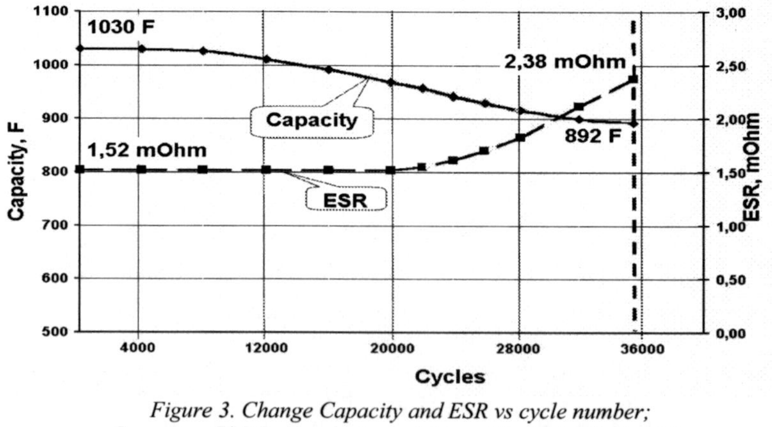

Figure 3. Change Capacity and ESR vs cycle number;
Capacitor C/NiOx, conductive additive: 25wt% 2939APH-RG;
Cycling regime: 1,400 cycles/day plus 650 hours of floating.

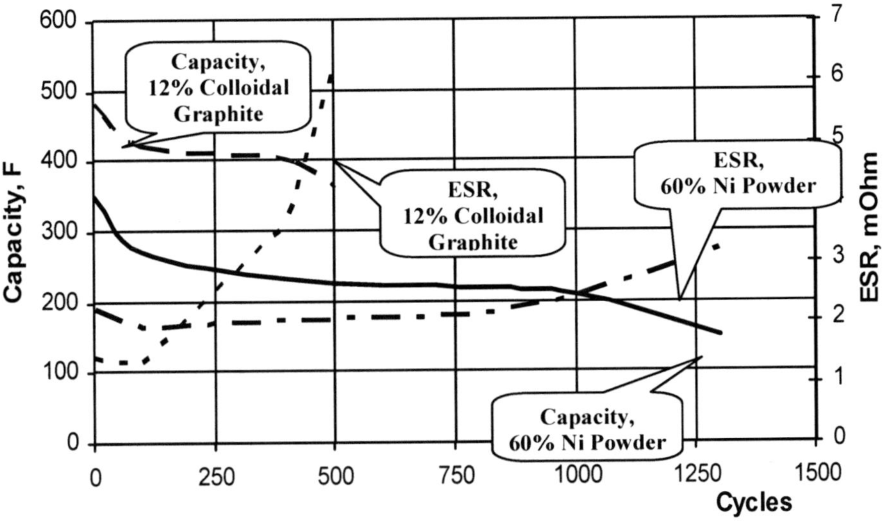

Figure. 4. Change of Capacity and ESR during cycling.

These parameters appear to be causing avalanche-type growth of contact resistances and accelerated failure of EC due to high ESR and capacity loss during operation. Instead, doping of flake graphite with Boron (2939APH-RG sample) appears to significantly improve oxidation resistance of conductive additive, thus resulting in superior cycle life of NiOx-carbon supercapacitors.

4. CONCLUSIONS

1. For electrochemical capacitors of the system carbon-carbon, in spite of the fact that the electrode body consists of conductive activated carbon, it is always necessary to use highly conducive additives, *preferably those selected from the group of carbon materials.* Exact type of additives plays a secondary role for this type of an EC, as it operates in the so-called "safe" voltage interval (1,12-1,24v), where limited to no oxidation of carbon takes place in this range of voltage.

2. For the electrochemical capacitors of Carbon-Ni Oxide system with aqueous KOH solution it is expedient to use carbonaceous graphite materials with expanded structure and modified surface. The best results were achieved with carbon surface doped with Boron, which makes this carbon superior than other conductive additives used in this study, due to its

increased resistance to oxidation. In this case, a number of contact resistances in the model presented by Figure 1 is decreased. Boron doping appears to lead to higher stability of conductive material under conditions of electrochemical oxidation (1,1-1,5v). Increased oxidation resistance enables long service life of an EC even in critical operation conditions.

3. Presently and for the near future, the greatest demand during development and production of electrochemical capacitors will lie in the sphere of optimization of the highly conductive micronized carbon structures, with modified surfaces against oxidation.

REFERENCES

1. B.E. Conway, *"Electrochemical Supercapacitors. Scientific Fundamentals and Technological Applications", 1999, Kluwer Academic /Plenum Publishers, NY.*
2. T. Extrom, S. Gordeev and V. Kuznetsov, *in Proceedings of the 6th International Seminar on DLC and Similar Energy Storage Devices, Dec. 9 – 11, 1996, Deerfield Beach, FL, USA.*
3. S. Dietz, V. Nquyen, *in Proceedings of the 11th International Seminar on DLC and Similar Energy Storage Devices, Dec. 3 – 5, 2001, Deerfield Beach, FL, USA.*
4. Lipka, D. Reisner, J. Dai, R. Ge, R. Cepulis, *in Proceedings of the 10th International Seminar on DLC and Similar Energy Storage Devices, Dec. 4 – 6, 2000, Deerfield Beach, FL, USA.*
5. D. Firsich, S. Lipka, *in Proceedings of the 10th International Seminar on DLC and Similar Energy Storage Devices, Dec. 4 – 6, 2000, Deerfield Beach, FL, USA.*
6. A. Beliakov, *in Proceedings of the 11th International Seminar on DLC and Similar Energy Storage Devices, Dec. 3 – 5, 2001, Deerfield Beach, FL, USA.*
7. A. Beliakov, *in Proceedings of the 12th International Seminar on DLC and Similar Energy Storage Devices, Dec. 9 – 11, 2002, Deerfield Beach, FL, USA.*

HYBRID SUPERCAPACITORS BASED ON α-MnO_2/CARBON NANOTUBES COMPOSITES

Volodymyr Khomenko, Encarnacion Raymundo-Piñero and François Béguin*

Centre de Recherche sur la Matière Divisée. CNRS-University, 1B Rue de la Férollerie, 45071 Orléans Cedex 02, France

1. INTRODUCTION

In recent years, attention has been focused on use of various metal oxides as supercapacitor electrodes for high-power applications[1-4]. The energy storage mechanism with these materials is based mainly on fast faradaic redox reactions, which occur at the interface between the oxide and the electrolyte[4-5], giving rise to the so-called "pseudo-capacitance". The great advantage in this case is that the specific capacitance which can be obtained is higher than for electrochemical capacitors where the charge is only stored in the double layer. The characteristics required for the use of a metal oxide as capacitor electrode are a pseudo-capacitive behavior, a large specific surface area, a high conductivity and electrochemical stability. Among the various metal oxides studied, amorphous and hydrated manganese dioxide (a-$MnO_2 \cdot nH_2O$) is the most promising owing to the low cost of the raw material. However, due to the low electrical conductivity of a-$MnO_2 \cdot nH_2O$, a conducting additive is required to realize a composite electrode for supercapacitors. In this study, taking into account that carbon nanotubes (CNTs) are strongly entangled, forming an interconnected conducting network, we propose to use them as alternative agent to carbon black. Since the electrochemical behavior of manganese oxide has been previously studied only with three-electrode cell systems and in the positive range of potential[6-12], we also propose to test real supercapacitors, i.e. two electrode cells, based on the a-MnO_2/CNTs composites. Therefore, for the first time, the problems of using manganese oxide in symmetric capacitors will be extensively discussed. Finally, the asymmetric construction, previously proposed by Hong *et al.*[13], will be optimized in order to allow both

* Corresponding author: E-mail: beguin@cnrs-orleans.fr

I.V. Barsukov et al. (eds.), New Carbon Based Materials for Electrochemical Energy Storage Systems, 33–40.

electrodes to operate in an optimal potential range, that broadens the voltage window of the device up to 2V in aqueous medium.

2. EXPERIMENTAL

Amorphous and hydrated manganese oxide (a-$MnO_2{\cdot}nH_2O$) was prepared by chemical co-precipitation from $KMnO_4$ and $Mn(OAc)_2.4H_2O$ in water medium[6-7]. For obtaining an homogeneous active material in view of the preparation of composite electrodes, predetermined amounts of carbon nanotubes (0, 15, 20, 25, 30 wt%) were added to the solution before α-$MnO_2{\cdot}nH_2O$ precipitation. The active material was also mixed with carbon black for comparison purposes. The high purity carbon nanotubes (CNTs) used in the present study were prepared by decomposition of acetylene over a powdered $Co_xMg_{1-x}O$ solid solution catalyst[14]. The amorphous character of the oxide was confirmed by XRD measurements. Applying the BET equation to its N_2 adsorption isotherms at 77 K, specific surface areas of around 250 m^2/g were obtained.

The capacitor electrodes were films formed by a mixture of the active material (a-$MnO_2{\cdot}nH_2O$/carbon composite, 90 wt%) with a binder solution (Teflon, 10%). The mixture was thoroughly homogenized in an agate mortar and dried at room temperature before being rolled into a thin film of uniform thickness. Electrode pellets were cut with a surface of 1 cm^2. Two electrode cell capacitors were built with a glassy fibrous separator and gold current collectors, using a Teflon Swagelok® type system. Different aqueous electrolytic solutions have been used: 1 $mol.L^{-1}$ of Na_2SO_4 or 2 $mol.L^{-1}$ KNO_3 with a pH of 6.4 or with a modified pH of 10 by the addition of NaOH. The values of capacitance were estimated by voltammetry (scan rate of potential from 2 to 10 mV/s) and galvanostatic charge/discharge cycling (VMP-Biologic, France). Additional three electrode cell experiments were done by using a Hg/HgO reference electrode.

3. RESULTS AND DISCUSSION

Table 1 contains the values of specific capacitance obtained when different amounts of carbon nanotubes are added to a-$MnO_2{\cdot}nH_2O$ for the preparation of the composite electrodes.

It is clearly demonstrated that adding carbon nanotubes improves the behavior of a-$MnO_2{\cdot}nH_2O$ as capacitor electrode. The specific capacitance values referred to the mass of a-$MnO_2{\cdot}nH_2O$ increase with the amount of CNTs. However, when the specific capacitance is referred to the total mass of the composite electrode material, to be realistic, it can be noticed that a carbon nanotubes loading higher than 10-15wt% does not improve the electrodes performance. Therefore, 10-15wt% of CNTs as conductive additive are enough to increase the capacitance of the a-$MnO_2 \cdot nH_2O$ based electrodes from 0.1 to ca. 140 F/g. Due to their entangled network forming open mesopores, the CNTs favor the ions accessibility from the electrolyte to an important part of the active a-$MnO_2 \cdot nH_2O$.

Table 1. Specific capacitance in 1 mol.L^{-1} Na_2SO_4 medium for a-$MnO_2 \cdot nH_2O$ loaded with different weight percentages of carbon nanotubes.

Carbon Nanotubes loading (wt%)	***Specific Capacitance ($F/g_{composite}$)***	***Specific Capacitance (F/g_{oxide})***
0	0.1	0.1
5	19	20
10	137	154
15	137	161
20	139	174
25	141	188
30	138	197
100	20	20

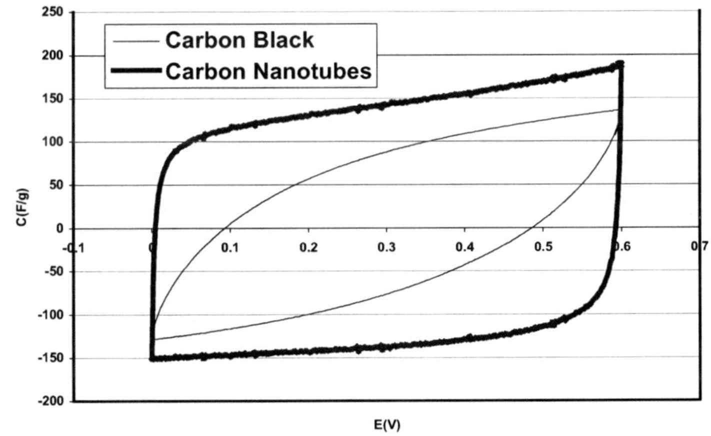

Figure 1. Cyclic voltammograms at 2mV/s in 2 mol.L^{-1} KNO_3 medium for two-electrode capacitors based on a-$MnO_2 \cdot nH_2O$ loaded with 15 wt% of carbon black or carbon nanotubes.

Figure 1 presents a comparative voltamperometric analysis of a-$MnO_2 \cdot nH_2O$ electrodes loaded with 15wt% of CNTs or 15wt% of carbon black, which is conventionally used as conductivity additive for electrodes. Using carbon nanotubes instead of carbon black produces a remarkable improvement of the capacitive behavior of the a-$MnO_2 \cdot nH_2O$ based electrode. The shape of the voltammetry curve demonstrates that the resistance of the electrode with CNTs is much lower than in the case of using the conventional carbon black, finding a performance closer to an ideal capacitor. An additional highly valuable effect of CNTs is the possibility to extract more energy from the a-$MnO_2 \cdot nH_2O$ electrodes, obtaining specific capacitance values of 140 F/g in comparison with only 70 F/g when carbon black is used.

Another important aspect for a material to be used as electrode for supercapacitors is its electrochemical stability. In Figure 2, presenting the specific discharge capacitance versus the cycle number for the optimized a-$MnO_2 \cdot nH_2O$/CNTs composite in 2 mol·L^{-1} KNO_3 (pH=6.5), it can be observed that the specific capacitance loss after 200 cycles is about 20%.

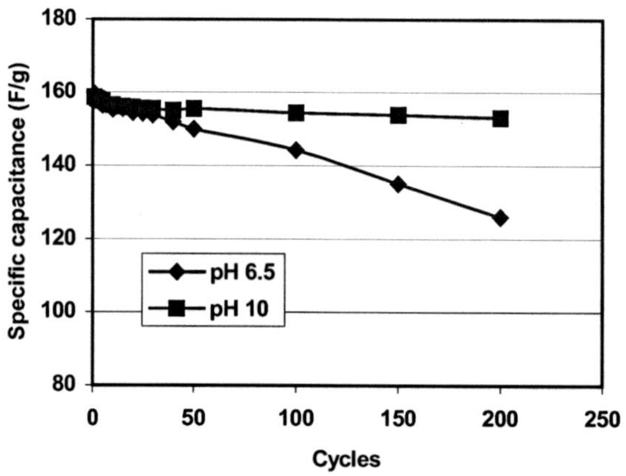

Figure 2. Specific discharge capacitance for the a-$MnO_2 \cdot nH_2O$ composite with 15 wt% of CNTs in 2 mol·L^{-1} KNO_3 medium at two different pH.

For explaining this decay, it has to be considered that the pseudocapacitance of hydrous oxides like a-$MnO_2 \cdot nH_2O$ is attributed to redox transitions with exchange of protons and/or cations with the electrolyte following the equation[12]:

$$MnO_a(OH)_b + nH^+ + ne^- \leftrightarrow MnO_{a-n}(OH)_{b+n},$$

where $MnO_a(OH)_b$ and $MnO_{a-n}(OH)_{b+n}$ indicate interfacial a-$MnO_2 \cdot nH_2O$ in higher and lower oxidation states, respectively. For the stability of the system these reactions must be reversible. After investigating the working potential range for the positive and negative electrodes, it is possible to observe that the negative electrode operates in the range of the irreversible Mn(IV) to Mn(II) reaction, which results in the dissolution of the electrode, being the reason of the poor cyclability. Taking into account that the potential for the equilibrium between Mn(IV) and Mn(II) depends on the electrolyte pH, the latter was increased to a value of 10 in order to avoid such an irreversible reaction. Figure 2 shows that the stability is remarkably increased, finding a loss of only 4 % of the initial specific capacitance after 200 cycles. However, even solving this problem, an important drawback of this kind of symmetric capacitor is the value of the working voltage which cannot be higher than 0.6 V.

The limitations of manganese dioxide in a two electrode capacitor were overcome by using activated carbon at the negative electrode. Such an asymmetric system was previously proposed[13], without sufficient explanation for the performance observed. In the present study, a deep study of the mechanism of charge storage for both electrodes allowed the system to be optimized.

Activated carbons (AC) are usually used for symmetric electrochemical double layer capacitors owing to their high specific surface area[5]. However, pseudocapacitance properties of AC in aqueous medium were shown when a negative polarization is applied[15]. Figure 3 presents the voltammetric analysis of AC in 2 mol.L^{-1} KNO_3 showing that high overpotential values for di-hydrogen H_2 evolution could be reached, up to ca. 0.5 V (the thermodynamic value in this medium is around –0.37 V vs NHE). Hydrogen produced by water decomposition at potentials lower than the thermodynamical values is immediately adsorbed in the carbon porosity[15], being the reason of the pseudocapacitive behavior.

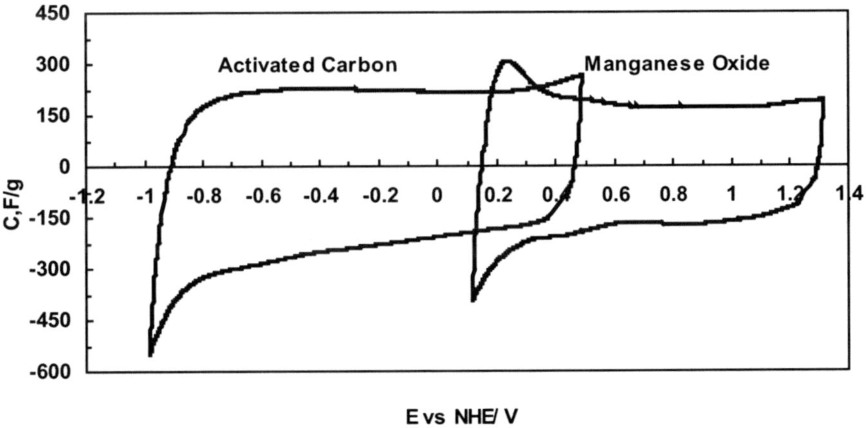

Figure 3. Cyclic voltammograms in three-electrode cells for activated carbon and for a-$MnO_2{\cdot}nH_2O$ loaded with 15wt% of carbon nanotubes in 2 mol·L^{-1} KNO_3 medium using Pt as auxiliary electrode.

Figure 3 also shows that manganese oxide has a very interesting behavior under positive polarization giving a high electrochemical reversibility up to 1.2 V. Therefore, by combining these two materials in an asymmetric system with activated carbon and amorphous manganese oxide as negative and positive electrodes, respectively, it is theoretically possible to obtain a working voltage window of 2.2 V in aqueous medium. Figure 4 presenting the voltammetry curve for this kind of asymmetric capacitor, confirms a quite rectangular shape characteristic of an ideal capacitive behavior up to 2.2 V.

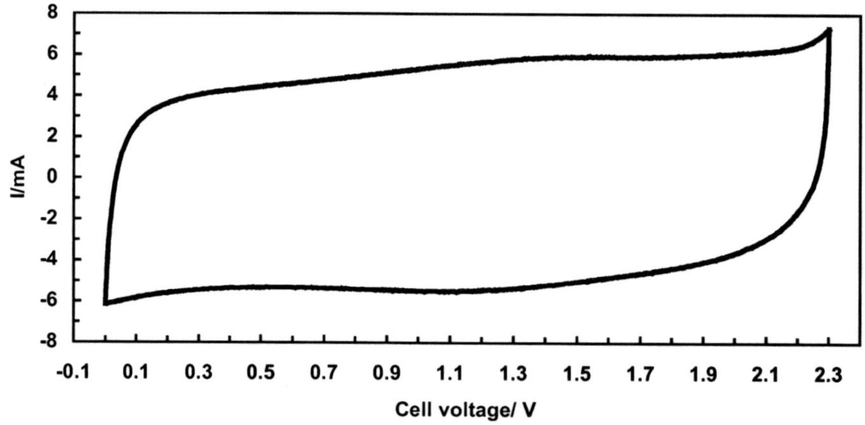

Figure 4. Cyclic voltammogram of an asymmetric capacitor with activated carbon and α-$MnO_2{\cdot}nH_2O$ as positive and negative electrodes, respectively, in 2 mol·L^{-1} KNO_3 medium.

In practicality, after optimizing the system, it was possible to obtain a working voltage window of 2 V in aqueous medium, while keeping an excellent cyclability. In these conditions, the energy density of the hybrid capacitor reaches 12.6 Wh.kg^{-1} that is an order of magnitude higher than in symmetric systems with activated carbon or manganese oxide in aqueous medium, and relatively close to the values of electric double layer capacitors working in organic electrolytes. Due to the various advantages of the aqueous electrolyte medium, it is worth to deeply investigate the systems based on this concept.

4. CONCLUSION

For the first time, a-$MnO_2{\cdot}nH_2O$ based composites have been studied in real two electrode capacitors. The a-$MnO_2{\cdot}nH_2O$/CNTs electrodes demonstrate an ideal capacitive behavior and high values of capacitance. Compared to the conventional carbon black, multi-walled CNTs are a very promising conductivity additive for capacitor or battery electrodes.

The performance of manganese oxide in a real two-electrode capacitor is limited by the irreversible reaction Mn(IV) to Mn(II) taking place at the negative electrode, which potential depends on the electrolyte pH.

A hybrid capacitor was optimized by combining two materials with a pseudocapacitive behavior in aqueous medium, i.e. the a-$MnO_2 \cdot nH_2O$/CNTs composite and an activated carbon as positive and negative electrodes, respectively. The optimized system gives a practical cell voltage of 2 V in aqueous medium, with energy densities close to the values obtained with electric double layer capacitors working in organic electrolytes.

ACKNOWLEDGMENTS

This research was supported by a Marie Curie fellowship of the European Community program "Improving Human Research Potential and the Socio-Economic Knowledge Base" under contract number HPMF-CT-2001-01453 and by a NATO Science for Peace grant number 973849.

REFERENCES

1. Zheng JP, Jow TR. A new charge storage mechanism for electrochemical capacitors. J. Electrochem. Soc. 1995; 142(1):L6.
2. Long JW, Swider KE, Merzbacher CI, Rolison DR. Voltammetric characterization of ruthenium oxide-based aerogels and other RuO_2 solids: The nature of capacitance in nanostructured materials. Langmuir 1999; 15(3):780-5.
3. Wu NL. Nanocrystalline oxide supercapacitors. Mat. Chem. Phys. 2002; 75:6-11.
4. Conway BE, Birss V, Wojtowicz. The role and utilization of pseudocapacitance for energy storage by supercapacitors. J. Power Sources 1997; 66:1-14.
5. Frackowiak E, Béguin F. Carbon materials for the electrochemical storage of energy in capacitors. Carbon 2001; 39:937-50.
6. Lee HY, Goodenough JB. Supercapacitor Behaviour with KCl Electrolyte. Journal of Solid State Chem. 1999; 144:220-3.
7. Lee HY, Kim SW, Lee HY. Expansion of active site area and improvement of kinetic reversibility in electrochemical pseudocapacitor electrode. Electrochemical and solid state letters 2001; 4(3):A19-A22.
8. Pang SC, Anderson MA, Chapman TW. Novel Electrode Materials for Thin-Film Ultracapacitors: Comparison of Electrochemical properties of Sol-Gel Derived and Electrodeposited Manganese Dioxide. J. Electrochem. Soc. 2000; 147(2):444-50.
9. Jiang J, Kucernak A. Electrochemical supercapacitor material based on manganese oxide: preparation and characterization. Electrochim. Acta 2002; 47:2381-6.
10. Hu CC, Tsou TW. Capacitive and textural characteristics of hydrous manganese oxide prepared by anodic deposition. Electrochim. Acta 2002; 47:3523-3532.
11. Jeong YU, Manthiram A. Nanocrystalline manganese oxides for electrochemical capacitors with neutral electrolytes. J. Electrochem. Soc. 2002; 149(11):A1419-A1422.
12. Hu CC, Tsou TW. Ideal capacitive behaviour of hydrous manganese oxide prepared by anodic deposition. Electrochem. Comm. 2002; 4:105-9.

13. Hong MS, Lee SH, Kim SW. Use of KCl aqueous electrolyte for 2 V manganese oxide/activated carbon hybrid capacitor. Electrochemical and solid state letters 2002; 5(10):A227-A230.
14. Delpeux S, Szostak K, Frackowiak E, Bonnamy S, Beguin F. High yield carbon nanotubes from the catalytic decomposition of acetylene on in-situ formed Co nanoparticles. J.Nanosci.Nanotec. 2002; 2: 481-84.
15. Jurewicz K, Frackowiak E, Béguin F. Towards the mechanism of electrochemical hydrogen storage in nanostructured carbon materials. Appl. Phys. A 2003; in press.

DEVELOPMENT OF SUPERCAPACITORS BASED ON CONDUCTING POLYMERS

V. Khomenko[1,3], E. Frackowiak[2], V. Barsukov[1*], F. Béguin[3]

[1]*Kiev National University of Technology and Design Nemorovicha-Danchenko 2, Kiev 02001, Ukraine*
[2]*Poznan University of Technology, ul. Piotrowo 3, 60-965 Poznan, Poland*
[3]*CRMD, CNRS-University, 1B rue de la Férollerie, 45071 Orléans Cedex 02, France*

Abstract

Electronically conducting polymers (ECPs) such as polyaniline (PANI), polypyrrole (PPy) and poly(3,4-ethylenedioxythiophene) (PEDOT) have been applied in supercapacitors, due to their excellent electrochemical properties and lower cost than other ECPs. We demonstrated that multi-walled carbon nanotubes (CNTs) prepared by catalytic decomposition of acetylene in a solid solution are very effective conductivity additives in composite materials based on ECPs. In this paper, we show that a successful application of ECPs in supercapacitor technologies could be possible only in an asymmetric configuration, i.e. with electrodes of different nature.

1. INTRODUCTION

The most common electrical energy storage devices are capacitors and batteries. Rechargeable batteries generally provide a high energy and a rather low specific power. Usually, capacitors can supply high specific power, but the amount of energy stored is very low. Electrochemical capacitors (often called supercapacitors or ultracapacitors) [1,2] can store the electric charge and the corresponding energy at high density with an excellent reversibility, as a regular capacitor, and hence can be operated at substantially greater specific power than most batteries. Supercapacitors can be classified basically into two different categories depending on the charge storage mechanism. Thus, there are the electric double layer capacitors (EDLCs) generally based on activated carbon materials and the so-called pseudocapacitors based on electronically conducting polymers (ECPs) [3-6] or metal oxides [7,8]. ECPs are promising materials for the realization of

* Corresponding author. E-mail: chemi@mail.vtv.kiev.ua

I.V. Barsukov et al. (eds.), New Carbon Based Materials for Electrochemical Energy Storage Systems, 41–50.

high performance supercapacitors, as they are characterized by high specific capacitances. Indeed, the charge storage with ECPs involves all the polymer mass and not only the surface, as in the case of double-layer capacitors based on activated carbons. Furthermore, the charge–discharge processes are generally fast with ECPs. However, compared to the activated carbons, which are commonly used as electrode materials in commercial supercapacitors, the ECPs still present some important drawbacks (e.g. poor stability), which preclude an industrial utilization. In order to make a progress towards the application of ECPs in supercapacitor technologies, it is necessary to improve the materials quality and their electrochemical properties.

In this paper, we report on the preparation of ECP composites based on carbon materials. In parallel with the development of the preparation processes and the electrochemical characterization of composites, we have performed an analysis of the supercapacitor cell design based on ECPs.

2. EXPERIMENTAL

Electronically conducting polymers (ECPs) such as polyaniline (PAN), polypyrrole (PPy), and poly(3,4-ethylenedioxythiophene) (PEDOT) and their composites have been chemically polymerized using an oxidative polycondensation of the corresponding monomers. Carbon nanotubes (CNTs) and commercial thermally exfoliated graphite (TEG) from Superior Graphite Co., Chicago, USA, have been used for making the composites. The carbon nanotubes were synthesized by catalytic decomposition of acetylene on a solid solution precursor [9]. ECPs were efficiently deposited by chemical polymerization of the monomer on the carbon support well dispersed in the reaction mixture under sonification, leading to homogenous composite materials. The proportion of the components was estimated by weighing the final product in the dried state. In some cases, acetylene black has been used for making the composite in order to determine if there is a beneficial effect of CNTs or TEG compared to this conducting additive. Scanning Electron Microscopy (SEM, Hitachi 4200) and Transmission Electron Microscopy (TEM, Philips CM20) have been used for a detailed observation of the composites.

The electrodes were 1 cm^2 pellets obtained by pressing the composite material. The pellets were pressed onto a gold current collector. The electrolytes were 1 M H_2SO_4 or 2M KNO_3 aqueous solutions, due to their high ionic conductivity. Two and three-electrode cells were used for the measurements. The two-electrode cells were built using a Swagelok® type construction, with a glassy fibrous separator and gold current collectors. In the case of the three-electrode cell, a Pt wire and Hg/Hg_2SO_4 were used as counter and reference electrodes, respectively. The electrochemical measurements were performed using a VMP multichannel potentiostat-galvanostat (Biologic, France).

3. RESULTS AND DISCUSSION

Scanning and transmission electron microscopies show that CNTs were coated with a homogeneous layer of ECPs (up to 50 nm). In the case of the composites based on TEG, ECP fibers with diameters up to 1 μm are observed on the TEG particles.

The electrochemical properties of all types of ECPs have been investigated using cyclic voltammetry, galvanostatic charge-discharge cycling and electrochemical impedance spectroscopy. It was established that an amount of carbon material ca. 20 wt% provides a good electronic conduction in the electrode when the polymer is in its isolating state. Acetylene black and especially TEG supply quite good electrochemical performance of the composites, however, CNTs as electrode component gave definitively a more homogenous dispersion of ECPs, that provided a better charge propagation. Also the CNT-based nanocomposites materials presented good mechanical properties, and swelling or shrinkage of the electroactive polymer during cycling did not lead to any electrode degradation.

Our previous investigations [10, 11] of the current-producing mechanism for ECP electrodes have shown that ECPs present a pseudo-faradaic behavior, at least within the potential range where they are at their low oxidation degree, while at higher oxidation degree, the introduced charged centers have a tendency to be delocalized, giving a metal-like character to the ECPs. In the latter case, the energy is stored in the conducting polymer by the accumulation of charges in the electrical double layer, mainly by electrostatic forces, without phase transformation in the electrode material. According to our data, it is possible to estimate an extremely high capacitance value (about 1000 F/g) for the potential range where the Faradaic process takes place, and a significantly lower value for the other potential range. Since the electrodes of a real supercapacitor work in a different potential range, the doped state of each electrode is different. Hence, the capacitance values are different for both electrodes. Taking into account the formula $1/C = 1/C_1 + 1/C_2$ for capacitors in series, one can see that the total capacitance of the device is imposed by the smallest value.

Figure 1 shows as an example the galvanostatic discharge of the PPy/CNTs composite in a three-electrode cell. It is easily seen that the slope of the curve strongly depends on the potential range. Based on the above equation for capacitors in series and taking into account the values of specific capacitance of the PPy composite materials according to Fig. 1, i.e. 250 F/g (C_1) calculated for the potential range from 0.2 V to –0.3 V and 903

F/g (C_2) for the negative range from -0.3 V to –0.6 V, the overall capacitance combining two electrodes working in these potential ranges will be the following:

$$C = \frac{C1 \times C2}{C1 + C2} = \frac{250 \times 903}{250 + 903} = 196F/g$$

This value of capacitance is well consistent with the experimental value of 192 F/g measured in a two-electrode cell. Hence, one must be extremely careful that the values of capacitance reported for ECPs in literature are generally obtained from three-electrode cell measurements. The above example shows clearly that in the case of a real two-electrode capacitor the capacitance values are always smaller than in a three-electrode cell configuration.

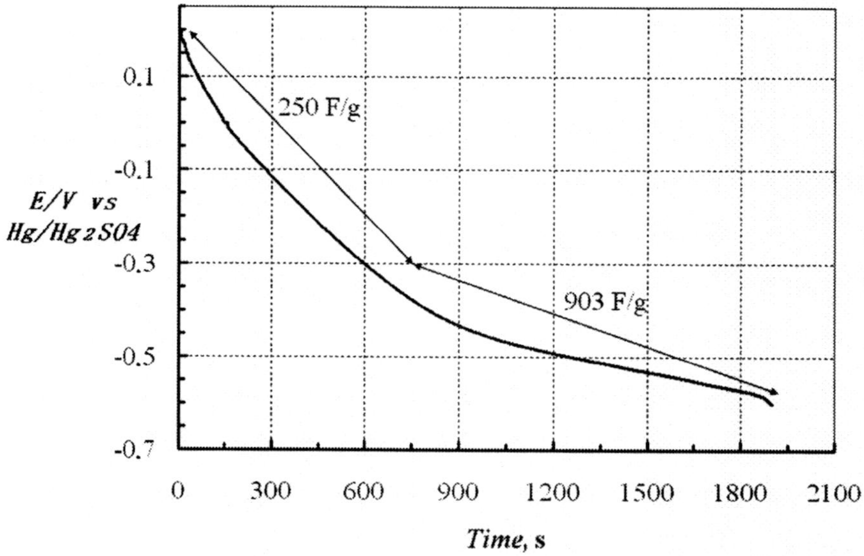

Figure 1. Galvanostatic discharge curve (three-electrode cell), at 2 mA, for a PPy/CNTs pellet electrode; m= 9.6 mg.

The cycling stability upon charge/discharge of a symmetric capacitor based on PPy composite electrodes was tested over several hundred cycles under constant current, and the variation of specific capacitance is illustrated in Fig. 2. These tests demonstrate that using the same material for both electrodes provides a poor cycling stability of the supercapacitor if the voltage range exceeds some limit. The cycleability of the PPy based capacitor is good when the cell voltage is 0.4 V during cycling. Upon cycling at U = 0.6 V, the capacitance is 20% less than the initial one after

500 cycles, and the loss reaches almost 50% when U=0.8V. A similar behaviour was observed for symmetric capacitors based on PANI or PEDOT.

Traditionally, many authors attribute the poor cycling life of ECPs based capacitors to the ECPs degradation process, mainly by oxidation. However, the analysis of the polyaniline electrochemical behaviour [11] proved that the redox processes taking place at the limits and beyond a given potential range can lead to the quick formation of thin passive layers of poor conductivity. Such new phases (e.g. leucoemeraldine salt, leucoemeraldine base etc.) rapidly inhibit and discontinue the electrochemical process playing the role of a switch limiting the whole supercapacitor performance. Therefore, a high cycleability (more than one million cycles according to some literature claims [12, 13] without significant change of the ECPs can be only realized using a three electrode cell in a strict potential range. By contrast, in an ECP-based two-electrode capacitor configuration, one cannot strictly control the potential of the electrodes, and the potential shift of one electrode to the region of ECP electrochemical unstability can provoke a capacity fade.

The galvanostatic tests, using a three electrode cell, on the negative electrode after the galvanostatic cycling of the PPy based supercapacitor at U=0.8 V (see Fig. 2), confirmed a noticeable capacity fade. Figure 3 (curve 1) shows that the capacitance of the negative electrode is about 160 F/g in the potential range from -0.1 V to -0.7 V vs. Hg/Hg_2SO_4 where this electrode works, while it was 610 F/g (curve 2 of Fig. 3) before cycling of the supercapacitor. It means that the potential of the negative electrode reached such a value that it achieved a full ECP undoping to the isolating state, being at the origin of the observed capacity fade due to the isolation of a noticeable part of the ECP from the charge/discharge process.

In order to solve this drawback, one could propose to recover the electrode capacitance by applying a more positive polarisation to the negative electrode that would allow the isolating phase to be converted into the conducting one. Figure 3 (curve 3) shows the galvanostatic charge of the negative electrode containing the isolating phase up to ca. 0.2 V. According to this curve the doping of the ECP mainly takes place (at least 80% of the total charge) at potentials above -0.1 V. Afterwards, the electrode was discharged, giving the curve 4 (Figure 4) with a capacity comparable to the initial state (curve 2), which confirms the re-doping of the electrode.

In a symmetric capacitor, where the cell voltage is imposed during galvanostatic cycling, it is impossible to realise such a potential shift for the negative electrode. Thus, the isolating phase accumulates in the negative electrode upon cycling of the supercapacitor. This process is strongly accelerated by an increase of the operating voltage, because the potential of the negative electrode is shifted to more negative values, what is experimentally supported by the data presented in Figure 2.

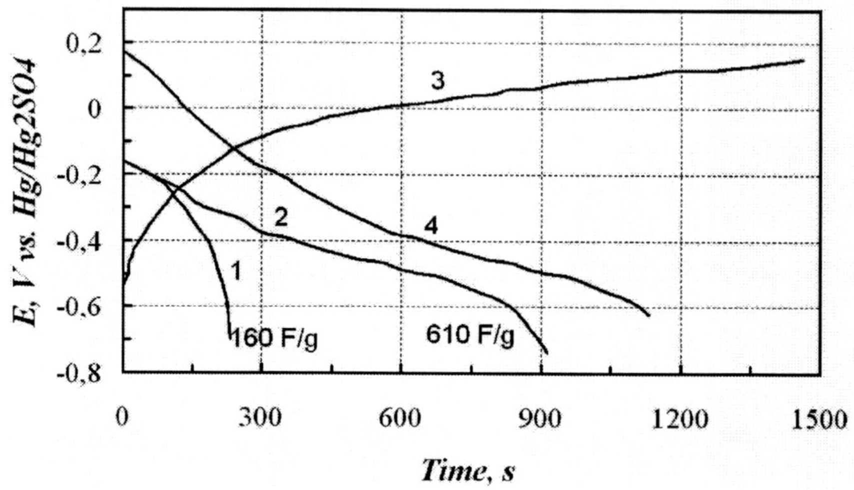

Figure 2. Cycleability of the PPy/CNTs composite at different cell voltages in 1M H_2SO_4; Current load I=2 mA; Mass of each electrode 8 mg; Two-electrode system.

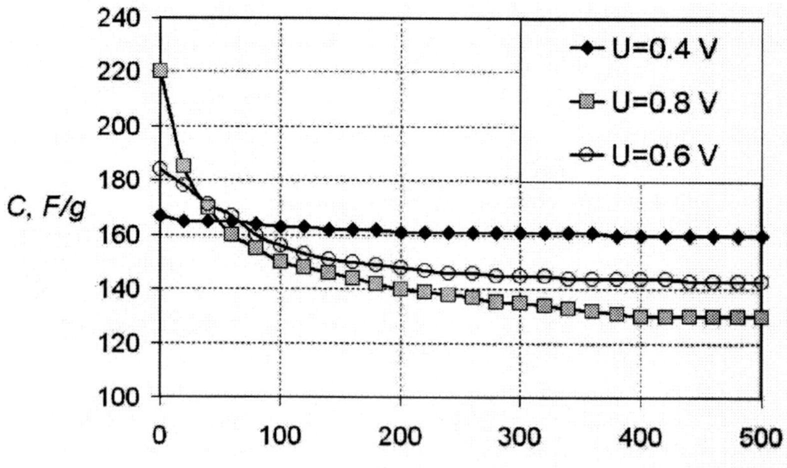

Figure 3. Galvanostatic discharge curves (three-electrode cell) at 2 mA of a PPy/CNTs pellet electrode (m=6.7 mg) before (2) and after (1) galvanostatic cycling in a symmetric capacitor (two electrode cell) at U=0.8 V. After the discharge (1), the electrode was charged up to 0.2 V (curve 3) and then discharged (curve 4).

It is interesting to note in Figure 2 that the value of the initial specific capacitance is more important at high voltage, but it rapidly decreases with cycling.

This suggests that the pseudocapacitance effects, which are at the origin of these high values gradually disappear with cycling (already after 50 cycles). Impedance spectroscopy was used in order to find an interpretation to this phenomenon, and the spectra of a supercapacitor based on PPy/CNTs composite electrodes before galvanostatic cycling and after 500 cycles at a voltage of 0.8 V are reported in Figure 4. The supercapacitor resistance at 10 kHz increased from an initial value of 0.35 Ohm to 1.43 Ohm after 500 cycles, that proves that the electrodes resistance also increased. Hence, one can conclude that irreversible redox transitions to the isolating state are at the origin of the high values of capacitance observed under high voltage during the first galvanostatic cycles.

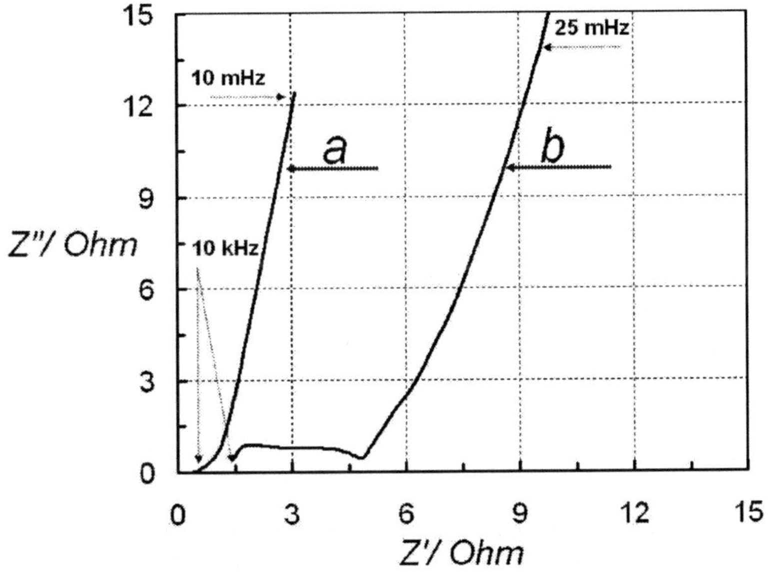

Figure 4. Impedance spectra of a supercapacitor based on the PPy/CNTs composite before (a) and after performing 500 galvanostatic charge/discharge cycles (b). Mass of the working electrodes: 8 mg.

The appearance of a semicircle at high frequency after cycling (Figure 3b) suggests a parallel existence of a charge transfer resistance, which is related with the already mentioned pseudofaradaic processes, i.e. with the formation of the isolating phase.

In our opinion, the above experiments proved that, for an optimal usage of the conducting polymer, the potential range of the negative electrode must be strictly controlled in order to prevent the ECP to reach its insulating state. In a symmetric supercapacitor this might occur only when the cell voltage is very low, that consequently limits the energy stored.

An asymmetric configuration, with two electrodes of different nature [4,5,14], is an excellent way to overcome the problems mentioned before. The main target, designing these systems, is to optimize the potential range of each electrode, taking into account the need to obtain a high cell voltage and a good cycling life of the supercapacitor as a whole. Thus, we have investigated the combination of different ECP electrodes or of an ECPs electrode and another electrode from an activated carbon or a metal oxide. For example, an asymmetric configuration was built where the ECPs/CNTs composite was the positive electrode and the commercial activated carbon (PX-21) was selected for the negative electrode. We have also tested amorphous manganese oxide, a-MnO_2, as positive electrode material in asymmetric supercapacitors where the ECPs/CNTs composites served as negative electrode. The advantages of a-MnO_2 are a pseudo-faradaic mechanism of charge storage allowing high capacitance values to be expected, a low cost and an environment friendly nature [7].

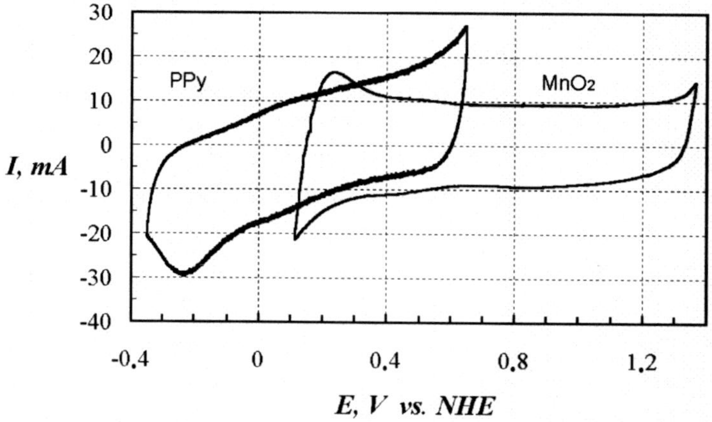

Figure 5. Cyclic voltammograms at 10 mV/s in three-electrode cells for PPy and for α-MnO_2 based electrodes in 2 mol·L^{-1} KNO_3 medium using Pt as auxiliary electrode.

Typical cyclic voltammograms of a-MnO_2 and PPy in 2 mol.L^{-1} KNO_3 are shown in Fig. 5. The electrochemical potential window of manganese dioxide is spread to more positive values than for activated carbons and ECPs, keeping high electrochemical reversibility up to 1.2 V. The electrochemical activity of PPy is in the range which should not include its redox undoping reaction at potentials below -0.2 V vs. NHE. Therefore, by combining these two materials in an asymmetric system, it is possible to obtain a working voltage window of 1.4V ($U=E_{MnO2} - E_{ppy}= 1.2 - (-0.2) = 1.4$), which will ensure a safe performance of both electrodes during long cycling. Thus, the above mentioned situation of deactivation of the negative

ECP based electrode of a symmetric capacitor would not occur when a-MnO_2 will be used as a positive electrode in an asymmetric configuration.

Table 1. Electrochemical characteristics of some asymmetric supercapacitors. U is the maximum available cell voltage. C_s is the specific capacitance of a pellet electrode calculated from $C_s = C_t*4/M$, where C_t is the capacitance of the asymmetric supercapacitor, M is the total mass of both electrodes.

Electrode materials		Supercapacitor characteristics	
Positive	**Negative**	**C_s (F/g)**	**U (V)**
PANI	PPy	280	0.6
PANI	PEDOT	160	0.8
PANI	Carbon (PX21)	330	1.0
α-MnO_2	PANI	160	1.4
PPy	Carbon (PX21)	220	1.0
MnO_2	PPy	140	1.6
PEDOT	Carbon (PX21)	110	1.2
MnO_2	PEDOT	120	1.8

The electrochemical performance of all types of asymmetric supercapacitors has been evaluated by galvanostatic charge/discharge tests, and the results are presented in Table 1. Combining an activated carbon for the negative electrode and the ECP/CNTs composite for the positive electrode significantly extends the practical voltage performance up to 1.2 V in aqueous electrolyte and provides a long cycling life of the supercapacitor. The most interesting results, which confirm our prediction presented above, were obtained with the newly designed hybrid electrochemical capacitor using the ECP/CNTs composite as negative electrode and manganese dioxide as positive electrode. Almost all of the charge residing on the ECP electrode can be used and the cell voltage reaches up to 1.8 V in aqueous electrolyte. Using such an asymmetric construction, the values of specific energy and specific power in aqueous electrolyte increase by more than 4 times compared with symmetric ECPs based capacitor. It should be additionally mentioned that the asymmetric supercapacitor was cycled over 1000 cycles without any noticeable decrease of the capacitance value.

4. CONCLUSION

Our data demonstrate that the symmetric configuration of ECP-based capacitor is not realistic, mainly because one cannot control the potential of the negative electrode during cycling. When the voltage range exceeds ca. 0.4 V, the negative electrode progressively shifts to its isolating state that leads quickly to a capacity fading of the system. Consequently, for a safe cycling of a symmetric ECP-based capacitor, the voltage should not exceed 0.4V, that makes the system of poor interest in term of energy stored.

In order to circumvent these drawbacks, we propose an asymmetric configuration with two different materials working in their optimal potential range for both electrodes. The combination of two different materials allows the operating voltage range to be extended up to 1.8 V in aqueous medium, keeping a good cycleability of the material without a noticeable aggravation of the capacitance values. Due to their environment friendly nature, these new high energy systems are quite promising in replacement of organic electrolyte capacitors.

ACKNOWLEDGMENTS

This work was supported by the NATO Science for Peace "Carbon" programme No. 973849.

REFERENCES

1. Conway B.E. in Electrochemical Supercapacitors. Kluwer Academic/Plenum Publishers, New York, 1999.
2. Frackowiak E., Béguin F. Carbon materials for the electrochemical storage of energy in capacitors. Carbon 2001; 39: 937-50.
3. Laforgue A., Simon P., Sarrazin Ch., Fauvarque J-F. Polythiophene-based supercapacitors. J Power Sources 1999; 80:142-8.
4. Mastragostino M., Arbizanni C., Soavi F. Polymer-based supercapacitors. J Power Sources 2001; 97-98:812-5.
5. Mastragostino M., Arbizzani C., Soavi F. Conducting polymers as electrode materials in supercapacitors. Solid State Ionics 2002; 148:493-8.
6. Jurewicz K., Delpeux S., Bertagna V., Béguin F., Frackowiak E. Supercapacitors from nanotubes/polypyrrole composites. Chem Phys Lett 2001; 347: 36-40.
7. Toupin M., Brousse T., Bélanger D. Influence of microtexture on the charge storage properties of chemically synthesized manganese dioxide, Chem Mater 2002; 14:3946-52.
8. Wu N.L. Nanocrystalline oxide supercapacitors. Mater Chem Phys 2002; 75:6-11.
9. Delpeux S., Szostak K., Frackowiak E., Bonnamy S., Béguin F. High yield of pure multiwalled carbon nanotubes from the catalytic decomposition of acetylene on in-situ formed cobalt nanoparticles. J. Nanosc. Nanotech. 2002; 2: 481-4.
10. Barsukov V., Chivikov S. The "Capacitor" concept of the current-producing process mechanism in polyaniline-type conducting polymers. Electrochim. Acta 1996; 41: 1773–79.
11. Barsukov V.Z., Khomenko V.G., Chivikov S.V., Barsukov I.V., Motronyuk T.I. On the faradaic and non-faradaic mechanisms of electrochemical processes in conducting polymers and some other reversible systems with solid-phase reagents. Electrochim. Acta 2001; 46: 4083–94
12. Kobayashi T., Yoneyama H., Tamura H. Polyaniline film-coated electrodes as electrochromic display devices J.Electroanal.Chem., 1984; 161: 429-23
13. Kobayashi T., Yoneyama H., Tamura H. Electrochemical reactions concerned with electrochromism of polyaniline film-coated electrodes. J.Electroanal.Chem 1984; 177: 281-91
14. Arbizzani C., Mastragostino M., Soavi F. New trends in electrochemical supercapacitors. J Power Sourc 2001; 100:164-170.

SUPERCAPACITORS: OLD PROBLEMS AND NEW TRENDS

Yurii Maletin*, Natalie Strizhakova, Vladimir Izotov, Antonina Mironova, Sergey Kozachkov, Valerie Danilin, Sergey Podmogilny

National Technical University of Ukraine "KPI", Department of Physical Chemistry, 37 Pobedy Prospekt, Kiev 03056, Ukraine

1. INTRODUCTION

Supercapacitors (SC) attract much attention as a good alternative or supplement to batteries. Over the past decade a considerable breakthrough took place in our understanding of how the supercapacitors function [1-3]. Though they are inherently behind the batteries by the specific energy stored, they are way ahead of them by their peak output power, efficiency and cycling life. As a result they can be used as pulse power sources, for load leveling the batteries, in hybrid power supply units (e.g., being combined with a fuel cell), etc. This paper is an attempt to analyze what are the prospects for SC to occupy their niche in the market of energy storage devices, what performance can be anticipated in near future, and how this performance might be achieved with the special emphasis on the SC power output and its dependence on the electrode properties.

2. PRINCIPLE OF OPERATION AND KEY PERFORMANCE CHARACTERISICS

Active electrochemical part of a typical SC device represents two conductive porous (normally carbonaceous) electrodes with a dielectric porous film (separator) interposed between them; the system being impregnated with a liquid electrolyte. The double electric layer forms at the electrode/electrolyte interface on charge, and this layer discharges during operation, making it possible for the electric energy to be stored and then to be delivered to the load in due time. The electrolyte can be either aqueous or organic aprotic one, and this differs SC of Generation 1 from Generation 2, if John Miller's [4] SC classification is used. As an alternative, hybrid devices

* Corresponding Author. E-mail: maletin@xtf.ntu-kpi.kiev.ua

I.V. Barsukov et al. (eds.), New Carbon Based Materials for Electrochemical Energy Storage Systems, 51–62.

can also be designed wherein one of carbonaceous "double layer" electrodes is replaced by an "electrochemical" electrode taken from the battery technology, e.g. NiO(OH) or PbO_2 in aqueous electrolyte (Generation 3 [4]) or $Li_4Ti_5O_{12}$ in aprotic electrolyte (Generation 4 [4]).

The specific energy stored and the specific power that can be delivered to the load are no doubt the crucial characteristics of a SC device along with others, such as its cycling life, self discharge current and efficiency.

Specific energy is proportional to the capacitance and voltage squared, and it should be noted that both of these parameters are close to their limit in modern SC devices comprising two carbonaceous electrodes and organic electrolyte. Double layer capacitance value is ca. 12(±2) μF/sq.cm for porous carbon materials [5], and their accessible specific surface area has practically reached the reasonable limit of ca. 1200 sq.m/g (readers should note that ultramicropores with the diameter less than 0.7÷0.8 nm are beyond the reach of organic electrolytes comprising relatively large species). This gives the specific capacitance of about 150 F/g for a carbon electrode in organic electrolytes, and this value has almost been reached in the best SC prototypes. As regards to the rated voltage, it can hardly be higher than 3 V for a SC unit cell with known electrolytes, and the value of 2.3÷2.5 V looks even more feasible if the long cycling life is required [6]. As a result, over a few past years the specific energy of the best SC devices with organic electrolytes has not exceeded 5 W.h/kg [7], and the value of 10 W.h/kg can be considered as the practical limit. For hybrid devices with dissimilar electrodes (*vide supra*) a bit larger specific energy can be obtained [4], though at the expense of lower power output and shorter cycling life.

High specific power value accompanied by long cycling life seems to be the most attractive advantage of SC devices since this is the performance which enables them to be effectively used for load leveling the batteries in electric or hybrid vehicle applications. However, to be competitive in the battery market, SC must demonstrate the specific power of about or higher than 10 kW/kg for a charge/discharge at ca. 95% efficiency. This power, p, can roughly be estimated using the equation (1) [7,8] as follows:

$$p = \frac{9(1-\eta)U^2}{16RM}, \qquad (1)$$

where η is the efficiency, U is the voltage in V, R is the equivalent series resistance in Ohm, and M is the mass in kilos.

This means the equivalent series resistance of SC should notably be reduced in order to increase their specific power. Hence, as can be shown, the electrode resistivity should be reduced significantly – down to 0.5

Ohm.cm^2 (of visible surface area) or even less. Some visible ways to lower the electrode resistivity are: (*a*) fabricating thin electrodes, (*b*) increasing the conductivity of electrolyte and electrode material, (*c*) matching the electrode porosity with the size of species in the electrolyte in order to facilitate the ion transport along the pores, and (*d*) assembling bipolar devices.

3. OPTIMIZATION OF THE ELECTRODE THICKNESS

In order to see how the electrode thickness might be optimized in order to provide the lowest electrode resistivity, we have developed a theoretical model to describe the charge/discharge processes in porous carbon electrodes. As a first approximation, let us consider an electrode having two sets of cylindrical pores, namely, nanopores (NP) of less than 3 nm in diameter and transport channels (TC) of more than 20 nm in diameter, with each nanopore having an exit to only one TC[‡].

The distribution of potential in TC is practically the same as that near the flat surface if the electrolyte concentration is about 1 mol/l [2]. So the discharge of TC may be considered as that of a double electric layer formed at the *flat electrode surface/electrolyte solution* interface, and hence, an equivalent circuit for the TC discharge may be presented as an ***RC*** circuit, where ***C*** is the double layer capacitance and ***R*** is the electrolyte resistance.

On the other hand, the gradient of potential is negligible within the NP [2]. That is why, here we will consider the process of NP charging or discharge as a diffusion process caused by the concentration gradient, which results from the concentration change within the diffused part of a double layer near the region where the NP face the surface of TC. As a result, an equivalent circuit for this electrode section may be represented by two ***RC*** circuits connected in parallel (see the corresponding portion of Fig. 2 below).

Now let us consider a model for a SC device that comprises two electrodes (anode and cathode), each of them being electrically connected to a current collector fabricated of Al foil. Let two of such collectors have a certain thickness of δ_{Al}. As an electrode material, an activated carbon powder is considered below. Anode and cathode are interposed with a separator of thickness δ_s. The electrodes and separator are impregnated with electrolyte. In this paper we mostly focus on the optimization of SC performance by varying the electrode thickness, while some other effects will briefly be considered in the next section.

As mentioned above, the electrode material comprises two types of pores – NP and TC, NP being mainly responsible for the electrode capacitance value. On the other hand, TC, which contribute far less to the

---

[‡] Such a choice of pore shape and distribution leads to a relatively simple mathematical formulation without loss of generality.

capacitance, are responsible for the ion transport through the electrode body up to the boundary between the electrode and separator. Let the electrode thickness be of δ_a or δ_c for anode or cathode correspondingly, and let the TC and NP be equally distributed within the electrode body.

Capacitance of a single pore is proportional to its surface area. Therefore, for a cylindrical NP it is proportional to $2\pi rl$, where r is the NP radius, and l is its length. Inner resistance of NP is proportional to l as well. The NP parameters depend on how the activated carbon powder was obtained but they are normally not affected by treating the powder when fabricating the electrode.

On the contrary, the shape and dimensions of TC depend on both the method for obtaining the porous carbon powder and the method for making electrode. Within the framework of our model, TC are considered to be the straight channels passing through an electrode such that their length is equal to the electrode thickness. As another model approximation, all the NP are assumed to face the TC forming a sort of tier structure, i.e. the NP exits are equally distributed along the TC circle and length – see Fig. 1. Let the distance between tiers be of $4r$. This approximation simplifies mathematical derivations significantly though does not distort a real physical picture. Number of NP, n, facing TC in a tier may be determined from the volumetric concentration of NP, ξ, and from the number of TC, μ, coming through the unit electrode section. Thus the n value can be determined as $n = 4r\xi/\mu$.

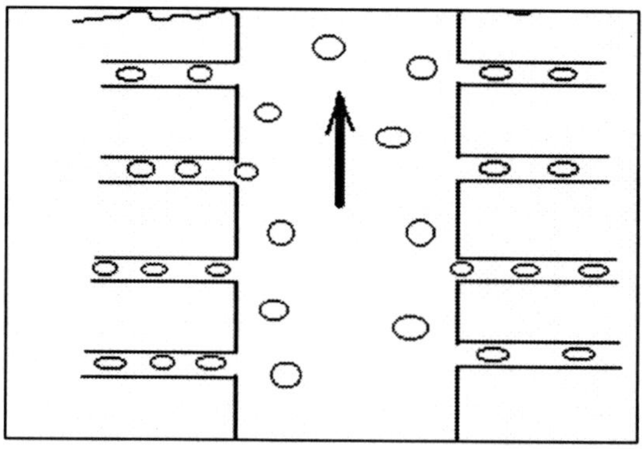

Figure 1. Model presentation of a few nanopore tiers facing a transport channel.

Now an equivalent circuit, which takes into account both the ion transport along the TC and the charge transfer through the carbon electrode material to the current collector, may be represented as in Fig. 2, wherein $N = \delta_{a(c)}/4r$, C_m and R_m are the total NP capacitance and resistance in a unit electrode volume (defined here as a product of a unit electrode area and the tier thickness), R_e is the electrical resistance of an electrode in the same unit volume, and R_i is the electrolytic resistance in TC located in the same unit volume. It should be noted that $R_m >> R_e$, R_i in real systems.

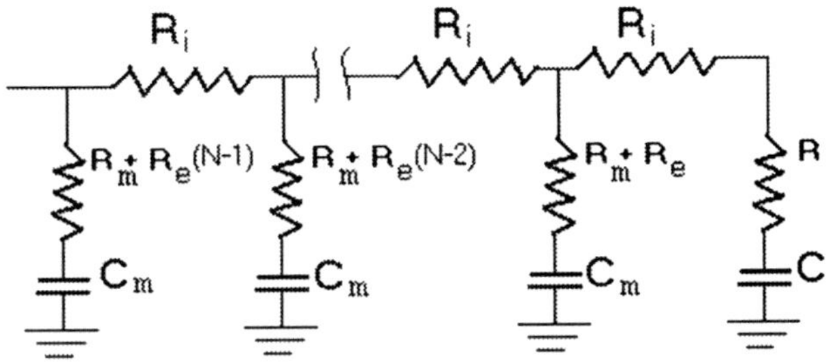

Figure 2. An equivalent circuit of a porous carbon electrode in SC.

Electrical parameters of the system under consideration have been estimated using an equivalent circuit presented by Fig. 2. Thus, the anode capacitance can be written as follows:

$$C_a = C_{m(a)}\ \delta_a/4r \tag{2}$$

However, it is not so easy to derive a similar functional relationship between the electrode resistance and its thickness that is why a computer simulation has been carried out to study the dependence of electrode resistance on its thickness. In the calculations, R_m was chosen as a unit resistance, and the distance between the NP tiers ($4r$) was chosen as a unit thickness. Figure 3 illustrates the results of calculations at various ratios between R_e, R_i and R_m, namely, $R_i = R_e = 0.002R_m$ (curve 1); $R_i = 0.004R_m$, $R_e = 0.0001R_m$ (curve 2). In these and other cases checked, the plots of resistance vs. thickness display a sharp drop followed by a slower rise after reaching a minimum. Let us consider these regions in more detail.

For two limiting cases – dip or rise of curve or, in other words, thin or thick electrode – the problem can be solved asymptotically. In case of a thin electrode, i.e. if $\delta << \frac{4rR_m}{R_i + R_e}$, the following expression for electrode resistance can be derived:

$$R = R_m \frac{4r}{\delta}. \tag{3}$$

In another limiting case, when $\delta >> \frac{4rR_m}{R_i + R_e}$, the dependence of R on the electrode thickness δ can be presented as follows:

$$R = \rho_{ef}\delta, \tag{4}$$

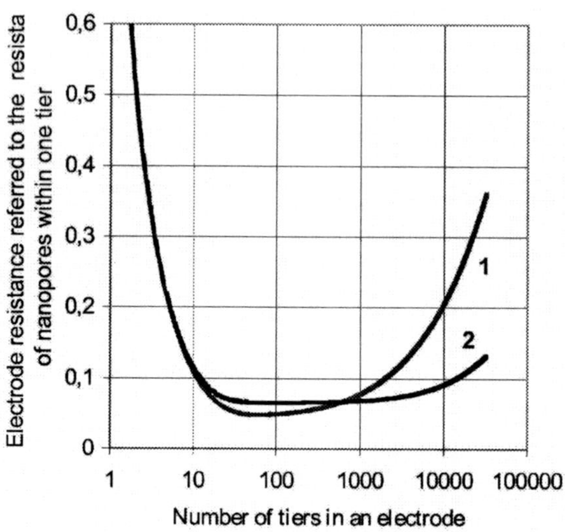

Figure 3. Plot of ESR of an electrode vs. its thickness measured in number of tiers.

where ρ_{ef} is the effective resistivity including the electrical resistivity of electrode material and the resistance of ion transport along the TC. These two conjugated components of ρ_{ef} cannot practically be separated. On the other hand, this is the ion transport, which normally limits the electrode resistance since not one porous electrode impregnated with electrolyte and charged, can then be discharged until counter-ions leave it. This means that conductivity of the electrolyte and its diffusion along the pores must play a crucial role for reducing the electrode resistivity (*vide infra*).

From the practical point of view, this is the discharge of a SC device under constant power conditions that is normally of the most interest. That is why the present work is aimed at determining the optimum electrode thickness that enables one to obtain the maximum energy, E, output (referred to as unit of volume or mass) if the discharge with a fixed power takes place. For the sake of simplicity we will speak about the energy density (E) and power density (p), but all the expressions derived below can easily be transformed to obtain the specific energy or power, if the volume is substituted by mass.

As a first step we have examined the discharge of SC having the volume of V, capacitance of C and inner resistance of R_{in}. For such a capacitor charged up to the voltage U, and then discharged to a nonlinear load to keep the constant power P until the voltage decreases to $U/2$ the expression (5) for the energy density ($\mathbf{E} = E/V$) and power density ($p = P/V$) can be derived [3].

In the equation (5) the inner resistance R_{in} includes not only the electrode resistance, but also the separator resistance R_s, defined as R_s =

$\rho_s\delta_s/S$, where ρ_s is the resistivity of separator to ion transport and S is the electrode (or separator) area. The quantities in Eq. (5) have the dimensions as follows: **E** in W.h/l, p in W/l, R_{in} in Ohm, C in F, and U in V.

$$\mathbf{E} = \frac{C(4R_{in}pV \ln(\dfrac{U-\sqrt{U^2-4R_{in}pV}}{\dfrac{U}{2}-\sqrt{\dfrac{U^2}{4}-4R_{in}pV}}) + U\sqrt{U^2-4R_{in}pV} + \dfrac{3}{4}U^2 - \dfrac{U}{2}\sqrt{\dfrac{U^2}{4}-4R_{in}pV})}{14400V} \qquad (5)$$

So far, this model differs from a real system because it does not take into account a decrease in capacitance with an increase in current (charge or discharge). As a result, it can be shown that calculations according to Eq. (5) start deviating from experimental Ragone plots in the range where the current values are high enough. Therefore, the dependence of capacitance on current was introduced into the model.

From experimental measurements it was found that there was a fairly good linear relation between the SC capacitance and current, the slope being dependent on the electrode material and electrolyte. If the dependence is known, it can be used to make the Ragone plot calculations more accurate. Therefore, the following method was employed in our evaluations.

If the power value is constant on discharge, the current varies evidently from the starting value i_s to the final one i_f. If one neglects the potential drop due to inner resistance of a SC, the i_s and i_f values may be determined as follows: $i_s = p/U$ and $i_f = 2p/U$. The starting and final values of capacitance ($C(i_s)$ and $C(i_f)$, correspondingly) may be evaluated from the experimental C-i plot, and then the average value $\bar{C} = (C(i_s) + C(i_f))/2$ may be put into Eq.(5). Some of the results of Ragone plot calculations using this equation and the experimentally determined C-i dependencies are presented by Figure 4.

As can be seen from Eq. (5), the maximum power density delivered to the load is:

$$p_{\max} = \frac{U^2 c}{16\tau} \qquad (6)$$

where c is the capacitance density, and τ is the SC time constant. For further calculations the following relationship may be used:

$$V = S(\delta_s + \delta_{Al} + \delta_a + \delta_c) \qquad (7)$$

For the sake of simplicity let $\delta_a = \delta_c = \delta$ below. Subsequently, expressions (2), (3) and (7) can be put into Eq. (5) and the maximum of **E** as a function of δ can be found, all the other quantities being constant. As a result it can be shown that for "thin electrodes" ($\delta << \frac{4rR_m}{R_i + R_e}$) and under the constant power discharge the maximum energy delivery can be achieved for the following electrode thickness:

$$\delta = \frac{U\sqrt{\delta_s + \delta_{Al}}}{4\sqrt{p\delta_s\rho_s}} - \frac{\delta_s + \delta_{Al}}{2} \tag{8}$$

In another limiting case ($\delta >> \frac{4rR_m}{R_i + R_e}$) the maximum energy can be delivered if the electrode thickness is as follows:

$$\delta = 0.3\frac{U^{2/3}(\delta_{Al} + \delta_s)^{1/3}}{(p\rho_{ef})^{1/3}} - \frac{\rho_s\delta_s + 2\rho_{ef}(\delta_{Al} + \delta_s)}{6\rho_{ef}} \tag{9}$$

4. VERIFICATION OF THE MODEL

If one substitutes now Eq. (6) into Eqs. (8) and (9) and then chooses the values[§] of $\delta_{Al} = 20$ μm, $\delta_s = 30$ μm, $\rho_s = 150$ Ω.cm, $\rho_{ef} = 60$ Ω.cm, $c = 15$ F/cc, $\tau = 0.2$-0.3 s for thin electrodes and $\tau = 1$-2 s for thick ones, the optimum electrode thickness δ might be evaluated within the range of 50-150 μm. That is why a series of experiments was carried out with the electrode thickness being varied within this range. As an electrolyte, 1.3 M Et_4NBF_4 in acetonitrile was used since, as was shown by various researchers and also in our experiments, acetonitrile-based electrolytes were able to provide the highest conductivity at room temperature as compared with other solvents - of ca. 60 mS/cm. Composite electrodes of various thickness were fabricated using the carbon powders produced by Skeleton Technologies as a result of TiC chlorination at high temperature, Al foil of 10 μm thick was used as a current collector, and the Celgard 2400 material of 30 μm thick was employed to separate the electrodes. The Ragone plot for a SC prototype having the electrodes of 120 μm thick is shown in Fig. 4. As it can be seen, the performance of the prototype is fairly good, and the suggested model is able to describe the SC characteristics with a reasonable accuracy.

[§] These values were estimated by us from preliminary experiments and from the known data.

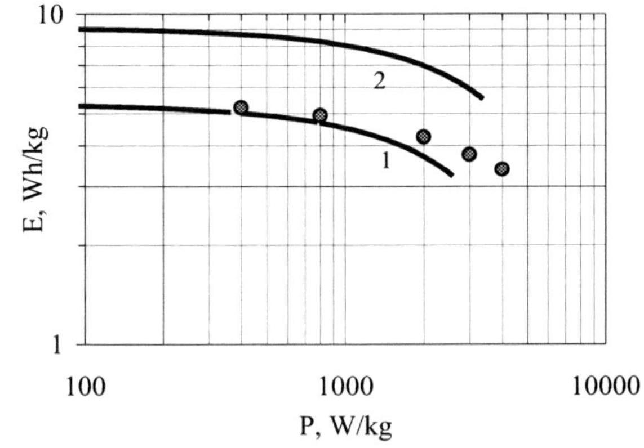

*Figure 4. Ragone plots calculated for 2.3 V (**1**) and 3.0 V (**2**); experimental dots were obtained in Dr. A. Burke's lab at 2.3 V.*

5. ELECTROLYTE TRANSPORT THROUGH THE CARBON ELECTRODES

As it has been mentioned above, this is the ion transport through the carbon electrodes that limits the rate of SC charge or discharge, and hence limits the specific power of a SC device.

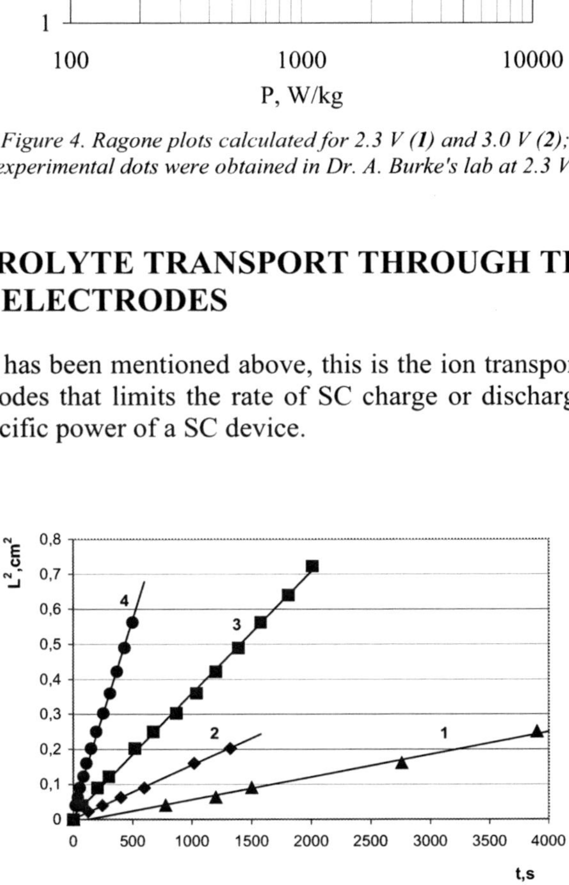

Figure 5. Length (squared) of the electrode impregnation with acetonitrile vs. time for various carbon materials produced by Skeleton Technologies: 1 – C/SiC; 2 – C/TiC-1; 3 – C/TiC-2; 4 – C/Mo_2C.

In accordance with the known equation: $\overline{L}^2 = 2Dt$, the mean ion displacement squared, $\overline{L}^2$, at time t is proportional to diffusion coefficient, D. Since we could not find any reliable data on diffusion coefficients for organic electrolytes in porous carbon materials, we have measured the electrolyte and/or solvent migration through the electrode body. With certain reservations the transport of liquid and transport of ions through the porous electrode body can be correlated similar to transport processes in porous membranes [9]. Therefore, we have developed a simple visual method to evaluate the rate of electrolyte transport through the electrode.

According to the method developed, a flat circular electrode is placed between two sheet glasses under the controlled pressure, one of the glasses having a tiny hole (a feeder) with a capillary near the center of the electrode. A solvent or an electrolyte can be fed to the electrode through the feeder, which is sealed to avoid the solvent evaporation. Then the migration of solvent or electrolyte (impregnation of the electrode) can be observed with microscope, and Fig. 5 illustrates the length of migration (squared) vs. time for different electrode materials.

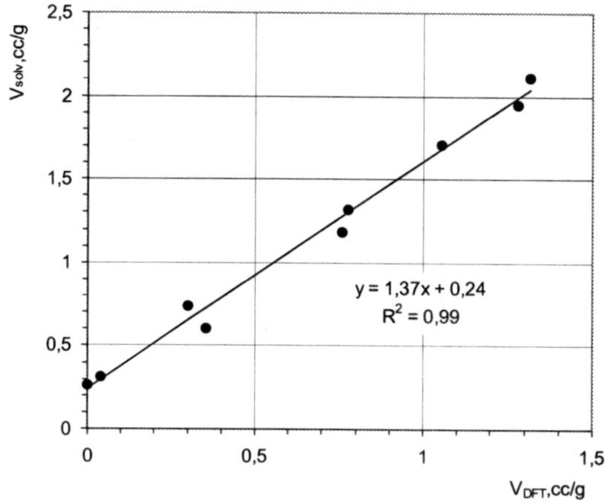

Figure 6. Plot of the specific electrode pore volume vs. the specific pore volume measured for the corresponding carbon powders.

As can be seen from Fig. 5, a fairly good linear correlation takes place for various carbon materials, the slope 2*D* being equidimensional with the diffusion coefficient and showing the rate of electrolyte migration through the electrode (more precisely, the volumetric rate of flow multiplied by the length *L*). In accordance with the Hagen-Poiseuille equation, for porous materials of the type presented in Fig. 1 the volumetric rate of flow must be proportional to the $\mathbf{pr}^2$ product, where **p** is the porosity of the material, and **r** is its mean pore radius [10]. The gravimetric porosity (the total pore volume per 1 gram of the material) was measured by two methods. The first method was employed for carbon powders and was based on the analysis of N_2 adsorption/desorption isotherms within the framework of the Density Functional Theory (DFT) using the ASAP2000M unit (produced by Micromeritics). The second method included the simple determination of solvent volume absorbed by the corresponding electrode using the same two glasses and a feeder as described above. The porosity values determined by the two methods for various carbon powders and electrodes made of them are plotted in Fig. 6, wherein the very left dot represents a non-activated anthracite. A reasonably good correlation indicates that electrodes possess about 0.24 cc/g of extra porosity due to macropores formed by carbon grains when the electrode is manufactured.

If we assume a quasi-cylinder pore structure of the electrode material as in Fig. 1, an average effective pore radius **r** can be evaluated from the relationship **r** = 2V/A, where V is the total pore volume, and A is the total pore surface that can also be determined using the DFT method (see also [5]). Then the migration coefficients determined as shown in Fig. 5 can be plotted vs. the **p**$\mathbf{r}^2$ product – see, e.g., Fig. 7 for five electrodes, which were made of various porous carbons produced by Skeleton Technologies.

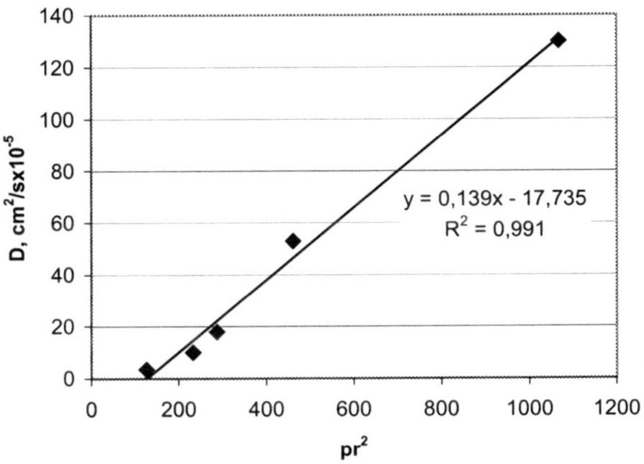

Figure 7. Plot of the migration coefficient (determined from the slope as in Fig. 5) vs. the **p**$\mathbf{r}^2$ *product of a carbon electrode.*

Figure 7 illustrates the dynamics of fluid migration through porous carbon electrodes to obey the Hagen-Poiseuille equation that is normally used to describe the transport through membranes having the pores of cylinder-like shape. Therefore, this method can probably be used for express analysis of the electrolyte dynamics in different porous carbon materials.

6. CONCLUSIONS

1. In order to be competitive in the market of energy storage devices SC must demonstrate the specific power output of about 10 kW/kg for a charge/discharge at 95% efficiency.

2. In contrast to battery technology, ultra thin electrodes cannot provide an increase in SC power density. Our theoretical model shows an optimum electrode thickness to exist in the range of 50-150 μm, the exact value depending on the porosity and nature of the electrode material.

3. This is the electrolyte mobility in pores that crucially influences the rate of electrode discharge and, hence, the SC power output.

4. A simple visual method to evaluate the rate of solvent/electrolyte migration through the porous electrode body has been developed. The method enables an express selection of carbon materials according to the mobility of electrolyte in their pore structure.

ACKNOWLEDGEMENTS

This work was supported by the Science-and-Technology Center in Ukraine (project STCU 1830). Financial support was also provided within the framework of INTAS 00-761 research project. Porous carbon materials derived from various carbides in accordance with methods, which were developed by Skeleton Technologies, were received from Central Research Institute for Materials (St. Petersburg, Russia).

REFERENCES

1. Conway, Brian, *Electrochemical supercapacitors: scientific fundamentals and technological applications*. New York: Kluwer/ Plenum, 1999.
2. Maletin YA, Strizhakova NG, Izotov VY, Mironova AA, Kozachkov SG, Danilin VV, Ekström T. Supercapacitor stacks of high energy & power density. Proceedings of the 7th International Seminar on Double Layer Capacitors and Similar Energy Storage Devices, Deerfield Beach, Florida, USA, Dec. 8-10, 1997.
3. Maletin YA, Strizhakova NG, Izotov VY, Mironova AA, Kozachkov SG, Danilin VV, Podmogilny SN. How to tune the EDLC components to improve its performance. Proceedings of the 10th International Seminar on Double Layer Capacitors and Similar Energy Storage Devices, Deerfield Beach, Florida, USA, Dec. 4-6, 2000.
4. Miller JR. The Evolution of Electrochemical Capacitor Technology: Future Performance Projections. 2003 International Conference on Advanced Capacitors, Kyoto, Japan, May 29-31, 2003.
5. Calas-Blanchard C., Comtat M., Marty J.-L., Mauran S. Textural characterisation of graphite matrices using electrolchemical methods. Carbon 2003; 41: 123-130.
6. Xu K., Ding S.P., Jow T.R. Towards Reliable Values of Electrochemical Stability Limits for Electrolytes. J. Electrochem. Soc. 1999; 146: 4172-4178.
7. Burke AF, Miller M. Characteristics and Applications of Advanced Ultracapacitors. Proceedings of the 12th International Seminar on Double Layer Capacitors and Similar Energy Storage Devices, Deerfield Beach, Florida, USA, Dec. 9-11, 2002.
8. Burke AF. Ultracapacitors: why, how and where is the technology. J. Power Sources 2000; 91: 37-50.
9. Mulder, Marcel, *Basic Principles of Membrane Technology*. Kluwer Acad. Publ., 1995 (Russian Transl., Moscow, Mir Publ., 1999).

MODELING POROSITY DEVELOPMENT DURING KOH ACTIVATION OF COAL AND PITCH-DERIVED CARBONS FOR ELECTROCHEMICAL CAPACITORS

Krzysztof Kierzek, Grazyna Gryglewicz, Jacek Machnikowski*

Institute of Chemistry and Technology of Petroleum and Coal
Wrocław University of Technology, Gdańska 7/9, 50 344 Wrocław, Poland

Abstract

High porosity carbons ranging from typically microporous solids of narrow pore size distribution to materials with over 30% of mesopore contribution were produced by the treatment of various polymeric-type (coal) and carbonaceous (mesophase, semi-cokes, commercial active carbon) precursors with an excess of KOH. The effects related to parent material nature, KOH/precursor ratio and reaction temperature and time on the porosity characteristics and surface chemistry is described. The results are discussed in terms of suitability of produced carbons as an electrode material in electric double-layer capacitors.

Keywords

KOH-activated carbons, porosity characteristics, chemical structure, electrochemical capacitors.

1. INTRODUCTION

Porous carbons are among the most attractive electrode materials for electric double layer capacitors (EDLC), where the charge accumulation occurs mainly by electrostatic attraction forces at the electrode/electrolyte interface [1-3]. Advantages of this class of materials include high surface

* Corresponding author. E-mail: jacek.machnikowski@pwr.wroc.pl

I.V. Barsukov et al. (eds.), New Carbon Based Materials for Electrochemical Energy Storage Systems, 63–72.

area, versatility in terms of porosity development and surface chemistry, availability and a reasonable price and a lack of negative environmental impact. In most general terms the target carbon material for EDLC should be characterized by an extensive surface area accessible by a given electrolyte and high specific capacitance per unit surface. The accessibility depends on both the pore size distribution and the affinity of a given electrolyte to the surface. Heteroatoms present on the surface can further improve the wettability by the electrolyte. The intrinsic capacitance of carbon surface is reported to vary from 15 to 50 $\mu F/cm^2$ [3]. Some authors [4] link the value with the structural aspect of carbon matter as directly related to the electric conductivity. Heteroatoms like oxygen or nitrogen are believed to enhance the charge accumulated at the unit surface due to pseudocapacitance effect.

While keeping in mind all these implications, the primary requirement in an attempt to store a huge charge based on the electrostatic forces seems to be high surface area of an activated carbon used. Among different ways of porosity development in carbons, the treatment with an excess of potassium hydroxide is most efficient in terms of microporous texture generation. Porous materials with BET surface areas in excess of 3000 m^2/g could be prepared using various polymeric and carbonaceous type precursors [5,6].

The recent reports [7,8] suggest that KOH activated carbons produced from very common precursors like bituminous coal and coal-tar pitch can be a suitable electrode material for EDLC. The excellent capacitance values approaching in some cases 300 F/g were measured for extremely high surface area (S_{BET}=2700-3200 m^2/g) materials produced using severe activation conditions [8]. No simple relationship between the BET surface area or pore volume and the electric capacitance was found. The carbons show so high capacitance despite the limited accessibility of the porous systems as indicated by a relatively low specific capacitance (calculated per BET surface area), being in the range of 7-11 $\mu F/cm^2$. For comparison, the maximum capacitance values of about 160 F/g and 16 $\mu F/cm^2$, respectively, were determined for a series of steam activated carbons, which were studied using exactly the same technique [9]. Moreover, a fairly good cycleability has been determined for pitch and activated carbon derived high surface area materials in contrast to those originated from coal [8]. Clearly, the electrode performance of KOH activated carbons depends on pore size distribution and intrinsic properties of the surface as related to the nature of precursor used. It is therefore anticipated that the porosity development and surface properties can be further tailored to the EDLC application by selection of parent material and adjusting process variables.

Hence, the objective of the present study is an advanced understanding of the effect of parent material nature and process variables on the porosity development and surface chemistry of KOH activated carbons with respect to their behavior as an electrode material in EDLC.

2. EXPERIMENTAL

2.1. Materials

Materials used in the activation with KOH include high volatile bituminous coal C, coal semi-coke CS, pitch mesophase PM, pitch semi-coke PS and commercial activated carbon AC. The semi-cokes CS and PS were produced by the heat treatment of corresponding parent materials at 520°C with a heating rate of 5°C/min and 2 hours soaking time. The preparation of mesophase PM comprised the soaking of coal-tar pitch at 450°C for 7 h with a continuous stirring. All the treatments were performed under argon in a vertical Pyrex retort of 45 mm diameter.

2.2. Activation

Raw materials were ground to a particle size less than 630 μm. Physical mixtures of solid KOH and a given precursor at 2:1, 3:1, 4:1 and 5:1 weight ratios were activated in the temperature range of 600-900°C for 1-5 h under argon flow of 15 dm^3/h. The treatments were carried out in a horizontal furnace using a nickel boat placed in a 36 mm diameter quartz tube. The heating rate was 10°C/min. The resultant materials were washed with 10% solution of HCl and distilled water and next were dried at 110°C for 6 h.

2.3. Analyses

Ash and volatile matter contents were determined according to the respective standard procedures. Elemental composition (CHNS) was analyzed using VarioEl elemental analyzer. Oxygen content was calculated by difference. The optical texture was evaluated using polarized light optical microscopy of polished surface of samples. N_2 adsorption/desorption isotherms at 77 K of the activated carbons were obtained using a gas adsorption analyser NOVA 2200 (Quantachrome). Samples were degassed overnight at 300°C in vacuum prior to the analysis. The relative pressure up to $p/p_o \leq 0.15$ was used for the calculation of the mean micropore width (L_D) by applying the Dubinin equation. The mesopore size distribution was evaluated using Kelvin equation.

3. RESULTS AND DISCUSSION

3.1. Parent Materials Characteristics

Characteristics of carbonaceous materials used in the activation with KOH are given in Table 1.

Table 1. Properties of the precursors used.

Sample	Ash wt %	Volatile matter wt%	Elemental analysis $(C/H)_{at}$	Elemental analysis (N + S + O) wt%	Optical texture
C	3.1	31.9	1.4	8.9	Isotropic
CS	3.8	15.7	2.2	8.8	Isotropic
PM	0.5	17. 4	2.1	2.9	Anisotropic
PS	0.3	6.6	2.4	2.8	Anisotropic
AC	7.1	1,7	21.2	5.6	Isotropic

The parent materials differ from each other in many aspects; the differences are being related both to their origin (coal or pitch) and heat treatment temperature. Clearly, coal should be classified as a polymeric type precursor while the others, such as carbonaceous precursors of relatively low, except for AC, carbonization degree. Specific of pitch-derived materials is distinctly lower mineral matter and heteroatoms content. Anisotropic appearance with predominating flow type texture proves the superior extent of structural ordering in pitch-derived materials.

3.2. Effect of Activation Variables on Porosity Development

The reagents ratio, reaction temperature, reaction time and inert gas flow have been reported as the variables of KOH activation process, which influence effectively the porosity development on the treatment of a given carbonaceous material with potassium hydroxide [6].

Figures 1-3 demonstrate the effect of KOH/precursor ratio, reaction temperature and reaction time, respectively, on porous structure parameters of carbon produced by KOH activation. While the presented relationships concern mostly carbonaceous mesophase, basically they are typical of all coal and pitch-derived materials of the study.

Varying KOH ratio in the mixture is a very effective way of controlling porosity development in resultant activated carbons. The trend in the pore volume and BET surface area increase seems to be similar for various precursors (Fig. 1a). It is interesting to note, however, a sharp widening of pores, resulting in clearly mesoporous texture, when a large excess of KOH is used in reaction with coal semi-coke (Fig. 1b). Increase in the reaction temperature within 600-900°C results in a strong development of porosity up to 700°C and rather moderate above this temperature (Fig. 2a).

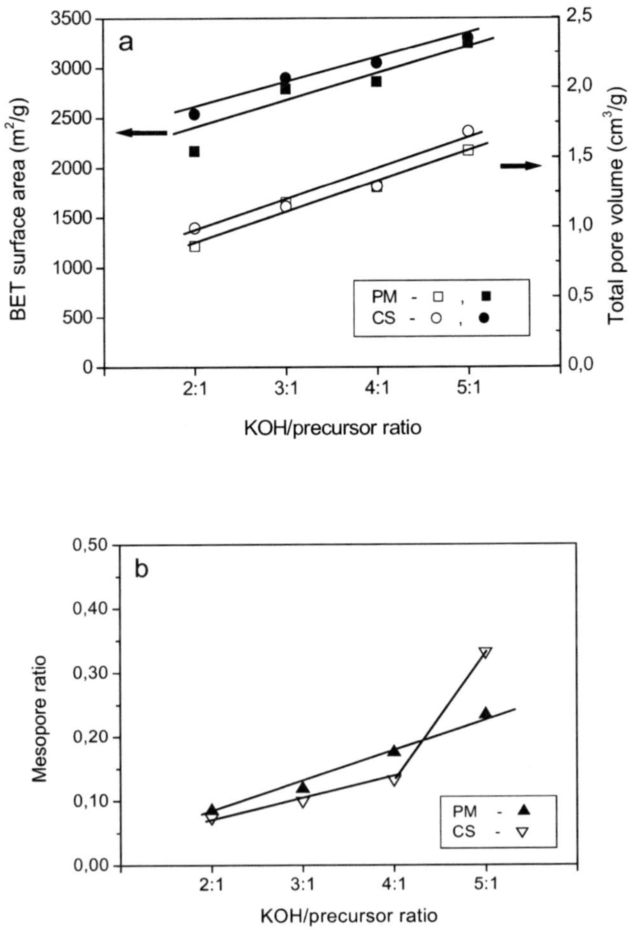

Figure 1. Effect of reagents ratio on pore volume and BET surface area (a) and mesopore ratio (b) in activated carbons produced on KOH activation of PM and CS at 800°C for 1 h.

The treatment of semi-coke or mesophase (3:1 mixture) for 2 h at 600°C gives a material of pore volume V_T about 0.7 cm^3/g and surface area S_{BET} about 1700 m^2/g which is typically microporous with rather narrow micropores (average size L_D below 1.2 nm). Activated carbons produced within the temperature range of 700-800°C have fairly similar porosity characteristics, V_T about 1 cm^3/g and S_{BET} near 2500 m^2/g. It is interesting to note that within the wide temperature range of 600-800°C the burn-off is at a reasonably low level of 20-23 wt%.

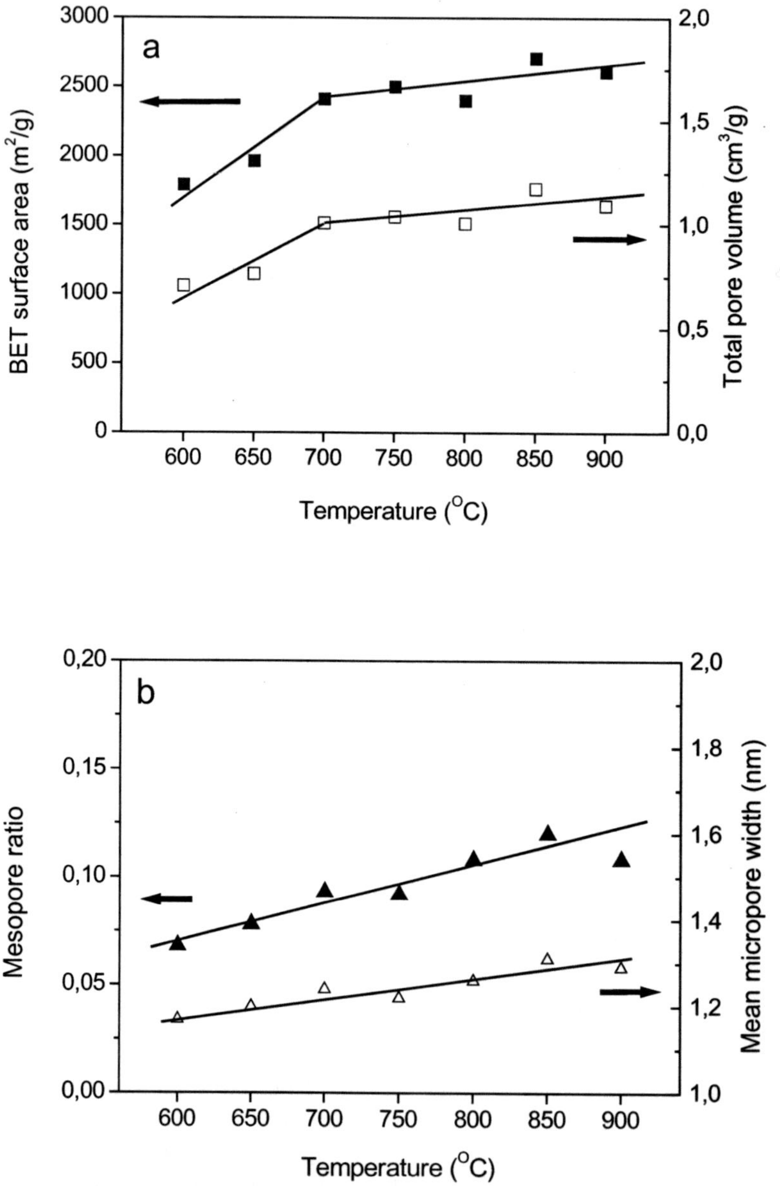

Figure 2. Variation of pore volume and BET surface area (a) and mesopore ratio and mean micropore width (b) with reaction temperature on the treatment of KOH/PM 3:1 mixture for 2h.

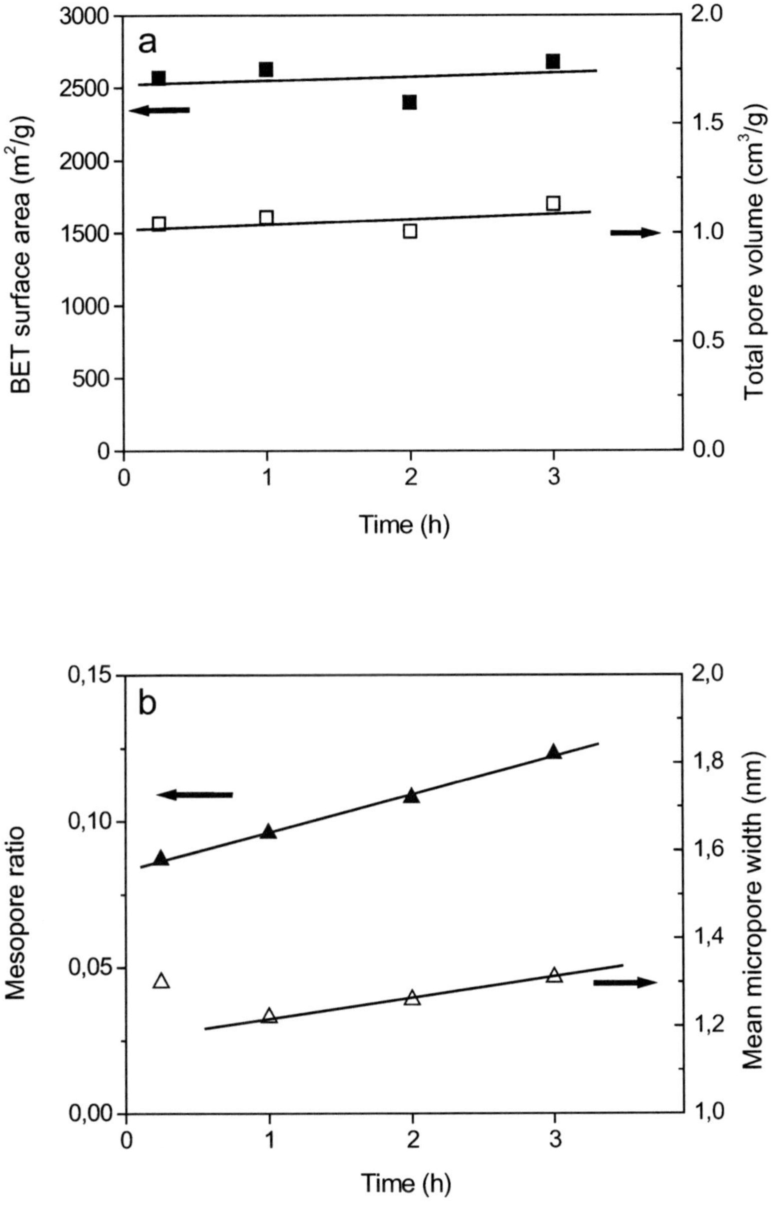

Figure 3. Variation of pore volume and BET surface area (a) and mesopore ratio and mean micropore width (b) with reaction time on the treatment of KOH/PM 3:1 mixture at 800 °C.

A sharp increase in burn-off above 800°C is associated with a limited enhancement of porosity. Pore widening (reflected by a slight but continuous increase in the mesopore ratio and micropore width), seems to be most characteristic effect of increasing reaction temperature (Fig. 2b).

The soak time seems to have a less pronounced effect on porosity development during treatment at 800°C. A considerable porosity (V_T = 1.04 cm^3/g and S_{BET} = 2570 m^2/g) is created within the early period of reaction and a limited change of the parameters occurs with prolonged heating to 3 h (Fig. 3a). Again, the widening of pores, as evidenced by an increase of both the mesopore ratio and the micropore width, is the most noticeable result of extended reaction time (Fig. 3b).

3.3. Effect of Parent Material on Porosity Development

In order to evaluate the potential to porosity development, which is intrinsically associated with a given precursor nature, all parent materials were treated using drastic conditions: reagent ratio 4:1, temperature 800°C, time 5 hours. Porosity parameters of resultant carbons are given in Table 2.

Nitrogen adsorption/desorption isotherms of all the activated carbons are of Type I, i.e. characteristic of basically microporous solids. There is a lack of adsorption/desorption hysteresis. More careful analysis permits to notice significant differences in the porous texture parameters depending on precursor origin.

Table 2. Porosity parameters of KOH activated carbons produced from different precursors under the same reaction conditions (800°C, 5 hours, KOH/precursor ratio 4:1)

Precursor	S_{BET} m^2/g	V_{total} cm^3/g	Pore volume distribution, %				Average pore size nm
			$V_{<2nm}$	V_{2-3nm}	V_{3-5nm}	$V_{5-50\,nm}$	
C	3150	1.612	73.8	15.0	6.1	5.1	2.04
CS	3190	1.445	81.9	13.9	2.6	1.6	1.82
PM	2660	1.209	83.6	10.9	3.3	2.2	1.82
PS	2750	1.227	85.4	9.5	2.7	2.4	1.79
AC	1900	1.051	68.6	12.3	8.3	10.8	2.21

At the exhaustive activation conditions, the polymeric-type precursor (coal C) gives an activated carbon of bigger pore volume and wider pore size distribution with a larger burn-off. The behavior can be attributed to a large proportion of non-carbon elements, which are readily removed from the system and a loose packing of structural units at the stage when the primary pyrolysis reactions are completed. Activation of low temperature carbons (mesophase PM, semi-cokes CS and PS) allows preparing carbons of very high S_{BET} (2700-3200 m^2/g) and rather similar pore size distribution. The commercial active carbon (AC) gives less porous material (S_{BET} = 1900 m^2/g) but with a considerable contribution of mesopores (above 30%), including those over 5nm. Apparently, the strange porosity characteristics appear as an effect of both a lower extent of reaction with alkali due to enhanced carbonization degree (Table 1) and overlapping of the original porous system with that generated by KOH activation.

3.4. Chemical Structure of KOH-Activated Carbons

Elemental analysis of activated carbons made from raw materials of this study under various reaction conditions is presented in Table 3.

Compared to the conventional steam activated carbons, characteristic of all the KOH activated carbons is the enhanced oxygen content. The content varies quite widely from one activated carbon to another. The data for a series of carbons activated for 5h at 800°C show rather surprisingly, the highest oxygen content in pitch derived materials. A lack of correlation in elemental composition between parent material and activation product suggests that the origin of the oxygen groups is the high temperature reaction with KOH and the post-activation treatment. A reduced oxygen content in carbons produced under severe treatment conditions is an effect of partial decomposition of the groups. Therefore, an enhanced oxygen content in the pitch derived activated carbons can prove that more thermally stable functionalities are created during reaction of KOH with material of higher structural ordering.

Table 3. Composition of activated carbons made by KOH activation of different precursors.

Precursor	Activation variables	C wt%	H wt%	N + S wt%	O_{diff} wt%
C	800°C, 5 h, 4:1	96.1	0.23	0.2	3.5
CS	800°C, 5 h, 4:1	92.8	0.20	0.3	6.7
PM	800°C, 5 h, 4:1	91.2	0.21	0.2	8.4
PS	800°C, 5 h, 4:1	91.3	0.18	0.2	8.4
AC	800°C, 5 h, 4:1	94.8	0.21	0.4	4.6
CS	800°C, 1 h, 3:1	86.1	0.36	0.7	12.8
PM	800°C, 1 h, 3:1	90.5	0.16	0.3	9.0
CS	600°C, 2 h, 3:1	79.3	1.02	1.0	18.7
PM	600°C, 2 h, 3:1	82.8	0.82	0.8	15.6

3.5. Potential of KOH-Activated Carbons as EDLC Electrode

Activation with KOH was recognized originally as an efficient way of producing microporous carbons with relatively narrow pore size distribution and extremely high surface area. The results of present study demonstrate a considerable flexibility of the process in terms of porosity development and, to some extent, surface properties.

With respect to the producing porous carbons for electrochemical capacitors the activation using potassium hydroxide represents several interesting aspects. Primarily, activated carbons of very high surface area and different mesopore ratio can be produced at a reasonable burn-off from a variety of precursors. In contrast to the agents used in conventional physical and chemical activation, KOH reacts efficiently with the whole range of polymeric and carbonaceous materials, including those of high extent of structural ordering as distinguished by optical anisotropy. This peculiarity can appear meaningful for improving the electric conductivity of electrode material. Another attractive aspect of KOH activated carbons is the presence of thermally stable oxygen functionalities, which can positively contribute to the electric capacitance by both enhancing wettability of surface by electrolyte and inducing pseudocapacitance effect. Oxygen content is controlled by both the parent material nature and reaction variables.

4. CONCLUSION

The activation with KOH of selected parent materials under appropriate process conditions (temperature, time, reagent ratio) can provide highly porous carbons of controlled pore size distribution and surface chemistry, also suitable for use as electrode materials in supercapacitors.

ACKNOWLEDGMENTS

Work was supported by the Polish Ministry of Scientific Research & Information Technology (Project T09B 13325), NATO SfP project 973849.

REFERENCES

1. Sarangapani S., Tilak B.V., Chen C.P., Materials for electrochemical capacitors – theoretical and experimental constraints, J. Electrochem. Soc. 1996; 143:3791-3799.
2. Qu D, Shi H. J. Power Sourc. 1998; 74:99-107.
3. Frackowiak E., Béguin F., Carbon materials for the electrochemical storage of energy in capacitors, Carbon 2001; 139, 937-950.
4. Qu D., Studies of the activated carbons used in double-layer supercapacitors, J. Power Sourc. 2002; 109:403-411.
5. Ahmadpour A., Do D.D., The preparation of active carbons from coal by chemical and physical activation, Carbon 1996; 34:471-479.
6. Lozano-Castello D., Lillo-Rodenas M.A., Cazorla-Amoros D., Linares-Solano A., Preparation of activated carbons from Spanish anthracite. Carbon 2001; 39:741-749.
7. Weng T.C., Teng H., Characterization of high porosity carbon electrodes derived from mesophase pitch for electric double-layer capacitors, J. Electrochem. Soc. 2001; 148:368-373.
8. Kierzek K., Frackowiak E., Lota G., Gryglewicz G., Machnikowski J., Electrochemical capacitors based on highly porous carbons prepared by KOH activation, Electrochim. Acta 2004; 49:515-523.
9. Gryglewicz G., Lorenc-Grabowska E., Lota G., Frackowiak E., Machnikowski J., Coal-based mesoporous activated carbons for double-layer capacitors, Carbon'03, International Conference on Carbon, CD-ROM Proc., paper 292, Oviedo 2003.

GENERAL PROPERTIES OF IONIC LIQUIDS AS ELECTROLYTES FOR CARBON-BASED DOUBLE LAYER CAPACITORS

A. Lewandowski* and M. Galinski

Faculty of Chemical Technology, Poznań University of Technology, PL-60 965 Poznań, Poland

1. INTRODUCTION

Activated carbons have been used for a long time in many industrial applications. One of the recent applications is using activated carbon as an active electrode material in double-layer capacitors (DLC). Electrochemical capacitors, based on the capacity of the double layer formed at the carbon/electrolyte interface, have received considerable attention as they can be used as high power-density energy-storage devices. Batteries have high energy density, but they suffer from low power density and low cyclability (usually <1000). On the other hand, electrochemical capacitors offer an order of magnitude higher power density and at least two orders of magnitude higher number of charge-discharge cycles (~100 000). However, the energy density is an order of magnitude lower in comparison to that characteristic for batteries. There are a number of electrochemical capacitor potential applications, but it seems that they are being investigated mainly for hybrid vehicles as a short-term power accumulator and supply. Therefore, the DLC research aims to develop an energy-storage device having energy density close to that typical of batteries and the power characteristic typical of capacitors. This goal can be obtained by optimisation of carbon materials as well as electrolyte. Carbon materials, for the use in double-layer capacitors, have been developed in various systems: powders, fabrics, tissues or aerogels [1-3]. It is possible to obtain activated carbons of high specific surface area at the level of 2500 m2/g. Similarly, many aqueous and organic liquid solutions of electrolytes as well as a limited number of polymer electrolytes have been developed for the use in carbon-based electrochemical capacitors [4].

* Corresponding author. E-mail: Andrzej.Lewandowski@put.poznan.pl

I.V. Barsukov et al. (eds.), New Carbon Based Materials for Electrochemical Energy Storage Systems, 73–83.

The ideal electrolyte should:

1. Produce high capacity of the carbon/electrolyte double electric layer;
2. Enter the pores of the high-surface area activated carbons;
3. Show possibly broad electrochemical stability window at an interface of carbon electrode;
3. Show high electrical conductivity, at the level of ≥10 mS/cm;
4. Be liquid in a broad temperature range;
5. Emit no volatile vapours.

One of the most important factors is the electrolyte resistance for electrochemical degradation (oxidation or reduction) at possibly high voltages applied to the capacitor. Let us assume, that the highest practical specific capacity of activated carbon in organic electrolytes is of the order of 160 F/g and that the device contains 25 wt.% of the carbon material. Therefore, the specific capacity is 40F/g. However, the capacity C of the device consisting of two identical electrodes is twice lower: (1/C = 1/C1 + 1/C2; C = Ci/2), and the final specific capacity of the capacitor equals ca. 20 kF/kg. The energy E which can be stored in the device is proportional to the voltage U in the second power: E = ½CU2. Figure 1 shows the energy accumulated by such a capacitor at different voltages applied to it (level of the specific energy of batteries is indicated along energy axis).

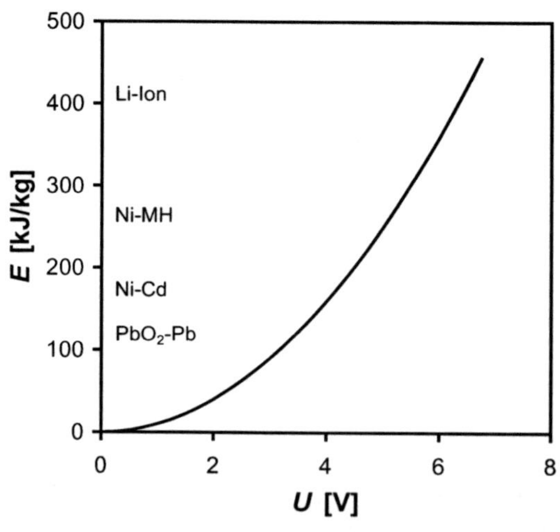

Figure 1. Energy accumulated in a capacitor having specific capacity of 20 kF/kg as a function of voltage (E = ½CU2).

It can be seen that an energy of ca. 150 kJ/kg, comparable to that accumulated in PbO_2-Pb or Ni-Cd batteries, can be obtained at voltages of 4V. Somewhat lower energy (100 kJ/kg) is accumulated at a voltage of 3V. Consequently, the searched system carbon/electrolyte should be characterised by: (i) specific capacity. ≥ 160 F per gram of activated carbon and (ii) electrochemical stability window at the level of ca. ≥3V.

The general aim of this work is to show the properties of activated carbon-ionic liquid interface (AC-IL) as well as performance of capacitors based on activated carbons as electrode materials and ionic liquids as electrolytes.

2. IONIC LIQUIDS

Conventional electrolytes applied in electrochemical devices are based on molecular liquids as solvents and salts as sources of ions. There are a large number of molecular systems, both pure and mixed, characterized by various chemical and physical properties, which are the liquids at room temperatures. This is the reason why they dominate both in laboratory and industrial scale. In such a case, solid salt is reacted with a molecular solvent and if the energy liberated during the reaction exceeds the lattice energy of the salt, the solid is 'liquified' chemically below its melting point, and forms the solution. Water may serve as an example of the cheapest and most widely used molecular solvent.

However, the lattice energy of the salt may be counterbalanced by providing heat to the solid; in such a case the solid is melted down at its melting point and forms a liquid phase, which does not contain any molecular solvent, but only ions and their charged and neutral combinations. Such a system is called ionic liquid or fused salt. Classical ionic liquids exist at relatively high, uncomfortable temperatures, determined by the melting point of salts. These temperatures can be lowered by an addition of other salts forming a eutectic. Even in such a case considerable heat is necessary to maintain the system in the liquid state. Aluminum electrolysis from Al_2O_3-Na_3AlF_6 eutectic fused at ca. 1000°C may serve as rather unique example of using fused salts in a large-scale industrial practice. Highly reactive metals, such as Li, Na or Mg are also produced from high-temperature fused salts. Ionic liquids have been used in thermally activated batteries, applied in military applications, when an extremely long storage time of the device, providing high power for a short time, is necessary. The battery is activated with the help of an explosive; the heat liberated during the explosion melts down the salt. Applying ionic liquids in commercial energy storage devices, such as batteries or electrochemical capacitors, has been not practically realized. On the other hand, salts characterized by a low melting point, and therefore being liquid at room temperature, or even below, can play a role of electrolytes in electrochemical systems. Room temperature ionic liquids (RTIL) are not new, but only within the past few years an increasing interest in their synthesis and application has been observed [5,6]. Their basic physicochemical properties are the same as high temperature molten salts, but the practical aspects of their application are different enough to merit a distinction. One of the first discovered RTIL (1914) seems to be $EtNH_3NO_3$, but the first extensively studied was 1-ethyl-3-methylimidazolium chloroaluminate ($EMImAlCl_4$) [5,6]. Ionic liquids based on the chloroaluminate ion seem to be a promising medium for reactions catalyzed by Lewis acids, such as for example Friedel-Crafts reaction. However, chloroaluminate ionic liquids are hydrolyzed by traces of moisture, which limits their possible practical application. Lately, ion-conducting systems based on nonchloroaluminate room temperature ionic liquids have been developed (Figure 2) for application as electrolytes [7-15].

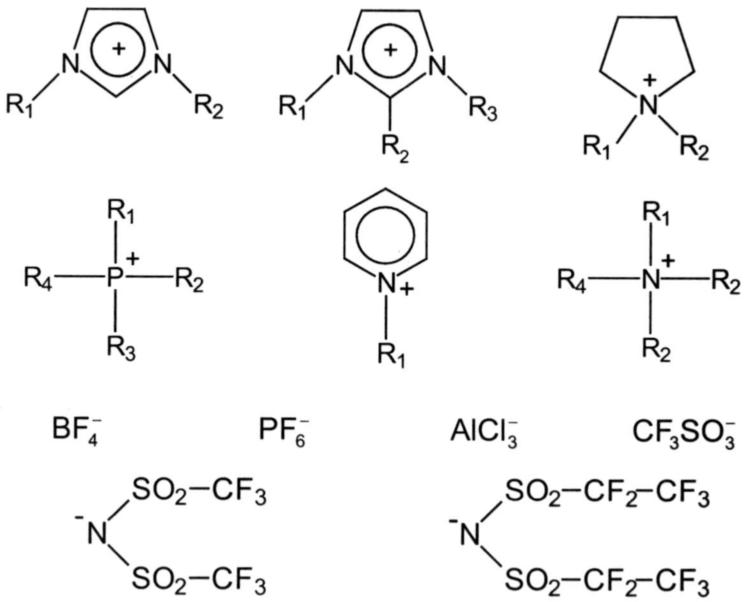

Figure 2. RTIL's typical cations and anions.

RTILs consist of large, unsymmetrical ions, such as 1,3-dialkyl-imidazolium, 1-alkylpyridinium, 1-alkylpyrazolium, tetralkylammonium or tetralkyl-phosphonium cations and tetrachloroaluminate, tetrafluoroborate, hexafluorophosphate, trifluoromethanesulfonate or bis((trifluoromethyl) sulfonyl)imide anions.

Table 1 shows examples of melting and freezing points of a number of organic salts, which may be regarded to be ionic liquids. As can be seen, the compounds are characterized by tendency for supercooling. The difference between melting and freezing points may be as high as 60°C. As a result, the same compound can be described in the literature as solid or liquid at room temperature. On the other hand, the term 'ionic liquid' is quite often extended to salts molten below 100°C. In such a case heat must be provided to keep the salt in the liquid state, and IL cannot be handled as an ordinary solvent.

Table 1. Freezing and melting points [°C] of some representative ionic liquids [9].

Ionic liquid	T_f	T_m	ΔT
$EMImBF_4$	-63	11	74
$EMImPF_6$	5	62	57
$EMImN(Tf)_2$	-50	-15	35
$DMPImPF_6$	18	78	60
$DMPIm\ N(Tf)_2$	-130	-81	49

Commonly is used a short term 'ionic liquid' instead of 'room temperature ionic liquid' or 'room temperature molten salt', which makes no distinction between salts liquid at room temperature and those liquid below 100°C.

3. CONDUCTIVITY

The room temperature specific conductivity of a number of ionic liquids is reported in Table 2, together with their viscosity. It can be seen from the table, that the conductivity varies in the range of 0.1–13 mS/cm. The most conductive IL in the series shown in the table is $EMImBF_4$.

Table 2. Concentration c, conductivity σ, viscosity η and Walden product $\Lambda\eta$ of some neat ionic liquids at 25 °C [9].

	c mol/dm^3	σ mS/cm	η cP	$\Lambda\eta$ cm^2cP/mol
$EMImBF_4$	6.5	13	43	86
$EMImPF_6$	5.7	5.2*	15	66*
$EMImN(Tf)_2$	3.9	8.4	28	60
$DMPImPF_6$	4.8	0.5	34	63
$DMPIm\ N(Tf)_2$	3.5	3.0	60	51
$HTEA\ N(Tf)_2$ [7]	-	0.7	17	-

*Data reported for the temperature of 80°C

The temperature dependence of the conductivity can be described by the classical Arrhenius equation $\sigma = \sigma^o \exp(-E^\#/RT)$, where $E^\#$ is the activation energy for the conduction process. According to the Arrhenius equation the $\ln\sigma$ versus 1/T plot should be linear. However, in numerous ionic liquids a non-linearity of the Arrhenius plot has been reported; in such a case the temperature dependence of the conductivity can be expressed by the Vogel-Tammann-Fuller (VTF) relationship: $\sigma = \sigma^o \exp\{-B/(T-T_o)\}$, where T_o represents the ideal glass transition temperature. The parameter B has no sense of activation energy. However, even when the $\ln\sigma = f(1/T)$ plot is not ideally linear, it is quite often treated according to the Arrhenius equation. Activation energies, determined from the Arrhenius plot, are usually in the range of 10-40 kJ/mol.

Table 3. Conductivity of 2mol/dm³ solutions of $EMImBF_4$ and $EMImPF_6$ in molecular liquids at 25 °C [9].

	$EMImBF_4$	**$EMImPF_6$**
-	13	5.2
PC	16, 18*	15
EC	23	22
DMF	25	33
Acetone	34	43
AN	47	60
Water	85*	-

*Data reported as quoted from ref. [16]

Table 3 shows conductivity of 2mol/dm³ solutions of $EMImBF_4$ and $EMImPF_6$ in a number of molecular solvents. A high increase of conductivity, in comparison to neat ionic liquids, can be observed after dilution with electrically neutral molecular liquids. However, solutions of ionic liquids in molecular liquids are simply conventional solutions of organic salts in nonaqueous solvents, and no distinction can be seen between them and commonly employed solutions of $(C_2H_5)_4NBF_4$.

4. POLYMER ELECTROLYTES WITH IONIC LIQUIDS

A polymer electrolyte may be generally defined as a system formed by dissolution of an electrolyte in a polymer matrix. The resulting composite shows ionic conduction and mechanical properties characteristic for polymeric materials. Polymer electrolytes may be prepared as thin foils showing rubber-like elasticity. However, the process may be reversed by dissolution of a polymer in ionic liquid. In such a case the polymer acts both as a diluent of the liquid salt, as well as the matrix providing mechanical strength to the system. The first polymer electrolytes incorporating ionic liquid were based on 1-butylpyridinium halides (chloride or bromide) and aluminum chloride [17,18]. More recently, ion conducting solid systems based on nonchloroaluminate room temperature ionic liquids have been synthesized [3]. Poly(vinylidene fluoride-hexafluoropropylene) (PVdF-HFP) [19,20] and NAFION [21] membranes were impregnated with room temperature, non-chloroaluminate ionic liquids. Ionic conductivity of a composite PVdF-HFP-ionic liquid was in the range of 3.2-0.06 mS/cm.

Polymeric conducting systems were also prepared by *in situ* polymerization of vinyl monomers in ionic liquids [22], with a conductivity of 1 mS/cm. A conductive polymer electrolytes were also prepared by polymerization in liquid $EMIm(HF)_nF$ leading to a composite poly(2-hydroxyethyl methacrylate)-$EMIm(HF)_nF$. Recently, polymer electrolytes were prepared in the form of thin foils, by incorporating ionic liquids in a polymer matrix [13-15]. Conductivities of polymer-IL or polymer-IL-solvent systems are collected in Table 4.

Table 4. Specific conductivity (25°C) of some polymer electrolytes based on ionic liquids.

Composition [wt.%]			σ
Polymer	**IL**	***TMS***	**[mS/cm]**
25.4% PEO	74.6% $BMImPF_6$	0.0 %	0.3
12.3% PAN	33.2% $BMImPF_6$	54.5%	6.3
67.2% PEO	32.8% $EMImBF_4$	0.0%	0.6
39.8% PVA	60.2% $EMImBF_4$	0.0%	1.6
26.6% PAN	73.4% $EMImBF_4$	0.0%	6.6
12.9% PAN	48.3% $EMImBF_4$	38.8%	15.0

It can be seen from the table that the system PAN-$EMImBF_4$-TMS, containing a plasticizer (TMS), shows the best conductivity of 15 mS/cm.

5. SPECIFIC CAPACITY OF CARBON/IL INTERFACE

It is difficult to compare the specific double-layer capacitance characteristic for capacitors based on highly porous activated carbons with that determined for non-porous electrodes (solid glassy-carbon and liquid mercury) as appropriate literature data on ionic liquids are limited to just a few studies [24,25]. Apart from a report on ionic liquid based on the chloroaluminate ion [25], one paper [24] about specific capacitance at the electrode/IL interface for glassy carbon (GC), mercury and a commercial carbon and four ionic liquids based on 1-ethyl-3-methyl imidazolium cation can be found in the literature. The integral capacitance of various ionic liquids at the dropping mercury electrode (DME) ranged between 10.6 and 12.4 $\mu F/cm^2$. For the $EMImN(Tf)_2$ electrolyte the capacitance was 11.7 $\mu F/cm^2$ at DME and 12.0 at glassy carbon, but only 3.7 $\mu F/cm^2$ at a high-surface SpectraCarb 2220 yarn (ca. 2000 m^2/g). This may indicate that the accessibility of the electrolyte to the carbon electrode surface limits the realizable specific capacitance of the device.

6. ELECTROCHEMICAL STABILITY WINDOW

It is essential from the point of view of high power-density to ensure the electrochemical stability of the system at possibly high voltages. Broad electrochemical stability windows are typical if ionic liquids, however, the practical voltage, which can be applied to capacitor carbon electrodes seems to be considerably lower. This is because the stability window is usually determined at glassy carbon electrodes while the capacitor electrodes are formed from different kinds of activated carbons, possessing functional groups at their surface. Moreover, a polymer used as an additive of the electrode, or being a component of a polymer electrolyte, may also undergo red-ox electrode reactions, which can additionally limit the stability window.

Table 5. Electrochemical stability window of ionic liquids (IL) at the glassy carbon (potentials [V] expressed versus Ag/Ag$^+$ 0.01M in DMSO reference) [26].

Ionic liquid	***Stability window***	
	cathodic	**anodic**
$EMImBF_4$	-2.1	2.1
$BMImBF_4$	-2.1	2.1
$BMImPF_6$	-2.1	2.0
$EMImN(Tf)_2$	-2.1	2.0
$BMPyN(Tf)_2$	-1.0	2.0

Table 5 shows cathodic and anodic limits of electrochemical stability windows of a number of ionic liquids. The cathodic limit of the stability window of the ILs based on the $EMIm^+$ and $BMIm^+$ cations, investigated at the glassy carbon electrode, is –2.1 V against the Ag/Ag$^+$ (0.01M in DMSO) reference. The $BMPy^+$ cation is reduced at the glassy carbon at considerably more positive potential, at ca. –1.0 V.

The anodic oxidation of the tetrafluoroborate anion occurs at potentials higher than 2.1 V and the remaining hexafluorophosphate and imide anions are oxidised at potentials higher than 2.0 V. Hence, the stability window of the $EMImBF_4$ and $BMImBF_4$ is 4.2 V. Ionic liquids $BMImPF_6$ and $EMImN(Tf)_2$ shows a similar stability window of ca. 4.1 V. However, the window of the $BMPyN(Tf)_2$, is considerably lower: ca. 3.0 V. This is consistent with data (ca. 4.1-4.2 V) found for a series of ionic liquids based on $EMIm^+$ and $DMPIm^+$ (1,2-dimethyl-3-propylimidazolium) cations [12].

Usually, an addition of a diluent (solvent), increases the conductivity, but also influences the stability window. It has been found, that among the most popular organic solvents, 2 M solutions in acetonitrile showed the highest conductivity level ($EMImN(Tf)_2$: 47 mS/cm and $EMImPF_6$: 60 mS/cm) and a broader stability window (ca. 4 V) [12]. Carbonate solvents decrease the cathodic limit by ca. 0.2-0.4 V, depending on the carbonate structure. It is worth noting that different authors give different stability windows for the same ionic liquid [10]; this is partially due to different electrode materials.

7. CARBON/IL CAPACITORS

Ionic liquids show a broad electrochemical stability window of ca. 4V, conductivity at a level of ca. 10 mS/cm as well as an acceptable specific capacity of ca. 10 μF/cm^2, which makes them promising candidates for electrolytes. However, there are only a few papers about capacitors based on carbon materials and ionic liquids. In 1997 some results on applying IL solutions in supercapacitors were reported [12]. Ionic liquids, based on $EMIm^+$ cation ($EMImBF_4$ and $EMImPF_6$) were dissolved in cyclic carbonates. Capacitors showed specific capacity, expressed versus mass of active carbon, of ca. 130 F/g and the electrochemical stability within the voltage range of ca. 3.5 V. One of the main problems in supercapacitors construction is a reduction of the equivalent series resistance (ESR) to low values. This may be done in two ways: by applying electrolytes of high conductivity or by developing a technology of ultra-thin electrodes. The best would be suppressing the ESR to very low values by combining high electrolyte conductivity with ultra-thin electrodes. In the paper [12], the conductivity was enhanced to values of ca. 26 mS/cm (2mol/dm^3 solution of $EMImPF_6$ in a mixture EC + DMC). However, solutions of ILs in molecular liquids are classical salt solutions, and they do not show any advantages in comparison to solutions of tetraalkylammonium salts in acetonitrile.

Table 6. Specific capacities of an activated carbon (2000 m^2/g) interface with 2mol/dm^3 solution of $EMImPF_6$ in cyclic carbonates [12]

	AC 2000 m^2/g	
Solvent	**F/g**	**μF/cm^2**
EC	123	6.2
PC	127	6.4
DMC	126	6.3
EMC	123	6.2
PC+EMC	129	6.5

Capacitors made of high-surface carbon with PVdF-co-HFP as a binder and filled with neat ionic liquids, without any molecular solvent, show high specific capacity, up to 180 F/g [13].

Table 7. Specific capacities of activated carbon interface with ionic liquids [13]

	AC 870 m^2/g		AC 2600 m^2/g	
Ionic liquid	**F/g**	**μF/cm^2**	**F/g**	**μF/cm^2**
$EMImBF_4$	55	6.3	178	6.8
$BMImBF_4$	50	5.7	155	5.7
$BMImPF_6$	45	5.2	160	6.1
$EMImN(Tf)_2$	51	5.9	180	6.9
$BMPyN(Tf)_2$	48	5.8	167	6.4

Similar or even better capacity was obtained in systems with polymer electrolytes based on ionic liquids. A series of electrochemical capacitors, based on activated carbons, applying polymer-IL solid electrolytes were prepared and tested [14]. Both polymer electrolytes as well as capacitor electrodes, in the form of thin foils, were prepared by a casting technique. All-plastic capacitors, having a coin-like shape, were constructed by sandwiching a polymer electrolyte between two electrodes. Ionic liquids: $EMImBF_4$, $BMPyN(tf)_2$ and $BMImPF_6$ served both as sources of ions as well as polymer plasticizers. In some systems sulpholane (TMS) was added to the system as an additional plasticizer. In other words, polymer was used as the ionic liquid diluent. Specific capacity was in the range of 4.2-7.7 $\mu F/cm^2$, calculated per total surface of the carbon materials, and up to 200 F/g expressed versus the mass of carbon material.

8. CONCLUSIONS

The main goal of the research on electrochemical capacitors is the development of an energy-storage device having energy density close as far as possible to that typical of batteries as well as the power characteristic and cyclability typical of capacitors. This can be obtained by optimization of carbon materials as well as electrolyte. Theoretical analysis shows that an electrochemical capacitor of specific capacity of 20 kF/kG and practical operation voltage of 3V would store energy comparable to that characteristic for the lead-acid battery (ca. 100 kJ/kg). The specific capacity of the device of ca. 20 kF/kG is equivalent to specific capacity expressed versus mass of activated carbon at the level of ca. 160 F/gC. Ionic liquids seem to be promising candidates for electrolytes in high energy capacitors, as they (i) produce high specific capacity at the AC/IL interface of the order ≥160 F/gC and (ii) they show broad electrochemical stability window determined at GC of the order of 4V. Till now, the literature of the subject is limited to a few papers on ionic liquid based capacitors.

REFERENCES

1. E. Frackowiak and F. Béguin, Carbon, 39 (2001) 937
2. G. Salitra, A. Soffer, L. Eliad, Y. Cohen, D. Aurbach, *J. Electrochem. Soc.*, 147 (2000) 2486
3. J. Gamby, P.L. Taberna, P. Simon, J.F. Favarque, M. Chesneau, *J. Power Sources*, 101 (2001) 109
4. B.E. Conway, Electrochemical Supercapacitors, Kluwer Academic/Plenum Publisher, New York, (1999)
5. J.S. Wilkes, *Green Chemistry*, 4 (2002) 73
6. T. Welton, *Chem. Rev.*, 99, (1999) 2071
7. J. Sun, M. Forsyth, D.R. MacFarlane, *J. Phys. Chem. B.* 102 (1998) 8858
8. D.R. MacFarlane, P. Meakin, J. Sun, N. Amini, M. Forsyth, *J. Phys. Chem. B.* 103 (1999) 4164

9. A.B. McEwen, H.L. Ngo, K. LeCompte, J.L. Goldman, *J. Electrochem. Soc.*, 146 (1999) 1687
10. R. Hagiwara, Y. Ito, *J. Fluorine Chem.*, 105 (2000) 221
11. T. Nishida, Y. Tashiro, M. Yamamoto, *J. Fluorine Chem.*, 120 (2003) 135
12. A.B.McEwen, S.F.McDevitt, V.R.Koch, *J. Electrochem. Soc.*, 144 (1997) L84
13. A. Lewandowski, M. Galiński, *J. Physics and Chemistry of Solids,* in press
14. A. Lewandowski, A. Swiderska, *Solid State Ionics*, 161 (2003) 243
15. A. Lewandowski, A. Swiderska, *Solid State Ionics*, in press
16. A. Lewandowski, S. Krajewski, in preparation for publication
17. M. Watanabe, S. Yamada, K. Saniu, N. Ogata, *J. Chem. Soc. Chem. Commun.* 929 (1993)
18. M. Watanabe, S. Yamada, N. Ogata, *Electrochim. Acta*, 40 (1995) 2285
19. J. Fuller, A.C.Breda, R.T. Carlin, *J.Electrochem Soc.*, 144 (1997) L67
20. J. Fuller, A.C.Breda, R.T. Carlin, *J. Electroanal. Chem.*, 459 (198) 29
21. M. Doyle, S.K. Choi, G.Proulx, *J.Electrochem Soc.*, 147 (2000) 34
22. A. Noda, M. Watanabe, *Electrochim Acta.* 45 (2000) 1265
23. T. Tsuda, T. Nohira, Y. Nakamori, K. Matsumoto, R. Hogiwara, Y. Ito, *Solid State Ionics*, 149 (2002) 295
24. C. Nanyunduiah, S.F. Mcdevitt, V.R., Koch, *J Electrochem Soc.* 144 (1997) 3392
25. R.J. Gale, R.A. Osteryoung, *Electrochim Acta,* 25 (1980) 1527
26. A. Lewandowski, M.Galiński upublished data

LIST OF ABBREVIATIONS

AC	activated carbon
AN	acetonitrile
$DMPIm^+$	(1,2-dimethyl-3-propylimidazolium) cation
DMC	dimethyl carbonate
DMF	NN-dimethylformamide
DME	dropping mercury electrode
$HTEA^+$	$(n\text{-}C_6H_{13})(C_2H_5)_3^+$ cation
EC	ethylene carbonate
EMC	Ethyl-methyl carbonate
$EMIm^+$	1-ethyl-3-methyimidaziolium cation
ESR	Equivalent series resistance
GC	glassy carbon
IL	ionic liquid
$N(tf)_2^-$	$^-N(CF_3SO_3)_2$ anion
PAN	poly(acrylonitrile)
PC	propylene carbonate
PEO	poly(ethyleneoxide)
PVA	poly(vinylalcohol)
PVdF-co-HFP	poly(vinylidenefluoride-co-hexafluoropropylene)
RTIL	room temperature ionic liquid
TMS	sulpholane (tetramethylene sulphone)

CHAPTER 2:

CARBON MATERIALS FOR GAS DIFFUSION ELECTRODES, METAL AIR CELLS AND BATTERIES

Chapter 2: Subject Overview

The following collection of papers focuses on the role of carbon materials in gas diffusion electrodes, currently used for the most part in Zinc-air batteries and cells. Such batteries use oxygen from ambient atmosphere to produce electrochemical energy. The air oxygen diffuses into the cell and is used as a cathode reactant. The cathode contains few forms of carbon: in the active material, as conductive additive, and as part of the current collector coating. A classic example of a household Zinc-air system is a so-called "hearing aid" cell. Besides that, these systems of power sources occupy several other niches of the energy market (discussed in length in this chapter). Editors believe that this chapter should be of interest to all: metal-air battery and most fuel cell system developers, due to obvious similarities between the construction of and requirements to the gas diffusion electrodes used by these chemistries. The chapter contains four articles.

The first paper by Ukraine's V. Barsukov *et al.* introduces a novel concept, which allowed constructing primary and rechargeable metal-air batteries of a new kind. The new chemistry uses a gas-diffusion electrode made out of a polyaniline (conducting polymer) / expanded graphite composite. The authors propose to put such an electrode against stabilized zinc or magnesium counter electrodes in the acidic solution. While featuring an apparent drawback of being a relatively low current density battery system, authors report achieving energy densities of 140 Wh/kg and 200 Wh/kg for the cells based on zinc and magnesium anodes, respectively. The data has been reported in fully functional AAA size cells.

The second and the third papers of this chapter were contributed by Bulgaria's National Academy of Sciences distinguished developers, Dr. A. Kaisheva and Professor I. Iliev. The two papers address magnesium-air and zinc-air chemistries, respectively. Concepts of mechanical recharging, new non-KOH based electrolytes, and the role of carbon materials in cell design have been thoroughly studied. Authors have reduced to practice their many developments. Systems created work(ed) in many specialty applications.

The last paper in this chapter is by Professor N. Korovin of Moscow Power Engineering Institute, Russian Federation. The author could not attend the NATO-CARWC in Chicago due to a last minute cancellation. His work, contributed to this volume is a comprehensive overview of various metal-air battery technologies, with a heavy focus on the role of carbon materials in these electrochemical systems.

While during the NATO-CARWC, the session on metal-air batteries featured other important contributions, it is, perhaps, by coincidence that only presenters from the former Eastern Block submitted their papers for this volume. In this regard, the editors wish to note, that papers below may be of special interest to the western reader, as they offer a rather unique insight into the science and application of the metal-air battery systems from the point of view of an Eastern European battery developer. Editors wish to respectfully refer the reader to [1]* for a "western" look into the subject.

Professor Ilia Iliev (In Memoriam)

Editors deeply regret that one of the authors, Professor Iliev passed away at the time of preparation of this volume. We have remembered his highly professional keynote talk, given during the NATO-CARWC, also his enthusiasm and high profile sense of humor. We hope that his laboratory in Bulgaria continues to operate and contribute to the world's body of electrochemical knowledge.

Below, we have published a memoriam to Professor Iliev by one of his former students, and now Assistant Professor at the Department of Chemical & Nuclear Engineering, University of New Mexico in Albuquerque, NM, USA, Dr. Plamen Atanassov.

Dr. Ilia Iliev was from the generation of Bulgarian scientists who entered the professional life in the late 50s. Broth in by the excitement of the physics in the mid-twentieth century and the growth of the post-WWII Eastern Europe, Ilia Iliev changed fields and become one of the key developers of the metal-air systems at the Central Laboratory of Electrochemical Power Sources, Bulgarian Academy of Sciences.

During the 70s he was engaged in the engineering of one of the largest electric vehicle propulsion programs being most contributive in the area of gas-diffusion electrodes of hydrophobic type. Associated hydrophobized carbon materials, electrode fabrication and device manufacturing were brought by his energy to a most successful fruition and market deployment. He contributed to development and commercialization of zinc-air power sources that lightened buys in the Baltic Sea, radioed messages from Mount Everest and circumnavigated the Earth.

Ilia Iliev embraced the new field of biosensors in the mid 80s and brought the most prominent Bulgarian contribution to it through his proven materials design approach. During the turmoil of the 90s he managed to sustain and grow these two technical venues continuing his association with the University of Sofia, where he taught electrochemistry to the new generation of Bulgarian chemical physicists. His contributions to the carbon-based materials and systems are well documented in many research publications and treatises.

His sudden death in 2004 became a shock. Professor Iliev will be remembered by all who knew him as a man of radiant energy, inventiveness and drive for innovation. Ilia Iliev is survived by his wife, Dr. Anastasia Kaisheva, collaborator and partner in all of the above research endeavors, and daughter Mila.

---

*[1] Handbook of Batteries / D. Linden, T.B. Reddy.-3rd ed. Chapt. 38, McGraw-Hill (2002).

NEW CONCEPT FOR THE METAL-AIR BATTERIES USING COMPOSITES: CONDUCTING POLYMERS / EXPANDED GRAPHITE AS CATALYSTS

V.Z. Barsukov[a*], V.G. Khomenko[a], A.S. Katashinskii[a] and T.I. Motronyuk[b]

[a] *Electrochemical Power Engineering & Chemistry Department, Kiev National University of Technologies and Design, 2 Nemorovich-Danchenko str., Kiev 02001, Ukraine*
[b] *National Technical University of Ukraine "KPI", 37 Pobedy prosp., Kiev 03056, Ukraine*

Abstract

The mechanisms and reasons of catalytic activity of polyaniline (PANI)-type conducting polymers toward oxygen reduction in acidic and saline solutions are investigated by electrochemical and quantum-chemical methods. The PANI/thermally expanded graphite compositions were developed for realization of fully functional air gas-diffusion electrodes. Principally new concept for creation of rechargeable metal-air batteries with such type of catalysts is proposed. The mockups of primary and rechargeable metal-air batteries with new type of polymer composite catalysts were developed and tested.

Keywords

Conducting polymers, polyaniline, catalytic activity, PANI/expanded graphite composites, metal-air batteries, primary & rechargeable cells.

1. INTRODUCTION

Oxygen reduction electrodes, where molecular oxygen is electrocatalytically reduced, are vital to metal/air batteries and fuel cells. Recently, noble metals such as platinum have played a dominating role as

* Corresponding author. E-mail: chemi@mail.vtv.kiev.ua

I.V. Barsukov et al. (eds.), New Carbon Based Materials for Electrochemical Energy Storage Systems, 89–104.

electrocatalysts for the oxygen reduction reaction. However a disadvantage of Pt is its high cost. There is a serious concern for wide commercial application of electrocatalysts based on Pt. Search for low-cost electrocatalysts for the oxygen reduction reaction has been an important goal of many investigations. Here we report a polyaniline (PANI) as a low-cost electrocatalyst for oxygen reduction in acidic and neutral solutions. To profoundly study and to understand the mechanism of the catalytic activity of PANI, we have carried out quantum-chemical calculations of the electronic structure of PANI and its adsorption complexes with molecular oxygen. The above catalytic activity phenomena has found a practical application for development of primary and rechargeable metal-airbatteries with low costs PANI/Expanded Graphite composite catalysts.

2. EXPERIMENTAL

The investigations in this work were mainly aimed at studying a polymer prepared chemically. The synthesis of PANI was carried out in a 1M solution of hydrochloric acid. The aniline concentration was not higher than 0.5M. Potassium dichromate was used as chemical oxidant. The polymer was extracted from the reaction medium by filtering the solution. The precipitate was then washed several times with distilled water. Final washing was performed in ethyl alcohol and acetone. PANI was dried at a temperature of no more than 100°C. To build a laboratory-scale model of polymer electrode, a polyaniline paste based on polytetrafluoroethylene (PTFE) emulsion was prepared. The dry PTFE residue content was 5% of PANI weight. The experimental PANI electrodes were made by applying the polymer paste to a current collector of graphite with $0.5cm^2$ area. The electrochemical measurements were made in a standard three-electrode cell using a silver chloride reference electrode and a platinum counter electrode. The measurements were made on a PI-50-1.1 potentiostat. All potentials in the paper are given with respect to a normal hydrogen electrode (NHE).

With respect to calculations of the electronic structure of molecular clusters and their complexes with oxygen, the PM 3 program of the computer-driven system called “MOPAC” was used.

3. RESULTS AND DISCUSSION

3.1. Assessment of PANI Catalytic Activity (the Experiment)

The catalytic activity of PANI for the oxygen reduction reaction was first characterized by using a thin film electrode as the working device in oxygen-and argon-saturated 1 M HCl solution.

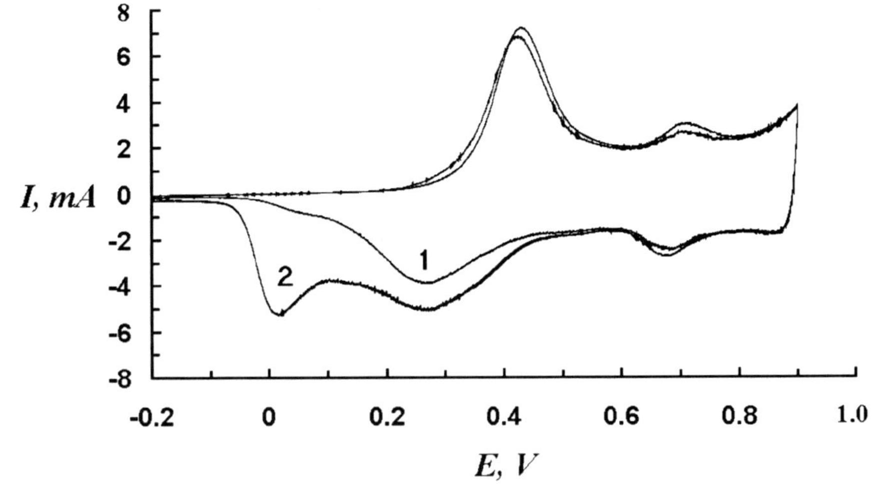

Figure 1. Cyclic voltamogramms (2 mV/s) of a PANI electrode in argon (1) and oxygen (2) saturated 1 M HCl solution.

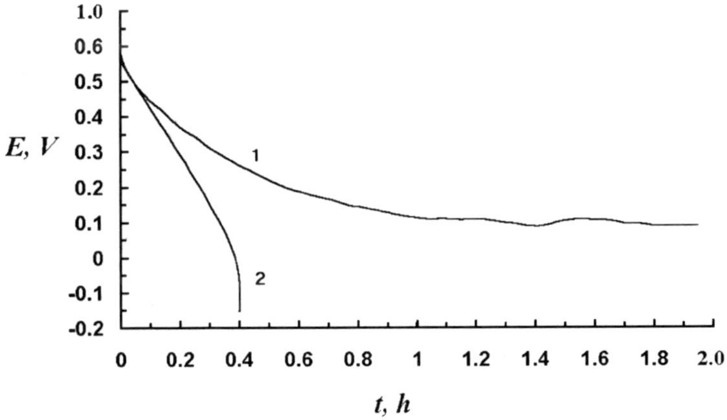

Figure 2. Galvanostatic discharge of a PANI electrode with current density at 20 mA/g (1) and 200 mA/g (2) in oxygen-saturated 1M HCl.

Figure 1 shows cyclic voltammograms in these two solutions. A reduction current is observed with two reduction peaks in the oxygen-saturated electrolyte. In the argon-saturated electrolyte, a single reduction peak at lower current density is observed. From the difference between the reduction currents in the two solutions, the reduction current in the oxygen-saturated solution must be due to catalyzed oxygen reduction reaction. On the other hand, the reduction of oxygen on PANI occurs at potentials that are close to the process of change of electrically conductive emeraldine (EM) into nonconducting leucoemeraldine (LEM). The formation of LEM leads to a sharp increase in electrode resistance and to loss of the electrochemical activity of the electrode. In the presence of oxygen, much electricity is spent on its electroreduction, and the process of nonconducting-phase formation slows down. Owing to this, the electrode loses its electroactivity at more negative potentials and is characterized by the appearance of an additional cathodic peak at E ~ 0 V.

Figure 2 shows that in the case of galvanostatic discharge of a PANI electrode with relatively low currents, a region of stable potential value (E ~ 0.1 V) appears (curve 1 in Figure 2). This region actually corresponds to the electroreduction of oxygen at the PANI electrode.

At the same time, the cathodic polarization of the electrode is accompanied by the formation of a nonconducting phase, leading to an increase in electrode resistance. Thus, at increased current densities, the region of stable potential value cannot be obtained as a result of electrode resistance increase and ohmic drop, as is shown by curve 2 in Figure 2.

3.2. Assessment of PANI Catalytic Activity (the Theory)

The above experimental data (Figs. 1 and 2) allow us to estimate the possibility for taking place the oxygen reduction process on PANI catalyst. If the special procedure is not taken to withdraw oxygen out of solution, PANI electrode in 1 M HCl usually demonstrates a potential of about 0.6 V (SHE). In compliance with the Fig. 1 the position of the reduction peak might be at 0. 28 V. From the data of the standard potentials for the following reaction:

$$O_2 + 2H^+ + 2e^- = H_2O_2, \quad E_0 = 0.682 \quad (1)$$
$$O_2 + H^+ + 2e^- = HO_2^-, \quad E_0 = 0.338 \quad (2)$$

it is possible to assume that oxygen reduction should occurs according to the above equations and includes an initial two-electron reduction of O_2 to H_2O_2 and HO_2^-. We have tested this approach by the qualitative reaction on hydrogen peroxide in the solutions with PANI electrode under cathodic polarization. The 0.5 M KI solution was used as an indicator. This qualitative reaction has proved the formation of H_2O_2 or HO_2^- in the solution.

In order to explain the catalytic activity of PANI, we have modeled the electronic structure of doped molecular PANI clusters and its adsorption complexes with oxygen and hydrogen. The geometric and electronic structures of PANI were modeled by a molecular cluster (Figure 3), which contained two phenyl rings in the polymer chain.

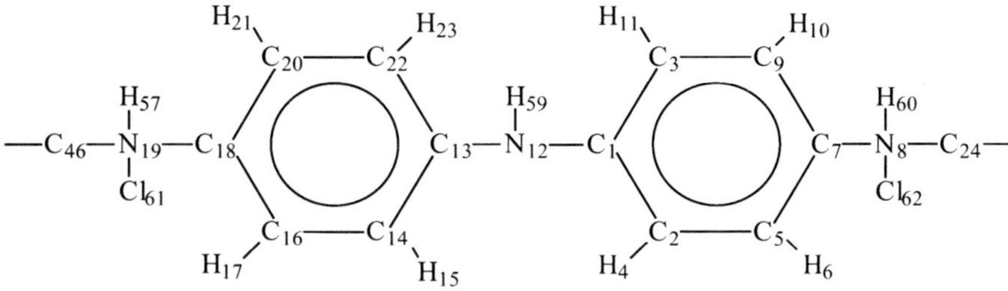

Figure 3. Fragment of a molecular cluster of PANI.

The model clusters were chosen in accordance with the structure of PANI proposed in Refs. [1, 3]. The present model of PANI also takes into account that under the influence of dopants (in this particular case, protons and anions (chlorine ions) which form bonds to PANI nitrogen), the spins of the highest occupied molecular orbital (HOMO) become unpaired, and PANI changes to triplet state. It should be noted that only in this state there is a considerable increase in PANI conductivity.

In constructing the initial model of PANI, standard values of C-H, C-C, and C-N valence bond lengths were originally adopted. The valence angles were taken to be 120 degrees. The results of calculations of PANI and AC's "PANI-oxygen" are listed in Tables 1 and 2.

The calculation of the electronic structure of PANI and adsorption complexes (AC) "PANI- oxygen" has been carried out with complete optimization of the geometric bond parameters.

Calculations show that between the dopant anion (Cl^-) and nitrogen atom there may form a weak ionic-covalent chemical bond (the bond order is no more then 0.3) with considerable contribution of the ionic component. As a result of doping, the coordination number (CN) and the positive-charge value of nitrogen atoms increase (Table 1). The large positive charges on the nitrogen atoms N8 and N19 at CN=4 suggest that under the influence of dopants, the nitrogen atom changes into cationic form. This conclusion agrees with the results of Ref. [2].

As it follows from Table 1, a π-charge of over 1e is concentrated on the carbon atoms. Therefore, each carbon atom can be in principle an adsorption site. The optimization of the position of oxygen atoms relative to the surface was performed at "frozen" geometric parameters of the PANI cluster. Calculations of the adsorption of molecular oxygen on PANI showed that the carbon atoms of the phenyl ring, which are shown by Figure 4, are the most stable surface adsorption sites (AS).

Table 1. Charges on atoms and π-electron density distribution on PANI and PANI-oxygen AC's (for the adsorption sites C_{13} and C_{22}).

№	Atom	PANI-O_2		PANI	
		Charge	**π-electron density**	**Charge**	**π-electron density**
1	C	-0.0412	1.0576	-0.0424	1.0545
2	C	-0.1547	1.0348	-0.1550	1.0356
3	C	-0.1441	1.0343	-0.1502	1.0326
4	H	0.1217	-	0.1267	-
5	C	-0.0202	1.0357	-0.0214	1.0316
6	H	0.1402	-	0.1421	-
7	C	-0.2203	1.0495	-0.2092	1.0506
8	N	0.4609	0.3320	0.4408	0.1031
9	C	-0.0630	1.0405	-0.0674	1.0330
10	H	0.1062	-	0.1055	-
11	H	0.1165	-	0.1136	-
12	N	0.1101	0.3484	0.1405	0.3767
13	C	0.1059	0.2882	-0.0572	1.0528
14	C	-0.1642	1.0294	-0.1439	1.0349
15	H	0.1241	-	0.1146	-
16	C	-0.0395	1.0180	-0.0279	1.0293
17	H	0.1495	-	0.1431	-
18	C	-0.1400	1.0564	-0.1964	1.0506
19	N	0.4261	0.2371	0.4298	0.1031
20	C	-0.1371	1.0320	-0.0763	1.0314
21	H	0.1209	-	0.1078	-
22	C	0.0424	0.1963	-0.1268	1.0330
23	H	0.1469	-	0.1263	-
24	C	-0.2176	1.0519	-0.2053	1.049
34	N	0.1211	0.2975	0.1188	0.2933
46	C	-0.1710	1.0629	-0.1747	1.0559
57	H	0.0506	-	0.0475	-
58	H	0.0537	-	0.0530	-
59	H	0.0726	-	0.0581	-
60	H	0.0449	-	0.0447	-
61	Cl	-0.4051	-	-0.3864	-
62	Cl	-0.3561	-	-0.3926	-
63	O	-0.2001	0.0557	-	-
64	O	-0.1806	0.0826	-	-

Moreover, calculation shows that molecular oxygen can be adsorbed on the PANI surface only when both oxygen atoms form bonds to surface atoms, i.e. a bridge model of adsorption is most probably (Figure 4).

The noticeable increase in the negative charge on O_2 (ads.) from 0 to – 0.3…-0.42 e (Table 2) indicates that the PANI surface in the PANI-O_2 AC is an electron density donor and O_2 (ads.) an electron density acceptor.

Table 2. Heats of formation, H (kcal), magnitude of electron density, Q (O_2), displace from the cluster to adsorbed oxygen molecules, equilibrium interatomic distances R (A-B) and bond orders P (A-B), corresponding to them, in PANI-O_2 adsorption complexes.

Bond	H, kcal	Q (O_2)	R (O_{63}-O_{64}), A^0	P (O_{63}-O_{64})
C2-C5	312.41	-0.3581	1.5059	0.9926
C3-C9	312.89	-0.3699	1.5015	0.9939
C5-C7	323.86	-0.3555	1.5110	0.9877
C7-C9	322.97	-0.3729	1.5052	0.9906
C14-C16	308.58	-0.4025	1.5045	0.9933
C20-C22	312.93	-0.4212	1.4982	0.9822
C13-C14	326.98	-0.3674	1.5036	0.9887
C13-C22	327.05	-0.3807	1.4969	0.9896

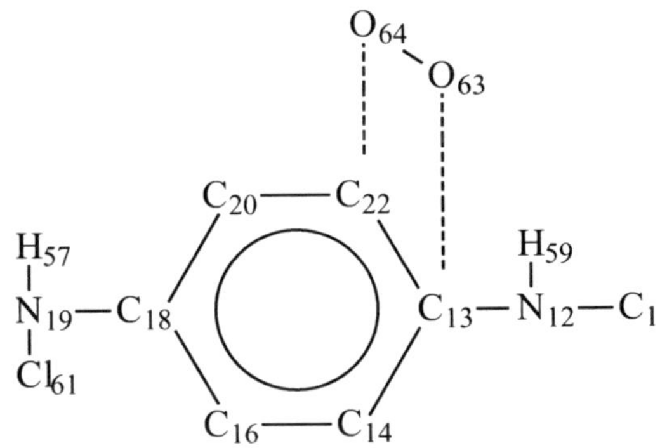

Figure 4. Bridge model of oxygen adsorption on PANI.

Electron density is transferred from PANI to O_2 mainly through the loss of electron density by adsorption sites. The latter (e.g. C_{13} and C_{22}) become electropositive (Table 1). The electron density displaced from the surface to O_2 during adsorption occupies the antibonding MO's of adsorbed oxygen molecule which is responsible for the destabilization of the O-O bond in O_2(ads.) during the chemisorptions of O_2 on PANI, π-bonds in O_2(ads.) and surface AS's undergo rupture. Instead of two rupture π - bonds, two donor-acceptor C-O σ-bonds are formed between PANI AS`s and the oxygen atoms that lie outside the phenyl ring plane.

Calculation of the heats of formation (H) of the bonds between oxygen atoms and PANI (62) AC atoms (Table 2) shows the bonds in which one of the AS`s (C1, C7, C13, C18) forms a bond to the nitrogen atom to be characterized by the highest H values. In this case, the stability of adsorption complexes is 10 - 15 kcal higher than that of the complexes formed on the carbon atoms not bonded to nitrogen. Thus, oxygen adsorption on pairs of AS`s which contain one of the carbon atoms C1, C7, C13, C18 is more probable than that on pairs of side chain sites.

Calculations show (Table 2) that the 0 - 0 bond orders in chemisorbed oxygen molecules are almost 30 % lower than those in free O_2 molecule. Besides, a noticeable increase (of more than 20 %) in O-O bond length takes place during adsorption. Thus, chemisorbed O_2 molecules have a fairly high degree of activation and can be easily reduced, which accounts for the catalytic activity of PANI.

3.3. PANI/Graphite (Carbon) Composites

The next important stage in the development of porous gas-diffusion electrode is an investigation of influence of thickness of PANI layer (or more easy controlled parameters like PANI mass or electrochemical capacity) on the local currents of O_2 electroreduction (table 3).

Analysis of data presented clearly suggests that a sharp increase of catalytic activity takes place in relatively thin layer of PANI, which corresponds to the mass of about 0.6...0.9 mg (or Q=108...164 mC) per geometric surface of electrode (0.5 cm^2). Further increase of PANI thickness (or mass) cannot increase enough the catalytic activity of porous electrode. That is the reason why the developed by us PANI/graphite composites have relatively high content of graphite (Table 4).

Table 3. Local currents of O_2 electroreduction at Graphite / PANI electrode (S=0.5 cm^2) with different mass (and electrochemical capacity) of PANI in 1M HCl solution.

Mass of PANI, M mg	Capacity of PANI, mC	I_1, μA	I_2, μA	I_3, μA
0.3	52	13	72	108
0.6	108	27	89	125
0.9	164	31	97	141
1.2	193	33	103	154
1.5	240	37	106	159
2.0	341	39	107	162

Notes for the Table:
The conditions of oxygen supply to the solution: I_1 - with low O_2 content (saturated by N_2); I_2 – with middle O_2 content (saturated by O_2); I_3 – with maximal O_2 content (during O_2 bubbling on the electrode surface).

Our experiments, as well as analysis of the proposed theoretical model for a generalized system of porous electrode "active material – carbon additive" proved that thermally exfoliated graphite (TEG) can be one of *the most effective conductive additive and structural support* for the different new and existing active materials of primary and secondary batteries, including "Air/PANI-Zn" batteries. The reason for such wide application of TEG is a following unique complex of TEG properties: low density, relatively high conductivity and stability to electrochemical oxidation during battery serviceability. Also it is interesting to note that particles of TEG are relatively hydrophobic. Thus it should be of a great benefit to use composite based on TEG for development of gas-diffusion electrodes.

The compositions of composites are different for primary and rechargeable variants of batteries. In rechargeable variant it is necessary to have considerable mass of PANI for charging of battery without oxygen evolution. In primary variant it is enough to have a relatively small amount of PANI for creation of its quite thin layer at the conductive and stable graphite collector. Thus we have developed two main types of composite materials based on TEG and PANI (Table 4).

Table 4. Composition of PANI/TEG composite electrodes.

Type of battery mockup	Composition of electrode, %	
	PANI*	TEG**
Primary battery	30	70
Rechargeable battery	90	10

Notes for the Table:
**Chemically synthesized Polyaniline, KNUTD (Ukraine)*
***Purified, thermally exfoliated natural flake graphite, Superior Graphite Co. (USA)*

3.4. New Concept for Development of Rechargeable Metal-Air Batteries

The main concept for development of metal-air batteries with new low-cost composite polymeric catalysts is to use *catalytic activity* of PANI/TEG composition towards the oxygen reduction during the discharge process of battery *side by side* with *non-Faradaic process* of anion doping during the charge process (please, see schemes below).

Schemes of Discharge & Charge Processes

1. Traditional Air-Met Battery (for instance, with carbon catalysts)

Discharge Process:

$$Met^0 - ne \Rightarrow Met^{n+} \quad (1)$$

$$O_2 + 4e + 2H_2O \Rightarrow 2OH^- \quad (2)$$

<u>Charge Process:</u>

$$Met^{n+} + ne \Rightarrow Met^0 \qquad (3)$$

$$4OH^- - 4e \Rightarrow 2H_2O + O_2 \qquad (4)$$

<u>Side reaction:</u> $C + O_2 \Rightarrow CO_2$ (5) - destruction of a catalyst

2. Rechargeable Air-Met Battery with PANI-based Catalysts

<u>Discharge Process:</u>

$$Met^0 - ne \Rightarrow Met^{n+} \qquad (1a)$$

$$2O_2 + 4e + 4H^+ \Rightarrow 2H_2O_2 \text{ (electrochemical)}$$

$$\underline{2H_2O_2 \Rightarrow O_2 + 2H_2O \qquad \text{(chemical stage)}}$$

$$O_2 + 4e + 4H^+ \Rightarrow 2H_2O \qquad (2a)$$

<u>Charge Process:</u>

$$Met^{n+} + ne \Rightarrow Met^0 \qquad (3a)$$

$$\underline{PANI - ne + nA^- \Rightarrow [PANI^{+n} \bullet nA^-] \; (4a)}$$

(intercalation)

Total: $Met^{n+} + PANI + nA^- \Rightarrow Met^0 + [PANI^{+n} \bullet nA^-]$

Thus, using PANI-type catalysis gives a possibility to exclude basically the typical side reaction of oxygen evolution during charging of battery, which usually conducts to destruction of catalytic active Air electrode.

3.5. Coin Cell Mockups of Rechargeable Metal-Air Batteries

In order to demonstrate the above concept in practicality, we have developed the coin cell mockups (active electrode area of 1 cm^2) of rechargeable metal-air battery. The composite containing 90% of PANI and 10% of TEG was used in rechargeable Zn-Air batteries. The aim of this investigation was to check the electrochemical behavior of PANI/TEG gas-diffusion electrode during the charging of battery. Hence, we have used thin Zn electrodes, which limited cell capacity (~70 mA·h). The galvanostatic charge/discharge curves of such battery are presented by Figure 5.

It is easy to conclude from the shape of discharge curves that the first stage of battery discharge is mainly related with the discharge of PANI component. On the second stage, a region of stable potential appears (U ~ 0.95 V). This stage actually corresponds to the electroreduction of oxygen at the PANI/TEG electrode. When approximately 65% Zn supply has been depleted, the voltage started to go down rapidly. After that, we switched battery to the charge mode of operation. The amount of capacity that could be recharged is limited by the amount of PANI. We have found that it is possible to utilize about 45% of initial capacity obtained on the first discharge at least during the 50-70 cycles. The limitations on cycling life of battery in a whole are related with well known problems of Zn electrode cycling (such as Zn dendrites formation) and could be overcome in the future by optimal development of separator and Zn electrode construction.

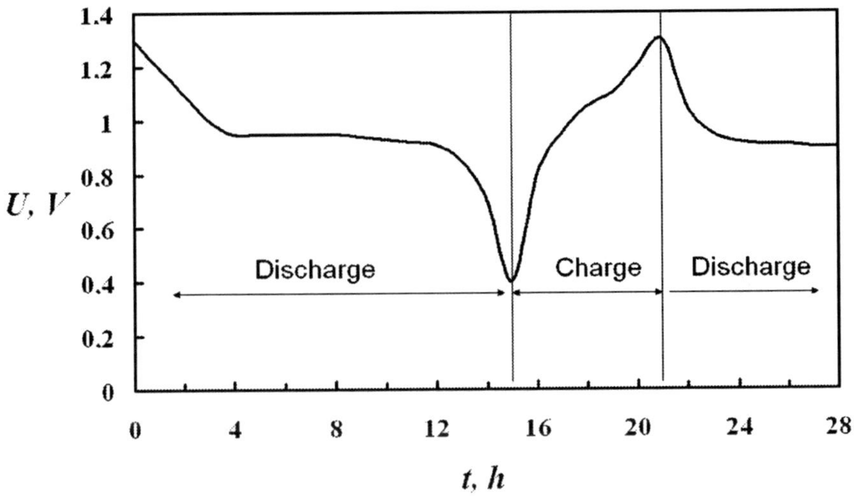

Figure 5. Galvanostatic charge/discharge curves of the coin cell mockups of Zn-Air battery ($I_{charge} = I_{discharge} = 3$ mA).

In this paper we would like to propose also an additional concept of utilization of rechargeable metal-air battery with PANI/ TEG catalysts. It is necessary to note the ability of PANI based electrode to self-charge in the presence of oxygen. A deep discharge of PANI electrode leads to the transformation of emeraldine (EM) to lecoemeraldine (LEM). It is well known that LEM is chemically non-stable at the presence of oxygen due to occurance of the following reaction:

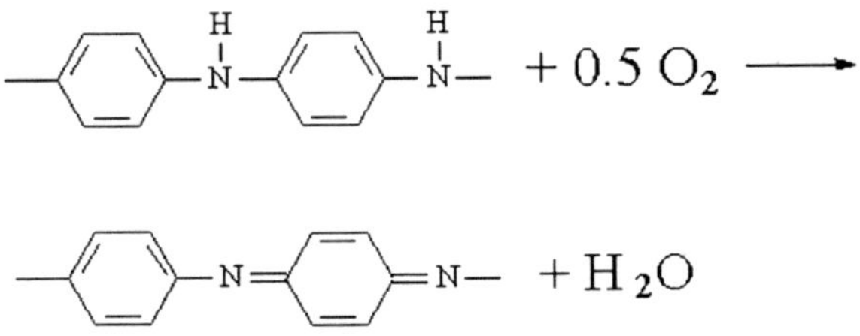

We have confirmed the above reaction by visual observation of EM to LEM transformation at the ITO glass. Thin film of PANI was electrochemically deposited on ITO glass. Then the glass electrode was placed in a standard three-electrode cell as a working electrode. After potentiostatic polarization at E= -0.2 V (vs. NHE) the green EM was reduced to LEM, which has a yellow color. The glass electrodes with yellow LEM were placed in the argon saturated solution. We observed that yellow color did not change during the few hours. On the other hand, the yellow color was changed to green at few minutes when electrode was placed in the oxygen-saturated solution. Thus the LEM can easy transformed to EM by reaction with oxygen.

For purposes of verifying of the concept of a self-discharge due to the LEM oxidation by air, we have designed a coin cell with a zinc electrode and a thin PANI/TEG cathode. The typical curves of voltage change for such electrochemical device are given by Figure 6.

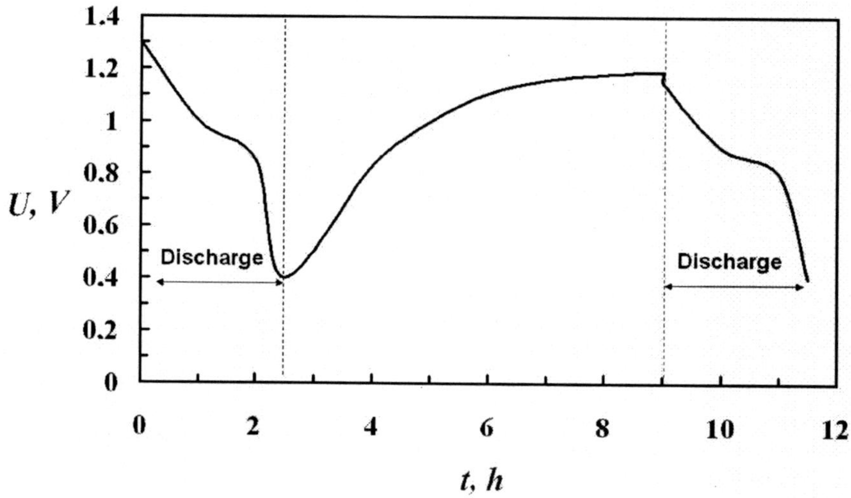

Figure 6. The galvanostatic discharge curves and self-charge curve for Zn-Air coin battery with PANI/TEG gas-diffusion electrode.

After the first galvanostatic discharge of battery up to Umin=0.4V a circuit was disconnected and computer system has registered an open circuit voltage. After reaching the Umax = 1,2V a circuit was connected again (the second discharge). It is interesting to note that during the second discharge a cell has demonstrated about 80 % of its initial capacity.

It was shown by our investigations that such cell can be cycled with self-charging at least during 10-15 cycles. After that, the internal resistance of such cell is incremented considerably due to the accumulation of discharge products on a zinc electrode. Thus, such battery can represent sufficient interest for some devices with a pulse mode of power consumption (for example, in signalling devices). It is necessary to note that by optimization of battery construction rather high value of a current in a pulse mode could be reached. Such battery represents some kind of the hybrid supercapacitor. PANI/TEG electrodes can be discharged with enough high rates and it gives possibility to use such electrodes also in supercapacitors.

3.6. The Full Size Cylindrical Metal-Air Batteries

The experimental samples of Air-Zn and Air-Mg batteries with PANI-TEG catalysts were realized in the standard AAA size of MnO_2-Zn primary batteries. The composite material based on 70% of TEG was used for construction of cylindrical cells of metal- air battery.

The discharge curves for all batteries (1, 2) have horizontal sections which correspond to the properties of air PANI/TEG composite electrode (Figure 7).

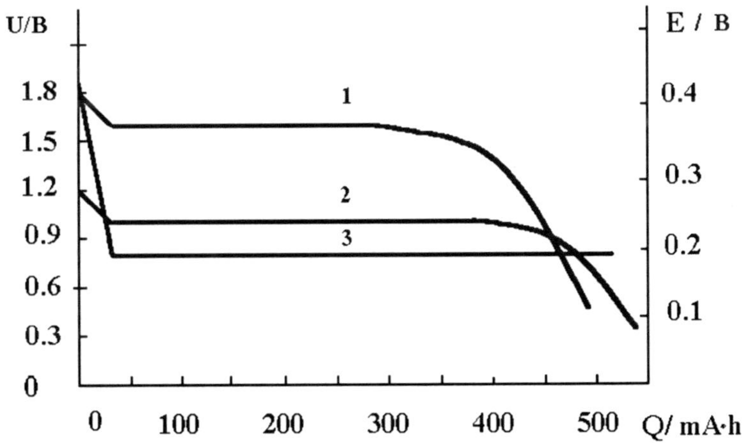

Figure 7. Discharge curves of AAA size Air/Metal cells (1, 2) with composite PANI catalysts, as well as discharge curve of separate Air PANI/TEG composite electrode (3). Discharge current I =5mA, temperature 18°C. ***Electrochemical Systems:*** *(PANI) Air /$MgBr_2$ /Mg (1); (PANI) Air /$ZnCl_2$ / Zn (2); (PANI) Air / $ZnCl_2$ / /AgCl /Ag (3).*

The discharge curve for Air electrode, which was measured using additional third Ag | AgCl reference electrode, is *ideally horizontal* since it is determined by the oxygen reduction potential in according medium

(Fig. 7, curve 3). The capacities of batteries are limited by the amount of Zn (or Mg) in batteries (Fig. 7, curves 1 and 2).

The main characteristics of cylindrical AAA size metal-air batteries with PANI/TEG catalysts, as well as standard Zn-MnO_2 battery have been gathered in the Table 5. Realization of all types of batteries in the same AAA size gives the possibilities for comparison of above electrochemical systems for some applications.

Table 5. Characteristics of cylindrical AAA size metal-air batteries with PANI/TEG catalysts, as well as MnO_2-Zn battery.

Performance Characteristics of cells	**(PAN) O_2|Zn**	**(PAN) O_2|Mg**	**MnO_2|Zn**
Discharge capacity, mA·h	420	340	380
Specific energy, W·h/kg	140	200	45
Weight of the cells, g	3	2.7	11
Dimensions, cm (H / d)	4.5 / 1.0	4.5 /1.0	4.5/1.0
Open circuit voltage, V	1.2	1.8	1.5
Average discharge voltage, V (at R_{DISCH} = 200 Ohm)	1.0	1.6	1.1-1.3
Short - circuit current, A	1.0	1.2	1.4
Self - discharge, % within a month	1.5-2.0	2.0-4.0	1.0-3.0
Service life, years	~ 5	< 1	< 2

The average weight for the experimental Air-Zn and Air-Mg batteries of AAA size is 3 g and 2.7 g correspondingly, vs that of 11 g for traditional MnO_2-Zn battery. That is the main reason of sharp increase of the specific energy of Air-Zn (140 W·h/kg) and Air-Mg (200 W·h/kg) batteries vs the traditional chemistry (45 W·h/kg).

The specific energy of Air-Zn battery based on slightly acidic electrolyte ($ZnCl_2$ with pH 5) is not very high, as compared with an ordinary Air-Zn battery with alkaline electrolyte, due to the relatively low average discharge voltage (1.0 V during discharge on the resistor of 200 Ohm). Nevertheless, the service life of such systems can be greater than that of ordinary alkaline systems owing to the absence of electrolyte carbonization. For example, some experimental samples of Air-Zn batteries have effectively demonstrated the service life of about 5 years.

Air-Mg batteries have larger average discharge voltage (1.6 V) and specific energy, but did not have a great service life and abilities for cycling due to the high activity of Mg electrode.

In general, we believe that new types of batteries with composite PANI/TEG catalysts can find their own diapason of practical applications depending on concrete conditions of application.

4. CONCLUSIONS

Polyaniline (PANI) was investigated as electrocatalyst for the oxygen reduction reaction in the acidic and neutral solutions. Galvanostatic discharge tests and cyclic voltammetry of catalytic electrodes based on polyaniline in oxygen-saturated electrolytes indicate that polyaniline catalyzes two-electron reduction of molecular oxygen to H_2O_2 and HO_2^-.

In order to study the possible reasons and mechanisms of the catalytic activity of conducting polymers (CPs), we have modeled the electronic structure of some molecular CPs clusters and its adsorption complexes with oxygen.

A specialized "MOPAC" computer software package and, in particular, its "PM3" quantum-chemical program has been successfully applied in calculations. The results of calculations have shown that both oxygen atoms form bonds with two more *active carbon atoms* of CP molecular cluster (so-called *"bridge" model* of adsorption). The total energy of system after a chemical adsorption at such *active atoms* is minimal.

In the CP-O_2^* complex, the CP surface is an electron density donor. For example, in the case of PANI, the bond orders in adsorbed O_2^* molecules decrease by about 30%, and the bond lengths L increase by about 24%. Thus, the adsorbed O_2^* molecules have a fairly high degree of activation and can readily interact with the protons. Hence, quantum-chemical analysis confirms the mechanism of O_2 electroreduction and gives possibility to understand the reasons of PANI catalytic activity.

The above phenomenon has found a practical application for development of metal-air battery mockups with low costs PANI/Expanded Graphite composite catalysts.

Specific energy, which could be attained as primary battery is of about 140 W·h/kg for Air/PANI-Zn and 200 W·h/kg for Air/PANI-Mg batteries. The discharge curves of such batteries is practically horizontal since there are determined by the oxygen reduction potential.

The effective service life of Air/PANI-Zn battery can be greater by several folds than that of conventional alkaline air-zinc batteries owing to the absence in principle of electrolyte carbonization ($pH \leq 7$).

In addition, the batteries can be rechargeable, the cyclability attained with them being limited now by the properties of Zn (Mg) electrodes and a separator.

A battery can be easy realized in any shape (prismatic, coin or cylindrical) and any size, including micro-size.

We believe that new type of conducting polymer / expanded graphite composite electrodes as gas-diffusion cathodes will find in perspective a practical application for some types of batteries and fuel cells.

ACKNOWLEDGEMENT

Authors would like to acknowledge Superior Graphite Co., Chicago, IL, USA for manufacturing and submission of various customized graphite samples to do the development work described in the paper. We like to also acknowledge NATO Science for Peace Programme for financial support of this work in framework of "Carbon" SfP 973849 project.

REFERENCES

1. Alan G. MacDiamid and Heeger, J. Semiconduction and Metallic Polimer: The Fourth Generation of Polymeric Materials. J. Phys. Chem. 2001; B105:8475-8491.
2. Bazong Zhao, Neon K.G., Liu F.T., at.al., Interactions Between Polianiline and Viologens. Synthetic Metals 2001;123:263-266
3. Alan G.MacDiamid, "Syntethetic <etals": A Novel role for Organic Polymers. Current Applied Phys 2001; 1: 269-279.

MECHANICALLY RECHARGEABLE MAGNESIUM-AIR CELLS WITH NaCl-ELECTROLYTE

Anastassia Kaisheva* and Ilia Iliev

Central Laboratory of Electrochemical Power Sources, Bulgarian Academy of Sciences, 1113 Sofia, Bulgaria

Abstract

Metal-air cells are developed with air gas-diffusion cathodes and Mg-anodes. Non-aggressive NaCl-solution is used as electrolyte. Carbon based catalysts for the oxygen reduction are selected and tested in the air gas-diffusion electrodes. Various Mg-alloys are tested as anodes. The V-A, power and discharge characteristics of the Mg-air cells are investigated.

Keywords

Electrochemical energy sources; Magnesium anode; Air gas-diffusion electrode; Neutral electrolyte.

1. INTRODUCTION

Metal-air batteries are characterised by a high energy density, a flat discharge voltage, long shelf life and relatively low price. A great advantage of the metal-air batteries is the use of a cheap and inexhaustible reactant – oxygen from the atmospheric air, which is reduced on an invariable air cathode [1].

Various metals can be used as an anode in metal-air cells, the amperhour capacity of the cell being determined by the type of metal anode.

Zinc is commonly used as an anode in metal-air batteries. Zinc-air batteries are commercially available and used as power supply for navigation buoys, communication systems, hearing aids, and etc. [2]. Mechanically rechargeable Zn-air batteries are developed recently for traction [3].

* Corresponding author. E-mail: kaisheva@bas.bg

I.V. Barsukov et al. (eds.), New Carbon Based Materials for Electrochemical Energy Storage Systems, 105–116.

One of the main inconveniences of the zinc-air batteries is the use of concentrated KOH-electrolyte. Strong alkaline solutions are chemically aggressive and their contact with the human body can cause serious injuries. The metal-air cell is an "open" system in which the electrolyte in the cell is in contact with the surrounding air through the pores of the air gas-diffusion electrode. There is a risk that drops of electrolyte may come out of the cell and cause accidents. The mechanical recharge of the zinc air cells poses also problems: the removal from the cell of the electrolyte and of the porous, flooded with electrolyte consumed zinc anodes is an operation with high chemical danger, which can be accomplished by trained people or robots only. In this respect the development of metal-air cells operating in non-aggressive electrolyte is of great interest. Promising results have been obtained with magnesium-air cells.

2. EXPERIMENTAL

Magnesium is a reactive metal with relatively high electrochemical equivalent of 2.2 Ah/g and low specific gravity 1.74 g/cm^3; its price is not high. This makes magnesium suitable for use as anode in metal-air cells.

A passive film is formed on the Mg electrode surface in alkaline solutions, which creates problems when the Mg-air cell is to be discharged. However Mg and especially some Mg alloys are suitable to be used in magnesium-air cells with NaCl-electrolyte [4], where the formation of the passive film on the magnesium surface is not observed.

The magnesium-air cells under investigation are electrochemical cells containing two air gas-diffusion electrodes (cathodes) and a magnesium anode. NaCl-solution is used as an electrolyte.

3. RESULTS AND DISCUSSION

3.1. Air Gas-Diffusion Electrode

The air gas-diffusion electrode developed in this laboratory [5] is a double-layer tablet (thickness ca.1.5 mm), which separates the electrolyte in the cell from the surrounding air. The electrode comprises two layers: a porous, from highly hydrophobic, electrically conductive gas layer (from the side of the air) and a catalytic layer (from the side of the electrolyte). The gas layer consists of a carbon-based hydrophobic material produced from acetylene black and PTFE by a special technology [6]. The high porosity of the gas layer ensures effective oxygen supply into the reaction zone of the electrode; simultaneously the leakage of the electrolyte through the electrode is completely avoided because of high hydrophobicity of the pores-surface of the gas layer. The catalytic layer of the electrode comprises uniformly distributed zones of the same hydrophobic material and carbon-based catalyst. An effective supply of oxygen in gas phase is achieved through the zones made from the hydrophobic material.

Various carbon-based catalysts for the electrochemical oxygen reduction have been tested in the air gas-diffusion electrodes [7]. The polarization curves of the air electrodes were measured when operating against an inert electrode in 2 N NaCl-solution. The potential of the air electrodes was measured versus saturated calomel electrode (SCE).

It is found that some types of active carbons possess enough catalytic activity to be used as catalysts in air electrodes operating at low c.d.

Catalysts from active carbon promoted with mixed Ni-Co- oxides are investigated [8]. These catalysts are produced by heat-treatment of active carbon impregnated with solution containing Co-and Ni-acetates. The concentration of the used Co and Ni- solutions is varied in order to optimise the content of the mixed Co-Ni oxides in the catalyst.

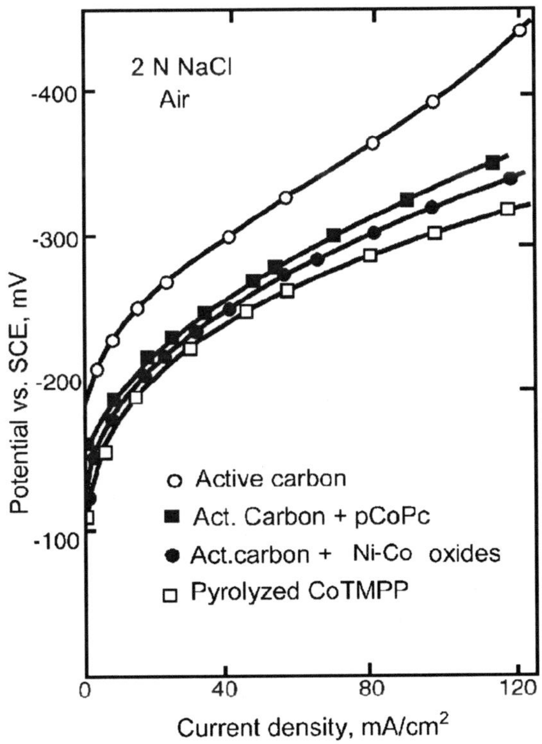

Figure 1. Polarisation curves of air gas-diffusion electrodes with various catalysts.

Catalysts from active carbon additionally activated with cobalt- or iron- phthalocyanines are also studied [7]. The results show that at current densities up to 50 mA/cm^2, the polarization of the air electrodes with catalyst from active carbon promoted with FePc is lower than that of the electrode with catalyst from active carbon promoted with CoPc. At higher current density the polarization of the electrode with catalyst from active carbon promoted with CoPc is lower, which is probably connected to the lower transport hindrances, due to the more favorable structure of this catalyst.

Pyrolyzed catalysts obtained by heat treatment in Argon of active carbon impregnated with solution of the compound Co-tetramethoxyphenylporphirine (CoTMPP) are studied [9]. Air gas-diffusion electrodes with this catalyst show low polarisation in a wide interval of current densities (up to 100 mA/cm^2) and stable long-term performance. These catalysts are more suitable for use in magnesium-air cells operating at high current drains, but unfortunately their price is comparatively high.

In Figure 1 we compare the polarization characteristics of the air gas-diffusion electrodes with the described catalysts.

One of the main difficulties in the design of air gas-diffusion electrodes operating in NaCl-solutions is to ensure the efficient current collection. Usually the current collection is achieved by means of a metal grid pressed on the catalytic layer of the electrode. However, it seems impossible to find a metal, chemically stable in contact with the saline solutions (at the working potentials of the air electrode). That's why the current collection in the air electrodes developed in this laboratory is achieved through a metal grid pressed into its hydrophobic, electrically conductive gas layer. In this case the current collector grid (a cheap Pb-plated iron grid is used) is not in contact with the NaCl-electrolyte and the corrosion problems are eliminated.

3.2. Magnesium Anodes

Sheets of magnesium or magnesium alloys were used as anodes in the developed magnesium-air cells.

Several types of magnesium alloys are tested: pure Mg, MgAM60, MgAM50 and MgAZ91 (Norsk Hydro Magnesium GmbH) and MA8M06 (Mg/TU190116/74) (Russia). The sheet of magnesium type MA8M06 is covered by the producer with a thin chromium protective layer, which was being dissolved immediately after its contact with the NaCl-solution.

The elemental compositions of the Mg-alloys used in this work (according the producers' product information data sheets) is presented in Table 1.

The current collection of the Mg anodes is achieved through the magnesium itself, without using additional current collectors. This type of current collection seems to be very easy but it shows some inconveniences. At the end of the discharge, the Mg anode is broken out and some pieces of magnesium are dropping into the electrolyte, so they cannot be discharged. This results in decrease of the cell capacity. In order to overcome the problem a special protective electrically conductive sheet (based on carbon) is used pressed between two magnesium sheets to form a "sandwich". The use of such "sandwich"-type magnesium anodes results in an improvement of the current collection of the magnesium anode, especially at the end of the discharge and in a more complete utilization of the magnesium.

Table 1. Composition of the Mg-alloys used as anodes.

Metal, %	Mg	Fe	Zn	Mn	Si	Al
pure Mg	99.99	-	-	-	-	-
MgAM60	94	-	-	-	-	6
MgAM50	95	-	-	-	-	5
MgAZ91	90	-	1	-	-	9
MA8M06	~97	0.02	0.12	1.3	0.01	0.12

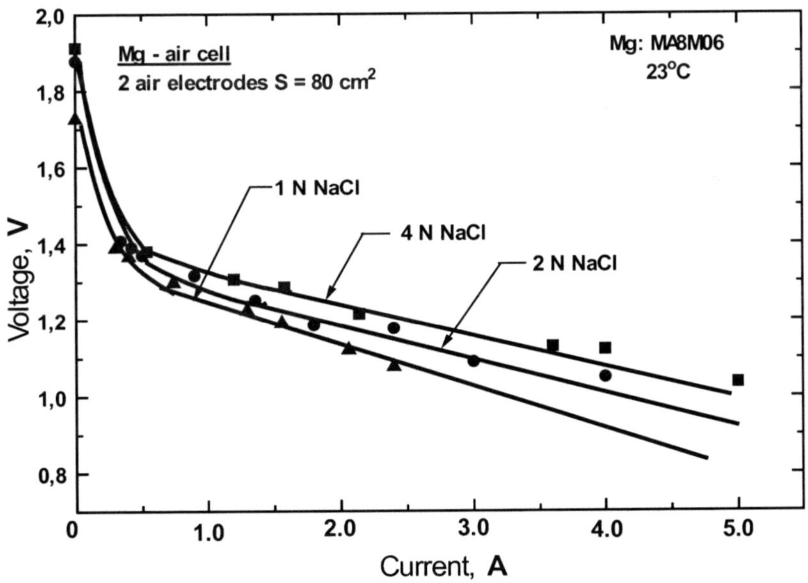

Figure 2. Typical current-voltage curves of Mg-air cell (S_{air} = 80 cm^2; Mg – MA8M06) operating in 1 N, 2 N and 4 N NaCl-electrolyte.

3.3. Experimental Mg-Air Cells

Several types of experimental magnesium-air cells were tested. These cells varied in their size (the working area of the air electrodes used) [10]. The current-voltage curves of an experimental Mg-air cell with two air electrodes (S_{air} = 80 cm^2) with pyrolyzed CoTMPP catalyst and sandwich-type Mg anode (MA8M06) operating in NaCl-electrolytes with different concentrations are presented by Figure 2.

The difference in the obtained curves is probably due mainly to the difference in the conductivity of the NaCl solutions with different concentrations. As expected, the performance in 4N NaCl is superior – currents up to 5 A can be reached at voltage higher than 1 V.

The discharge curve of the same cell at constant load ca. 0,3 Ω is presented in Figure 3.

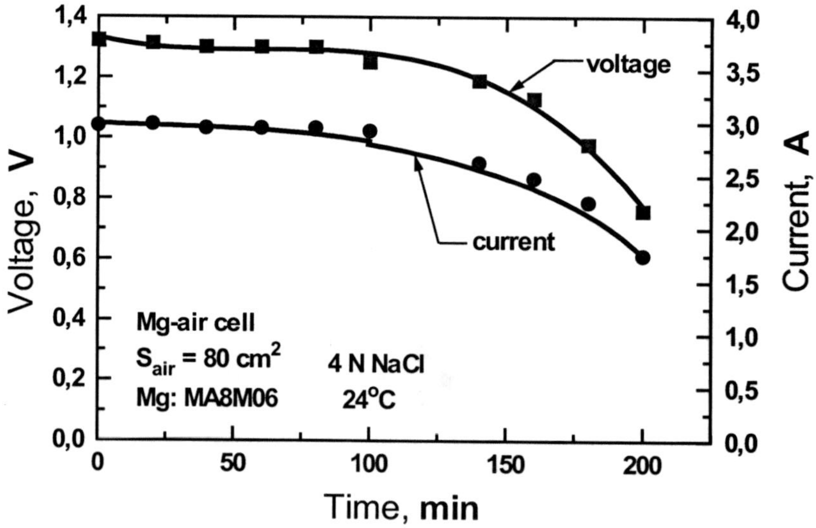

Figure 3. Discharge curve of the Mg-air cell from Fig. 1 at constant load 3 Ω.

The testing of the different magnesium anodes is performed in magnesium-air cells with two air electrodes with pyrolyzed CoTMPP catalyst (S = 65 cm^2). One-sheet Mg anodes were used in 4N NaCl media.

In Figure 4 we compare the current-voltage characteristics of the investigated Mg-air cell with different Mg-anodes.

The comparison of the data from Figure 4 shows that the pure magnesium (99.99%) is the most suitable material to be used as anode in magnesium-air cells with NaCl-electrolyte. The magnesium alloy type

MA8M06 (Russia) which is produced specially for use in water activated Mg-AgCl batteries is also suitable for use in Mg-air cells.

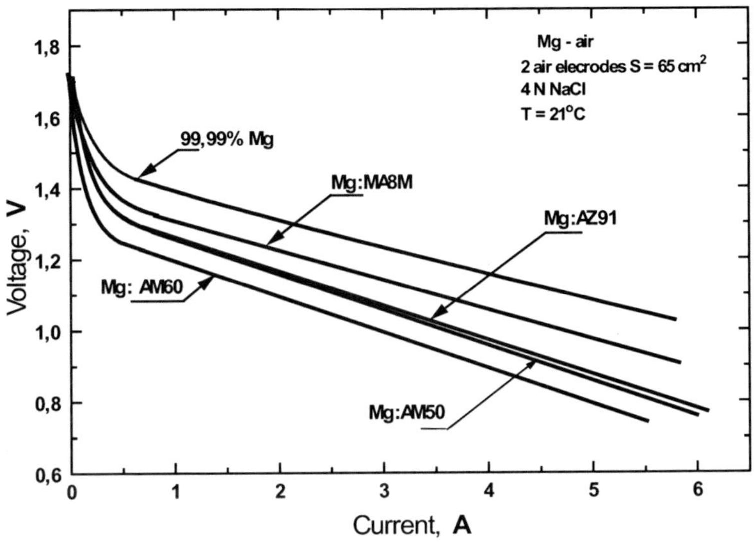

Figure 4. Current-voltage characteristics of Mg-air cells with different Mg- anodes.

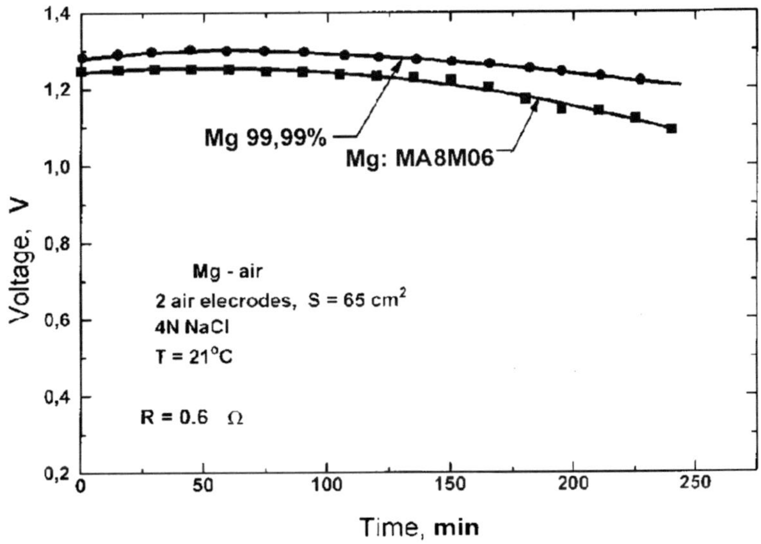

Figure 5. Discharge curves of Mg-air cells with anodes from Mg 99.99% and from Mg-alloy MA8M06.

In Figure 5 we compare the discharge curves at a constant load of 0,6 Ω of similar magnesium-air cells with anodes from Mg 99.99% and from Mg-alloy MA8M06.

3.4. Mechanically Rechargeable Mg-Air Cells

Mechanically rechargeable magnesium-air cells were designed and investigated. The most important feature of these cells is that the cell-case with the air electrodes can be used many times.

Figure 6. Experimental mechanically rechargeable Mg-air cell.

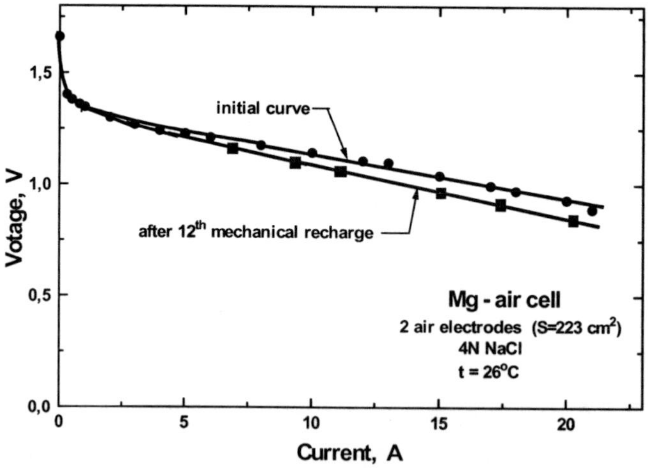

Figure 7. Current-voltage characteristics of one and the same Mg-air cell: initial and after 12th mechanical recharge.

Once the Mg-anode has been discharged, it was removed out of the cell together with the electrolyte and new Mg electrode and fresh NaCl-electrolyte were introduced, by which the cell is operational again. The electrolyte in the cell was also changed.

The initial current-voltage curve and the current-voltage curve after 12th mechanical recharge of one and the same Mg-air cell are presented in Fig. 7. The results show that the Mg-air cell can be recharged mechanically several times without any significant loose in its performance.

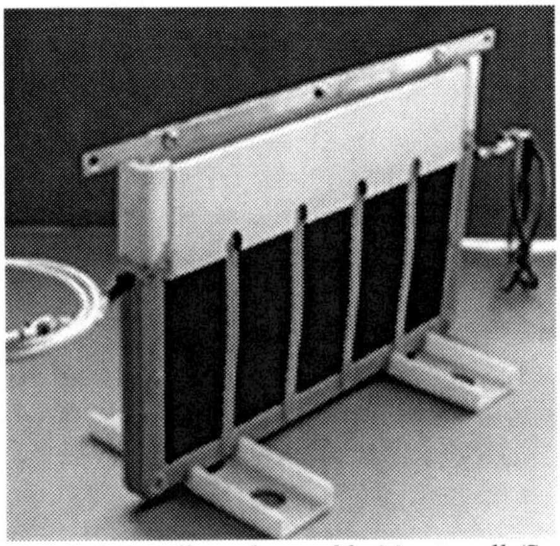

Figure 8. Mechanically rechargeable Mg-air cell (S_{air} = 223 cm^2).

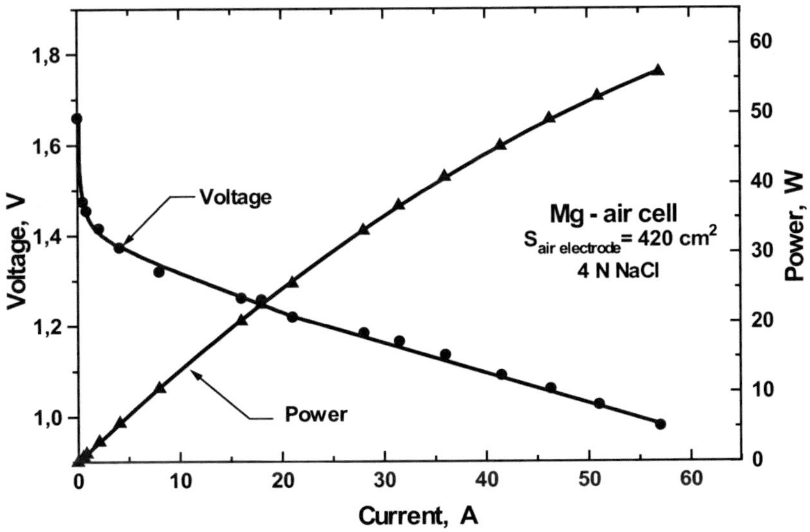

Figure 9. Current-voltage and power characteristics of the Mg-air cell from Fig. 8.

Two types of mechanically rechargeable Mg-air cells for operation at high nominal currents are designed. The air electrodes in these cells are with pyrolyied CoTMPP catalyst. Mg anodes from the alloy MA8M06 were used in this design.

In Figure 8 we have presented a picture of the mechanically rechargeable Mg-air cell with total working area of the air electrodes 420 cm^2. It is known [4] that a large amount of heat is released during the discharge of Mg-air cells. This heat promotes an evaporation of the electrolyte from the cell, which can be overcome by providing an adequate electrolyte-reservoir. That's why the upper part of the Mg-air cell from Fig. 8 is designed as a reservoir in which an additional amount of electrolyte is available.

In Figure 9 we presented the current-voltage and the power characteristics of the Mg-air cell form Fig. 8. It is seen that currents up to 50A could be obtained from this cell at voltage higher than 1 V. The specific power of this cell is ca. 50 W/kg.

Even more powerful Mg-air cell was constructed with total working area of the air electrodes S = 660 cm^2. Current loads up to 100 A could be reached with this cell at a voltage higher than 1 V.

The current-voltage and the power characteristics of this cell are presented in Figure 10.

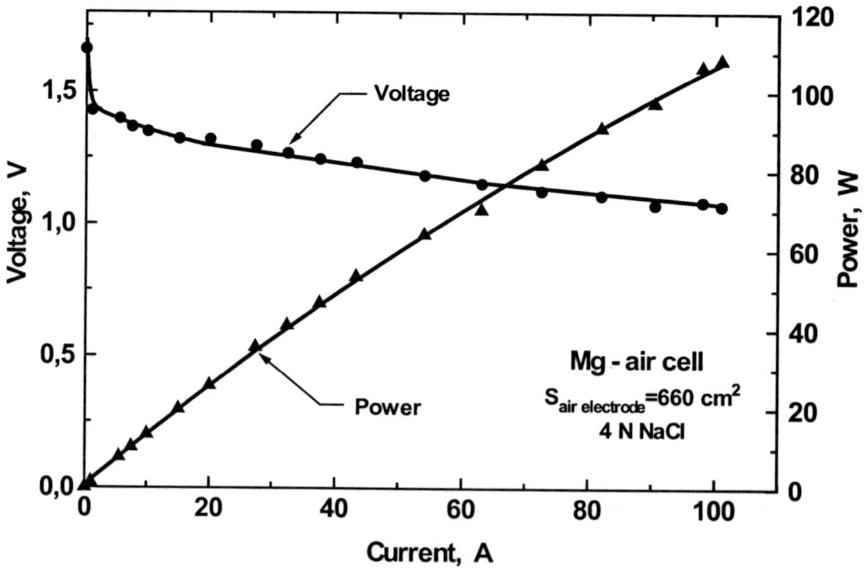

Figure 10. Current-voltage and power characteristics of the mechanically rechargeable Mg-air cell (S_{air} = 660 cm^2).

4. CONCLUSIONS

Magnesium-air air cells with NaCl-electrolyte were developed and investigated. The current-voltage and the discharge characteristics of the cells with were studied. Air gas-diffusion electrodes suitable for operation in NaCl-electrolytes were designed. Various carbon-based catalysts for the electrochemical reduction were tested in these air electrodes. Magnesium alloys suitable for use as anodes in Mg-air cells were found.

Mechanically rechargeable Mg-air cells for operation in middle and high nominal current were designed.

The main advantage of the magnesium-air cells is the use of NaCl-electrolyte. This makes the magnesium-air cells environmentally friendly. They are clean, no toxic emission; are produced in a way that is safe for user, even for children because the NaCl-electrolyte is not chemically aggressive and magnesium is not toxic. The mechanical recharging of these cells can be accomplished without any chemical danger.

Something more: the magnesium-air cells possess an infinite shelf life because the electrolyte can be removed easily before storage. When power is needed, the electrolyte is poured back into the cell. No other electrochemical energy sources do exist in which the electrolyte can be removed, stored and reused by the consumer.

From the presented results it can be concluded that the magnesium-air cells feature substantially increased power output. They can be used as primary, mechanically rechargeable and emergency energy sources for the residential, commercial and industrial markets.

ACKNOWLEDGEMENTS

This work has been partially supported by ZOXY Energy Systems AG, Germany.

REFERENCES

1. Kim Kinoshita, *Electrochemical oxygen technology*, John Willey & Sons, Inc. New York, Chichester, Brisbane, Toronto, Singapore.
2. Mantell C.L. *Batteries and Energy Systems*, New York, Mc-Graw-Kill, 1970.
3. Iliev I., Kaisheva A., Stoynov Z. Pauling H.J, Mechanically rechargeable Zinc-air cells, Proceedings of the 3th International Battery Recycling Congress, July 2 – 4, 1997, Noordwijk Aan Zee, the Netherlands.
4. Gregory D.P., *Metal-air batteries*, Mills & Boon Limited, 1972.
5. Iliev I. Air electrodes for primary metal-air batteries, 160th Meeting of the Electrochemical Society, Denver, Colorado, Oct. 1981, Extended Abstracts, p. 268-269.
6. US Patent No. 4031 033, 1972.

7. Iliev I., Air electrodes for aluminium- air batteries, Bull.Soc.Chim.Beograd, 1983; 47 (Supplement): S317-S338.
8. Kaisheva A., Gamburzev S., Iliev I. Carbon air gas-diffusion electrodes for operation in neutral electrolytes containing mixed metal oxides, Elektrokhimija 1981; 17:1362-66
9. Kaisheva A., Gamburzev S., Iliev I., Characteristics of air gas-diffusion electrodes from active carbon and pyrolyzed CoTMPP. Acid and neutral electrolytes, Electrochimija 1982; 18:139-43.
10. Kaisheva A., Iliev I., Milusheva J. Mechanically rechargeable magnesium-air cells with non-aggressive electrolyte. International Congress for Battery Recycling, July 3 – 5, 2002, Vienna, Austria.

APPLICATION OF CARBON-BASED MATERIALS IN METAL-AIR BATTERIES: RESEARCH, DEVELOPMENT, COMMERSIALIZATION

Anastassia Kaisheva* and Ilia Iliev

Central Laboratory of Electrochemical Power Sources, Bulgarian Academy of Sciences, 1113 Sofia, Bulgaria

Abstract

Metal-air cells have been developed, comprising a carbon air gas-diffusion electrode providing low polarization and stable long-term operation and metal anode with low self-discharge, operational at high current densities. The mechanism of gas-transport in the hydrophobic gas layer of the air electrode is investigated. Various carbon-based catalysts for the electrochemical oxygen reduction were tested. Methods for diagnostic of the activity and transport properties of the catalysts are proposed. A range of metal air cells developed in this laboratory is presented.

Keywords

Gas-diffusion electrode; metal-air cells; gas-transport in porous media; carbon-based catalysts.

1. INTRODUCTION

Metal-air batteries combine a metal anode (similar to that used in the conventional primary batteries) and an air gas-diffusion electrode (cathode) similar to that used in the fuel cells. During operation the metal anode is electrochemically oxidized for the expense of the oxygen from the air, which is reduced on the air gas-diffusion electrode.

The air gas-diffusion electrode possesses two advantages over the metal-oxide cathode in the conventional primary batteries: infinite charge

* Corresponding author. E-mail: kaisheva@bas.bg

I.V. Barsukov et al. (eds.), New Carbon Based Materials for Electrochemical Energy Storage Systems, 117–136.

capacity and low weight independent of the capacity. The amper-hour capacity of the metal-air battery is determined by the amount of the active metal in the anode, but not by the air electrode.

The specific characteristics (such as Ah/kg and Ah/l) of the metal-air batteries is significantly higher than that of the classical electrochemical systems with the same metal anode.

The theoretical data [1] shows that Li and Ca possess very high energy density (13172 and 4560 Ah/kg respectively) but these metals are not suitable to be used as anodes because of their instability in aqueous electrolytes. The theoretical energy densities of Mg and Al are also high (6846 Wh/kg and 8212 Wh/kg). It is shown that some alloys of Mg and Al can be successfully used as anodes, especially in metal-air cells with neutral electrolytes. The theoretical energy density of Zn is much lower than that of Li and Ca, but the self-discharge of Zn can be effectively suppressed by the use of suitable inhibitors. That's why the zinc-air batteries with KOH electrolyte are the first metal-air system brought into service.

2. EXPERIMENTAL

The cathode in all types of metal-air cells and batteries is an air gas-diffusion electrode. The air gas-diffusion electrode is a porous thin, light plate, which serves as a wall of the metal-air cell and separates the electrolyte in the cell from the surrounding air. This poses some conflicting requirements to the air electrode: it must be highly porous and permeable for the gaseous oxygen and simultaneously the leakage of the electrolyte through the porous electrode must be completely avoided. The electrode must be electrically conductive and must possess enough mechanical strength to withstand the hydrostatic pressure of the electrolyte and eventual hydrodynamic shocks. The air electrode must contain an active catalyst for the electrochemical reduction of oxygen in contact with the electrolyte. Stable operation of the air electrode with time is also needed.

3. RESULTS AND DISCUSSION

3.1. Air Gas-Diffusion Electrode

The air gas-diffusion electrode developed in this laboratory [2] is a double-layer tablet comprising a porous hydrophobic gas layer and a catalytic layer in which the catalyst is placed. The gas layer of the electrode is prepared from hydrophobic material: carbon black modified with PTFE by a special technology [3]. The catalytic layer contains a porous catalyst.

3.1.1. Hydrophobic Gas Layer

The hydrophobic gas layer of the air electrode [4] possesses high porosity (ca. 0,9 cm^2/g), such that an effective oxygen supply through this layer is obtained. From the experimental porogrames measured by both mercury and 7 N KOH-porometry the contact angle Θ_{eff} of the hydrophobic material with water electrolytes is obtained (Θ_{eff} =116° – 118°). Because of its high hydrophobicity, the gas layer prevents the leakage of the electrolyte through the electrode.

The investigation of the pore size distribution (Fig. 1) shows that nano-size pores (radius ca. 20 nm) predominate in the gas layer from this hydrophobic material.

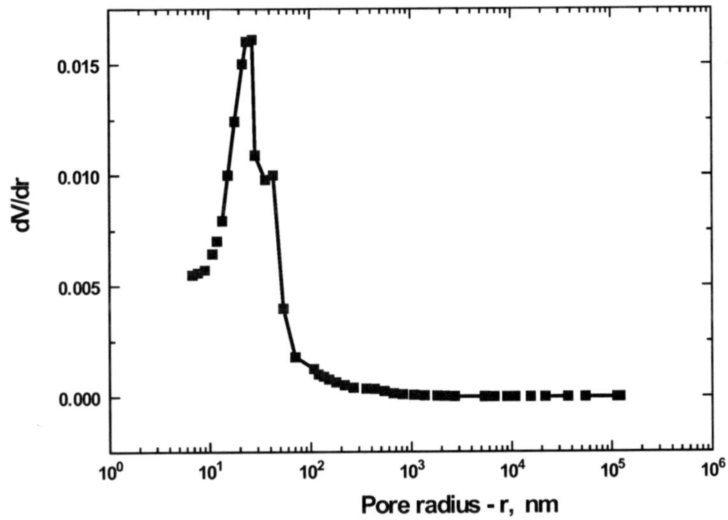

Figure 1. Distribution of the volume porosity as a function of the pore radius.

The mechanism of gas transport through the porous hydrophobic layer of the developed air gas-diffusion electrodes is theoretically and experimentally investigated [5].

The gas-permeability coefficient **K** of the material is obtained by measuring the gas flow **J** through a flat sample with thickness **L** under a total pressure difference **ΔP**. The gas-permeability coefficient **K** is a two terms sum:

$$\mathbf{K} = \mathbf{D_{1K}} + \mathbf{K_\eta} \qquad /1/$$

viscous term $\mathbf{K_\eta}$ accounting the viscous flow of the gas (accepted as a viscous fluid) and Knudsen term $\mathbf{D_{1K}}$ accounting the Knudsen diffusion mechanism (which takes into account the influence of the bouncing of the gas molecules with the walls of the porous system).

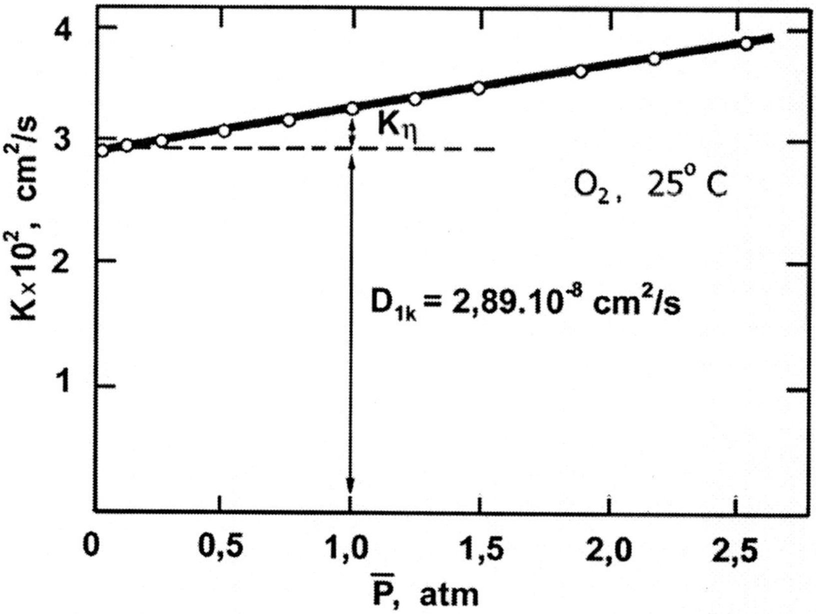

Figure 2. Permeability coefficient as a function of the mean oxygen pressure.

In Figure 2 we presented the permeability coefficient **K** of oxygen as a function of the mean gas pressure experimentally obtained for a sample of porous material from acetylene black modified with 35% PTFE. The experimental linear dependence is obtained. The intercept with the abscissa corresponds to the Knudsen term D_{1K}. The value obtained is $2{,}89.10^{-2}$ cm^2/s. The slope of the straight line is small, so that the ratio K_η / D_{1K} at mean gas pressure 1 atm. is small ($\approx$ 0.1) which means that the gas flow is predominantly achieved by Knudsen diffusion and the viscous flow is quite negligible. At normal conditions (1 atm, 25°C) the mean free path of the air molecules ($\lambda \approx$ 100 nm) is greater than the mean pore radii in the hydrophobic material ($r \approx$ 20 nm), so that the condition ($\lambda >> r$) for the Knudsen-diffusion mechanism of gas transport is fulfilled.

Gas diffusion in the nano-porous hydrophobic material under partial pressure gradient and at constant total pressure is theoretically and experimentally investigated. The "dusty-gas model" is used in which the porous media is presented as a system of hard spherical particles, uniformly distributed in the space. These particles are accepted as gas molecules with infinitely big mass. In the case of gas transport of two-component gas mixture (**i = 1,2**) the effective diffusion coefficient $(\mathbf{D_i})_{\mathbf{eff}}$ of each of the components is connected with the corresponding Knudsen diffusion coefficient $\mathbf{D_{i,K}}$ and with the effective binary diffusion coefficient $\mathbf{(D_{1,2})_{eff}}$ by the Bosanquet relation:

$$(D_i)_{eff}^{-1} = D_{iK}^{-1} + (D_{1,2})_{eff}^{-1} = (D_i)_{eff}^{-1} + \left[\left(\frac{\varepsilon}{q}\right).D_{1,2}^{(1)} \right]^{-1}.P \qquad /2/$$

in which $\mathbf{(D_{1,2})_{eff} = (\varepsilon/q).D_{1,2}}$, where $\boldsymbol{\varepsilon/q}$ is the porosity/tortuosity factor of the porous media and $\mathbf{D_{1,2}}$ is the binary diffusion coefficient. $\mathbf{D_{1,2}}$ is inversely proportional to the gas mixture total pressure $\mathbf{P}$ $\mathbf{(D_{1,2} = D_{1,2}^{(1)}/P)}$, where $\mathbf{D_{1,2}^{(1)}}$ is binary diffusion coefficient at total pressure 1 atm.

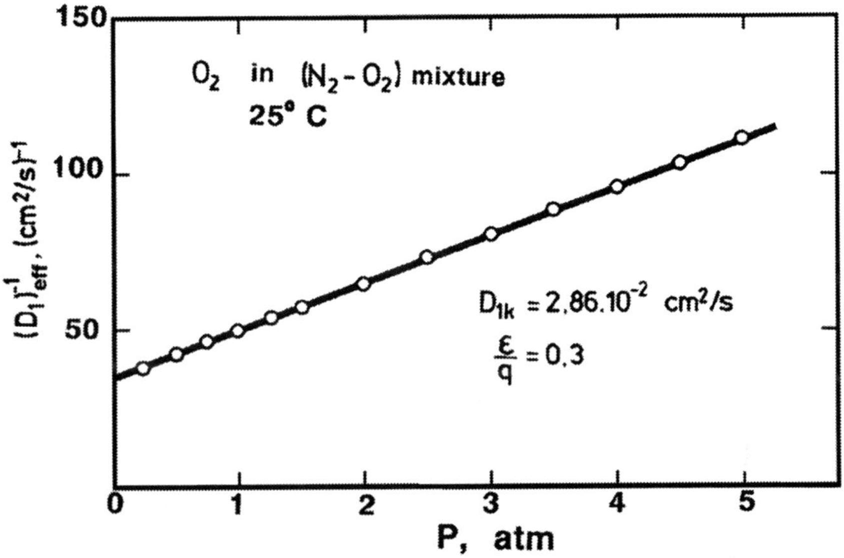

Figure 3. Effective diffusion coefficient of O_2 in O_2-N_2 mixture as a function of gas pressure.

In Figure 3, we have presented the experimentally obtained reciprocal values of $\mathbf{(D_1)_{eff}}$ of oxygen in a sample of the nano-porous hydrophobic material as a function of the total pressure $\mathbf{P}$ of gas mixture (O_2-N_2) when the oxygen concentration in the mixture is 21%. From the intercept of the straight line with the ordinate the value of the Knudsen diffusion coefficient can be also determined. It must be underlined that the value of Knudsen diffusion coefficient obtained by these diffusion measurements ($2,86.10^{-2}$ cm^2/s) is in very good coincidence with the value obtained by the gas permeability measurements.

It must be noted that the effective diffusion coefficient $\mathbf{(D_1)_{eff}}$ is obtained by electrochemical measurements of air gas-diffusion electrodes with sufficiently thick gas layer so that the limiting process is the gas transport in this layer. The polarization curves of these air electrodes are measured when operating in a close vessel with a gas cocktail consisting of electrochemically active (oxygen with partial pressure $\mathbf{P_{10}}$) and electrochemically non-active (nitrogen) gases. From the polarization curves the diffusion limited current $\mathbf{I_{1L}}$ is obtained which is connected with the diffusion limited flow of oxygen $\mathbf{J_{1L}}$. The effective diffusion coefficient of the oxygen $\mathbf{(D_1)_{eff}}$ is calculated according the equation:

$$J_{1L} = (kT)^{-1}.(D_1)_{eff}.(\frac{P_{10}}{L}) \quad /3/$$

It can be concluded that the predominating mode of gas transport in the investigated nano-porous hydrophobic material is Knudsen diffusion, so that the diffusion is the main mechanism of gas transport in electrochemical systems based on such material and operating with gaseous reactants.

3.1.2. Catalytic Layer

The catalytic layer of the air electrode is made from a mixture of the same hydrophobic material and porous catalyst [2]. It comprises hydrophobic zones through which the oxygen is transported in gas phase and zones containing catalyst where the electrochemical reduction of oxygen is taking place. It must be noted that the overall structure of the electrode is reproducible when various kinds of carbon-based catalysts are used.

One of the main problems in the development of air gas-diffusion electrodes for metal-air cells is to find active and stable catalysts for the electrochemical reduction of oxygen. Carbon-based catalysts are mostly used, because of their highly developed surface area and capability for adsorption of O_2, suitable morphology, chemical stability, good electric conductivity and comparatively low price.

Various carbon-based catalysts were tested in the investigated air gas-diffusion electrodes: pure active carbon [6], active carbon promoted with silver [7] or with both silver and nickel. Catalysts prepared by pyrolysis of active carbon impregnated with a solution of the compound Co-tetramethoxyphenylporphyrine (CoTMPP) are also studied [8].

It is well known that the performance of the air gas-diffusion electrode is influenced not only by the activity of the catalyst, but also by all transport processes taking place in its porous structure. In addition, the transport hindrances in the electrode are function not only of its overall structure, but also of the porous structure and the surface properties of the catalyst. Methods for diagnostic of the activity and the transport properties of air gas-diffusion electrodes were proposed [9].

It was experimentally found that the polarization curves of the investigated air gas-diffusion electrodes in a semi-logarithmic scale at low current densities (below 10 mA/cm^2) are straight lines, which can be treated as Tafel plots. At these low current densities the transport hindrances in the air electrode are negligible so that activation hindrances only are available.

In Figure 4 we have presented the experimental Tafel plots of air electrodes with catalysts from pure active carbon and from active carbon promoted with different amounts of silver. The obtained curves are straight lines with identical slopes. It must be underlined that the investigated electrodes possess identical gas layers and catalytic layers, which differ in the type of catalyst used only. Therefore, the differences in the observed Tafel plots can be attributed to differences in the activity of the catalysts used. The current density ***a*** at potential zero (versus Hg/HgO), obtained from the Tafel plots of the air electrodes is accepted as a measure of the activity of the air gas-diffusion electrodes: the higher value of ***a*** corresponds to higher activity of the air electrode.

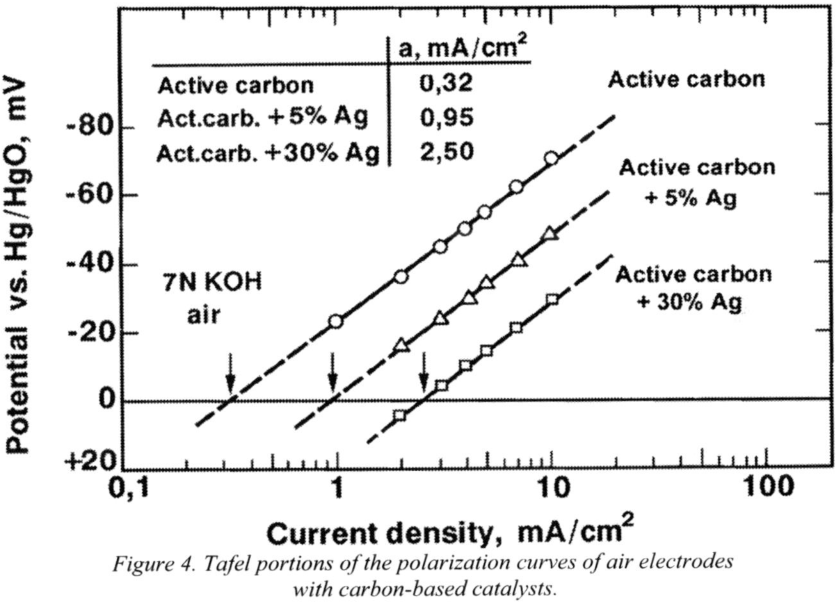

Figure 4. Tafel portions of the polarization curves of air electrodes with carbon-based catalysts.

From Figure 4 it is visible that the electrodes with silver in the catalyst are more active than that with pure active carbon catalyst. Moreover, the increase of the amount of promoting silver in the catalyst results, as expected, in a higher activity of the electrode.

Nevertheless, the comparison (Figure 5) of the polarization curves of the same air electrodes show that the polarization at the high current density range of the most active electrode is much higher that that of the electrodes with lower activity. This fact can be connected with the higher transport hindrances in the catalyst containing high amount of silver.

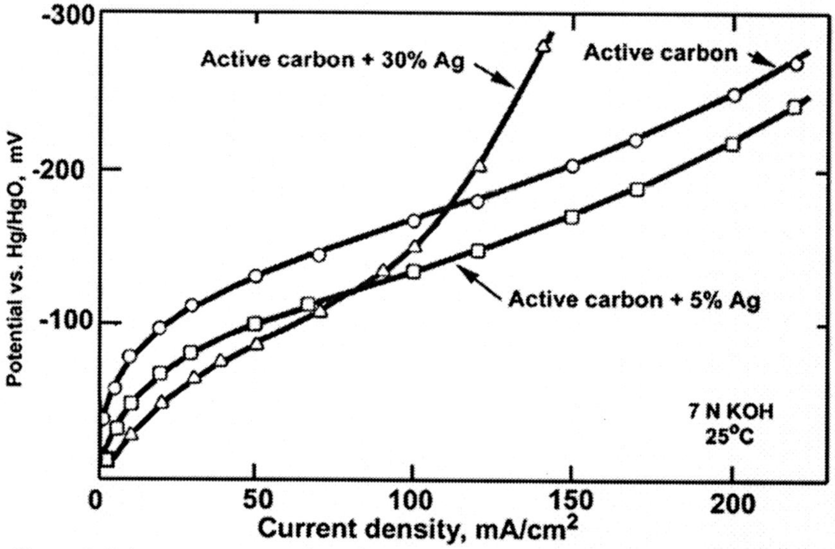

Figure 5. Polarization curves of air electrodes with catalysts from active carbon and with active carbon promoted with 5% and 30% of silver.

Generally, the hindrances in the transport processes, which take place in the porous structure of the air gas-diffusion electrodes, influence strongly the electrode performance and are of prime importance for their practical application.

An experimental method is proposed for estimation of the transport hindrances in air gas-diffusion electrodes. As a measure of the transport hindrances in the air gas-diffusion electrodes is introduced the difference ΔE between the potentials of one and the same electrode when operating with air and with pure oxygen at one and the same current density. The difference ΔE can be theoretically described by the equation:

$$\Delta E = E_{O_2}(I) - E_{air}(I) = \frac{RT}{\alpha zF} \ln \frac{P_{oxygen}}{P_{air}} + \frac{RT}{\alpha zF} \ln \frac{f_{oxygen}(I)}{f_{air}(I)} \qquad /4/$$

which comprises two terms.

The first term /5/ takes into account the partial pressures of oxygen only

$$\Delta E_0 = \frac{RT}{\alpha zF} \ln \frac{P_{oxygen}}{P_{air}} \qquad /5/$$

and is independent of the current density. The second term depends on the current density through the efficiency factors of the electrode when operating with air $\mathbf{f_{air}(I)}$ and with pure oxygen $\mathbf{f_{ox}(I)}$ [9]. The efficiency factors being defined as the ratio of the currents generated at one and the same potential by the real electrode and by the hypothetical electrode with the same structure and catalyst in which there are no transport hindrances.

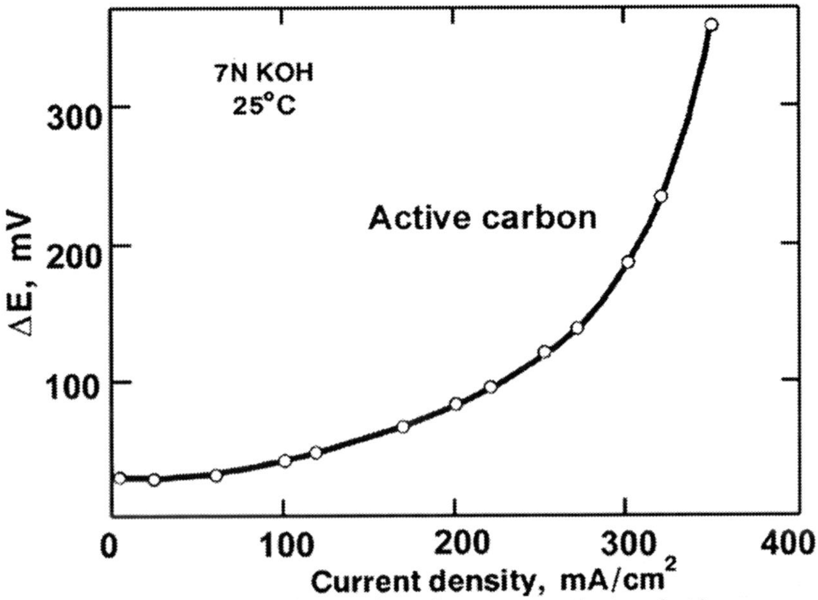

Figure 6. Polarization curve of an air electrode operating with air and with pure oxygen.

With the increase of the current density, the second term in the equation /4/ is increasing, because the efficiency factor of the electrode operating with air $\mathbf{f_{air}(I)}$ decreases more rapidly with the increase of the current density than the efficiency factor of the same electrode operating with pure oxygen $\mathbf{f_{ox}(I)}$. According to equation /4/ the value of $\mathbf{\Delta E}$ will increase with the increase of the current density.

Figure 6 presents the polarization curves of one and the same electrode with active carbon as catalyst when operating with air and with pure oxygen.

The value of $\mathbf{\Delta E}$ is obtained by the extraction of the potential of the two curves at one and the same current density, so it is free from the IR drop between the air electrode and the reference electrode.

The obtained values of ΔE are presented in Fig. 7 as a function of the current density. It is seen that the values of $\mathbf{\Delta E}$ increase with the current. At low current densities, where the transport hindrances in the electrode are negligible low, $\mathbf{\Delta E}$ is practically independent of the current density. This is visible in Fig. 8, where are presented data from Fig. 7 for low current densities (up to 10 mA/cm^2).

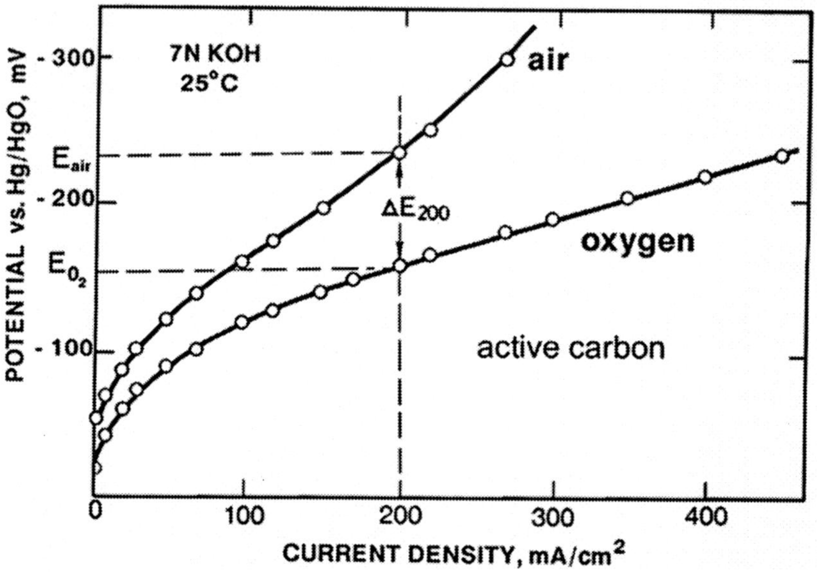

Figure 7. ΔE – I curve obtained from Figure 6.

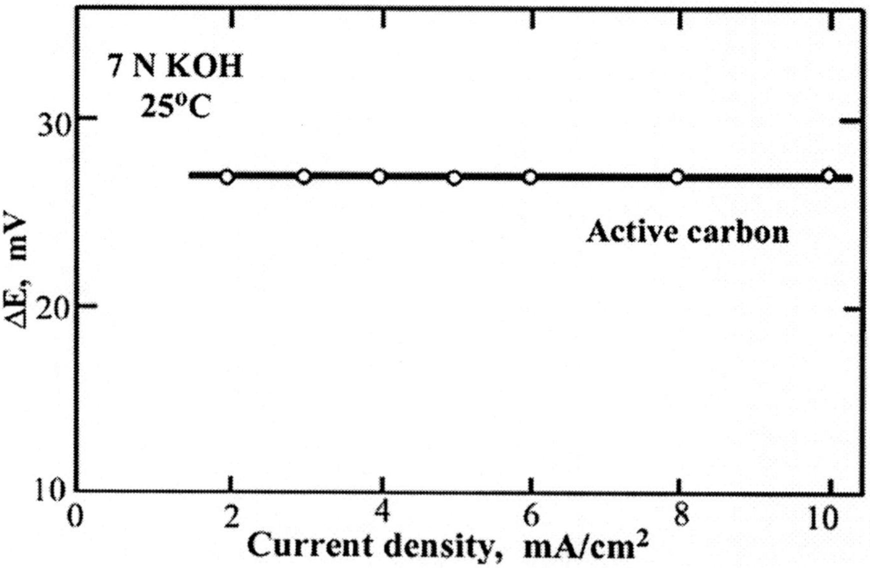

Figure 8. ΔE – I curve at low current densities obtained from Figure 6.

Experimental **ΔE – I** curves can be used for comparison of the air electrodes with respect to the transport hindrances. In order to illustrate this possibility in Fig. 9 are presented the **ΔE – I** curves for air electrodes with identical catalytic layers and gas layers differing in their thickness only. Apparently the hindrances in the transport of molecular oxygen will be higher in the electrodes with thicker gas layers.

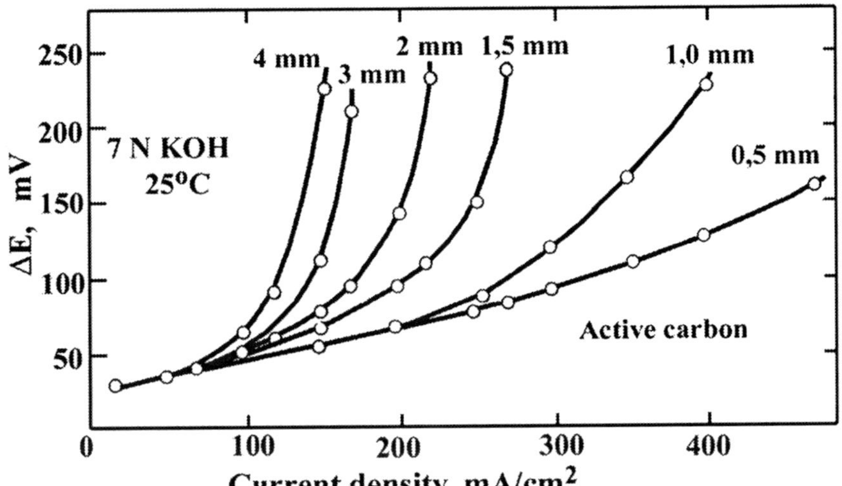

Figure 9. ΔE – I curves of air electrodes with identical catalytic layer and gas layer with different thickness.

From Figure 9 it is seen that the value of **ΔE** increases more rapidly with the current density for the electrodes with thicker gas layer which is due to the higher transport hindrances.

The estimation of the transport hindrances by the use of ΔE can be performed in two ways: by comparison of the shape of the **ΔE – I** curves, or by comparison of the values of **ΔE** at a constant current density.

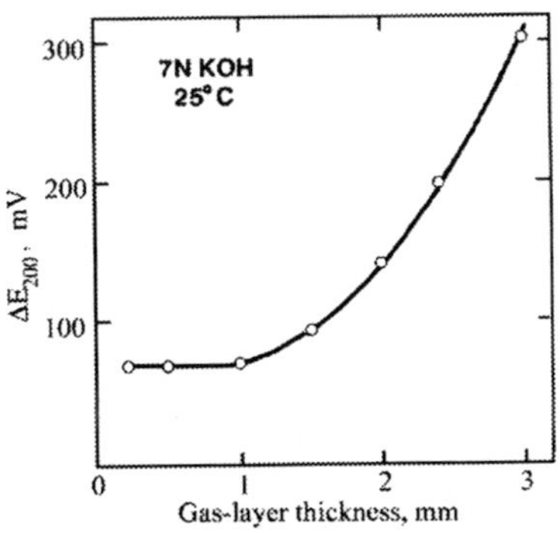

Figure 10. ΔE at c.d. 200 mA/cm² as a function of thickness.

In Figure 10, we presented the value of **ΔE** at current density 200 mA/cm^2 as a function of the thickness of the gas layer of the electrodes from Figure 10.

It is seen that the transport hindrances remain practically constant at thickness of the gas layer up to 1 mm. The further increase of the gas layer thickness results in significant increase of the transport hindrances.

The described method for the diagnostic of the activity and the transport hindrances in air gas-diffusion electrodes is very useful in the research of porous catalysts for air electrodes. The comparison of the activity and the transport hindrances of air electrodes with catalysts from various types of active carbon allow a proper selection to be accomplished.

The diagnostic method can also be successfully used in the assessment of promoted catalysts.

As it has been previously shown (Figure 4), the activity of catalysts from active carbon promoted with Ag increases with the increase of the content of the promoting silver in the catalyst. But the transport hindrances are strongly influenced by the content of silver in the catalyst.

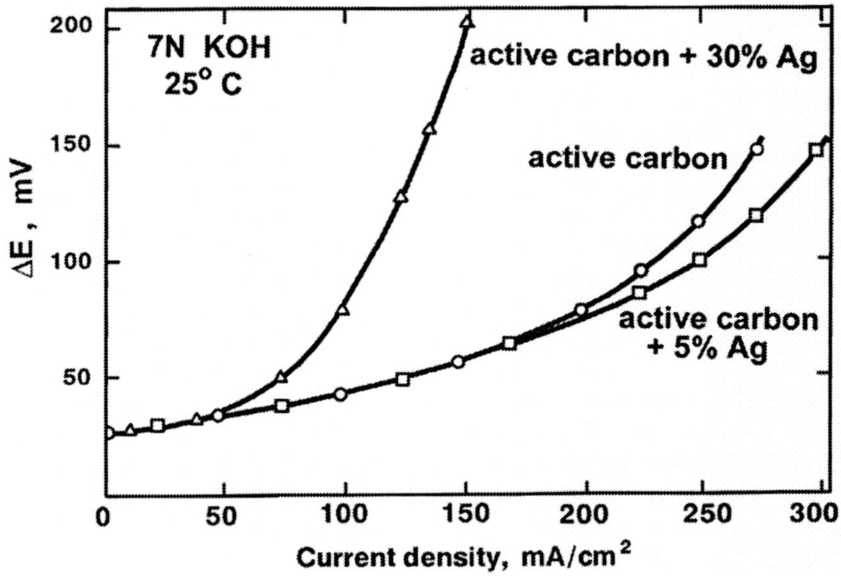

Figure 11. ΔE – I curves of air electrodes with catalysts from active carbon and active carbon promoted with different amounts of silver (from Figure 5).

In Figure 11, we have presented the **ΔE – I** curves of air electrodes with catalysts from active carbon and active carbon promoted with different amounts of silver (from Figure 5). It is seen that the transport hindrances in the electrodes with catalysts from pure active carbon and with active carbon promoted with 5% of silver are near to each other. The transport hindrances in the air electrodes with catalyst containing 30% of silver are much higher. That's why catalysts containing large amount of silver are suitable to be used in air electrodes operating at comparatively low current densities.

Using the method for diagnostic, optimal amounts of Ni in catalysts from active carbon promoted with both Ag and Ni were found. Catalysts are investigated from active carbon containing one and the same amount of Ag (5%) and different amounts of Ni. It is found that the activity of these catalysts increases with the increase of the Ni content up to 7.5%. The further increase of the Ni-content from 7.5 to 25% does not influence the activity of the catalyst. The investigation of the ΔE values of air electrodes with these catalysts shows that the transport hindrances remain practically constant at Ni content up to 10%. At higher Ni contents the transport hindrances increase significantly. It can be concluded that the optimal content of promoting metals in the catalysts is ca. 5%Ag + 7%Ni.

The study of catalysts produced by heat treatment in argon of active carbon impregnated with the compound C-tetramethoxyphenylporphyrine (CoTMPP) is of great interest. The polarization curves of air electrodes with

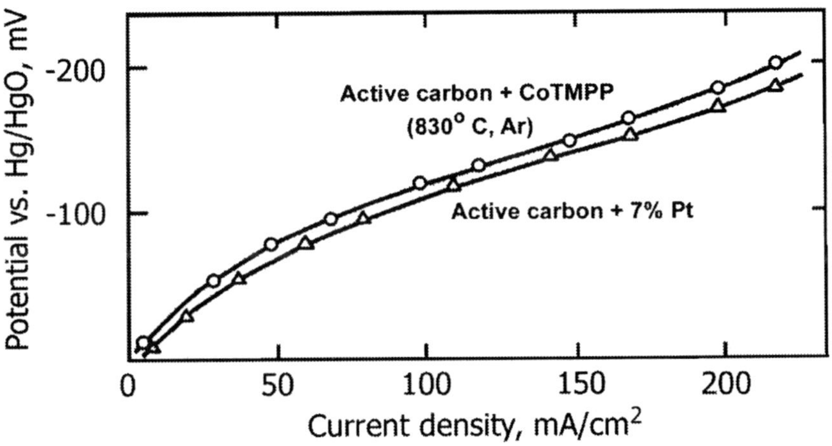

Figure 12. Polarization curves of air electrodes with catalysts from active carbon promoted with 5% Pt and with pyrolyzed CoTMPP catalyst.

this catalyst are comparable to that of air electrodes with catalyst containing 5% Pt (Fig. 12). Something more on electrodes with catalyst containing pyrolyzed CoTMPP. These show stable performance: they can operate continuously for more than 8000 hours at current density 100 mA/cm^2.

The activity of air electrodes with pyrolyzed CoTMPP catalysts is investigated as a function of the heat treatment temperature. In Figure 13 we have presented the initial curve, and the curves obtained after long-term operation of the electrode at c. d. 100 mA/cm^2. It is seen that a maximum in the activity is obtained at pyrolysis temperature ca. 800°C, the activity being practically preserved with the time of operation of the air electrode.

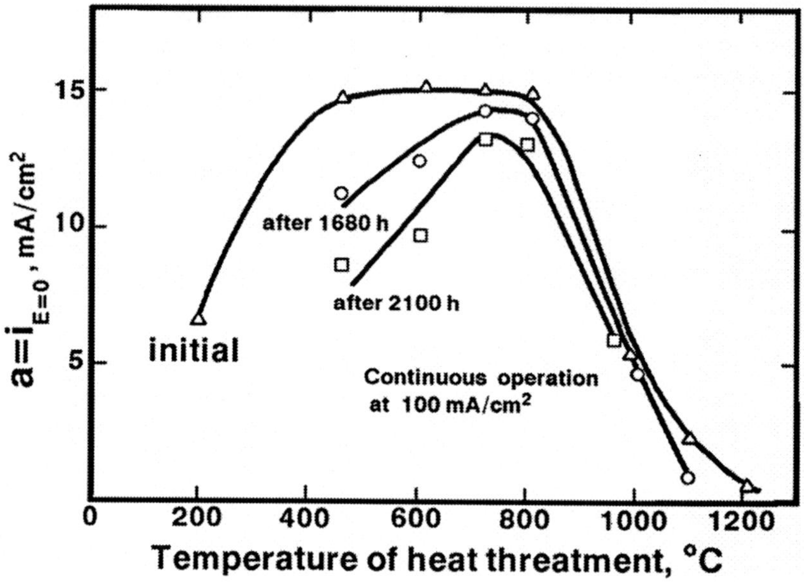

Figure 13. Activity of air electrode with pyrolyzed CoTMPP catalysts as a function of the pyrolyzis temperature.

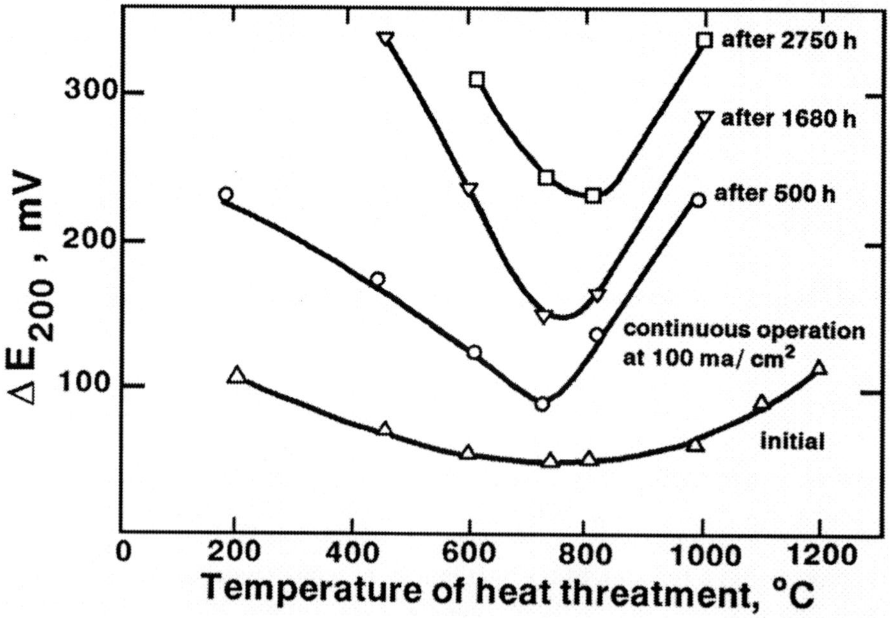

Figure 14. Transport hindrances in air electrodes with pyrolyzed CoTMPP catalysts as a function of pyrolysis temperature.

The transport hindrances (Figure 14) show a minimum at pyrolysis temperatures 700-800°C, but the transport hindrances in the air electrode increase significantly with the time of operation. It can be concluded that the main cause for the aging of the investigated air electrodes is the increase of the transport hindrances with the time of operation.

It is theoretically predicted [10] and experimentally proven [11] that the existence of a network of very tiny hydrophobic pores in the carbon-catalyst grains play an important role for the effective operation of the electrodes and is of prime importance for their long exploitation life.

It was found that the H_2O_2 formed by the electrochemical reduction of oxygen oxidizes the carbon surface, by which parts of the pores in the carbon-catalyst loose their hydrophobicity and is gradually filled with electrolyte. As a result the hindrances in the transport of oxygen through the carbon catalyst grains increase. This seems to be the main cause for the aging of the electrodes during long-term operation. One way to impede this effect is the use of catalyst with high activity not only for the electrochemical reduction of oxygen but also for the heterogeneous H_2O_2 decomposition (such as pyrolyzed CoTMPP catalyst).

The investigations of various types of carbon-based catalysts allow suitable air electrodes to be developed for use in the large variety of metal-air cells and batteries designed in this laboratory.

3.2. Metal-Air Cells

Various types of metal-air cells are developed using the described air gas-diffusion electrodes. In the most cases zinc is used as an anode [13].

Porous zinc electrodes with highly developed surface are elaborated in this laboratory, which show extremely high discharge rate capability in wide temperature range (down to -40°C). The self–discharge of the zinc is well suppressed without the use of any unacceptable mercury and a high amper-hour capacity of these zinc electrodes is achieved.

A series of primary zinc-air cells are developed with capacity ranging from 100 Ah to 3300 Ah and nominal currents ranging from 2 A to 50 A. Technologies for the production of the air electrodes, porous zinc electrodes and for the assembling of the cells are elaborated. Some of the developed zinc-air cells are in production since 1981.

The high capacity, low power zinc-air cells from the types ZV500 and ZV3000 are used as a power supply of navigation buoys in Baltic see for more than 10 years. Batteries from these cells operate for more than 1 year continuously. Zinc-air cells from the same types are successfully used in Bulgarian Himalayan expeditions (Lhotze 1981, Everest 1984 and Anapurna 1986) as power supply of the telecommunication system and electric light at the base camp. It must be underlined that in these cases the zinc-air cells operate at extreme condition – altitude 5000 m above the sea level (low oxygen concentration of in the air) and temperatures down to - 30°C.

Active carbon promoted with small amount of silver is used as catalyst in the air electrodes of these cells. In Figure 15 we presented the discharge curve of the zinc-air cell ZV3000 at constant current 1 A.

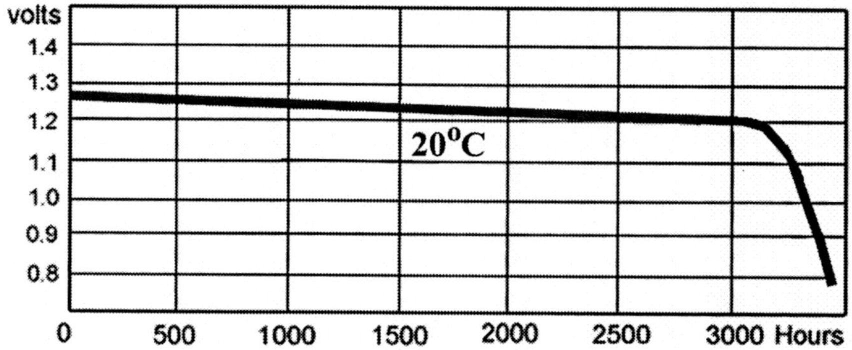

Figure 15. Typical discharge curve of a zinc-air cell ZV3000.

High power primary zinc-air cells are also developed with nominal current 20A ÷ 40A and capacity in the range 100 Ah ÷ 320 Ah. Typical application of these powerful zinc-air cells is as easy activated reserve power supply. These cells also can be successfully used for traction.

In order to achieve comparatively high nominal current of the cells, pyroyzed CoTMPP catalyst is used in the air electrodes.

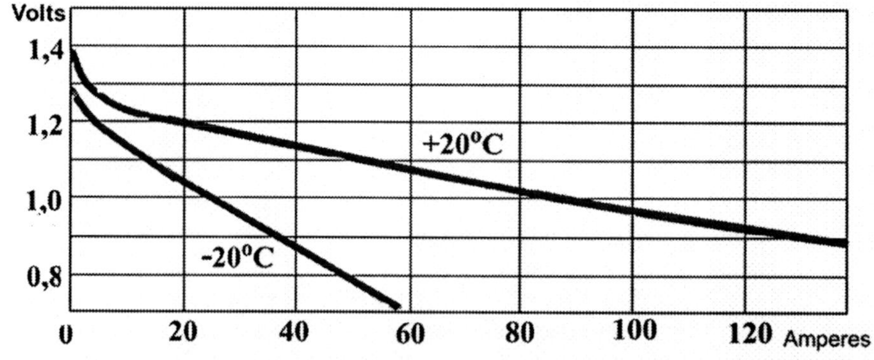

Figure 16. Polarization curves of a zinc-air cell ZV320 at temperatures 20 ℃ and -20 ℃
Zinc-air cells from the types ZV280, ZV300, ZV320 and others are constructed and tested.

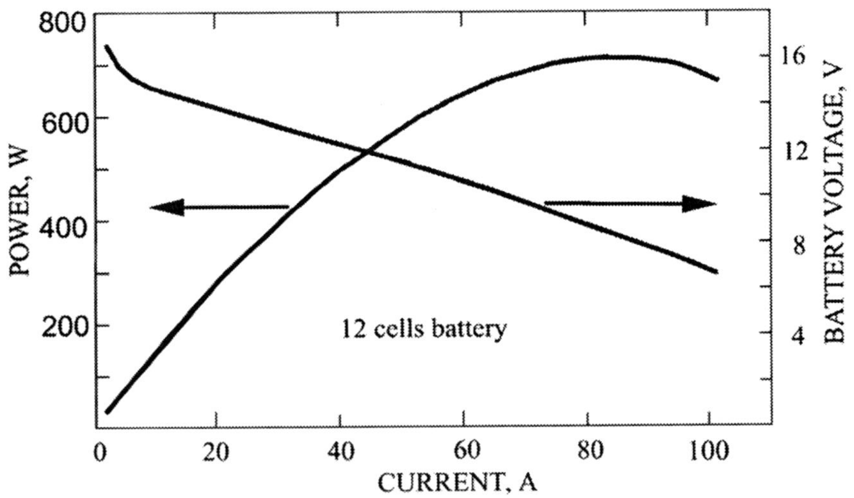

Figure 17. Voltage and power of a 12 cell battery, consisting of zinc-air cells ZV300 as a function of the current.

In Fig. 17 we have presented the current-voltage and the power characteristics of 12-cells battery from the cells type ZV300. It must be reminded that, when the zinc anode in the primary zinc-air cells is discharged, the cell is out of operation. By this, the air electrodes, which in principle can operate for a long time, are lost. In order to avoid this inconvenience, mechanically rechargeable zinc-air cells were developed [14]. The essential feature of the mechanically rechargeable zinc-air cell is that the air electrodes are used many times.

Once the mechanically rechargeable zinc-air cell is discharged, the spend (oxidized) zinc electrode is removed from the cell together with the electrolyte and new zinc electrode and fresh electrolyte are introduced by which the cell is operational again.

The removed from the cell discharged anode, containing the products of the discharge of zinc is put in a separate cell, where it is recharged electrochemically against an inert electrode. During this charging process zinc is produced again from the discharge products. The experiments have shown that such a procedure of recharging of the zinc anodes can be repeated several times without significant losses of the amperhour capacity of the zinc electrode.

Several types of mechanically rechargeable zinc-air cells are developed in this laboratory, operating at nominal current 20A ÷ 40 A with capacity ca.300 Ah.

Batteries for such cells are tested in experimental electric cars and scooters.

Another type of zinc-air system is designed and investigated – the so-called zinc-air fuel cell. It consists of a cell with two air electrodes and an inert anode. Slurry from zinc power dispersed into the electrolyte is circulating through the cell, by which the zinc particles are discharged on an inert electrode.

In Table 1 we have summarized data for some Zn-air cells developed in CLEPS.

Table 1. Data for some zinc-air cells developed in CLEPS.

Type/purpose	**Capacity Ah**	**Current A**	**Voltage V**	**Power W**	**Specific energy Wh/kg**
ZV-1 (Electric truck)	160	40	1,0	40	160
ZV-2 (Electric car)	250	80	1,0	80	114
Zn-Air Fuel cell	-	150	1,0	150	40
$Zn-O_2$	280	20	1,18	23	200
ZV320	320	40	1,17	46	180
ZV500	500	2	1,2	2,4	250
ZV3000	3300	2	1,2	2,4	270
ZVMR1 (Electric car) Mechanically rechargeable	500	30	1,1	33	180
ZVMRII (Electric car) Mechanically rechargeable	300	20	1,1	22	150

Air gas-diffusion electrodes with similar overall structure suitable for operation in neutral electrolytes are developed. On the basis of these air electrodes metal-air cells are developed using aluminum and magnesium as an anode.

An aluminum-air battery has been designed (2500 Ah, 2A, 12V) with sea water as an electrolyte (in cooperation with the University of Belgrade). The battery [15] is designed for use in a small yacht. It is known that during the discharge of aluminum a gelatinous aluminum hydroxide is obtained. It could form an insulator layer on the air electrode surface by which its further operation will be troubled. In order to avoid this effect, the electrolyte in the aluminum-air battery is changed every 8 hour. After the use of the battery during the night, the electrolyte (sea water) is pumped out of the battery and a fresh sea water is pumped in the battery when power supply is needed.

Several types of magnesium-air cells [16] with neutral electrolyte (NaCl solution) are also developed and investigated.

4. CONCLUSIONS

Air gas-diffusion electrodes were developed, suitable for use in metal-air cell with alkaline or with saline electrolytes. A variety of carbon-based catalysts are used in these air electrodes. Methods for the estimation of the activity and the transport hindrances are proposed and used successfully for the optimization of the carbon-based catalysts.

A variety of primary zinc-air cell and batteries are designed with capacity ranging from 100 Ah to 3300 Ah, operating at nominal currents from 2 to 40A at temperatures in the range +40 ÷ -40°C. Mechanically rechargeable zinc-air cell are also developed and tested in experimental electric cars and scooters.

Al-air and Mg-air cells with saline electrolytes have also been designed.

REFERENCES

1. Gregory D.P., *Metal-air batteries*, Mills & Boon Limited, 1972.
2. Iliev I., Air electrodes for primary metal-air batteries, Prroceedings of the 160th Meeting of the Electrochemical Society, Oct. 1981, Denver, Colorado, p. 268-269.
3. US Patent No. 4031 033, 1972.
4. Kaisheva A., Doctor Thesis, 1984.
5. Kaisheva A., Iliev I. Gas-transport in nano-porous hydrophobic media, Procedings of the Int. Symposium Micro/nanoscale energy conversion and transport, 2002, April 14 – 19, Antalya, Turkey, p. 107.
6. Iliev I., Mrha J., Gamburzev S., Kaisheva A., On the effect of various active carbon catalysts on the behaviour of carbon gas-diffusion air electrodes, Journal Power Sources 1976/77; 1: 35-46.
7. Iliev I., Gamburzev S, Kaisheva A., Mrha J. Influence of mass transport on the performance of carbon gas-diffusion electrodes in alkaline solutions, J.Appl. Electrochem., 1975; 5 :291-297.
8. Iliev I., Gamburzev S., Kaisheva A., Gas-diffusion electrodes with transition metal macrocyclic catalysts for electrochemical reduction of oxygen, Proceedings of the 31 ISE Meeting; 1980 Sept. 22 – 26, Venice, Italy, Vol. I, p. 286-288.
9. Kaisheva A., Iliev I., Gamburzev S., Comparative methods fo estimation of the activity and the transport hindrances of air gas-diffusion electrodes, J.Power Sources, 1984; 13:181 – 195.
10. Iliev I., Gamburzev S., Kaisheva A., Optimization of the pyrolysis temperature of active carbon-CoTMPP catalyst for air electrodes in alkaline media, J.Power Sources, 1986; 17: 345 352.
11. Iliev I., Structure and transport processes in carbon air electrodes,Proceedings of the Fall Meeting of Electrochemical Society, 1983 Oct. 9-14, Washington, Vol. 83-2, p. 618-19.
12. Gamburzev S., Iliev I., Kaisheva A., Steinberg G., Mokrousov L., Behaviour of carbon wetproofed electrodes during long-term operation in alkaline electrolyte, Elektrokhilija, 1980; 16:1069-72.
13. Budevski E., Iliev I., Varbev R., Zinc-Luft Elemente mit hohem Energievorrat – Probleme und Eisatzmoeglichkaiten, Z. Phys. Chem. Leipzig, 1980; 261:716-21.

14. Iliev I., Kaisheva A., Stoynov Z., Pauling H.J., Mechanically rechargeable zinc-air cells, Proceedings of the 3rd Int. Battery Recycling Congress, 1997 July 2 – 4, Noordweejk Aan Zee, the Netherlands.
15. Budevski E., Iliev I., Kaisheva A., Despic A., Krsmanovic K., Investigation of a large capacity medium power saline aluminium-air battery, J. Appl. Electrochem., 1989; 19:323-330.
16. Kaisheva A., Iliev I., Milusheva J., Mechanically rechargeable magnesium-air cells with non-aggressive electrolyte, Proceedings of the Int. Battery Recycling Congress, 2002 July 3 – 5, Vienna, Austria.

METAL – AIR BATTERIES WITH CARBONACEOUS AIR ELECTRODES AND NONMETALLIC CATALYSTS

Nikolay Korovin*

Moscow Power Engineering Institute (Technical University (9)
14 Krasnokasarmennaya, Moscow, 11250, Russian Federation

Abstract

Problems of the metal-air batteries and air electrode are considered. Results of research, design work and development of the air – Zn primary and air-Al mechanically rechargeable batteries, as well as electrically rechargeable electrodes are presented.

Keywords

Batteries; air electrodes; catalysts; carbon.

1. INTRODUCTION

In the batteries with air electrodes, the oxidizer is oxygen from the atmosphere. This fact assures many advantages of these batteries, namely: high energy density, long shelf life, environmental compatibility, relatively low cost and flat discharge curve. Theoretical parameters of batteries with air electrodes (please see, for reference, Table 1) are much higher than theoretical parameters of most traditional, newly developed and emerging battery systems.

Obviously, the practical performance of batteries with air electrodes naturally is lower than the theoretical values indicated by Table 1, and is a function of parameters of the reagents making up these batteries.

Sucessful application of the air electrode requires solving some key problems: the air electrode catalyst, the alkaline electrolyte carbonization, the oxygen reaction with anode materials, an influence of an air humidity on an electrode behavior.

*E-mail: nikolay.korovin@mtu-net.ru

I.V. Barsukov et al. (eds.), New Carbon Based Materials for Electrochemical Energy Storage Systems, 137–149.

Table 1. Theoretical parameters of various battery chemistries based on air electrodes (Po_2=21kPa) at 298K.

Anode	EMF, V	Specific capacity, Ah / g	Specific energy, Wh/g	Energy density, kWh/l
Al	2.55	2.98	7.6	22.11
Mg	3.16	2.20	6.97	12.12
Zn	1.64	0.82	1.34	9.6
Fe	1.26	0.96	1.21	9.52
$LaNi_5H_6$	1.2	0.37	0.44	2.64

Of those types mentioned in Table 1, main attention has been paid to the development of batteries with Zn, Mg and Al anodes [1-7]. Recently battery air-metal hydride was proposed and investigated [3, 8].

It is possible to divide the batteries with air electrodes into three following groups: primary, mechanically and electrically rechargeable. In these batteries alkaline, saline and seawater electrolytes are used (Table 2).

Parameters and behavior of these batteries depend on activity and stability of air electrodes, thus it is necessary to consider processes at the air electrode, problems of air electrode and ways of their solution, as a complex and multifunctional task.

Table 2. Types of batteries with air (oxygen) electrodes and applicable electrolytes.

Anode	Primary Batteries	Mechanically Rechargeable Batteries	Electrically rechargeable batteries
Zinc	Alkaline	Alkaline	Alkaline
Aluminum	Seawater	Alkaline Saline	-
Magnesium	Seawater	Saline	-
Iron	-	-	Alkaline
Metal hydride	-	-	Alkaline

2. AIR ELECTRODES

2.1. Processes

The active component of an air electrode is oxygen. The reaction of oxygen reduction (4-electron reaction) can be written by equation (1)

$$O_2 + 2H_2O + 4e^- = 4OH^-;\ E^0_1=0.401\ V \qquad (1)$$

It is possible to attain the theoretical potentials of the reaction (1) only at the special conditions of the electrodes, and provided the deep purification of electrolytes. Stationary electrode potentials are usually lower than the theoretical potentials by 100-500 mV. The reason for that effect is occurance of parallel and side reactions in the cell.

Many scientists deeply investigated the processes of oxygen reduction in metal-air cells [9 - 11]. The process in question is very complicated. Products of the oxygen reduction usually contain hydrogen peroxide. Therefore, the oxygen reduction can proceed by a two-electron scheme:

$$O_2 + 2H_2O + 2e^- = HO_2^- + OH^-;\ E^0_2 =-0.076\ V \qquad (2)$$

Hydrogen peroxide then can be reduced or decomposed in water and oxygen. Thus, oxygen reduction can proceed by two parallel ways: (a) the four-electron reaction of conversion into water and (b) the two-electron reaction of transformation into hydrogen peroxide. The four-electron reaction can proceed, if oxygen adsorbs with a rupture of chemical bonds in O_2:

$$O_2 \rightarrow 2O_{ads}. \qquad (3)$$

Adsorbed oxygen (O_{ads}) from process (3) then reduces, e.g. by reaction (4):

$$O_{ads} + H_2O + 2e^- = 2OH^-. \qquad (4)$$

Energy of the oxygen adsorbtion on the pyrographite, active carbon and some other electrodes is not enough for the rupture the bonds in oxygen. In this case the reaction proceeds by the electron addition to the adsorbed oxygen molecule, which usually limits all process of oxygen reduction.

2.2. Catalysts

The oxygen electrode catalysts accelerate the adsorption step, or the step of the adsorbed oxygen reduction or the hydrogen peroxide decomposition. A catalytic activity is shown by: platinum metals, silver, oxides (MnO_2, NiO +Li_2O, Co_2O_3), spinels ($NiCo_2O_4$, $MnCo_2O_4$, $CoAl_2O_4$), perovskites ($LaCoO_3$, $La_{1-x}Sr_xCoO_3$, $La_{1-x}Sr_xMnO_3$, $La_{1-x}Ca_xCoO_3$), activated carbon, polyaniline, macrocycles, their polymers and pyropolymers [9-14]. Properties of some of them are shown in the Table 3.

Table 3. Catalysts for the oxygen reduction in the alkaline solution.

Catalysts	Activity	Stability	Cost
Pt,Rh,Pd	Good	Very good	Very high
Silver	Very good	Good	High
MnO_2	Good	Middle	Acceptable
Spinels (AB_2O_4)	Middle	Middle	Acceptable
Perovskite (ABO_3)	Middle	Middle	Acceptable
Activated carbon	Low	Middle	Low
Polymer macrocycles	Good	Low	Acceptable
Pyropolymers	Good	Good	Acceptable

The exchange current density of Pt-metals is relatively small, but they have high stability. Very high cost does not permit to use them in the batteries of wide application. Noticeably higher activity, very good stability and lower costs are demonstrated by silver. The most inexpensive catalyst is activated carbon that has very high surface area. This type of catalyst is used in some batteries. Activity of carbon electrode can be improved by additive of oxide (e.g. MnO_2) or pyropolymers.

2.3. Porous Electrodes

Oxygen reduction can be accelerated by an application of electrodes with high surface area, e.g. the porous electrodes [9, 13]. The porous electrodes usually consist of catalysts, hydrophobic agent (polytetrafluoroethylene-PTFE) and conductive additive. Electrode kinetics on the porous electrodes is complicated by the mass and charge transfer in the pores and is called "the macrokinetics of electrode processes".

An analysis of the equations of macrokinetics of oxygen reduction on the porous hydrophobic electrodes gave some conclusions [15, 16]:

1. The dependences of current density on weight percent of PTFE and on porosity go through a maximum;

2. Current density increases with decrease of the catalysts and PTFE particles size, and with the temperature increase;

3. The dependence of current density on potential is very complicated;

4. Effective Tafel coefficient can changed from b to 4b.

5. The characteristic length of processes L is determined by the relationship between the constants of transfer and reaction velocity.

Partial pressure of oxygen is 5 times lower than that of the air pressure. Inert nitrogen changes the mechanism transfer of oxygen in the gaseous phase [17]. In pure oxygen, the Stephanov flux takes place, because pressure in the reaction zone is lower, than pressure of air. Therefore, in gaseous phase the oxygen diffusion does not limit process. Inert component (nitrogen) suppresses Stephanov flux, because difference between pressure in the air and in the reaction zone is very small. Concentration of oxygen in the reaction zone is lower than that in air. Therefore, it is necessary to have special thin diffusion layer for the oxygen of air transfer.

3. PRIMARY AIR-ZINC BATTERIES

3.1. Processes and Problems

The overall reaction in zinc-air battery with alkaline electrolyte at the beginning of discharge may by written as follows:

$$2Zn + O_2 + 2H_2O + 2OH^- = 2Zn[(OH)_4]^{2-}. \qquad (5)$$

Zincate-ions can dissociate and after the solubility exceeds the equilibrium level precipitation of zinc oxide proceeds:

$$Zn[(OH)_4]^{2-} = ZnO + H_2O + 2OH^-. \qquad (6)$$

Thus, after partial discharge, the reaction (5) can be interchanged with reaction (7):

$$2Zn + O_2 = 2ZnO. \qquad (7)$$

The cathode can of an air –zinc cell has a hole for the air access (the respiratory hole). The substitution of the cathode oxides, which are used in traditional batteries, with air increases capacity and energy density of such a battery. The battery voltage depends first of all on the air potential, which at constant current density and temperature is also constant; hence the voltage of an air-zinc battery at the discharge practically does not change.

There are several recognized problems in air-zinc batteries. Among these are: relatively high cathode polarization, the anode passivation, the zinc corrosion, the electrolyte aging, the carbonization of electrolyte and the air humidity influence on battery performance parameters. The cathode polarization is decreased by the porous electrode with catalyst (activated carbon, pyropolymers and other) use. The anodic passivation arises owing to the zinc oxide formation at high current density. For its prevention it is necessary to decrease current density, for example by the zinc powder electrode application. Some additives, e.g. Li^+ or SiO_3^{2-} slow down the aging of electrolyte. The zinc corrosion proceeds as with the hydrogen evolution.

$$Zn + 2H_2O = Zn^{2+} + H_2 + 2OH^-, \qquad (8)$$

and as with the oxygen reaction: (5) and (7). For purposes of decreasing of the corrosion rate, zinc of high purity is applied, and also the additives in zinc (e.g. Pb) and in electrolyte (e.g. amine) are used.

If battery is open to the air, carbonization of alkali takes place in accordance with process (9). For purposes of prevention of electrolyte carbonization, the respiratory holes should be sealed until after the beginning of battery use.

$$CO_2 + 2OH^- \leftrightarrow CO_3^{2-} + H_2O. \qquad (9)$$

The air humidity influences the performance parameters of the air-zinc batteries. At high air humidity the electrode flooding is possible. On the other hand, at low air humidity a concentration of the electrolyte increases, the air electrodes get too dry.

3.2. Portable Batteries

These types of batteries are available in button and prismatic forms. Their main application is as power sources for hearing aids. Other applications include various specialty uses in the notebook computers, electronic pagers, portable battery chargers, various medical devices, the wireless crew communicator systems [18, 19].

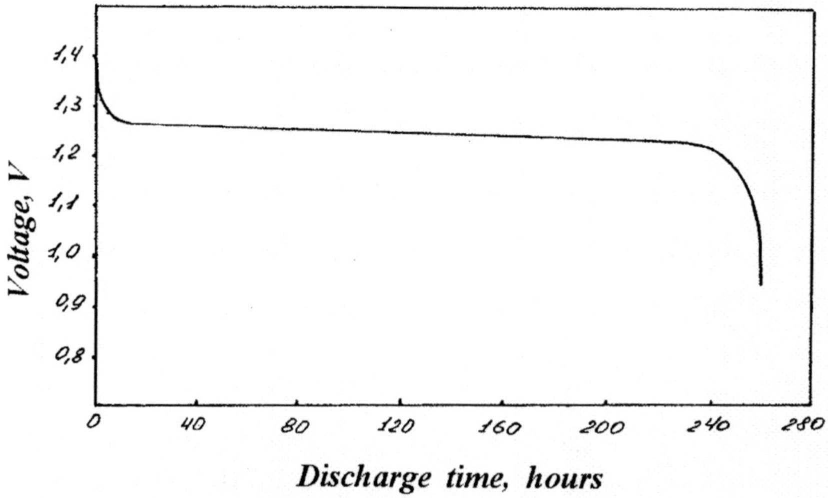

Figure 1. Discharge curve of a primary button air – zinc cell PR-44 at the resistance 625Ω and at 293 K.

Button cells consist of cathode and anode cans (used as the terminals), powdered zinc anode, containing gelled electrolyte and the corrosion inhibitor, separator with electrolyte, thin (0.5 mm) carbon cathode with catalyst and PTFE, waterproof gas-permeable (teflon) layer and air distribution layer for the even air assess over the cathode surface. Parameters of battery depend on the air–transfer rate, which is determined by quantity and diameters of air access holes or porosity of the gas-diffusion membrane. Air-zinc batteries at low rate (J=0,002-0,01C at the idle drain and J= 0,02-0,04C at the peak continuous current) have flat discharge curves (typical curve is shown by Figure 1).

With the air-transfer acceleration, current density increases (or the cathode polarization decreases), but at very high air velocity, electrode can be flooded (at high air humidity of more than 60% at 293K), or get too dry and fail (at low air humidity). Some batteries have the devices for the air-transfer control. Capacity of batteries decreases with temperature decrease (e.g. by a factor of two with corresponding temperature decrease from 294 to 273K), as well as with current density increase. Now the button batteries with capacity from 50 to 6300 mAh are being produced. Some of them were developed at MPEI(TU) [2, 20] and are manufactured at the battery plant 'Energiya", c. Eletz, Russia. Typical useful service-life of these batteries are 2-3 months, specific energy –200-400 Wh /kg and 400-800 Wh /L. Air-zinc batteries are the best power sources for hearing aids. Constant voltage of air-zinc batteries provides also best compatibility of these batteries with hearing aids from the medical point of view.

4. AIR–ALUMINIUM MECHANICALLY RECHARGE-ABLE BATTERIES WITH SALINE ELECTROLYTE

Air-Al batteries with saline electrolyte (NaCl solution) are simple in maintenance and low cost [4, 7, 19, 20].

4.1. Processes and Problems

Cathodic reaction in this battery chemistry is described by equation (1). Anodic reaction may be written in a form of reaction (10):

$$Al - 3e^- \rightarrow Al^{3+}. \qquad (10)$$

In saline solutions, Al is also exposed to corrosion, which can be decreased (down to 10^{-6} mm/h) by additions of small quantities of Ga, Sn and Pb to alumimum. Real processes in batteries are very complicated. The Al^{3+} ion is exposed to hydrolysis with decrease of pH (reaction (11)).

$$Al^{3+} + H_2O \leftrightarrow AlOH^{2+} + H^+. \qquad (11)$$

Chloride-ions participate in the anodic processes. Activation of Al by Cl^- and pH decrease accelerate anodic dissolution of Al. At the cathodic department concentration of OH^- increases, and as a result the complex ions, e.g. $[Al(OH)_4]^-$, are formed. In solution more complicated ions can be formed, for example: $[Al(OH)_xCl_y]$. With the chloride concentration increase, share of hydroxide ions in the complex ions decreases and pH solution grows, as it was shown in our experiments [7, 20]. The dependence of voltage of air-Al cell on NaCl concentration has a maximum at concentration of 15wt%. That type of dependence is connected mainly with influence of NaCl concentration on the anodic polarization. For our batteries we choose 15wt% solution of NaCl. Reaction products are most of all are in the colloidal condition. Experiments show that in non-stirred solution it is possible to receive an energy density of up to 100Ah/liter. Intensive stirring increases said performance parameter.

4.2. Battery Performance Parameters

In Moscow Power Engineering Institute (TU) portable air–aluminum batteries with saline electrolyte were developed [7, 18, and 20]. In our devices, the air electrodes consist of two layers. Diffusion layer contains PTFE, carbon black and metal screen; active layer consists of activated carbon and PTFE. At 293 K and the range of current density 2-25 mA/ cm^2 dependence of cathode potential E (in H-scale) upon current density J (Figure 2) may by written by the Tafel equation (12).

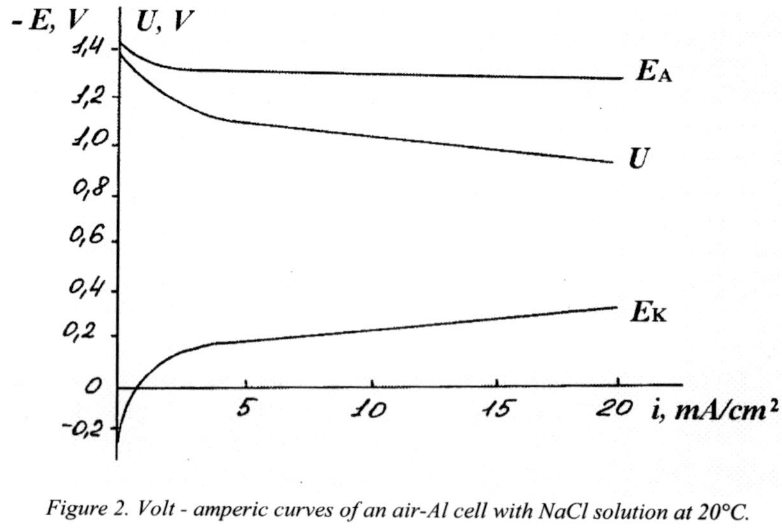

Figure 2. Volt - amperic curves of an air-Al cell with NaCl solution at 20°C.

$$E = 0.03 - 0.18 \lg J. \qquad (12)$$

Dependence of cell voltage U on current density (Figure 2) also is expressed by equation: U = 1.5-0.25 lgJ. The discharge voltage at a current density of 20 mA/cm^2 increases in the beginning and after that changes only a little (Figure 3).

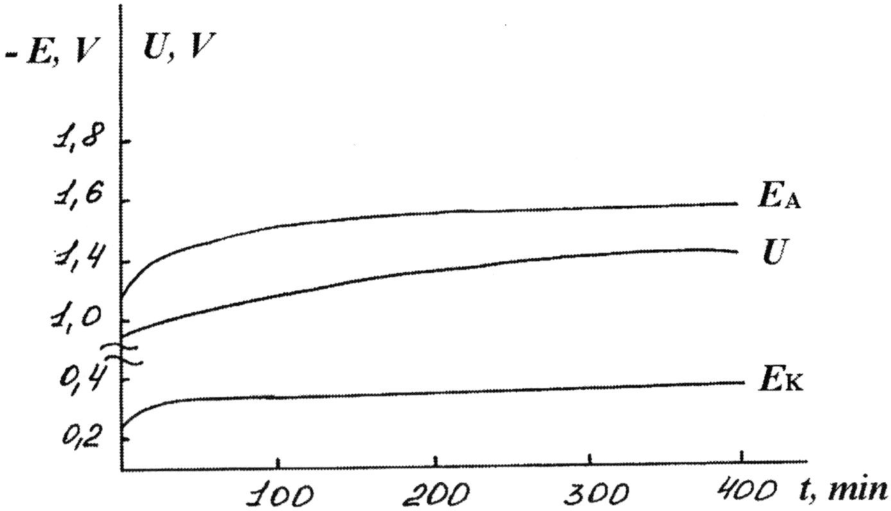

Figure 3. Discharge curves of air-Al cell with NaCl solution at 20 mA/cm^2 and 20^{0}C.

Coulombic efficiency of Al is 80-90 %. Batteries with capacities in the range of 50-1000 Ah (at repeated exchanges electrolyte after 6-8 hours of work and anodes after 50-100 hours of work) have been developed. Specific energy is 80-150 Wh/ kg without anode exchange and 200-400 Wh/kg with anode exchange. Batteries, having thermal insulation, can work in the low surrounding temperatures (down to 236K). Specific costs of these batteries is in the range of 100-120 $/kWh.

4.3. Application

The main fields of application for this battery chemistry include: portable TVs, radio receivers, lamps, flashlights, electric shavers, barrier lightning, instruments, batteries for portable rechargers, emergency power supplies, small refrigerators, power sources for tourists, hunters, geologists, shepherds and so on. Parameters of some batteries, assembled from the modules, are summarized in Table 4.

Table 4. Parameters of some air-Al batteries with saline electrolyte.

Voltage, V *Uocv, U(disch)*	**Dimensions, cm**	**Weight without electrolyte, kg**	**Application**
3.7 (2.4)	14x8x13	0,5(with insulation)	Lamp, 1.9W
6.1 (4.0)	10x10x10	0,4	Flashlight, 4W
9.1 (7.0)	16x12x10	0.9	Lamp, 7W
12.2 (8.0)	20x10x10	0.8	Tape-recorder, 12 W
18.3 (12.0)	30x10x10	1.2	Portable TV, 12W

5. ELECTRICALLY RECHARGEABLE AIR ELECTRODE

The most attractive batteries with air electrodes are the so-called electrically rechargeable batteries, because they have very high specific energy, simple maintenance, environmental compatibility and relatively low cost. But there are many problems of rechargeable air electrodes. Aluminum and magnesium can not recharge in water solution. Therefore zinc, iron and metal hydride can be used as rechargeable anode. Main attention is paid to the air-zinc batteries. Very complicated problem is development of the rechargeable air electrode.

Reversible reaction on air electrode may be written by equation (13):

$$O_2 + 2H_2O + 4e^- \leftrightarrow 4OH^-;\ E^0 = 0.401\text{V} \qquad (13)$$

Limiting steps of the oxygen reduction and the oxygen evolution are different. The catalysts of direct reaction are usually not the best catalysts of reverse reaction. Process of the oxygen evolution is accelerated by nickel and some oxides (spinels, RuO_2-TiO_2, IrO_2, RhO_3). That process proceeds at high anodic potentials, at which many catalysts and carriers are not stable. The graphitized cloth, the carbon black, treated at high temperature and nickel can be used as the catalyst carriers and current collectors.

Three approaches of solution of the reaction (13) catalyst problem are used: application of bifunctional catalysts, two catalysts in one layer and two catalysts in two layers.

The bifunctional catalysts, which were proposed are: silver, rhodium and iridium oxides [3], $Na_xPt_3O_4$ [21], pyrochlores $A_2B_2O_7$ (A=Pb, Bi; B=Ru, Ir) [22], perovskites –ABO_3 (A=La, Ca; B=Co, Ni, MnO_2) [3], pyropolymers [24, 25], NiO_xCoO_x.

Comparative activity parameters of catalysts and electrodes based on these catalysts are summarized in the Table 5.

Table 5. Differences in the potentials of oxygen evolution and reduction from air at the current density of 100 mA/cm² at 298K for some catalysts [3, 21-25].

Catalyst	ΔE, mV	Catalyst	ΔE, mV
$Pb_2Ru_{1.42}Pb_{0.58}O_7$	392	$LaNiO_3$	620
$Na_xPt_3O_4$	532	$La_{0.6}Ca_{0.4}CoO_3$	637
$Pb_2Ir_{2-x}Pb_xO_7$	557	Electrode with Ni-PP*	850
RhO_3	587	Electrode with PP*	950
NiO-CoO	1200	Electrode with Ag	910

Note for the Table: PP-pyropolymer.*

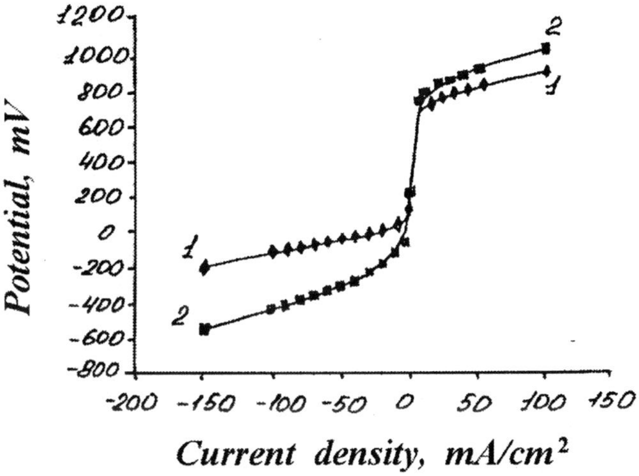

Figure 4. Polarization curves of the rechargeable air electrode at 293 K. Electrode with (1) and without (2) pyropolymer catalyst.

As it follows from Table 5, many catalysts contain metallic platinum. We have developed bi-layer porous hydrophobic air electrodes, which do not contain platinum metals, are active and can be cycled [24, 25] (Figures 4-6). These bifunctional catalysts are pyrolized Co – macrocyclic compounds. Said catalyst has high catalytic activity for the oxygen reduction and also features acceptable stability, however its activity for the oxygen evolution is not high enough.

Our recent our works show that even higher activity and stability can be demonstrated by the three-layer electrodes with nickel layer, active in the oxygen evolution, middle layer with catalyst, active in the oxygen reduction (MnO_2, pyropolymer or a perovskite), and a diffusion (waterproof) layer, needed for the air transfer. The potential difference of oxygen evolution and reduction on such electrodes was 800-900mV (Table 5). Haas with co-workers developed rechargeable electrode with a bifunctional catalyst $La_{0.6}Ca_{0.4}CoO_3$ [23]. The potential differences of oxygen evolution and reduction at current densities of 100 mA/cm^2 were 900-955 mV. Electrodes are believed to be stable and permit to perform more than 500 charge-discharge cycles.

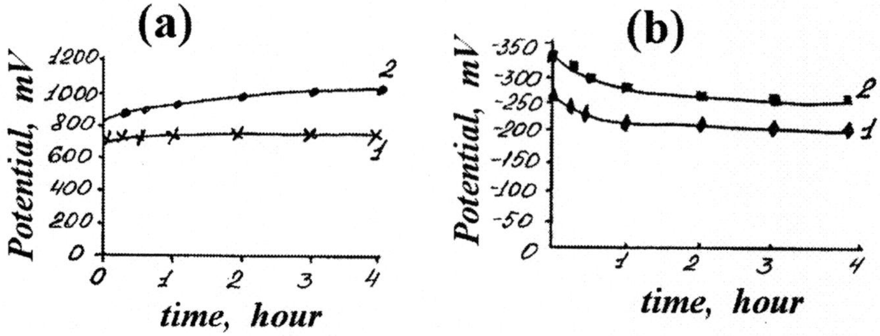

Figure 5. Charging and discharging curves of an air electrode with (1) and without (2) pyroplymer catalyst at 293 K.
The current density (mA/cm^2) is 50 (charging) and 100 (discharging), respectively.

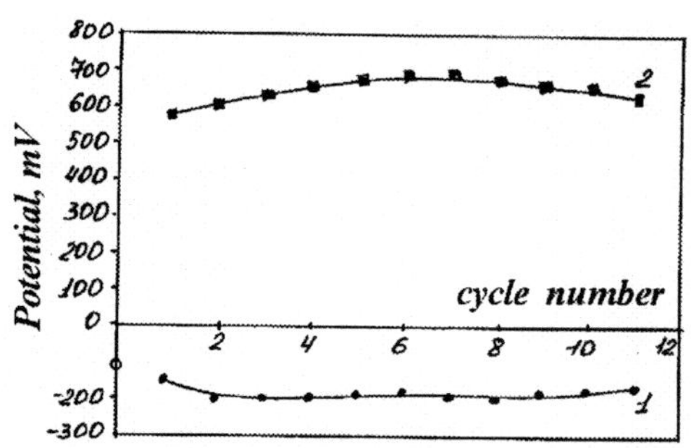

Figure 6. Change of potentials of the electrode with pyropolymer upon cycling at the discharge (1) and at the discharge (2) at 293K, with the current density of 50mA/cm^2.

6. CONCLUSIONS

Air-Zn primary and air–Al mechanically rechargeable batteries with carbonaceous air electrodes were designed and tested. Air–Zn batteries are mostly used in the hearing aids, while air-Al batteries find wider range of applications in the portable devices. Electrically rechargeable air electrodes with pyropolymer catalyst have been developed.

REFERENCES

1. Bagotzky V.S., Skundin A.M., *Batteries*. Moscow: Energoatomizdat. 1981, (Russian).
2. Korovin N.V. "Advanced half-gas systems for rechargeable batteries". In *New Promising Electrochemical Systems for Rechargeable Batteries (NPESRB)*, V.Z. Barsukov, F. Beck, ed. Dordrecht: Kluver Academic Publisher. 1996; 171-180.
3. Jorissen L., Garche J. "Application of bifunctional air electrodes", In *(NPESRB)*, V.Z. Barsukov, F. Beck, ed. Dordrecht: Kluver Academic Publisher. 1996; 159-169.
4. Korovin N.V., *New Batteries*. Moscow: Energiya.1978. (Russian).
5. Blurton K.F., Sammells A.F. Metal/air batteries: their status and potential – a review. J. Power Sources. 1979; 4: 263.
6. Budevsky E., Iliev I., Varbev R., Zinc – Luft Elemente mit hohen Energievorrat-Probleme und Eisatzmoeglichkaiten, Z. Phys. Chem. Leipzig, 1980; 261: 716-21.
7. Korovin N.V., Kleimenov B.V., Dyachkov E.V., Voligova I.V. Aluminium-air batteries for portable devices.Ext Abstr.1997 Joint. Intern. Meeting ES and ISE. Paris.1997.P. 88.
8. Gamburtzev S., Velev O.A., Danin R., Srinivasan S., Appleby A.J. "Performance of an improved design of metal hydride/air rechargeable cell". In *Batteries for portable application and electric vehicles*. C.Holmes, A.Landgrebe ed. Pennington: Electroch. Soc, 1997, 726-33.
9. Korovin N.V. *Electrochemical Power Sources*. Moscow: Energoatomizdat. 1991.(Russian).
10. Wiesener K., Garche J., Schneider W. *Electrochemische Stromquellen*. Berlin: Akademie Verlag. 1981.
11. Hoar J.P. *The electrochemistry of oxygen*. N.Y: Interscience Publ. 1968.
12. Korovin N.V., Kasatкin E.V. Electrocatalysts of electrochemical devices. Russian Electrochemistry. 1993; 39: 448.
13. Iliev I., Gamburzev S., Kaisheva A., Mrha J. Influence of mass transport on the peformance of carbon gas-diffusion electrode in alkaline solutions. J. Appl. Electrochemistry. 1975; 5: 291-97.
14. Barsukov V., Khomenko V., Chivikov Santonenko P. Physico-chemical base of development of the air-meral batteries with the catalysts on anyline base. Electrochemical Power Sources. (Russian). 2001; 1: 24 – 30.
15. Chismadjev Yu.A., Markin V.S., Tarasevich M.R., Chirkov Yu.G. *Macrokinetics of processes in porous electrodes*. Moscow: Nauka, 1971 (Russian).
16. Korovin N.V., Luzhin V.K. Knudsen diffusion in oxygen and air electrodes. Russian Electrochemistry. 1979; 15: 360 - 62.
17. Korovin N.V., Kicheev A.G. Luzhin V.K. Influence of the inert components on mass-transfer in the gas phase of porous electrode. Russian Electrochemistry. 1972;.8: 146 –49.
18. Korovin N., Kleimenov B., Agaphonov N., Guryanov M. Air-metal batteries Ext. Abstr. 46th ISE Meeting; 1995; Xiamen, China. 1995. 1-5-12.
19. Hamlen R.P. "Metal/air batteries." In *Handbook of batteries,* D. Linden, ed. N.Y: McGrow-Hill Inc., 1995, 38.1-38.45.
20. Voligova I.V., Korovin N.V., Kleimenov B.V., Dyachkov E.V. " Air-Al batteries with saline electrolyte." In *Fundamental problems of electrochemical power sources (FPEPS)*, I Kasarinov, ed. Saratov: SGU, 1999, 169.
21. Swette L., Kackley N., McCatty S.A. Oxygen electrodes for rechargeable alkaline fuel cells, J. Power Sources. 1991; 36: 323 - 39.
22. Horowitz H.S., Longo J.M., Horowitz H.H. Oxygen electrocatalysis on some oxide pyrochlores. J. Electrochem. Soc. 1983; 130: 1851- 59.
23. Muller S., Striebel K., Haas O. Electrochimica Acta. 1994; 39: 1661.
24. Korovin N.V., Voligova I.V., Kleimenov B.V. Bifunctional catalysts for the air electrodes. Ext. Abstr. 50th ISE Meeting.; 1999; Pavia: 1999, 1.19.
25. Korovin N.V. Batteries with air electrodes. Electrochemical power sources. 2001; 1: 16-23. (Russian).

CHAPTER 3:

CARBON ANODES FOR LITHIUM-ION BATTERIES

Chapter 3: Subject Overview

Currently, lithium-ion and lithium-ion polymer cells use graphitic carbons as the active material of the negative electrode. The phenomenal rate of growth of the lithium-ion battery chemistry has resulted in a steadily increasing demand for anode grade carbons. Thus, for reference, the actual worldwide sales volumes of graphitic carbon powders into this market amounted to approximately 3,000 tons in 2001, propelled up to 6,000 tons in 2003, and well exceeded 11,000 tons in 2004. Due to fierce competition among carbon manufacturers and also the lithium-ion battery becoming commodity chemistry, the prices for anode grade carbons declined from the levels of $50/kg to approximate levels of $20/kg within the same time frame. Still, one may estimate that as of the end of 2004, the market of anode materials for lithium-ion batteries amounted to a minimum of USD 220 million per year. A collection of 11 papers in this chapter seeks to address latest issues associated with development, synthesis, characterization and use of new advanced carbonaceous materials for the rapidly growing segment of electrochemical energy storage market in question.

This chapter begins with an overview paper by T. Takamura and R. Brodd. The authors broadly discuss various uses of carbon materials in the main types of power sources. Besides heavy focus on carbon materials for lithium-ion batteries, authors also review a recent work by U.S. manufacturers aimed at incorporation of expanded graphite into the cathodes of alkaline primary batteries. The role of irreversible capacity loss in the design of lithium-ion grade carbons was also discussed. Placing surface coatings and control of morphology of the graphite assists in minimizing the irreversible capacity loss. In addition, the editors wish to comment that spherodizing of lithium-ion grade graphite is a necessary, but not sufficient step, when designing a carbon with low irreversible capacity loss.

The second paper by H.J. Santner *et al.* is a derivative of an overview presentation given during the NATO-CARWC by Professor M. Winter of Graz Technical University, Austria. This interesting work summarizes over 10 years of research and development investigations by the University in the search for the most efficient electrolyte additives and solvents, which are the working environments for graphite anodes when in lithium-ion batteries. This fundamental paper could be quite useful, in particular, to those battery developers, who seek to understand the phenomena of formation of protective layers on graphite. The authors clearly differentiate between the types of layers, which are being created on graphite during the process of formation of lithium-ion batteries.

In the third paper, M. Walkowiak *et al.* report on findings of Central laboratory of batteries and Cells (CLAiO) in Poland, as related to the electrochemical performance of spherodized purified natural graphite and boron-doped carbons in lithium-ion batteries. While it is noteworthy that these authors are still working on debugging of their electrochemical devices, the fact that full cells are being used to qualify new materials at CLAiO deserves a complement, as it is less common that full cells are used for such purposes, with most academia scientists doing individual electrode studies, often being unaware of the real requirements of the battery industry.

The article by Professor D. Aurbach *et al.* offers a comprehensive analysis of the reasons for why the synthetic and natural graphite, and mesocarbon microbeads fail in PC-containing organic electrolytes (PC is a highly attractive solvent for lithium-ion battery technology due to its low cost and inherent ability to work at temperatures below –40 °C). These authors (from the Bar-Ilan University in Israel) suggest that the failure mechanism of graphite/carbon electrodes in PC-based solutions is due to particle cracking, followed by their electrical isolation by surface films, rather than due to complete graphite exfoliation due to co-intercalation of PC molecules with Li^+ into graphite lattices. Though not widely accepted by the scientific community, the concept appears to be very innovative. Authors report to have collected lots of evidence, proving the above theory to be valid. Among other advanced analytical methods used in their work, readers are encouraged to review the presented data utilizing atomic force microscopy (AFM).

One of the obvious values of paper by F. Henry *et al.* is the fact that it comes from a U.S. commercial manufacturer of the industrial carbon and graphite. The reader will get an unusual insight into the evolution in understanding of what were the properties, which needed to be tailored in order to create a graphite line suitable for lithium-ion battery application. It is noteworthy that reversible and irreversible capacities, which are being usually looked at in the academia as key qualifying properties, carry a secondary role, when compared to the practical needs of having excellent adhesion to the copper substrates, increased packing densities and acceptable abuse tolerance in various lithium-ion battery environments.

A fundamental work by Professor F. Beguin *et al.* addresses the mechanisms of reversible and irreversible intercalation and BET adsorption in lithium-ion grade carbons. The authors discuss, among other phenomena, an interesting concept that values of the active surface area (ASA) on carbon correlate with irreversible capacity loss. This is a departure from traditionally accepted belief that regular BET single point values of carbon are in proportional relationship with the irreversible capacity loss. It is noteworthy that editors also have seen the later theory to be highly inaccurate, and thus, consider the approach taken by authors as valuable. Readers should be aware that the method of determining the ASA currently is being actively promoted by Micromeritics. So far, it has received limited acceptance in the electrochemical labs dealing with fuel cell research, but the method remains relatively unknown to the battery research community.

The seventh paper in this chapter (R. Yazami *et al.*) has focused on fundamental aspects of understanding the transition between graphite intercalation stages #1 and #2. Authors note peculiar "anomalies", apparently not described in the literature up until now (even though the subject of investigation of graphite staging upon its lithiation has been around for decades). The paper offers a fundamentally new look into the stage transition phenomena.

In the article by the renowned Accumulator Research and Design Institute "Istochnik" of St. Petersburg, Russia, basic performance of various types of carbon materials in the coin and pouch-type lithium-ion cells is described. On a side note, detailed descriptions of various component suppliers furnished by the authors, suggest that Russia is slowly but surely establishing an internal supply chain of anode, cathode, separator, current collector materials and components for the needs of the lithium-ion technology, which is being actively developed internally as well. Authors reported performance of the following carbon materials: pyrolytic carbon; a version of nanotubes; expanded graphite; "spectrally pure" graphite; special "RIECP graphite" (derived from natural flake) – all are Russian-made grades. Authors also tested thermally purified spherodized flake graphite SLA1020 and surface coated spheroidal flake graphite SLC1115 from Superior Graphite (USA). According to conclusions by the authors, the best for Li^+ intercalation are: SLA, SLC and "Spectral Pure"carbon materials. Editors, however, have no doubts, that many inquiries will be made to "Istochnik" after publishing their comments from testing of Russia's Astrin Co. nanotubes. What could not be effectively achieved by other "classic" nanotube developers thus far, apparently could be achieved with a particular grade of Russia-made nanotubes. The authors report observing good stability of these nanomaterials under conditions of prolonged cycling in lithium-ion cells. The editors wish to comment that the authors did not discuss the nature of these nanotubes, but it may be speculated that such nanotubes might be rather coarse, and partially-to-well-graphitized nanoparticles of untraditional nature.

In the next paper by R. Yazami, I. Goncharova and N. Plakhotnik, a joint work between the U.S./France and Ukraine researchers is presented. Authors investigate mechanisms of BF_4^- and PF_6^- anion intercalation into the single walled carbon nanotubes (SWCNT), produced by pulsed laser ablation technique. Authors conclude that BF_4^- and PF_6^- anions can intercalate into SWCNT structures, and that this process has some reversibility. It is noteworthy that the only electrochemical method used by the authors to arrive with the above conclusion was a method of cyclic voltametry (CV). It is known that CV is a highly sensitive method when it comes to studying the surface processes occurring on working electrodes, while the limitation of the method is its inability to detect what is happening in the bulk electrode. In the mean time, intercalation is a process, which is happening in the bulk of carbon. Intercalation processes are also well studied by the technique of galvanostatic cycling, a constant current technique that was not used in the present paper.

A fundamental review paper by Professor A. Churikov *et al.* of Saratov State University, Russia focuses on cons and pros of lithium intercalation into hard and soft carbons, as related to application of these materials as lithium-ion battery anodes. Besides offering what the editors consider a good tutorial on various forms of carbon used for intercalation, the authors arrive with the conclusion that an optimum carbon for lithium-ion battery should be a physical blend of soft and hard carbons. While being an interesting concept, editors wish to note that North American and Pacific Rim lithium-ion battery industries already use an enhanced version of this concept, e.g. soft or hard carbon coated graphitic carbons. Authors also report observing a synergistic capacity increase effect from combining two types of carbons. As shown by Figure 2 in the subject paper, the resultant capacity, surprisingly goes above 372 mAh/g, which is the theoretical limit for graphite formulated as LiC_6.

A paper by Dr. J. Liu *et al.* of Argonne National Laboratory in the U.S. describes some of the analytical procedures used to qualify candidate carbon materials for their use as anodes of emerging high power lithium-ion batteries for hybrid electric vehicle (HEV) application. This work in the United States is being supported by the U.S. Department of Energy's FreedomCar program, which, besides conducting the development work, enables cooperation between all members of the battery production chain with three major car manufacturers. The particular paper describes performance of thermally purified, spheroidal natural graphite, SL-20, and also, that of surface coated by amorphous carbon, thermally purified, spherodized natural graphite SLC1015 (both grades are the products of Superior Graphite Co., a Chicago, IL, USA graphite/carbon manufacturer). Carbon coating and reduction of particle size were seen to improve high rate, pulse performance and abuse tolerance of natural graphite to the extent of making it a suitable candidate material for the HEV battery application.

We hope you will find these papers interesting and useful in your work.

CARBONACEOUS MATERIALS FOR BATTERIES

Tsutomu Takamura[1] and Ralph J. Brodd[2*]

[1]*Department of Applied Chemistry, Harbin Institute of Technology, Harbin, 150001, China*
[2]*Broddarp of Nevada, Inc., 2161 Fountain Springs Dr., Henderson NV USA*

1. INTRODUCTION

Practically every battery system uses carbon in one form or another. The purity, morphology and physical form are very important factors in its effective use in all these applications. Its use in lithium-ion batteries (Li-Ion), fuel cells and other battery systems has been reviewed previously [1 – 8]. Two recent applications in alkaline cells and Li-Ion cells will be discussed in more detail. Table 1 contains a partial listing of the use of carbon materials in batteries that stretch across a wide spectrum of battery technologies and materials. Materials stretch from bituminous materials used to seal carbon-zinc and lead acid batteries to synthetic graphites used as active materials in lithium ion cells.

Table 1. Partial Listing of the Uses of Carbon and carbon materials in Batteries.

Battery System	Use/Function
Carbon-Zinc	Sealant, cathode conductive carbon black matrix
Alkaline Zn-MnO_2	Sealant, cathode conductive graphite matrix
Zinc-Air	Sealant, cathode catalyst support and current collector
Li- CFx	Cathode active mass
Li-SO_2	Cathode polymer bonded conductive matrix
Li-Ion	Anode active mass, cathode conductive diluent
Lead Acid	Anode Conductive diluent
Ultracapacitors	Anode and cathode active matrix
Fuel Cells	Cathode and anode catalyst support Intercell connectors

* Corresponding author. E-mail: dbrodd@ix.netcom.com

I.V. Barsukov et al. (eds.), New Carbon Based Materials for Electrochemical Energy Storage Systems, 157–169.

It is estimated that over 14 billion alkaline $Zn\text{-}MnO_2$ cells and 30 billion carbon-zinc cells were made worldwide in 2003. This compares to the about 500 million lead acid automotive batteries that were made in the same period.

2. ALKALINE CELLS

Today, the alkaline $Zn\text{-}MnO_2$ cells set the standard for performance for primary cells. The need for higher performance alkaline cells was met in part by the use of "expanded" graphite from Superior Graphite. This type of graphite is produced from natural graphite by treatment in nitric acid followed by heating to cause exfoliation of the graphite. Photomicrographs of the expanded graphite and natural graphite are shown in Figure 1. These expanded particles mixed with MnO_2 form a matrix with higher conductivity than that produced by from the same formulation using natural graphite.

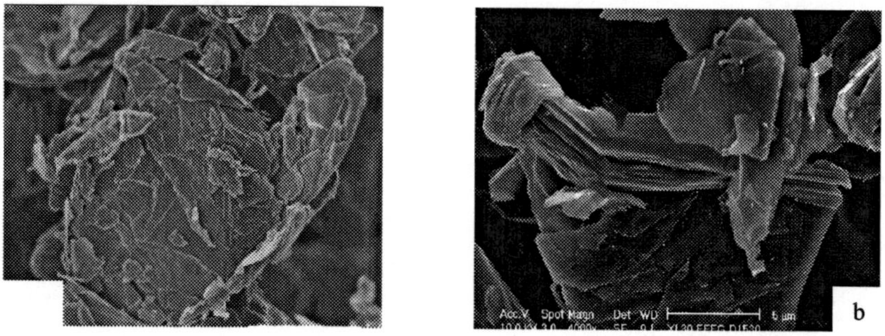

a b

Figure 1. Purified expanded (a) and natural (b) graphite. (Courtesy of Superior Graphite).

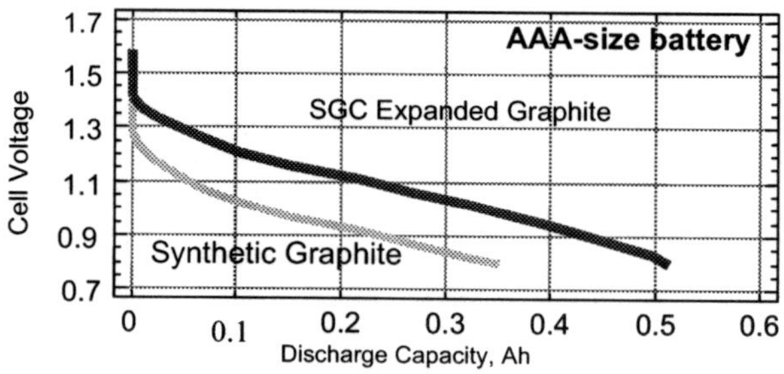

Figure 2. Continuous Discharge of AAA-Size Alkaline Cells at 500 mA. (Courtesy of Superior Graphite).

This permits a reformulation of the cathode to include more active material and a lower internal resistance, as noted in Figure 2. The resultant lower internal resistance and higher active mass content produced a 30% increase in cell performance at high rate discharges.

The proper mixing of the graphite with the MnO_2 is critical for high rate performance. It is necessary to "put work into the mix" to uniformly coat the manganese dioxide particles with a thin layer of graphite. This provides a low resistance path from the manganese to the cell terminals to minimize the internal resistance of the cell. A similar effect can be expected if used in cathode formulations for Li-Ion cells.

2. LI-ION CELLS

The mobile electronic devices, including the 3rd generation of portable phones, portable personal computers and personal digital assistants (PDAs) have an insatiable appetite for "evermore" power and energy (capacity). The power sources of choice for these devices are Li-Ion batteries. There are several excellent reviews of recent work in this area [9 – 11]. The working voltage of Li-Ion batteries is about 3.5 volts, which is over two times that of conventional portable batteries, and yields batteries that are capable of delivering very high specific energy density. Li-ion has become the system of choice for portable electronic devices and shows promise to become the next generation universal rechargeable battery system.

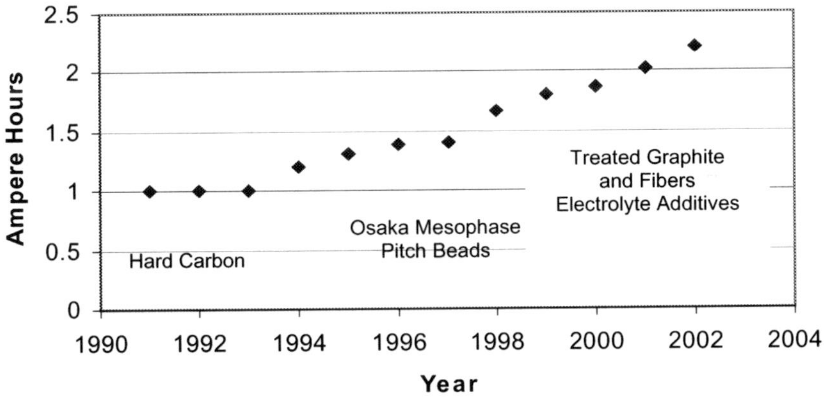

Figure 3. Capacity improvement in the ICR 18650 cell and the anode composition.

The capacity of the system has more than doubled since it was introduced in 1991. The improvements in performance were due in large part from the use of new carbon anode materials, as noted in Figure 3. Over 850 million Li-Ion cells were manufactured worldwide in 2002.

The Li-Ion system was developed to eliminate problems of lithium metal deposition. On charge, lithium metal electrodes deposit moss-like or dendrite-like metallic lithium on the surface of the metal anode. Once such metallic lithium is deposited, the battery is vulnerable to internal shorting, which may cause dangerous thermal run away. The use of carbonaceous material as the anode active material can completely prevent such dangerous phenomenon. Carbon materials can intercalate lithium into their structure (up to LiC_6). The intercalation reaction is very reversible and the intercalated carbons have a potential about 50mV from the lithium metal potential. As a result, no lithium metal is found in the Li-Ion cell. The electrochemical reactions at the surface insert the lithium atoms formed at the electrode surface directly into the carbon anode matrix (Li insertion). There is no lithium metal, only lithium ions in the cell (this is the reason why Li-Ion batteries are named). Therefore, carbonaceous material is the key material for Li-Ion batteries. Carbonaceous anode materials are the key to their ever-increasing capacity. No other proposed anode material has proven to perform as well. The carbon materials have demonstrated lower initial irreversible capacities, higher cycle-ability and faster mobility of Li in the solid phase.

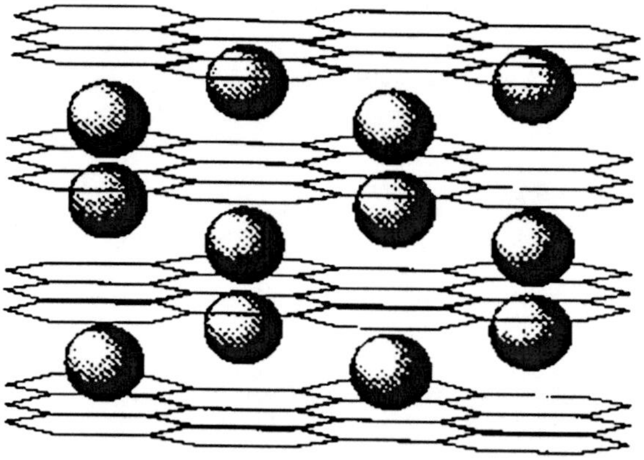

Figure 4. Depiction of lithium intercalated into the carbon/graphite lattice.

The heart of the carbon anode utility is its structure, shown in Figure 4. Lithium ions in the organic electrolyte approach the surface of the graphite where they exchange an electron and enter the carbon structure (intercalate) along the basal plane. The lithium atoms settle into the center of the hexagonal carbon structure in the amount of up to one lithium to six carbon atoms (LiC_6). As the lithium content of the graphite increases, the graphite approaches the lithium potential. With graphite, there is a small (5%) volume change as the electrode reaches full charge. As the electrode approaches the lithium potential a thin film forms on the surface from the reaction with the organic solvent to form a protective passive layer called the Solid Electrolyte Interphase (SEI) layer. This reaction uses lithium from the cathode and is called "first cycle loss." It is irreversible and related to the surface and type of carbon/graphite. Electrolyte additives can control the losses related to the formation of the SEI. In a well formed film, there is no capacity loss associated with solvent reduction and SEI layer formation after the first cycle.

The reaction at the anode in Li-Ion cells is given in Equation 1. During charge the lithium ions approach the surface of the carbon where they accept an electron and enter the lattice. On discharge, the opposite reaction occurs. The electrochemical reaction is thought to occur on the edge planes and not the basal plane of the carbon/graphite particles.

$$Li^+ + e^- = Li \text{ (in C)} \qquad (1)$$

3. KINDS OF CARBONACEOUS MATERIALS, WHICH CAN BE USED AS ANODES IN LI-ION CELLS

These kinds of carbon materials include:

1) Highly Oriented Pyrolytic Graphite (HOPG)
2) Natural graphite
3) Artificial graphites
4) Hard carbons
5) Mesophase low temperature fibers and MCMB carbons

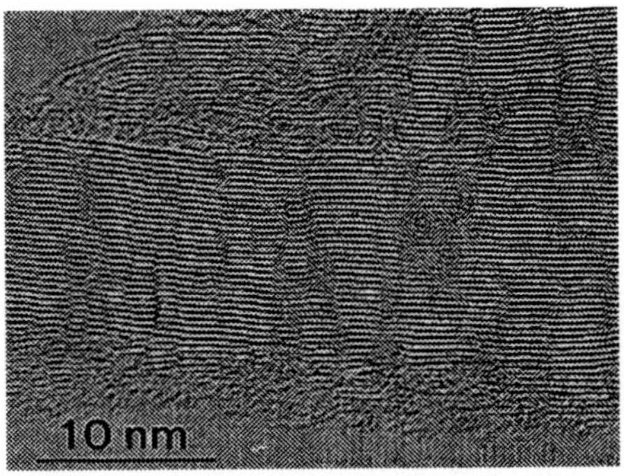

Figure 5. TEM photograph of natural graphite [12].

While it is possible to prepare square centimeter, and larger, pieces of HOPG. the performance of HOPG is very similar to that of natural graphite but much more expensive and it will not be considered further here. A TEM photograph of natural graphite is shown in Figure 5 in which the movement of lithium in the carbon skeleton is very rapid along the horizontal (basal plane) direction. The discharge curve is very flat and the charge/discharge reversibility after the first cycle is essentially 100 % (Figure 6). The first cycle loss for typical 5 m^2/g natural graphite is about 10% to 12%. The high power rate, however, is restricted because the graphite tends to form leaf-like platelets with very little edge plane exposed and a long diffusion path for the lithium reaction.

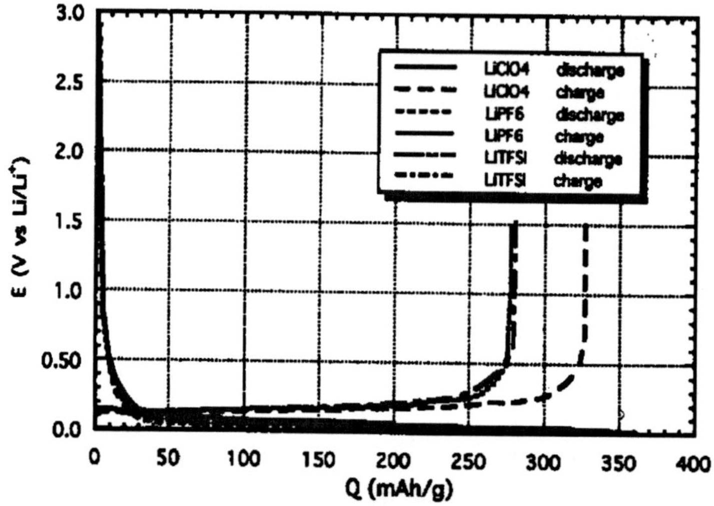

Figure 6. Low rate charge discharge performance of natural graphite electrode [13].

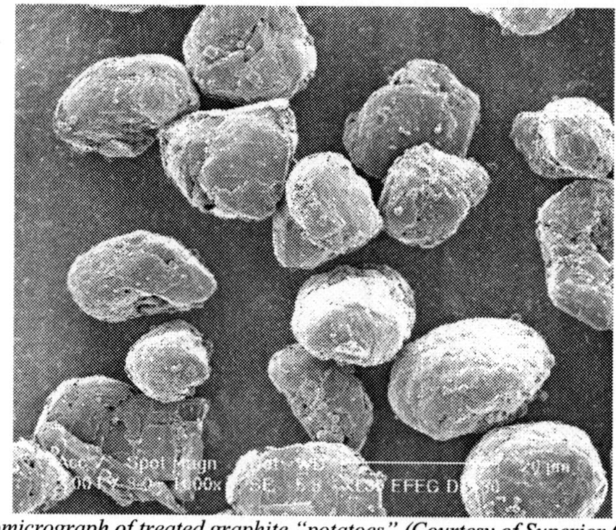

Figure 7. Photomicrograph of treated graphite "potatoes" (Courtesy of Superior Graphite).

When the natural graphite powders are coated on the current collector, they tend to be aligned to be parallel to the collector surface, which raises the ohmic resistance perpendicular to the electrode surface, which is important for the high rate charge/discharge. Surface treated natural graphites, called potatoes, shown in Figure 7 can avoid these surface orientation related problems to yield a high capacity material with a low first cycle loss of about 5%. Since the lithium transfer is through the edge plane to the electrolyte, the electrode reactions may be hindered by the structure of the SEI layer at the electrolyte-graphite interface. If the graphite particles can align with the basal plane perpendicular to the current collector, the reaction is facilitated. Graphite materials presently used in the practical commercial batteries include mesophase carbon beads and fibers (MCF), artificial graphites having the potato-like shape and surface treated natural graphite.

3.1. Mesophase Carbon Fibers

Petoca, Ltd. prepared mesophase carbon fibers by melting the mesophase pitch. The fiber is formed by blowing molten pitch through tiny holes in a die into a heated chamber. The fibers are then heated at about 300°C in an atmosphere of low-pressure oxygen and water vapor, followed by heating at 500°C in nitrogen to produce a thin hard protecting skin. Finally, heating up 3000°C produces a well-graphitized bead or fiber. The formed graphite fiber has a radial structure of the crystal as shown in Figure 8 (a) and (b). The fiber is cut into short lengths before the graphitization.

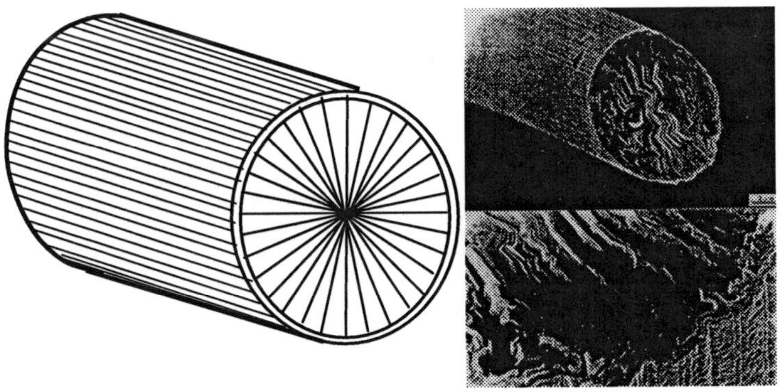

(a) **(b)**

Figure 8. (a) Schematic structure of carbon fibers and (b) an SEM of a Melblon 3100 fiber showing the fiber and internal structure of a fracture.

This procedure produces fibers, such as Melblon 3100, which have 1) a radial crystal orientation which can accept Li ions from all the side wall areas and 2) the side wall is covered with a porous thin graphen skin which protects the fiber from the shape change during the Li intercalation and de-intercalation (swelling and shrinkage). Fibers as thin as 7 μm can be prepared to facilitate their use in electrode coating operations. The anode discharge curves are shown in Figure 9, where (a) is for the normal product and (b) is for the boron-doped product. [14]. In addition, surface coating of the fibers with silver, zinc, etc., improves the transfer across the interphase boundary to yield very good high rate performance [15].

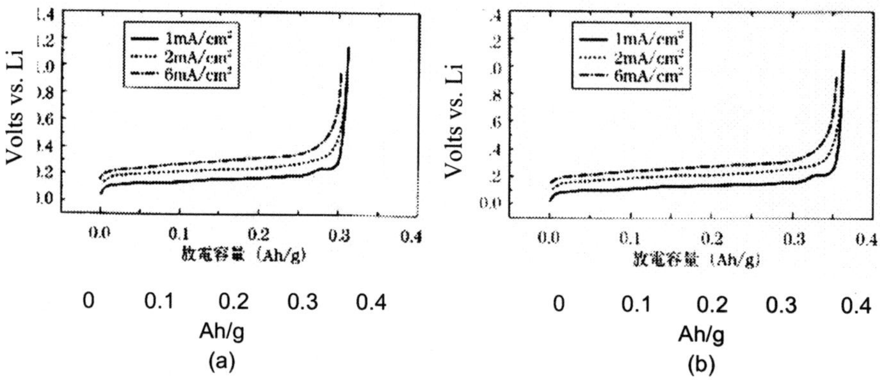

Figure 9. Discharge curves of (a) normal and (b) boron doped Melblon carbon fiber at 1, 2, and 6 mA/cm^2.

Toshiba, among others, have succeeded to produce a very thin (as thin as 1 mm thick for the 7 mm batteries that have superior performance, as compared to conventional Li-Ion batteries. The success is based on the use of the Petoca boron doped MCF (BMCF), new electrolytes containing $LiBF_4$ dissolved in EC/MEC/γ-BL, and an Al laminate plastic sheet container. The low temperature performance is superior to that of the conventional Li-ion battery. The carbon fiber is electrically highly conductive but it is still strongly recommended to coat the milled carbon fiber with an appropriate amount of conductive additives, such as natural graphite powder or Timcal synthetic graphite powder. The particle size of the powder should be 5μm, or less, to maintain good contact with the fibers.

Acetylene/carbon black is also quite effective but has an initial irreversible capacity that cannot be ignored. The amount of irreversible loss for acetylene black component ranges up to 20%. The particle size of conductive additives is recommended to be less than 5 microns. The addition is very effective to improve to improve: 1) cycle life, 2) high power capability, and 3) the initial charge efficiency (reduce the initial irreversible capacity loss). The use of the expanded graphite can serve the same purpose without the high irreversible loss associated with carbon blacks. The coating know-how is very important in producing high performance electrode structures. The know-how is not restricted to the slurry formulation but the control of coating machine operations are equally, or more, important for thickness control and particle orientation.

3.2. MCMB (Mesophase Carbon Micro Beads)

When the mesophase pitch is heated, tiny spherical beads having a radial orientation of the graphite embryo form and grows in the mother pitch. When the size of the beads in the pitch reaches an appropriate size (10-30 μm), the pitch is cooled down and the beads can be collected by dissolving the pitch matrix with an appropriate solvent. The beads are then heat-treated in an oxygen/water gas mixture to produce the protective skin, as for the MCF, and then graphitized, as depicted in Figure 10.

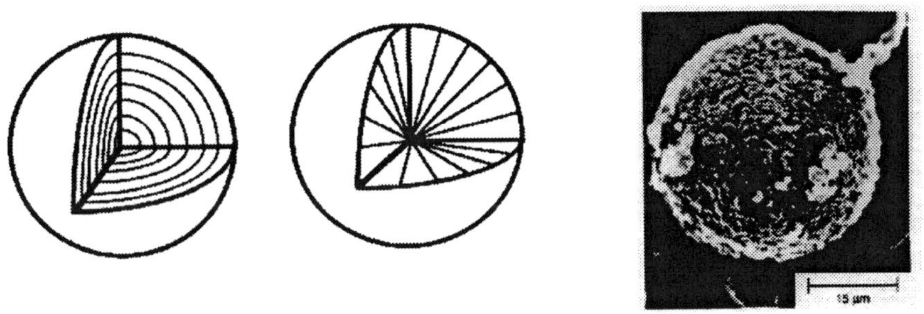

Figure 10. Depiction of the (a) Onion structure, the (b) Radial structure, and (c) Actual product MCMB, [16].

MCMB produced by Osaka Gas Co. has very good performance and is easily coated on the Li-Ion anode current collector (Cu). These materials are used widely throughout the world. The price is expensive, and cannot be reduced. This production process is inherently expensive due to the large volume of solvent required to be wash out and recover the beads from the pitch matrix.

3.3. Hard Carbons

When thermosetting resins, such as, phenol resin, epoxy resin, cellulose, sugars or proteins are heated in an inert gas or vacuum, the materials decompose to release water, and other gases to form carbon. The carbon particles are inherently joined together in a complicated structure and cannot reorient to form graphite. The resulting carbon is very hard and has no preferred crystallite orientation but has a randomly oriented micro-structure of graphite and voids or cavities. Therefore, the hard carbons have a higher capacity than that of graphite.

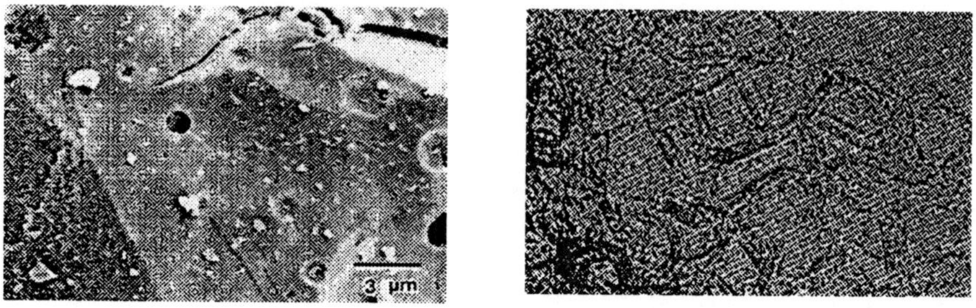

a b

Figure 11. (a) Photomicrograph of Kureha Carbotron P, (b) TEM of 2800 C heat-treated Sugar carbon.

These materials have an advantage that they do not decompose propylene carbonate (PC) so that the low cost PC electrolytes, having very good performance, can be used. In contrast, PC tends to decompose vigorously when it is in contact with lithiated graphite. These are two advantageous points of hard carbon as compared with graphite. In addition, hard carbons have very little change in volume during the charge/discharge reaction. However, these carbons have several demerits: 1) they have very poor high rate performance related to the slow diffusion of lithium in the internal carbon structure; 2) the discharge curves is not flat but gradually slopes down to the end-of-charge voltage; 3) large initial irreversible capacity, as noted in Figure 11, of 20% or more. A comparison of the three materials is given in Table 2.

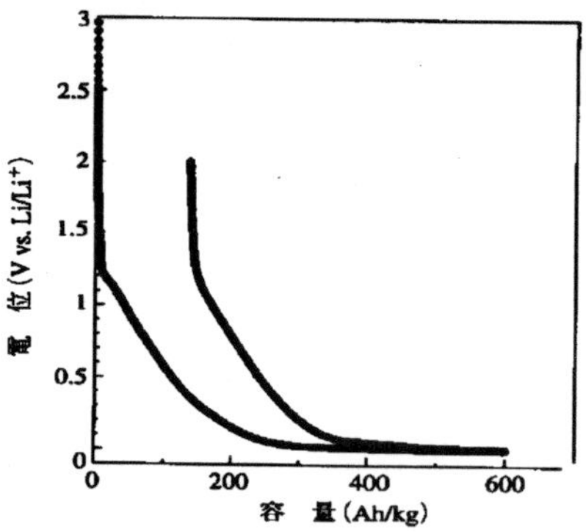

Ah/kg

Figure 12. Charge and discharge curves of Carbotron P.

Table 2. Comparison of the anode performances of three carbonaceous materials.

Materials / Performance	Natural graphite	Artificial graphite	Hard carbons
Capacity	**	**	***
High rate capability	—	***	*
High working voltage	***	***	*
Flatness of discharge	***	***	□
Price	***	*	**
Low temp. perform.	**	***	**
Use PC based electrolytes	— —	— —	**
Electric conductivity	***	***	**

Sony has stuck to their hard carbon for the anode because of its long cycle life and still makes a line of hard carbon anode cells. Sony has moved to graphite because they have recognized that hard carbon cannot be improved to deliver the high power and high capacity requirements of current portable electronic devices. Some hard carbons have demonstrated very high capacities but are considered unusable in their present form because of their high initial irreversible capacity and slow rates of Li doping and undoping (intercalation). It should be noted that recently electrolyte additives have been developed that allow the use of natural graphite in low cost propylene carbonate (PC) electrolytes (17, 18). These materials control the initial SEI layer formation, improve storage and cycle life.

Li-Ion batteries are poised to enter a new era of high performance. Figure 13 depicts the segmentation into two market segments. One path will continue the drive for higher performance but will require new, more expensive materials [19]. The other path will emphasize low cost materials with essentially equivalent performance to serve new developing markets such as power tools and hybrid vehicles. This path puts the Li-Ion system in a position to challenge the Ni-Cd and Ni-MH for power tools and the hybrid automotive applications. It also raises the option to compete with primary alkaline Zn-MnO_2 cell for low-end portable electronic applications. One key to meeting these challenges is the availability of low cost electrolyte and electrode materials. Solution additives have lowered the first cycle loss and opened the possibility to use lower cost electrolyte compositions and eliminate the use of $LiPF_6$ solutes. The other key is the increased safety realized by the use of low cost, stable cathode materials.

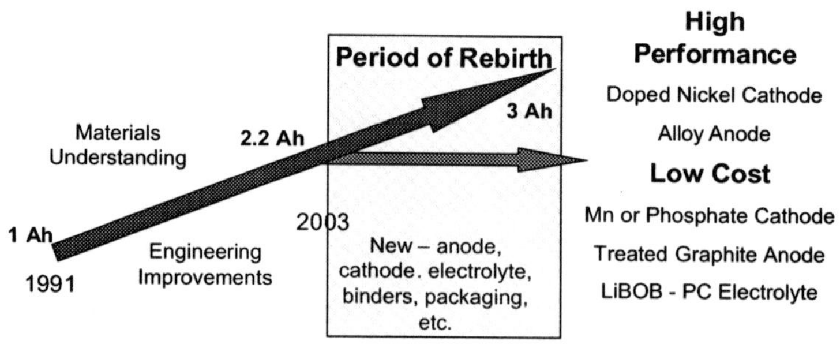

Figure 13. Depiction of the development of the Li-Ion battery system.

4. ULTRACAPACITORS

The carbon-carbon ultracapacitor has become the main memory protection device in appliance applications. These devices store charge in the electrical double layer that forms at all electrode-solution interfaces. High surface area carbon blacks (1000 m^2/g) are the material of choice. Since no electrochemical reactions are involved, these devices have very long life (10^5 cycles or more). This field has been very active as it is possible that the ultracapacitor will find application in hybrid vehicles. Recently, Okamura reported that devices with 45 Wh/kg and 60 Wh/l are possible with carbons where development of a high internal surface area results from intercalation and removal of potassium [20]. The potassium opens channels of nanoporosity with high surface area. Asymmetric capacitors, with a carbon anode and a battery electrode (nickel, lead, etc.), have been developed with very long cycle life. In another development, Naoi deposited π-stacked supramolecule redox couples onto carbon nanotubes. He reported long cycle life and reasonable energy density [21].

5. CLOSING REMARKS

The future remains bright for the use of carbon materials in batteries. In the past several years, several new carbon materials have appeared: mesophase pitch fibers, expanded graphite and carbon nanotubes. New electrolyte additives for Li-Ion permit the use of low cost PC based electrolytes with natural graphite anodes. Carbon nanotubes are attractive new materials and it appears that they will be available in quantity in the near future. They have a high ratio of the base plane to edge plain found in HOPG. The ultracapacitor application to deposit an electronically conductive polymer on the surface of a carbon nanotube may be the wave of the future.

REFERENCES

1. *The Electrochemistry of Carbon*, S. Sarangapani, J. Akridge and B Schuum, Eds., The Electrochemical Society, Pennington, NJ, 1984, PV84-5.
2. *K. Kinoshita, Carbon Electrochemical and Physicochemical Properties, J. Wiley, New York, 1988.*
3. M. Winter and J. Besenhard in *Handbook of Battery Materials*, J. O. Besenhard (Ed.) Wiley-VCH, Weinheim (1999).
4. Mitio Inagaki, *Carbon Material Engineering*, Nikkan Kogyo Publ. (Japanese) (1985).
5. M. Inagaki and Hishiyama, *New Carbon Material*, Gohodo, (1994) (Japanese).
6. *Denchigijutu,* (No 6) 1994-(No. 10) 1999 Battery Committee Japan (Japanese).
7. *Handbook of Batteries and Fuel Cells*, D. Linden and T. B. Reddy, Eds., McGraw-Hill, New York, 3rd Edition, 2001.
8. *Handbook of Battery Materials*, J.O. Besenhard, Ed., Wiley-VCH, Weinheim, 1999.
9. M. Yoshio and A. Kozawa, *Lithium ion Batteries*, Nikkan Kogyo, Tokyo, 2nd edition (2000).
10. *Lithium Ion Batteries*, M. Wakihara and O. Yamamoto, Wiley VCH Weinheim, 1998.
11. *Advances in Lithium-Ion Batteries*, W.A. van Schalkwijk and B. Scrosati, Klewer Academic/Plenum Publishers, 2002.
12. X.Y. Song, Xi Chu and K. Kinoshita, in *New Sealed Rechargeable Batteries and Supercapacitors*, MRS Symposium Proceedings Volume **393**, Materials Research Society, Warrendale, 1995, p321.
13. S. Mori, H. Asahina, H. Suzuki, A. Yonei and K. Yokoto, *J. Power Sources*, **68**, 59 (1997).
14. T. Tamaki, T. Kawamura, and Y. Yamazaki, in *Materials for Electrochemical Energy Storage*, MRS Symposium Volume **496**, D.S. Ginley, D.H. Doughty, B. Scrosati, T. Takamura and Z. Zhang, Eds., Materials Research Society, Warrendale, 1998, p.569.
15. J. Suzuki, O. Omae, K.Sekine and T. Takamura, Solid State Ionics, **152-153**, 111(2002).
16. M.Nishizawa□R. Hashitani, T. Ito, T. Matsue and I. Uchida, *Electrochem. and Solid State Letters*, **1**, 10 (1998).
17. D. Aurbach In *Proceedings 1st PBFC at Jeju Island, Korea*, June 1 – 6, 2003, J. Power Sources, in press.
18. M. Broussely, S. Herreyre, P. Biensan, P. Kaztejna, K. Nechev and R. J. Stanowitz, J Power Sources **97-98**, 13 (2001).
19. T. Takamura, Solid State Ionics, **152-153**, 19 (2002).
20. M. Takeuch, T. Maruyama, K. Koike, A. Morgami and M. Okamura, Electrochemistry, 66, No. 12 (1998).
21. K. Naoi, In *Proceedings 1st PBFC at Jeju Island, Korea*, June 1 – 6, 2003, J. Power Sources, in press.

ANODE-ELECTROLYTE REACTIONS IN Li BATTERIES: THE DIFFERENCES BETWEEN GRAPHITIC AND METALLIC ANODES

H. J. Santner, K. C. Möller, W. Kohs, C. Veit, E. Lanzer, A. Trifonova, M. R. Wagner, P. Raimann, C. Korepp, J. O. Besenhard, and M. Winter**

Institute for Chemical Technology of Inorganic Materials, Graz University of Technology, A-8010 Graz, Austria,

Abstract

An overview about more than 10 years of R&D activities on solid electrolyte interphase (SEI) film forming electrolyte additives and solvents at Graz University of Technology is presented. The different requirements on the electrolyte and on the SEI formation process in the presence of various anode materials (metallic lithium, graphitic carbons, and lithium storage metals/alloys are particularly highlighted.

Keywords

Solid electrolyte interphase (SEI), electrolyte additive, lithium ion battery, Li metal, graphite, lithium alloy.

1. METALLIC LITHIUM AS ANODE MATERIAL IN RECHARGEABLE LITHIUM BATTERIES

Considering thermodynamic reasons for the selection of a metallic anode material, light metals like Li, Na, or Mg are favored as they combine outstanding negative standard potentials with low equivalent weights, with lithium showing the lowest weight and most negative potential. The strong reducing power of metallic lithium leads to its thermodynamic instability in most known battery electrolytes. Fortunately, in suitable electrolytes, films of electrolyte decomposition products spontaneously form after the

* Corresponding author. E-mail: martin.winter@tugraz.at

I.V. Barsukov et al. (eds.), New Carbon Based Materials for Electrochemical Energy Storage Systems, 171–188.

immersion of Li into the electrolyte solution. In the ideal (but so far not fully realized) case, the films still act as a membrane for the active charge carrier, the lithium cation, but are electronically insulating and are impermeable to other electrolyte components and thus hinder further electrolyte reduction [1]. In other words, these films behave as a solid electrolyte interphase (SEI) [2]. The observation of the kinetic stability of lithium in a number of nonaqueous electrolytes induced a rapid proliferation of research in the field of lithium batteries beginning in the 1950s, and the commercialization of primary lithium batteries followed quickly. The commercial breakthrough of rechargeable batteries using metallic lithium, however, failed, as upon recharge in liquid electrolytes lithium plating occurs simultaneously with Li corrosion and passivation (formation of the SEI). The SEI is inhomogeneous, both with regard to composition and thickness, and lithium deposition through the SEI therefore does not proceed uniform and smooth but locally very different, that dendritic highly reactive metal depositions can be observed (cf. Fig. 1). The dendrites are substrates for further inhomogeneous SEI formations, which results in further dendritic Li deposition, etc. In other words, the dynamic lithium surface during cycling results in continuous SEI formation, which again promotes the dynamics of the surface. As a consequence of this "vicious cycle", Li and electrolyte losses occur in each cycle and thus the cycle life of the cell is drastically reduced. Even worse, the dendrites may grow to filaments which locally short-circuit the cell. Due to the low melting point of metallic Li (about 180°C) the local overheat can trigger a disastrous thermal runaway [3].

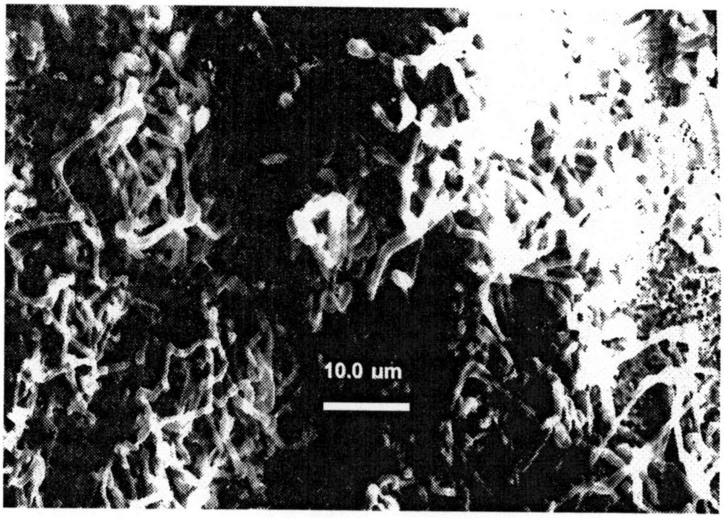

Figure 1. Lithium deposition on Ni performed in 1 M $LiClO_4$ in propylene carbonate (PC) as electrolyte, $Q = 2.7$ C, $I = 1$ mA/cm^2, magnification: x 1000 [4].

2. CARBON REPLACES METALLIC LITHIUM IN RECHARGEABLE LITHIUM BATTERIES

The market breakthrough of rechargeable lithium batteries was achieved when the performance and safety limited metallic lithium electrode had been replaced by layered carbons (favorably graphitic carbons), the graphite being an intercalation host for lithium cations. The main difference to metallic lithium is that the positions for charge storage and protective SEI formation are different. The SEI is formed on the surface of graphite, and the lithium cations are transported through the SEI into the graphite, where they are accommodated into the host structure. Thus, the surface of graphite in contact with the electrolyte remains stable and so does the SEI (Figure 2). As a consequence and at contrast to metallic lithium, SEI formation on graphites and other carbons takes place in a confined period of time, i.e., mainly in the first cycle. Therefore, SEI formation on graphite can be (but also has to be) controlled during the 1st charge process [3, 5, 6].

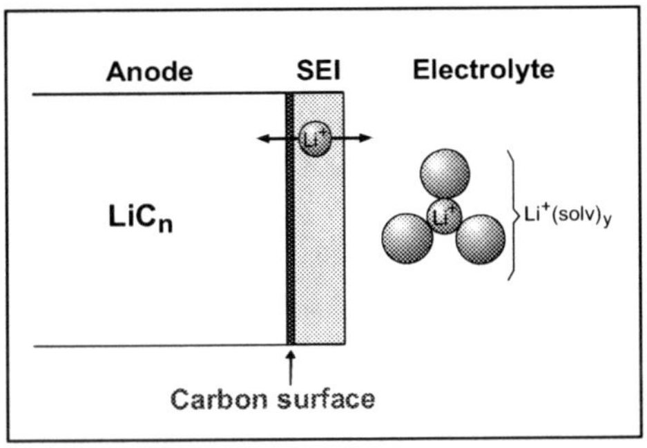

Figure 2. SEI film formation on the carbon surface (for details, cf. text).

3. FILM FORMING ELECTROLYTE COMPONENTS AND ADDITIVES

SEI formation control is the key to good performance and the safety of the whole lithium ion battery, as not only anode operation but also cathode properties are strongly affected by the SEI formation process (the cathode is the lithium cation source of lithium ion cells). Apart from control of the graphite (surface) properties, an appropriate composition of the electrolyte is usually helpful for creation of an effective SEI.

Currently, graphite-based lithium ion batteries use mixed solvent electrolytes containing highly viscous ethylene carbonate (EC) and low viscosity dilutants such as dimethyl carbonate (DMC) or diethyl carbonate (DEC) as main solvents. EC is indispensable because of its excellent filming characteristics. DMC and/or DEC are required to get the low temperature performance of the electrolyte at least reasonable. The performance limitations of EC/DMC or EC/DEC based electrolytes, e.g., at lower and higher temperatures, under abuse conditions, or with regard to the 1st cycle irreversible charge losses, stimulated various R&D efforts for improved electrolytes, either by addition of certain electrolyte additives or even by replacement of certain solvent components, e.g., either EC or DMC/DEC by other solvents and additives [3, 5, 6].

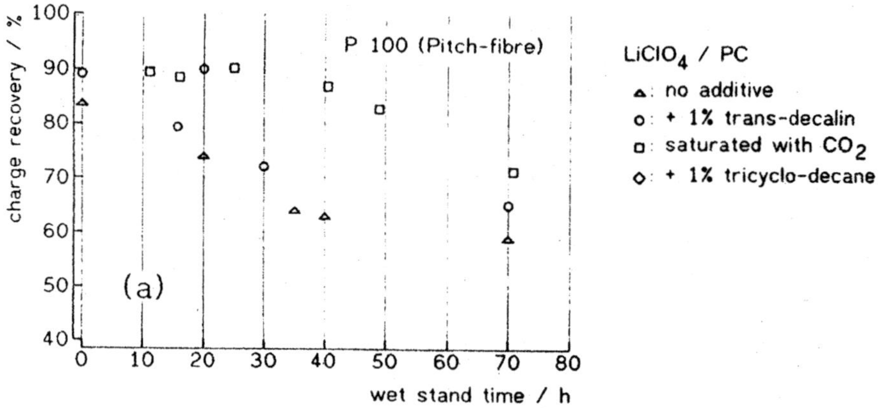

Figure 3. Charge recoveries vs. wet stand time of LiC_n in PC electrolytes without and with electrolyte additives. Electrolyte: 0.5 M $LiClO_4$ in PC, Carbon: Highly graphitic carbon fiber P100 (Amoco), 1st charge with -50 μA mg^{-1}. [7].

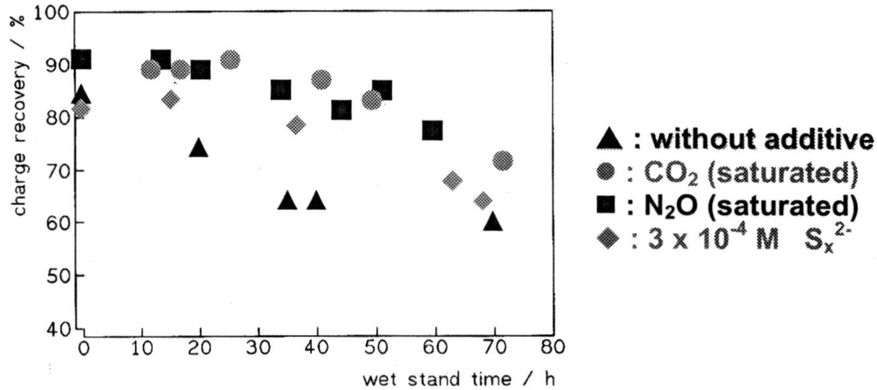

Figure 4. Charge recoveries vs. wet stand time of LiC_n in PC electrolytes without and with electrolyte additives (CO_2, N_2O, S_x^{2-}). Electrolyte: 0.5 M $LiClO_4$ in PC, Carbon: Highly graphitic carbon fiber P100 (Amoco), 1st charge with -50 μA mg^{-1} [4].

The first promising results have been achieved with inorganic additives such as CO_2 [4, 7, 8] (Figs. 3-5), and N_2O [4, 8] or S_x^{2-} [4, 8] (Figs. 4-5) added to organic solvent-based electrolytes, which in comparison to the additive-free electrolyte showed some improvements of the SEI formation process. However, most surprisingly also inorganic component based electrolytes, e.g., SO_2-based electrolytes [4, 9] (Fig. 6) and (with some limitations) also $SOCl_2$-based electrolytes (Fig. 7) [4, 9] showed some compatibility with graphitic carbons.

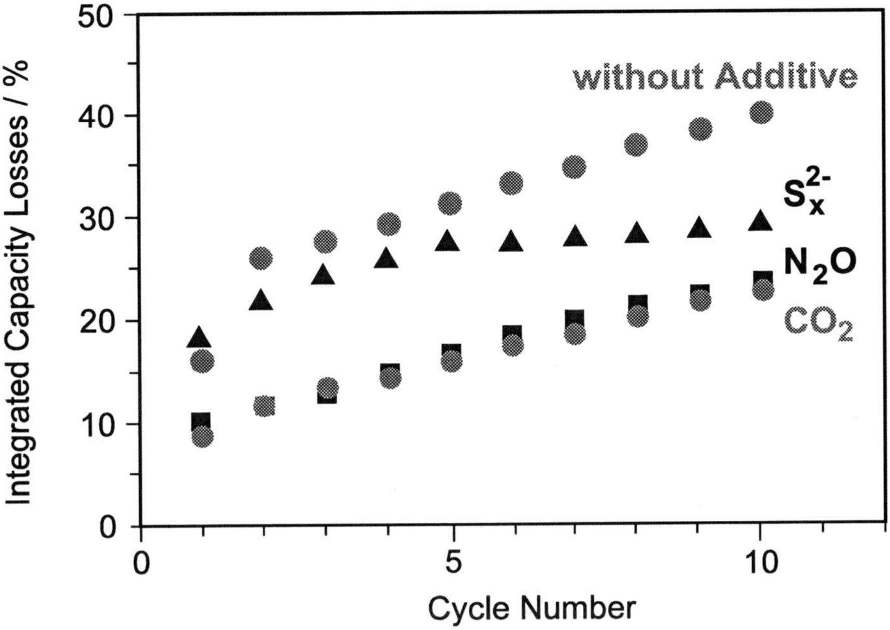

Figure 5. Integrated capacity losses of LiC_n in PC electrolytes without and with electrolyte additives (CO_2, N_2O, S_x^{2-}). Electrolyte: 0.5 M $LiClO_4$ in PC, Carbon: Highly graphitic carbon fiber P100 (Amoco), $i = 50\ \mu A\ mg^{-1}$, cut-off: 0-2 V vs. Li/Li^+ [4].

The beneficial effect of CO_2 on the SEI formation process and the cycling performance of graphitic carbons have been reported by many groups [4, 7, 8, 10, 11]. It is most nicely demonstrated in γ-butyrolactone based electrolytes obviously, CO_2 suppresses the decomposition reaction of γ-butyrolactone at approx. 0.8 V vs. Li/Li+ (Figs. 8, 9). Independent of the used electrolyte salt ($LiClO_4$ (Fig. 10, top, left), $LiBF_4$ (Fig. 10, top, right), $LiPF_6$ (Fig. 10, bottom, left), and $LiN(SO_2CF_3)_2$ (Fig. 10, bottom right), the cycling performance of the γ-butyrolactone based electrolytes improves when CO_2 is present. Nevertheless, the significant differences in the electrochemical performance indicate that the electrolyte salt anion plays a strong role in the SEI formation process, i.e., that the filming behaviour of the anion may interfere with that of the additive or with that of γ-butyrolactone [12, 13].

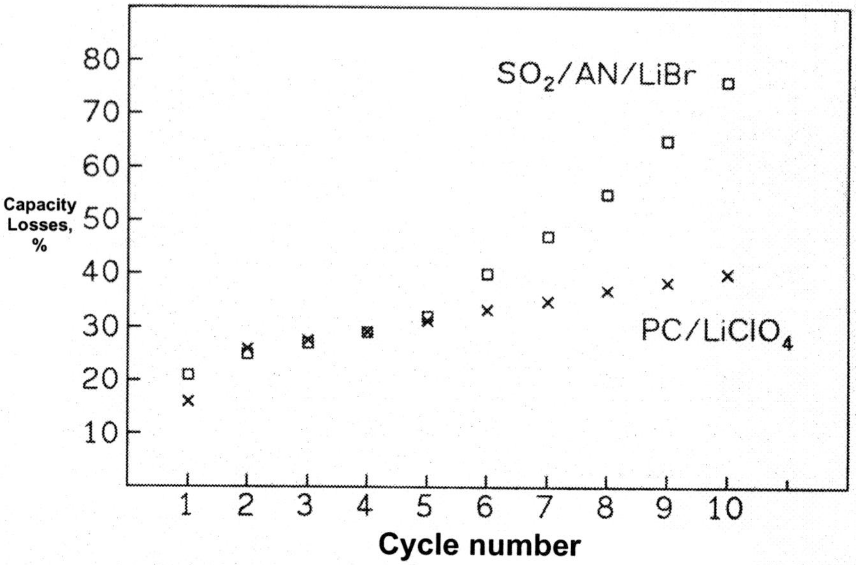

Figure 6. Integrated irreversible capacities of LiC_n in 1 M LiBr in SO_2/acetonitrile (AN) and 0.5 M $LiClO_4$ in PC as electrolytes. Carbon: Highly graphitic carbon fiber P100 (Amoco), $i = 50\ \mu A\ mg^{-1}$, cut-off: 0-2 V vs. Li/Li$^+$ [4].

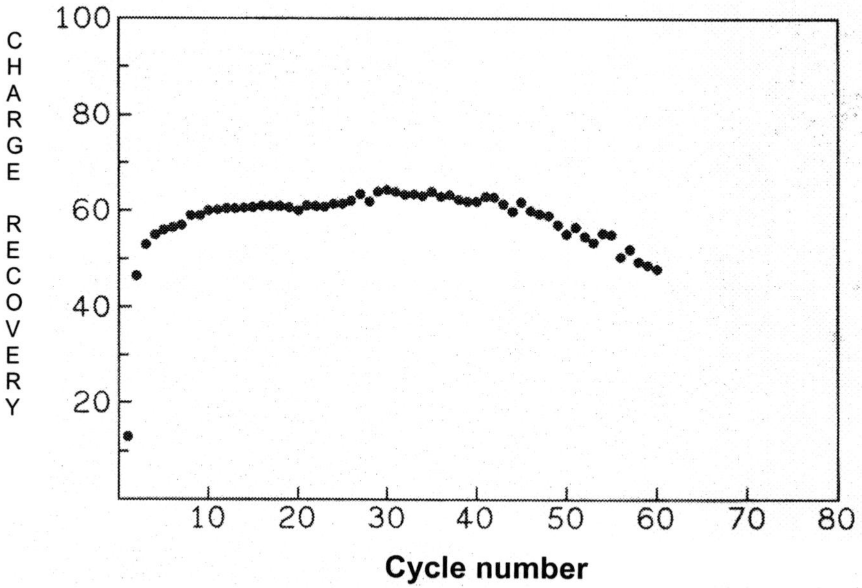

Figure 7. Constant current charge/discharge cycling of highly graphitic carbon fiber P100 (Amoco) in 1.8 M $LiAlCl_4$ in $SOCl_2$ electrolyte, $i = 300\ \mu A\ mg^{-1}$, cut-off: 0-2V vs. Li/Li$^+$ [4].

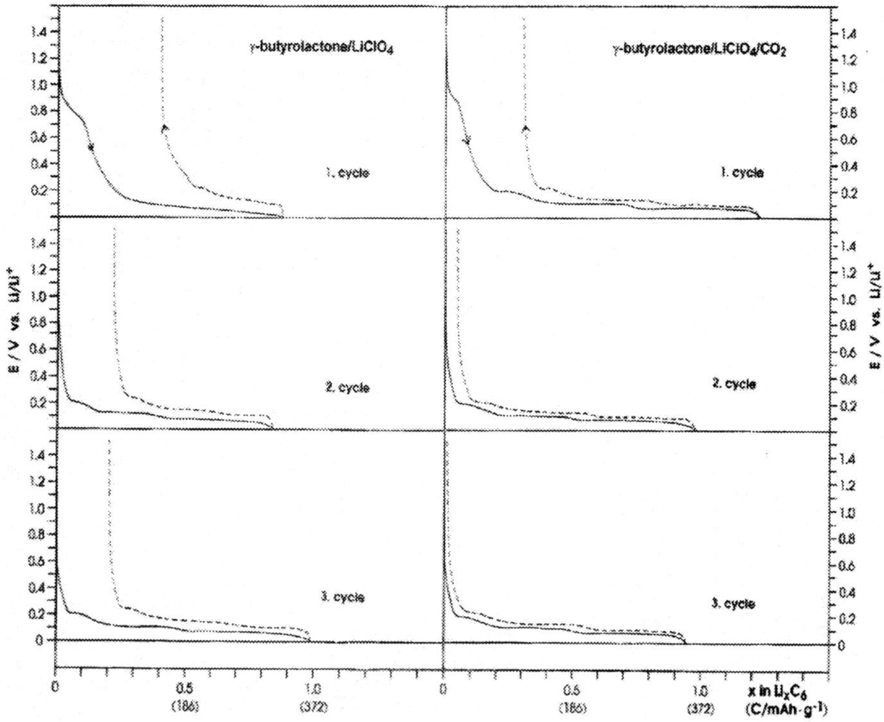

Figure 8. Constant current charge/discharge cycling (1.-3. cycles) of graphite (Lonza KS44 synthetic graphite) in 1 M $LiClO_4$ in γ-butyrolactone as electrolyte without and with CO_2 (saturated in electrolyte) as electrolyte additive, i~10μA mg^{-1}, cut-off: 0-1.5V vs. Li/Li^+ [12,13].

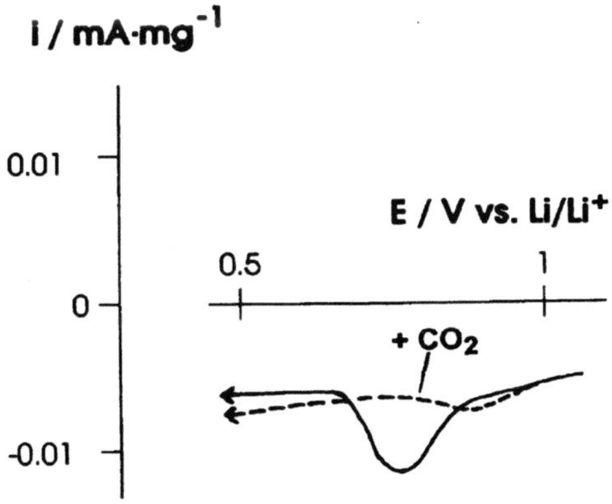

Figure 9. Excerpts of the peaks due to "electrolyte decomposition" recorded by voltammetry in 1 M $LiClO_4$ in γ-butyrolactone as electrolyte without and with CO_2 (saturated in the electrolyte) as electrolyte additive, scan rate: 0.01 mV s^{-1} [12, 13].

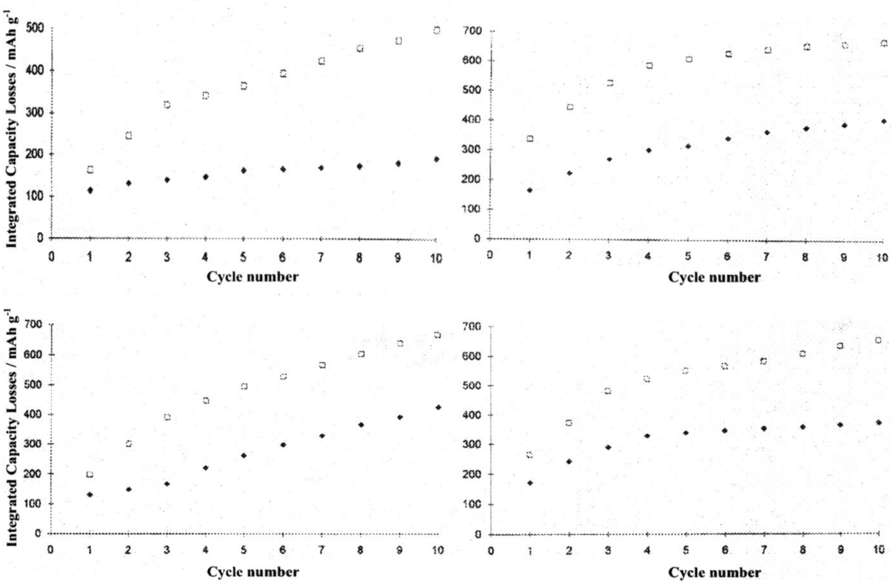

Figure 10. Integrated irreversible capacities of LiC_n in γ-butyrolactone based electrolytes without (full symbols) and with (open symbols) CO_2 as electrolyte additive using various electrolyte salts: $LiClO_4$ (top, left), $LiBF_4$ (top, right), $LiPF_6$ (bottom, left), $LiN(SO_2CF_3)_2$ (bottom, right). Carbon: Lonza KS44 synthetic graphite, $i = \sim 10\ \mu A\ mg^{-1}$, cut-off: 0-1.5 V vs. Li/Li^+ [12].

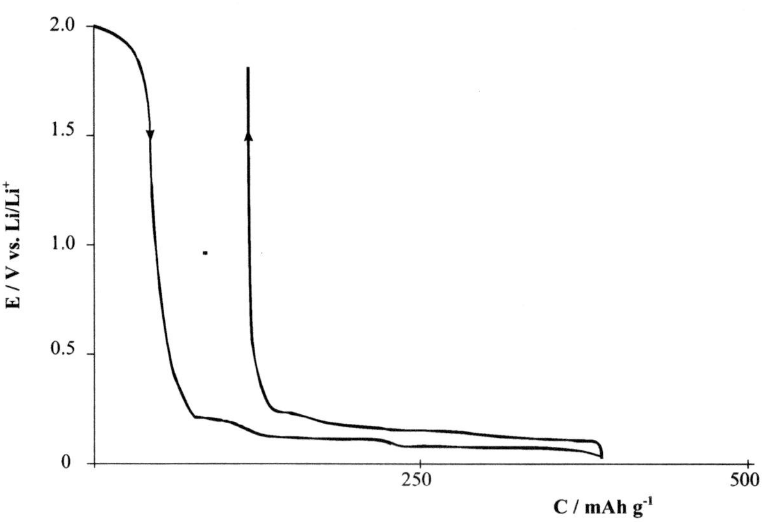

Figure 11. First cycle constant current charge/discharge curves of synthetic graphite TIMREX® SFG 44 using 1 M $LiClO_4$ in PC:PS (propylene sulfite) (95:5 by volume) as electrolyte. $i = \pm 20\ mA\ g^{-1}$, cut-off = 1.8/0.025 V vs. Li/Li^+.

Due to easier handling, liquid electrolyte additives and solvents have found more interest than gaseous additives. Among these components, organic sulfites [14-19] (Fig. 11), partially fluorinated aprotic solvents [12, 20-25], and vinylene additives [24, 26-32] have found particular attention.

Partially fluorinated components can be used either as electrolyte solvents (Fig. 12) or as electrolyte additives (Fig. 13). In many cases they show much superior SEI forming capabilities compared to their non-fluorinated counterparts. Moreover, fluorinated solvents are in general much less flammable as less hydrogen is available, which might contribute to cell safety [12, 23, 25].

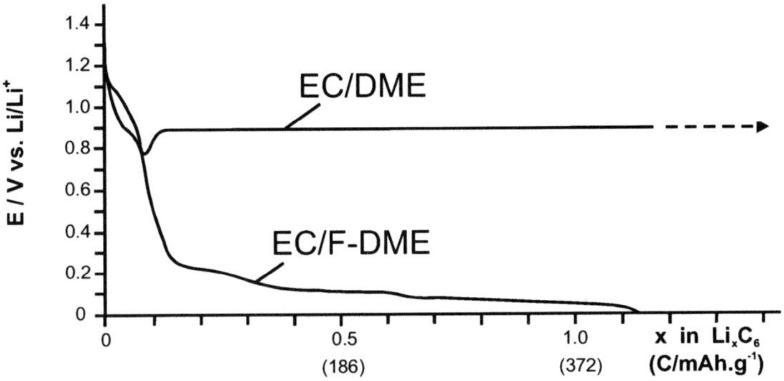

Figure 12. 1st cycle constant current charge curves of synthetic graphite LONZA KS 44 (i) using 1 M $LiN(SO_2CF_3)_2$ in EC:DME (dimethoxy ethane, $CH_3OCH_2CH_2OCH_3$) (3:2) as electrolyte. The measurement was stopped when the graphite was exfoliated, (ii) using 1 M $LiN(SO_2CF_3)_2$ in EC:F-DME (partially fluorinated dimethoxy ethane, $CH_3OCF_2CF_2OCH_3$) (3:2) as electrolyte. $i = \pm 20$ mA g^{-1}, cut-off = 0.0 V vs. Li/Li^+ (adapted from [12]).

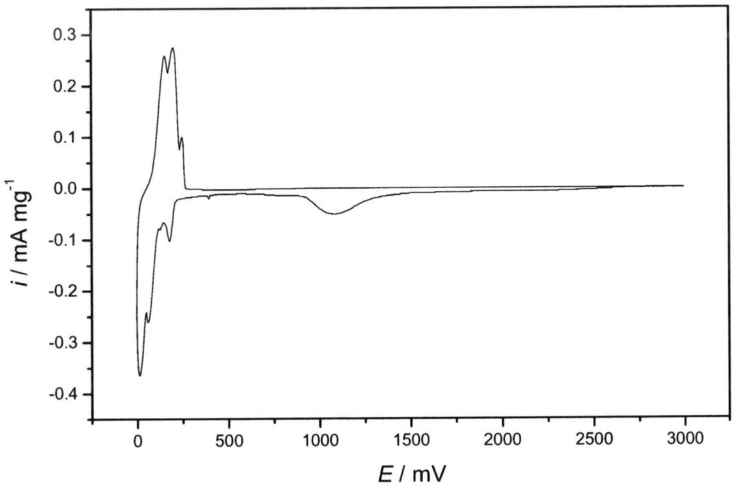

Figure 13. Cyclic voltammogram of graphite (Lonza KS6, synthetic graphite) in 1 M $LiClO_4$ in PC/novel fluorinated additive (90:10, v:v); scan rate 30 $\mu V\ s^{-1}$, Potentials vs. Li/Li^+.

Vinylene additives are very attractive at the moment both from an academic and industrial point of view [24, 26-32]. Usually, vinylene additives can be very effective even when they are used in very low amounts (≤ 1 vol.-%), in particular when the first charging step is properly adjusted (Fig. 14).

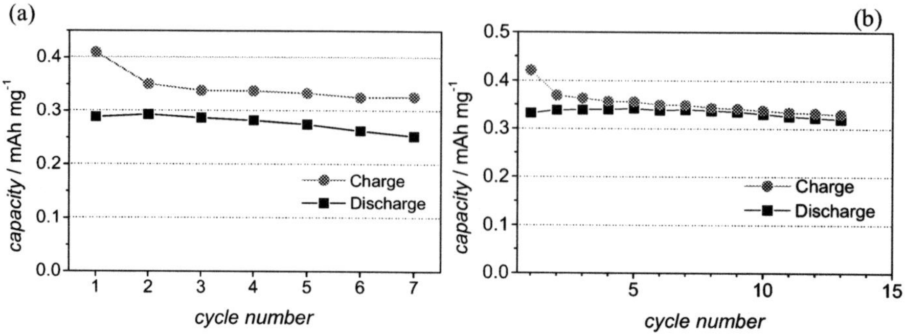

*Figure 14. **(a)** Constant current charge cycling of graphite (TIMREX SFG 44) in 1M $LiClO_4$ in PC/AAN (99:1, v:v) with a constant current of ±20 mA g^{-1}, cut-off: 0.02/1.5 V vs. Li/Li^+. The efficiency was 70% in the 1st cycle, and <85% in the following cycles. **(b)** Same graphite and electrolyte, but different charging process: Potentiodynamic ramp from 3.0 to 0.5 V vs. Li/Li^+ in the 1st charge, then constant current cycling at ±20 mA g^{-1}, cut-off: 0.02/1.5 V vs. Li/Li^+ in the rest of the 1st cycle and subsequent cycles. The efficiency was >80% in the 1st cycle and >>90% in the following cycles. The applied combined potentiodynamic/ galvanostatic 1st cycle charge process might is not the ultimately best one, but illustrates the effect of the charge conditions on the performance.*

4. WHAT IS SPECIAL ABOUT THE SEI FORMATION ON GRAPHITE?

With regard to SEI formation, a main difference to lithium metal (above) and lithium storage metals/alloys (below) is that the structure of layered graphite gives rise to basically two kinds of surfaces, prismatic (edge) surfaces and basal plane surfaces. Ideal (= defect and contaminant free) basal plane surfaces are homogeneous and "smooth" and consist only of carbon atoms. In contrast the prismatic surfaces are heterogeneous and "rough" and apart from carbon may contain various, mostly oxygen-containing surface groups [33]. It is well known that the prismatic and basal plane surface areas of graphite show a different SEI formation behaviour [33, 24]. For example, in propylene carbonate (PC) based electrolytes ternary solvated graphite- intercalation compounds take part via the prismatic surfaces in the electrolyte reduction and SEI formation mechanism.

In fact, crystalline graphites usually cannot be operated in PC electrolytes, unless effective film forming electrolyte additives are used (see above) as propane gas evolution [35], creation of solvated graphite intercalation compounds (sGICs) [36], and graphite exfoliation take place. Recently [37, 38], it was found that propylene evolution is observed at graphite, while absent at lithium active metallic anodes, e.g., Sn and SnSb.

The compared to other anode materials unique propane gas evolution process at graphite, can be explained by decomposition of the only kinetically stable sGICs, as only graphite, but not metallic Li and lithium storage metals/alloys are vulnerable to solvent co-intercalation (cf. experiment, in Fig. 15). Apparently, the decomposition products of the sGICs, which are created apart from propane, do not form an effective SEI hindering further co-intercalation. Thus, even when only a limited graphite area may be subject to solvent co-intercalation (which may be not visible by XRD), gas evolution takes place inside graphite and the formed gas could act as strong lever enabling exfoliation and mechanical destruction of graphite [39, 40] (Fig. 16).

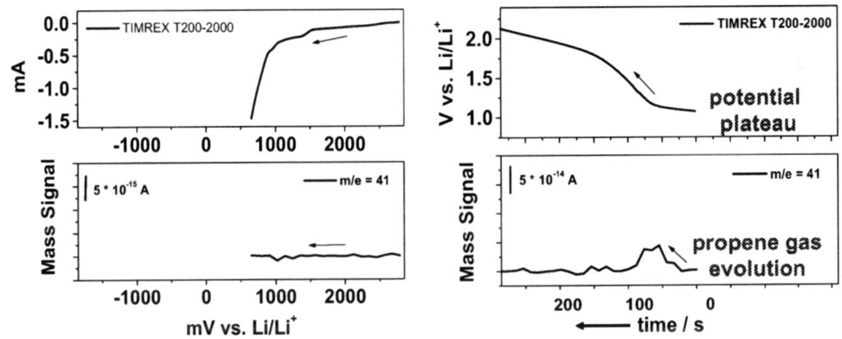

Figure 15. In order to monitor gas evolution in parallel to the electrochemical experiment we took use of on-line mass spectrometry. The experiment was performed as follows: Following a fast potentiodynamic step with 10 mV s^{-1} from ca. 3.0 to 0.65 V vs. Li/Li$^+$, the electrode (TIMREX® T200-2000) has immediately been kept at open circuit. At open circuit, the potential and the mass signal of propylene have been recorded vs. time. At these fast scan rates, only small areas of the graphite react. Thus only small, but still significant mass signals can be recorded: No strong propylene evolution is observed during the voltammetric scan (Fig. 15 left). At open circuit the graphite potential jumps from 0.65 V to >1 V vs. Li/Li$^+$, where a potential plateau develops (Fig. 15, right). This plateau potential is typical for solvated graphite intercalation compounds (sGICs). The formed sGICs are only kinetically stable and gradually decompose during time, which is visible by a gradual shift of the open circuit potential to more positive values. This is accompanied by propene gas evolution (m/e = 41), the gas being a decomposition product of the solvated intercalates [39].

At the prismatic surfaces, the primary reduction reaction is the reduction of those solvent(s) co-intercalated inside graphite, whereas on the basal plane graphite sites, reduction reactions may proceed in an excess of all electrolyte species (including electrolyte salt anions and solvents that do not co-intercalate). Therefore, the electrolyte decomposition products formed on the basal plane surface should have a different composition to those formed at the prismatic surfaces [38]. In addition, it should be noted, that due to the fact, that lithium intercalation and solvent reduction occur in parallel at the prismatic surface sites, the SEI at the prismatic surfaces contains a higher fraction of lithium compounds than the SEI at the basal plane sites.

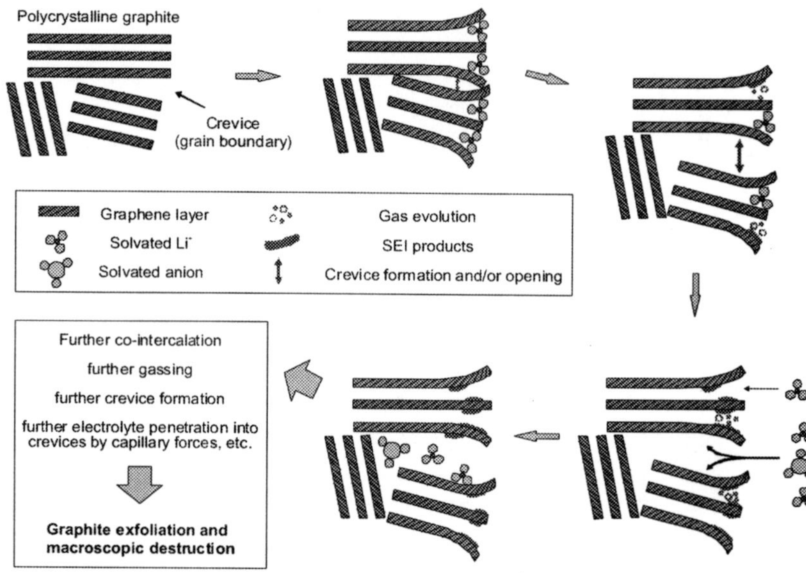

Figure 16. Model: In PC based electrolytes, solvent co-intercalation, gas formation and crevice formation in polycrystalline graphite materials are inter-related reactions. In fact, there is a subsequence of reactions: (1) PC co-intercalation, (2) gas formation, (3) crevice formation ultimately resulting in exfoliation and macroscopic destruction of graphite [40].

It should be noted here, that not only the (chemical and morphological) composition of the protective layers at the basal plane surfaces and prismatic surfaces is different, but that these layers also have completely different functions. At the prismatic surfaces, lithium ion transport into/from the graphite structure takes place by intercalation/de-intercalation. Here the formed protective layers of electrolyte decomposition products have to act as SEI, i.e., as transport medium for lithium cations. Those protective layers, which have been formed on/at the basal plane surfaces, where no lithium ion transport into/from the graphite structure takes place, have no SEI function. However, these "non-SEI layers" still protect these anode sites from further reduction reactions with the electrolyte.

Though there is a distinct difference between the functions of the SEI and non-SEI layers, it is accustomed in the literature to designate the protective layers as SEI, independent of their function. However, a most important impact of the differentiation between SEI- and non-SEI layers on graphite is that during processes such as SEI aging, formation, or dissolution, any measurable changes of the non-SEI layers at the basal plane surface sites can not be directly related to the electrochemical data recorded in parallel, such as charge, discharge or self-discharge as these electrochemical processes take place only at the prismatic sites of graphite (Fig. 17).

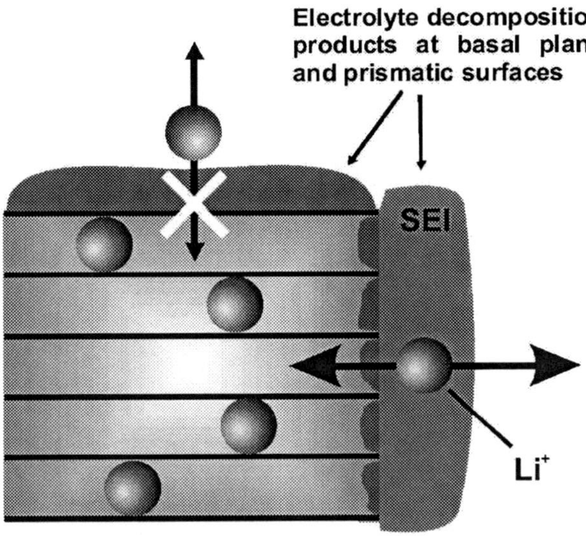

Figure 17. The basal plane and prismatic surfaces of graphite have different functions with respect to lithium intercalation and de-intercalation (= charge, discharge, self-discharge, etc.). As a consequence, only the electrolyte decomposition product layers at the prismatic surfaces have SEI function. Any processes related with electrolyte decomposition product layers at the basal plane surfaces (= non-SEI layers) therefore can not be directly related to electrochemical data such as charge, discharge, self-discharge, etc. The situation is even more complex as the SEI composition and morphology at the basal and prismatic surface planes varies [40].

5. SEI FORMATION AND SEI STABILITY ON LITHIUM STORAGE METALS AND ALLOYS

Electrolyte effects on the cycling stability of lithium storage metals and alloys indicate the importance of SEI formation in this case, too. Very early measurements suggest that additives such as CO_2 do not only improve the cycling stability of metallic lithium [41] and graphitic carbons (see above), but also that of lithium storage metals (Fig. 18), which may be related with the electrical properties of the SEI (Fig. 19) [13].

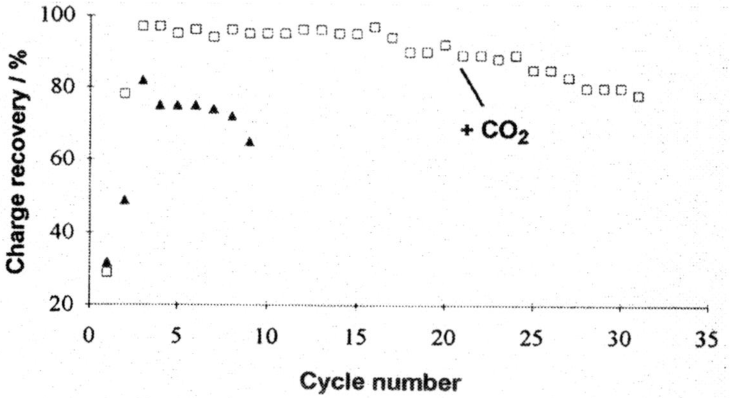

Figure 18. Cycling of metallic Sn (2.5 μm) on a Cu-substrate in 1 M $LiClO_4$/PC in absence and presence of CO_2 (saturated at ambient pressure). $i_c = i_d = 0.025$ mA cm^{-2}, capacity limited charge of 1.7 Li/Sn, discharge cut-off: 0.8 V vs. Li/Li$^+$ [13].

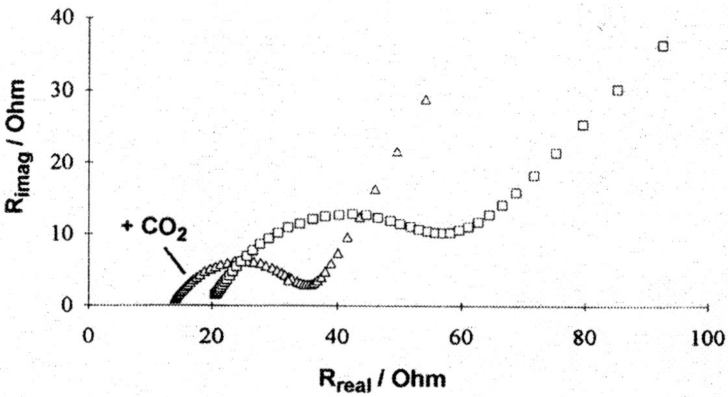

Figure 19. Impedance spectra of "$Li_{1.5}Sn_{0.72}Sb_{0.28}$" after storage for 40 days in 1 M $LiClO_4$/PC in absence (f_{apex} = 341 Hz) and presence (f_{apex} = 674 Hz) of CO_2 (saturated at ambient pressure) [13].

Recent studies on the electrolyte decomposition processes on graphite and lithium storage metals however revealed that the reduction mechanisms and SEI compositions as well as the requirements on the SEI properties are different (see above). Because of the strong volume changes of the metallic hosts during lithium uptake and removal in lithium alloy electrodes, the SEI is subject to continuous mechanical stress. This may be relieved by the formation of a more flexible SEI, whereas graphites require a dense SEI suppressing solvent co-intercalation. From the point of view of surface dynamics, the SEI stability on lithium storage metals/alloys may be rather compared with that on metallic lithium than that on graphites (Fig. 20) [42].

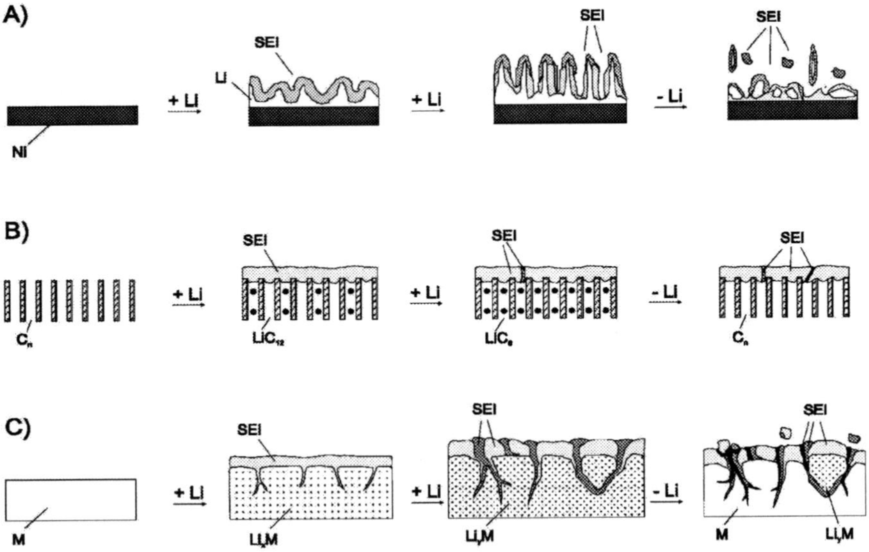

Figure 20. SEI formation on different anodes for rechargeable Li batteries: (A) lithium metal, (B) graphitic carbon, and (C) metals and intermetallics. Different colors of the SEI indicate SEI products formed at different stages of charge and discharge (and do not indicate different composition) [42].

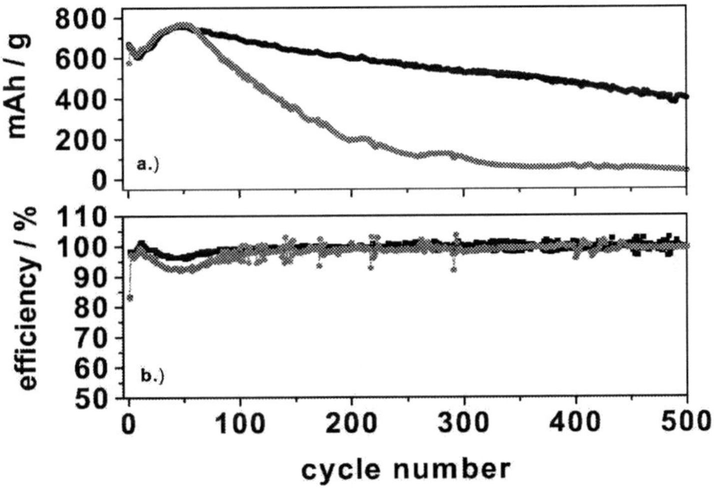

Figure 21. Effect of adapted electrolyte on the cycling performance of electroplated thin film Sn_2Co anodes with regard to: a.) discharge capacity and b.) efficiencies. (●) 1 M $LiClO_4$ in EC/PC (1:1), (■) 1 M $LiClO_4$ in EC/PC (1:1) + 20 vol.-% new electrolyte component. Charge procedure: potentiodynamic charging step: 2.5 - 0.8 V vs. Li/Li^+, followed by constant current charge/discharge cycling at 250 $\mu A\ cm^{-2}$, cut-off: 20-1200 mV vs. Li/Li^+.

Figure 1 presents results on the beneficial effect of a new electrolyte solvent on the cycling behavior of the lithium storage alloy Sn_2Co. The influence of the adapted electrolyte may become even more clearly by taking a look at the cumulated charge losses (Fig. 22). It should be noted that this new electrolyte component is not able to suppress solvent co-intercalation into graphite [37], emphasizing the above-mentioned different requirements on the electrolyte.

As long as lithium storage metal and alloys do not show a dimensional and structural stability similar to that of graphite, the graphite anode in a lithium ion cell may not be simply replaced by the lithium storage metal/alloy without adapting the electrolyte solution. Moreover, the fact, that graphitic carbons and lithium storage metals and alloys may require different electrolyte compositions may be an obstacle for the application of graphite/lithium storage metal composites.

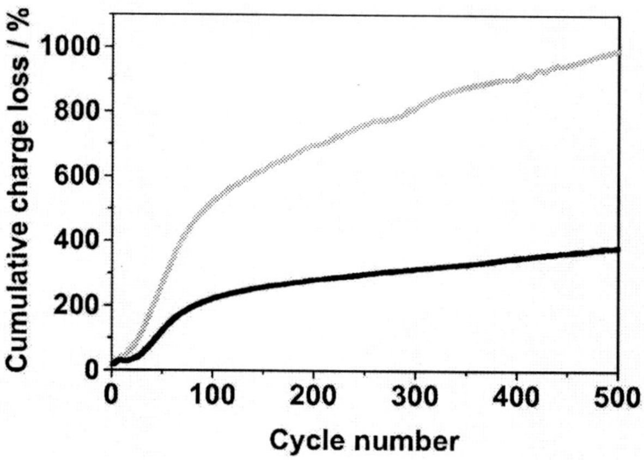

Figure 22. Effect of adapted electrolyte on the cumulative charge loss of thin film Sn_2Co anodes. (●) 1 M $LiClO_4$ in EC/PC (1:1), (■) 1 M $LiClO_4$ in EC/PC (1:1) + 20 vol.-% new electrolyte component.

ACKNOWLEDGEMENTS

Support by the Austrian Science Funds through the special research program "Electroactive Materials" is gratefully acknowledged.

REFERENCES

1. A. N. Dey, Electrochem. Soc. Fall Meeting 1970, Abstr. 62.
2. E. Peled, in: Lithium Batteries (Ed.: J.-P. Gabano), Academic Press, London, 1983, Ch. 3.

3. M. Winter, J. O. Besenhard, M. E. Spahr, P. Novak, Adv. Mat., 1998, 10, 725.
4. M. Winter, Diploma Thesis, University of Münster, Germany, 1993.
5. M.Winter, J. O. Besenhard: in: Handbook of Battery Materials (Ed.: J. O. Besenhard), Wiley-VCH, Weinheim, 1999, 383.
6. M. Winter, K. C. Moeller, J. O. Besenhard (2003)Science and technology of advanced lithium batteries (Eds: G. A. Nazri, G. Pistoia), Kluwer Academic Publishers, New York, 2003, Ch. 5.
7. J. O. Besenhard, P. Castella, M. W. Wagner, Mater. Sci. Forum, 1992, 91-93, 647.
8. J. O. Besenhard, M. W. Wagner, M. Winter, A. D. Jannakoudakis, P. D. Jannakoudakis, E. Theodoridou, J. Power Sources, 1993, 43-44, 413.
9. J. O. Besenhard, M. Winter, J. Yang, W. Biberacher, J. Power Sources 1995, 54, 228.
10. O. Chusid, Y. Ein-Ely, D. Aurbach, J. Power Sources, 1993, 43-44, 47.
11. B. Simon, J. P. Boeuve, M. Broussely, J. Power Sources, 1993, 43-44, 65.
12. M. Winter, Doctoral Thesis, University of Münster, Germany, 1995.
13. J. O. Besenhard, M. Winter, J. Yang, International Workshop on Advanced Batteries (Lithium Batteries), Osaka 1995, 129.
14. G. H. Wrodnigg, J. O. Besenhard, M. Winter, J. Electrochem. Soc., 1999, 146, 470.
15. G. H. Wrodnigg, T. M. Wrodnigg, J. O. Besenhard, M. Winter, Electrochem. Comm., 1999, 1, 148.
16. G. H. Wrodnigg, C. Reisinger, J. O. Besenhard, M. Winter, ITE Batt. Lett., 1999, 1(1) 111.
17. H. Ota, T. Sato, H. Suzuki, T. Usami, J. Power Sources, 2001, 97-98, 107.
18. H. Gan, E. S. Takeuchi, Eur. Pat. Appl. No. 513428, 2000.
19. G. H. Wrodnigg, J. O. Besenhard, M. Winter J. Power Sources, 2001, 97-98, 592.
20. J. O. Besenhard, K. v. Werner, M. Winter, US Pat. No. 5916708, 1997.
21. W. K. Appel, J. O. Besenhard, S. Pasenok, M. Winter, G. H. Wrodnigg, Ger Pat No. DE 19 724 709, 1998.
22. W. K. Appel, J. O. Besenhard, L. H. Lie, S. Pasenok, M. Winter, Ger. Pat. No. DE 198 58 924, 1998.
23. K.-C. Möller, T. Hodal, W. K. Appel, M. Winter, J. O. Besenhard, J. Power Sources, 2001, 97-98, 595.
24. K.-C. Möller, S. C. Skrabl, M. Winter, J. O. Besenhard, in: Proc. 3rd Hawaii Battery Conference, Hawaii, USA, 2001, 228.
25. L. H. Lie, T. Hodal, K.-C. Möller, G. H. Wrodnigg, W. K. Appel, J. O. Besenhard, M. Winter, ITE Batt. Lett., 1999, 1(1), 105.
26. C. Jehoulet, P. Biensan, J. M. Bodet, M. Broussely, C. Moteau, C. Tessier- Lescourret, in Batteries for Portable Applications and Electric Vehicles (Eds.: C. F. Holmes, A. R. Landgrebe), The Electrochemical Society, Pennington NJ, 1997, PV97-18, 974.
27. M. Fujimoto, M. Takahashi, K. Nishio (Sanyo) US Pat. No. 5352548, 1994; B. Simon, J.-P. Boeve (SAFT) US Pat. No. 5626981, 1997; J. Barker, F. Gao (Valence) US Pat. No. 5712059, 1998; Y. Naruse, S. Fujita, A. Omaru (Sony) US Pat. No. 5714281, 1998.
28. K.-C. Möller, H. J. Santner, J. O. Besenhard, M. Winter, in: Proc. 4th Hawaii Battery Conference, Hawaii, USA, 2002, 238.
29. M. Yoshio, H. Yoshitake, K. Abe, T. Umeno, K. Fukuda, 11th International Meeting on Lithium Batteries, Monterey CA, 2002, Abstract No. 9.
30. H.J. Santner, K.-C. Möller, J. Ivančo, M.G. Ramsey, F.P. Netzer, S. Yamaguchi, J. O. Besenhard, M. Winter, J. Power Sources, 2003, 119-121, 368.
31. K.-C. Möller, H.J. Santner, W. Kern, S. Yamaguchi, J.O. Besenhard, M. Winter, J. Power Sources, 2003, 119-121, 561.
32. C. Korepp, H. Santner, J. O. Besenhard, M. Winter, K. C. Möller, J. Anal. Bioanal. Chem., in press.
33. J. P. Olivier, M. Winter, J. Power Sources, 2001, 97-98, 151.
34. D. Bar-Tow, E. Peled, L. Burstein, J. Electrochem. Soc., 1999, 146, 824.
35. A. N. Dey, B. P. Sullivan, J. Electrochem. Soc., 1970, 117, 222.

36. M. Winter, G. H. Wrodnigg, J. O. Besenhard, W. Biberacher, P. Novák, J. Electrochem. Soc., 2000, 147, 2427, and references therein.
37. M. R. Wagner, P. Raimann, K. C. Möller, J. O. Besenhard, M. Winter, Electrochem. Solid-State Lett., 2004, in press.
38. M. R. Wagner, P. Raimann, H. J. Santner, A, Trifonova, C. Korepp, K.-C. Möller, J. O. Besenhard, M. Winter, Lithium-Battery-Discussion 2003, September 14-19, 2003, Bordeaux-Arcachon (France), Abstract No. Inv. 7.
39. H. J. Santner, M. R. Wagner, G. Fauler, P. Raimann, C. Veit, K. C. Möller, J. O. Besenhard, M. Winter (2003). An Overview on SEI Formation Processes of Lithium Battery Anodes in Organic Solvent Based Electrolytes, Taipei Power Forum and Exhibition (TPF2003), December 1-3, 2003, Taipei (Taiwan) Invited lecture.
40. H. J. Santner, M. R. Wagner, G. Fauler, P. Raimann, C. Veit, K. C. Möller, J. O. Besenhard, M. Winter (2003), Taipei Power Forum and Exhibition (TPF2003), December 1-3, 2003, Taipei (Taiwan), Proceedings Volume.
41. S. B. Brummer, V. R. Koch, R. D. Rauh, in: Materials for Advanced Batteries (eds.: D. W. Murphy, J. Broadnead, B. C. H. Steele, Plenum, New York, 1980.
42. M. R. Wagner, J. O. Besenhard, K.-C. Möller, M. Winter, Proceedings of the 7th International Symposium on Advances in Electrochemical Science and Technology (ISAEST-VII), November 27-29, Chennai (India), in press.
43. M. Winter, W. K. Appel, B. Evers, T. Hodal, K.-C. Möller, I. Schneider, M. Wachtler, M. R. Wagner, G. H. Wrodnigg, J. O. Besenhard, Chem. Monthly, 2001, 132, 473.

PERFORMANCE OF NOVEL TYPES OF CARBONACEOUS MATERIALS IN THE ANODES OF CLAiO'S LITHIUM-ION BATTERY SYSTEMS

Mariusz Walkowiak[1*], Krzysztof Knofczynski[1], Daniel Waszak[1], Maciej Kopczyk[1], M. Rusinek[1], and Jacek Machnikowski[2]

[1] *Central Laboratory of Batteries and Cells, Forteczna 12, 61-362 Poznan, Poland*
[2] *Wroclaw University of Technology, Institute of Chemistry and Technology of Petroleum and Coal, Gdańska 7/9, 50-344 Wrocław, Poland*

Abstract

The electrochemical performance of two anode materials has been investigated in CLAiO's fully functional lithium-ion battery mockups. Our findings show that the spherodized purified natural graphite SL20 possesses sufficient reversible capacity and low irreversible capacity loss, and thus has been qualified as a suitable candidate material for application in lithium-ion batteries. Another carbon material investigated in this work was a Boron-doped carbon (25B2) synthesized by WUT via co-pyrolysis of coal tar pitch with pyridine borane complex. Its irreversible capacity was seen to be higher than that of SL20, while reversible capacity is promising. The unique electron acceptor properties, which are typical in boron-doped graphites did not manifest themselves in the investigated sample. Ball milling of boron-doped carbon may be an effective way of improving the WUT precursor and thus designing a suitable lithium-ion grade anode material.

1. INTRODUCTION

Lithium-ion (shuttle-cock, rocking-chair, swing) battery is widely considered as the most advanced power source for consumer electronics and is regarded as the most promising battery technology for a variety of other applications, such as electric vehicles, medicine and space exploration. One of the most critical factors in designing successful Li-ion cell is the choice of

* Corresponding author. E-mail: mw@plusnet.pl

I.V. Barsukov et al. (eds.), New Carbon Based Materials for Electrochemical Energy Storage Systems, 189–196.

proper active anode materials. For many years carbonaceous materials have been attracting particular attention of lithium-ion battery scientists and engineers due to the possibility of reversible hosting lithium ions in their structures [1-3]. The reaction of lithium insertion/deinsertion is the basic process taking place in the anodes of Li-ion cells. Among carbons tested as potential anode materials for Li-ion cells were natural and synthetic graphites, non-graphitised carbons (both soft and hard), carbon fibers, carbon composites, nanotubes and others. Recently doped carbons (especially boron-doped) have been considered as promising new anode materials due to their unique properties. Since boron atoms introduce electron acceptor levels the materials have been expected to have enhanced capacities.

As the end-user in the NATO SfP project "Carbons as materials for the electrochemical storage of energy" Central Laboratory of Batteries and Cells does research and development works on the application of novel carbonaceous materials to the Li-ion technology. The general idea of these works is to build prototypes of cylindrical Li-ion cells on the basis of materials produced in the cooperating laboratories. The aim of this paper is to examine the applicability of selected commercial and non-commercial carbon materials (with special attention devoted to boron-doped carbons) to the construction of a practical cylindrical Li-ion cells.

2. EXPERIMENTAL

Two different natural graphites manufactured by Superior Graphite Co. (SL-20 and LBG-73) were tested as received on the possibility of using as anode of a cylindrical Li-ion battery. For comparison, typical synthetic graphite KS-15 from Lonza was examined.

Innovatory boronated carbons (manufactured in the Institute of Chemistry and Technology of Petroleum and Coal, Wroclaw University of Technology, Poland) were obtained by co-pyrolysis of coal-tar pitch with a pyridine-borane complex. In the first stage of pyrolysis (520°C) the so-called semi-coke is obtained. Further carbonization at 2500°C leads to obtaining boron-doped carbonaceous material (sample labeled 25B2).

Grinding was performed using Retsch ball-mill. Particle size distribution was measured using Malvern (Mastersizer) laser particle sizer. BET surface area measurements were carried out using ASAP (Micromeritics) surface area analyzer.

Basic electrochemical parameters of the tested materials were obtained using coin-type half-cells with metallic lithium as the counter electrode and 1M $LiPF_6$ in EC/DEC (Merck) as electrolyte. The cells were galvanostatically charged/discharged at the current density 20 mA per gram of active mass. Full cylindrical Li-ion cells were manufactured in a laboratory scale using AA-size cases. Anode active masses after mixing with a binder (PVdF, Fluka) were coated in the form of a viscous slurry onto a 20 μm thick copper foil using a hand-operated laboratory coating tool. Similarly, cathode active mass (manganese spinel, Merck) mixed with a binder and a conductive additive was coated onto a 20μm thick aluminum foil. After sandwiching with a 25μm thick polypropylene separator (Celgard) the electrode set was wound and inserted in the case. Finally the cell was dozed with electrolyte and closed. The most critical operations were carried out in a glove box.

3. RESULTS AND DISCUSSION

Construction of lithium-ion battery is a complex problem involving many critical stages. Among these the choice of proper active anode material is one of decisive importance. The ideal candidate for anode should be characterized not only by excellent electrochemical parameters but also low cost and good processability.

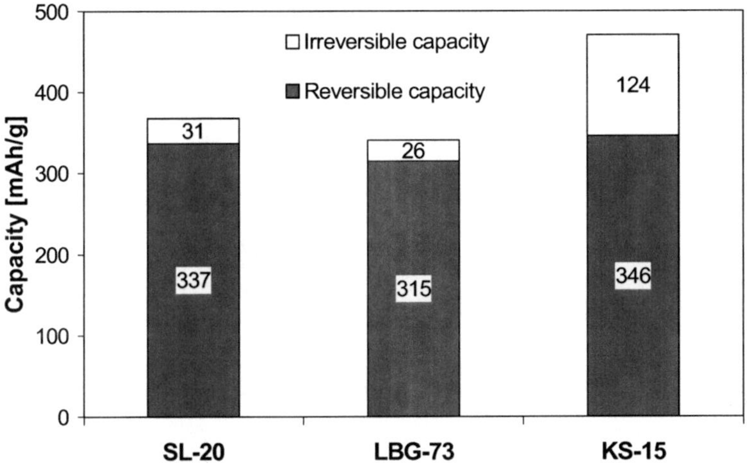

Figure 1. Comparison of the reversible and irreversible capacities of commercial graphites.

Three commercial graphite materials were evaluated as possible candidates for anodes of a Li-ion cell: SL-20 and LBG-73 (Superior Graphite Co.) and KS-15 (Lonza). The first two materials represent a wider range of graphites specially designed by the SGC researchers to meet the stringent requirements of Li-ion technologies. Prior to the actual construction of a cylindrical Li-ion battery all materials were subjected to electrochemical tests in half-cells in order to select the best candidate for anode material in terms of reversible and irreversible capacities. These values are also necessary in the actual battery designing process, which involves the stage of establishing the anode/cathode mass balance. The bar chart on the Figure 1 shows the respective capacities for all the commercial graphites tested.

From the data presented it appears that all the tested materials exhibit reversible capacities distinctly exceeding 300 mAh/g, which is typical for graphites. Graphite grades SL-20 (SGC) and KS-15 (Lonza) are close to the maximum theoretical reversible capacity accessible for graphitic materials (372 mAh/g). However the irreversible capacity for KS-15 is unacceptably high (124 mAh/g, which means 26 % of the total electrical charge delivered to the electrode upon first charging). The corresponding values in the case of the SGC graphites are 31 mAh/g (8 %) and 26 mAh/g (8 %) for SL-20 and LBG-73 respectively. Such low values of irreversible losses together with high reversible capacities make the SGC materials good candidates for the anodes of practical Li-ion batteries.

The model cylindrical Li-ion battery (AA-size) was manufactured using SL-20 graphite as anode active material. The general appearance of the cells is shown by Figure 2; for more detailed description of the cells see the experimental part of the paper.

Figure 2. The general view of the coin and cylindrical cells.

One important conclusion from the works on the construction of the cell is that the material (SL-20) can be described as one having excellent process properties. By this, one should among others understand perfect adhesion to the copper foil and the ability to form smooth and uniform layers. The discharge capacities for the first ten cycles are presented in the Figure 3, together with the discharge profile for the first cycle.

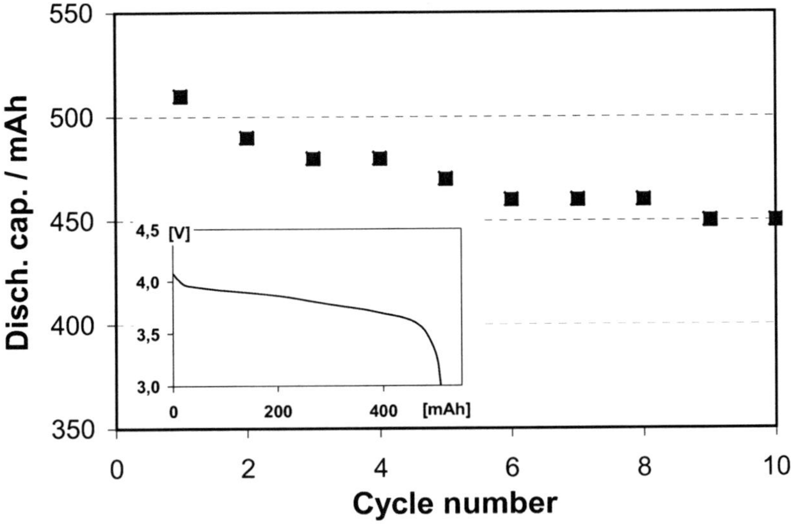

Figure 3. Discharge capacities for he first ten cycles for the Li-ion cell SL-20 anode. The inset: discharge curve for the first cycle.

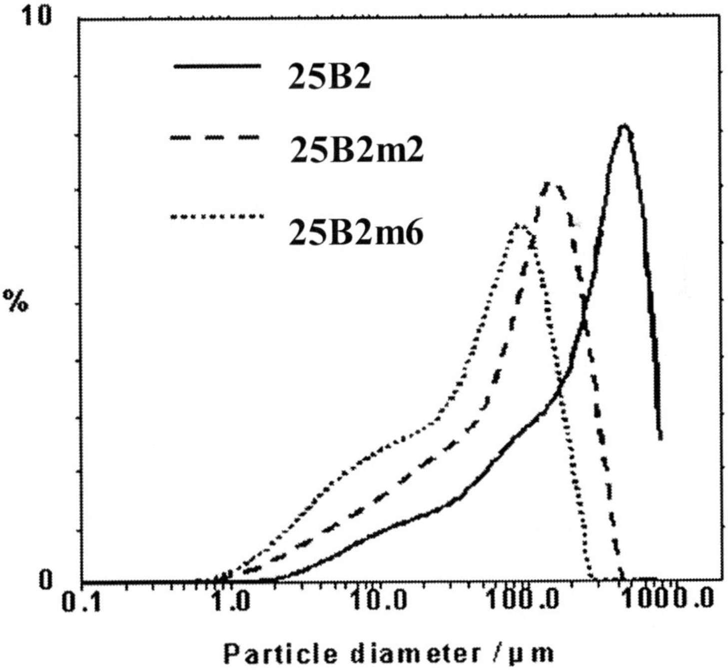

Figure 4. Particle size distributions for the original and ground B-doped carbon from WUT.

The capacities obtained initially exceed 500 mAh (which is ca. 70 % of those exhibited by commercial products), but decline in the subsequent cycles to reach a fairly stable value of ca. 450 mAh in the tenth cycle.

Boron-containing carbons synthesized by co-pyrolysis of coal-tar pitch with pyridine-borane complex (series 25Bn) have already been considered as hosts for lithium insertion [4]. Unlike the commercial graphites described above, the boron-doped carbon 25B2 (WUT) as received was not suitable for direct use in the cylindrical cell due to very large and hard particles. This feature makes the coating process very difficult.

To improve the processing properties, the material was ground in a ball-mill for two hours (sample designated as 25B2m2) and separately for 6 hours (sample designated as 25B2m6). In order to check the effect of milling the particle distribution of all the samples was determined (see Fig. 4).

The curves give information about the distribution (in weight percentage terms) of particle sizes in a given sample. While the majority of the original material (25B2) concentrates around 500 μm (see the peak of the 25B2 curve, note that the particle size axis is a logarithmic one), the particles of the milled samples are markedly reduced in size (the peaks are located around 150 μm and 80 μm for 25B2m2 and 25B2m6 respectively). Prolonged grinding makes the WUT material applicable for the Li-ion technology. Together with the reduction of particle sizes an increase in the BET surface areas is observed (0.4 m^2/g, 5.8 m^2/g and 8.5 m^2/g for 25B2, 25B2m2 and 25B2m6 respectively). What is interesting, the XRD pattern analysis proves that the crystal structure of the material was not affected by grinding.

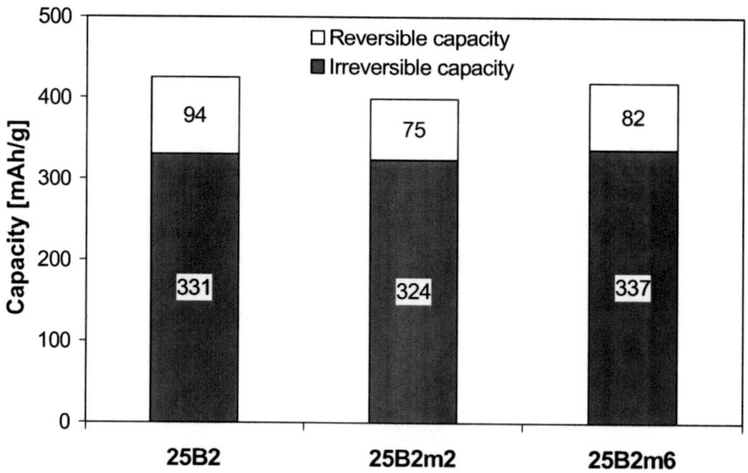

Figure 5. Comparison of the reversible and irreversible capacities of the original and ground B-doped carbon from WUT.

Electrochemical tests in half-cells allow the preliminary assessment of the WUT carbon as well as of the impact of grinding on the electrochemical performance. The data from the chart on the Figure 5 indicate that the material has high reversible capacity (similar to the capacities of the commercial graphites described earlier).

This capacity is practically not affected by grinding. This finding is in contradiction to many literature reports stating that excessive ball milling can have negative impact on the reversible capacities due to introduction of turbostratic disorder. Also irreversible capacities do not change significantly upon grinding. However the ratios of irreversible charge consumptions are higher than in the case of SGC graphites (e.g. 20% for 25B2m6). To sum up, ball milling of the original B-doped carbon (WUT) creates material with satisfactory electrochemical parameters and having greatly improved processing properties. Figure 6 shows the discharge capacities for the first ten cycles for the AA-size Li-ion battery with the 25B2m6 anode. The insert on this picture shows the first discharge profile for this cell.

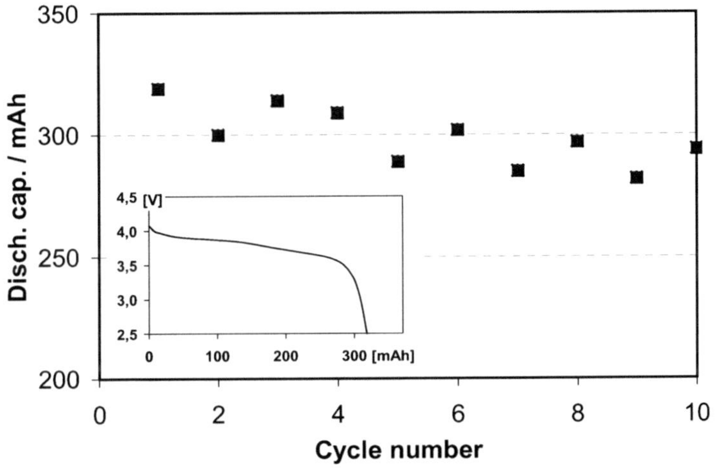

Figure 6. Discharge capacities for he first ten cycles for the Li-ion cell SL-20 anode. The insert: discharge curve for the first cycle.

4. CONCLUSIONS

Commercial and non-commercial carbons were tested for their applicability as anode of lithium-ion battery. It was found that Superior Graphite Co's materials are characterized both by high reversible capacities and low irreversible capacities and thus can be regarded as good candidates for practical full cells. Cylindrical AA-size Li-ion cells manufactured using laboratory techniques on the basis of SL-20 anode had initial capacities over 500 mAh (volumetric energy density ca. 240 Wh/dm^3). Boron-doped carbon

25B2 (WUT) exhibits high reversible capacity but its irreversible capacity is slightly greater than that of SGC graphites. Moreover, very large particles hinder direct application of this material. It was found that ball milling could be considered as an effective method of improving the usability of this carbon as anode of Li-batteries.

ACKNOWLEDGEMENTS

The authors would like to thank Superior Graphite Co. for delivering graphite samples. Special thanks for NATO SfP "Carbon" 973849 project consortium members for valuable material and intellectual assistance, in particular to Prof. François Beguin, Prof. Elżbieta Frąckowiak, Prof. Jacek Machnikowski, Prof. Andrzej Lewandowski and Prof. Vyacheslav Barsukov.

REFERENCES

1. S. Flandrois, B. Simon, Carbon, 37 (1999) 165-180.
2. M. Endo, C. Kim, K. Nishimura, T. Fujino, K. Miyashita, Carbon, 38 (2000) 183-197.
3. F. Cao, I. V. Barsukov, H. J. Bang, P. Zaleski, J. Prakash, J. Electrochem. Soc., 147 (2000) 3579-3583.
4. J. Machnikowski, E. Frąckowiak, K. Kierzek, D. Waszak, R. Benoit, F. Beguin, J. Phys. Chem. Solids, 65 (2004)

WHY GRAPHITE ELECTRODES FAIL IN PC SOLUTIONS: AN INSIGHT FROM MORPHOLOGICAL STUDIES

Doron Aurbach*, Maxim Koltypin, Hnnan Teller and Yaron S. Cohen

Department of Chemistry, Bar-Ilan University, Ramat-Gan 52900, Israel

Abstract

This paper deals with capacity fading mechanisms of graphite electrodes. Three types of graphite particles were used: synthetic flakes, natural flakes and mesocarbon microbeads (MCMB). We used a probe solution, EC-PC/$LiClO_4$, in which electrodes comprising different types of graphite particles behave very differently from each other. The tools for this study included in-situ AFM imaging, scanning electron microscopy, FTIR spectroscopy, XRD and standard electrochemical techniques. The morphology of the edge planes of the graphite particles plays an important role in their ability to develop passivating surface films when graphite electrodes are polarized cathodically in the electrolyte solutions. Another critical factor that determines the passivation of the graphite electrodes is the cohesion and adhesion of the solution reduction products. A major failure mechanism of graphite particles in PC solutions seems to be cracking of the particles because of reactions in crevices, which lead to a build-up of internal pressure. Such cracking processes can be clearly observed by AFM imaging. The unexpected dependence of the irreversible capacity of some types of graphite electrodes on the particle size (increases with larger particle size) is also evident for the above failure mechanism. Factors that determine stabilization and failure of graphite electrodes are demonstrated and discussed.

Keywords

Graphite electrodes, ethylene carbonate, propylene carbonate, capacity fading, surface films, morphology, AFM, passivation.

* Corresponding author. E-mail: aurbach@mail.biu.ac.il

I.V. Barsukov et al. (eds.), New Carbon Based Materials for Electrochemical Energy Storage Systems, 197–211.

1. INTRODUCTION

There is no question that the development and commercialization of lithium ion batteries in recent years is one of the most important successes of modern electrochemistry. Recent commercial systems for power sources show high energy density, improved rate capabilities and extended cycle life. The major components in most of the commercial Li-ion batteries are graphite electrodes, $LiCoO_2$ cathodes and electrolyte solutions based on mixtures of alkyl carbonate solvents, and $LiPF_6$ as the salt.[1] The electrodes for these batteries always have a composite structure that includes a metallic current collector (usually copper or aluminum foil/grid for the anode and cathode, respectively), the active mass comprises micrometric size particles and a polymeric binder.

Graphite intercalates electrochemically and reversibly with lithium via phase transitions in four stages up to a stochiometry of LiC_6 (corresponding to the capacity of 372 mAh/gr).[2] Since Li insertion into graphite electrodes occurs at very low potentials, very close to that of Li deposition (between 0.3 and 0.01 V *vs.* Li/Li^+), all the relevant nonaqueous solvents and salts, in which electrochemical Li intercalation into graphite can proceed, are reduced on the cathodically polarized graphite electrodes at potentials that can be as high as 1.5 V above lithium insertion potentials.[3] Most of these reduction processes precipitate surface films on the graphite particles, which are comprised mostly of organic and inorganic Li salts (similar to the Li electrodes that are covered by surface films in all nonaqueous solutions in which lithium is apparently stable).[4] These surface films formed on graphite electrodes, which are usually ion conducting and electronically insulating, passivate these electrodes, and were widely explored in terms of structure, properties, etc.[5-10] It appears that their properties are critical to the stability and reversibility of graphite electrodes in repeated Li insertion/deinsertion cycling. Hence, the behavior of lithiated graphite electrodes depends strongly on the composition of electrolyte.

The failure mechanisms of lithiated graphite electrodes in commonly used Li salt solutions relevant to the field of Li ion batteries are still in the focus of study by a number of research groups.[11,12] A common opinion is that in cases where the passivation by the surface films is not sufficient, solvent molecules may cointercalate with Li ions. Consequently, the fragile lattice structure of graphite may be destroyed by exfoliation of the graphene planes that simply fall apart from each other, thus causing an amorphization of the graphite particles.[11-13] Another possibility raised was that solvent molecules cointercalate with the Li ions even at potentials as high as 1 V *vs.* Li/Li^+, and are reduced within the graphite, close to the surface of the particles, thus blocking further Li-ion insertion into the lattice.[14]

There is an ongoing interest in the study of failure mechanisms of graphite electrodes. It is very important to understand the source of the irreversible capacity of graphite electrodes[15,16] (related to the irreversible reduction of solution species) and how to reduce it.[17,18] In these studies, issues such as morphological aspects, crystal structure (phase composition), surface area and solution composition are considered, and their possible impacts on the irreversible behavior of graphite electrodes are explored.[15-20]

Figure 1 provides several electrochemical windows of important, relevant processes, including the reduction of alkyl carbonates, ethers, Li insertion into graphite, and Li metal deposition. Recent studies revealed two major failure mechanisms of graphite electrodes in repeated Li insertion/ deinsertion processes:[21]

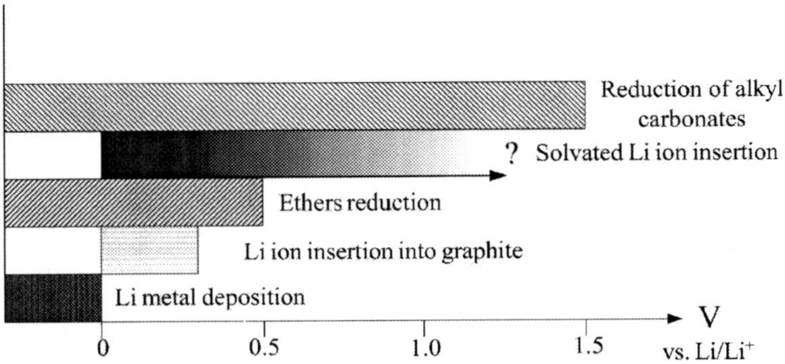

Figure 1. Schematic representation of the electrochemical windows of the processes, occurring during cathodic polarization of graphite electrodes in nonaqueous solutions.

1) When the electrolyte solutions are not too reactive, as in the case of ethereal solutions, there is no massive formation of protective surface films at potentials above Li intercalation potential, and most of the solvent reduction processes may occur at potentials lower than 0.3 V *vs.* Li/Li^+. Hence, the passivation of the electrodes is not sufficient to prevent cointercalation of solvent molecules. This leads to an exfoliation of the graphite particles into amorphous dust (expholiated graphene planes). This scenario is demonstrated in Figure 2a as the reduction of the 002 diffraction peak[21] of the graphite electrode, polarized cathodically in an ethereal solution.
2) In the case of solutions based on a solvent such as propylene carbonate (PC), the failure of graphite electrodes is attributed by some researches to the exfoliation of the graphite particles due to cointercalation of PC molecules with the Li ions.[14,22] The difference between ethylene carbonate (EC) and PC in this respect may, according to this approach, be attributed to the higher ability of PC molecules to solvate Li ions.[23] Hence, cointercalation of PC molecules takes place because their desolvation from Li ions, which migrate from solution phase to the intercalation sites in the graphite,

is more difficult than in the case of EC molecules.[14,22,23] As demonstrated in Figure 2b, the failure of graphite electrodes in PC solutions does not necessarily involve a complete exfoliation of the graphite particles and destruction of their 3D structure.[21] Cumulative studies, based on scanning electron microscopy (SEM), X-ray diffraction (XRD) and standard electrochemical techniques, converged to the conclusion that the failure mechanism of graphite electrodes in PC solutions in more likely due to cracking of the particles and their electrical isolation by surface films rather than due to the complete exfoliation (on the nanometer scale).[24,25]

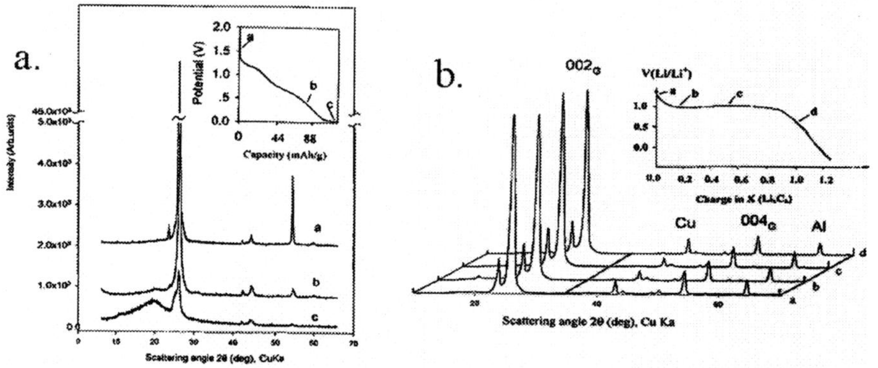

Figure 2. XRD patterns measured ex-situ from graphite particles and chronopotentiograms of graphite electrodes polarized galvanostatically from OCV to low potentials in CH_3-OCH_2CH_2-OCH_2CH_2-OCH_3 (diglyme)/$LiClO_4$ solution (a) and in PC/$LiAsF_6$ solution (b). The letters near the V vs. capacity curves correspond to the relevant XRD patterns, which have been measured after different periods of galvanostatic polarizations. Notice that in the diglyme solution (a) the XRD patterns clearly indicate progressive destruction of the active mass, while in the PC solutions (b) the XRD patterns show that the active mass remains pure graphite.

This article discusses the origin of the failure mechanisms of graphite electrodes in PC solutions, focusing on the electrodes' irreversible capacity, mainly from a morphological point of view. The EC-PC/$LiClO_4$ electrolyte solution was chosen as the major probe solution, because it was found that graphite electrodes of different types and morphologies behave very differently in these solutions in terms of the irreversible capacity measured during a first cathodic polarization of the electrodes and the potentials at which the major reduction of solution species takes place, which are lower by more than 0.5 V compared with those in the EC-DMC/$LiAsF_6$ solution. Nevertheless, after stabilization, the electrodes behave reversibly in this probe solution, which means that after the first few cycles, the behavior of these systems stabilizes.[25]

2. EXPERIMENTAL DESCRIPTION

The experimental tools for this research were chronopotetiometry (galvanostatic cycling),[25] atomic force microscopy (AFM),[26,27] scanning electron microscopy (SEM), and X-ray diffraction (XRD).[21,25] It should be mentioned that the AFM imaging was conducted *in-situ* under potential control and in a special homemade glove box filled with highly pure argon atmosphere. This system has been already described in detail in the literature.[28]

This article concentrates on three types of graphite particles, which differ in their morphology, based on a rigorous examination of SEM micrographs:[25]

1) Synthetic graphite flakes, obtained from Timrex Inc., whose morphology has been characterized by a high level of crevices in the facets perpendicular to the basal planes, through which lithium ions are inserted into the graphite lattice (edge planes).
2) Mesocarbon microbeads (MCMB) particles, obtained from Osaka Gas Co., which are round graphite particles. They are also far from having a smooth morphology and have crevices on their surface.
3) Natural graphite (NG) particles, which have smoother facets and a lesser amount of crevices in their edge planes.

3. RESULTS AND DISCUSSION

3.1. On the Proposed Scenarios of Stabilization and Failure of Graphite Electrodes

The surface reactions of graphite electrodes in many nonaqueous solutions have been investigated intensively,[29,30] and the major reaction paths in a variety of alkyl carbonate solutions seem to be quite clear. Both EC and PC decompose on graphite electrodes, polarized cathodically, to form solid surface films with $ROCO_2Li$ species as major components,[31] and ethylene or propylene gases, respectively, as co-products.

Figure 3 illustrates schematically the processes of graphite electrodes in standard solutions (such as EC-DMC/$LiAsF_6$). During the first cathodic polarization, highly passivating surface films are formed, which stop the surface reactions. Since the reduction of EC is quick enough and forms sufficiently passivating films, they protect the fragile graphite structure from cointercalation of solvent molecules and exfoliation. As Li is inserted, there is a volume expansion of the graphite, and hence, the surface films on the edge planes (through which the Li insertion takes place) are stretched. As Li is de-intercalated, the graphite's volume contracts back to its initial size.

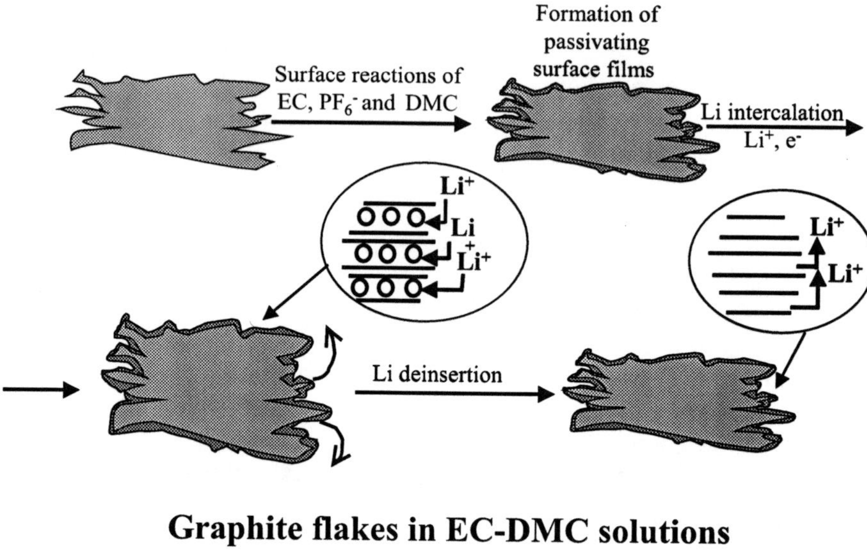

Figure 3. An illustration of morphology, surface processes and changes during a Li insertion-deinsertion cycle of an electrode comprising synthetic graphite flakes in EC-DMC solutions, in which the electrodes behave reversibly.

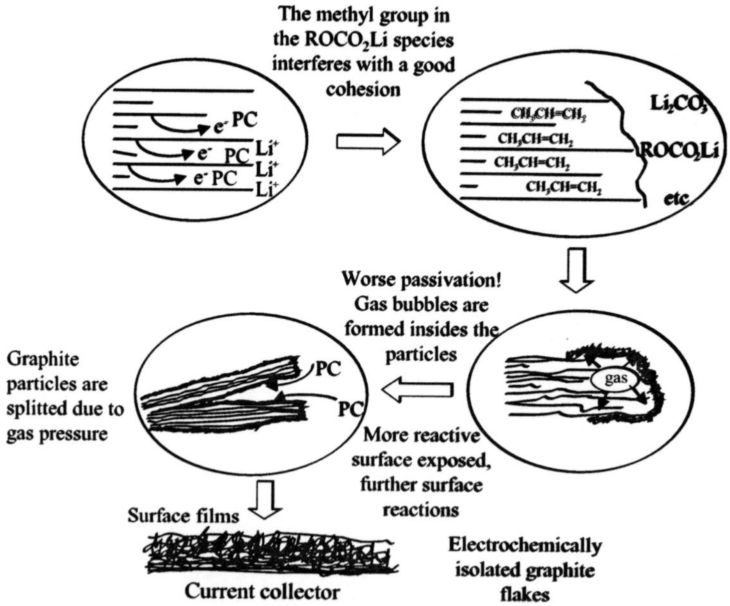

Figure 4. A scheme of the proposed mechanism of surface reactions of PC on graphite flakes, which contain crevices.

The passivation of the graphite particles depends on the accommodation of the surface films to these volume changes, and to the extent that their stability and integrity are maintained (when the graphite is fully lithiated). It is assumed that upon the repeated volume changes of the graphite particles during the Li insertion-deinsertion cycling, passivation by the surface films on the edge planes may be damaged, and thus reactive Li-graphite compounds are exposed to the solution, thus allowing a further small-scale reduction of solution species.[32] These minor surface reactions thicken the surface films during prolonged, repeated cycling, and hence, explain the increase of impedance of these electrodes observed during prolonged charge-discharge cycling.[33]

Figure 4 presents schematically the surface reactions in PC based solutions, especially when the particles' morphology is extremely rough. At potentials below 1.5 V, PC is reduced on any metallic or carbon electrode, in the presence of Li ion.[34] The major products are propylene gas and $CH_3CH(OCO_2Li)CH_2OCO_2Li$.[35] The former solid product is insoluble in PC, and hence precipitates on the active mass of the electrode as a surface film. When the carbon particles have deep crevices in which the surface reactions take place, pressure may be developed inside the crevices by gaseous products such as propylene, as the solid products that precipitate may block the outward gas diffusion. The developed pressure inside the crevices may then mechanically crack the particles very easily, thus exposing more reactive surfaces of the particles to the solution and allowing further surface reactions to take place. In the following sections we present experimental evidences that support the above failure mechanism.

3.2. Experimental Evidence: Irreversible Capacity vs. Surface Area and Particle Size

Figure 5 shows the dependence of the irreversible capacity on the size and the specific surface area of synthetic graphite particles in EC-PC solutions compared to EC-DMC solutions. It is expected that as the average size of the graphite particles decreases, and their surface area is higher, the irreversible capacity should also be higher, as was indeed reported previously for EC-DMC solutions.[10,11,14,31] Nevertheless, in the EC-PC solutions used herein, as the particle size is higher, the opposite trend is observed and the irreversible capacity is higher as well. Hence, surprisingly, the irreversible capacity is lower in these systems as the surface area of the particles is higher.

These results are in contrast to the approach that assumes that as the graphite flakes are bigger, the average energy needed to separate the graphene layers by solvated ions is higher, and thus the level of exfoliation may be lower. However, these results are in agreement with the mechanism presented above in which graphite particles which contain deep crevices and have rough morphology are cleaved because of development of internal pressure in the crevices. A fresh surface of active mass is then exposed to the solution species, thus allowing further surface reactions, and hence, increasing irreversible capacity. As the particles are larger, the freshly exposed surface of active mass due to cracking (along the basal planes) is larger (compared to the surface area of the original facets), and thus the irreversible capacity is higher. Thereby, the irreversible capacity behaves inversely with the specific surface area of the particles (the specific surface area is higher as the particles are smaller).

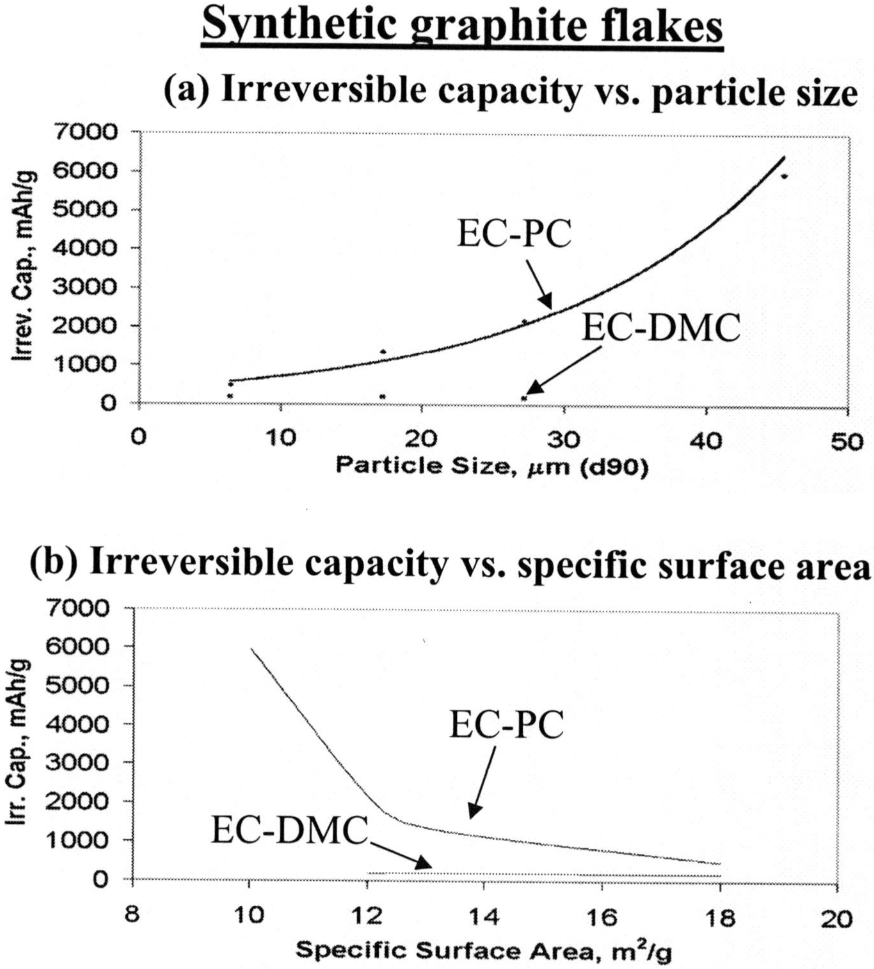

Figure 5.Irreversible capacity vs. particle size (a) and vs. specific surface area (b) of graphite electrodes comprising of synthetic flakes in EC-PC and EC-DMC solutions, as indicated.

In the case of EC-DMC solutions, since the surface species are deposited quickly and form very compact passivating films, the passivation of the active mass is obtained before products such as ethylene gas have the chance to be accumulated in crevices and an internal pressure to grow. Indeed, in these solutions the irreversible capacity depends inversely on the size of the particles (as expected).

3.3. Experimental Evidence: Irreversible Capacity vs. Discharge Rate

It was found that in the case of EC-PC solutions the irreversible capacity depends on the discharge rate of the graphite electrodes. As the discharge rate is higher, the irreversible capacity is higher, as presented in Figure 6. This result reflects changes in the balance between the kinetics of particles cracking and the development of a proper passivating films, at different polarization rates: as the discharge rate is higher, the surface reactions occur more intensively and hence products such as propylene gas accumulate more quickly, while the solid products precipitate fast as well and block its way out (thereby, a fast build-up of internal pressure cracks the graphite particles). As a consequence, fresh surfaces of active mass are exposed to the solution, and hence, surface reactions occur intensively.

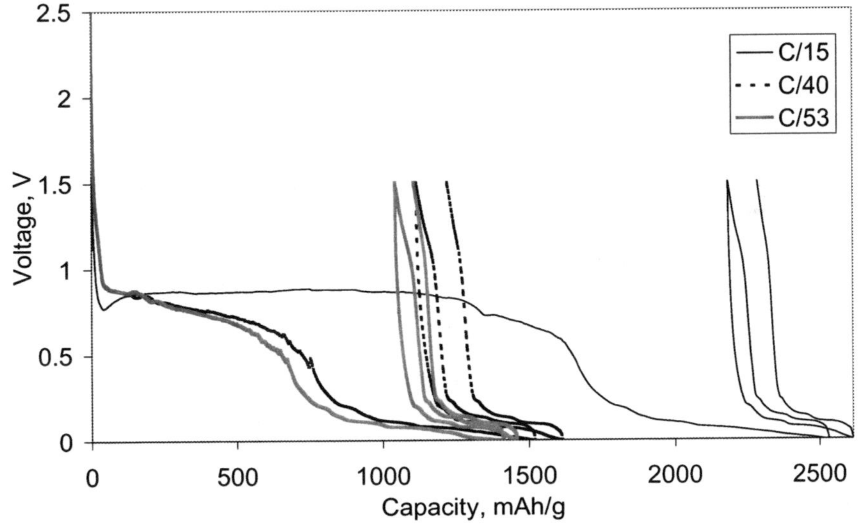

Figure 6. Potential vs. capacity curves obtained from cycling tests of synthetic graphite flakes in EC-PC/$LiClO_4$ solutions in different discharge rates. Notice that as the discharge rate decreases – the irreversible capacity decreases accordingly.

As the discharge rate is low enough, gaseous products are formed slowly, and have the chance to escape from the crevices before internal pressure builds up and cracks the particles.

3.4. Experimental evidence: AFM Imaging and Surface Chemistry

The main issue of this paper relates to the morphological aspects of the graphite particles. As described above, the three types of graphite particles, which are the subject of this study, differ in their pristine morphology, mainly in the roughness and the level of crevices in their edge planes (perpendicular to the basal planes).[25]

Figure 7 shows typical images obtained by *in situ* AFM measurements, which were conducted with electrodes comprising different types of graphite particles.[26] These measurements nicely demonstrate cracking of the graphite particles upon cathodic polarization of the electrodes, as was described above. This figure presents images of synthetic graphite, MCMB and natural graphite (NG) particles in an EC-PC solution at OCV, before polarization and deposition of surface films, and after the formation of surface films, during the first cathodic polarization of the electrodes.

Images "a" and "b" in Figure 7 belong to synthetic graphite flakes in the probe solution. As the electrodes are polarized from OCV (3V vs. Li/Li^+, Figure 7a.) to potentials below 1V, pronounced morphological changes are observed: the surfaces of the particles at their boundaries are increasingly lifted upwards. This is clearly seen when comparing images 7a and 7b while focusing on a gap between two particles. It is clear that the electrodes retain their integrity, and after the irreversible morphological changes observed by the AFM imaging, they can function electrochemically and reversibly in repeated Li insertion-deinsertion processes.

Pictures 7c and d are AFM images measured *in situ* during a first galvanostatic polarization of an electrode comprising MCMB (round) particles. The images obtained at OCV (not seen in the figure) reflect initially a rough surface. Image 7c relates to the carbon surface covered by surface films, whose morphology differs considerably from that of the pristine material. The surface films cover the entire surface and, in fact, hide the initial rough structure. As polarization proceeds, the imaging shows the clear formation of cracks that deepen and widen during polarization, during the irreversible surface processes (corresponding to the potential plateau around 0.8 V *vs.* Li/Li^+ that is seen in the galvanostatic curves). Upon further polarization, the cracks are filled by solution reduction products, which are formed within the cracks. Imaging electrodes comprised of MCMB particles provide a clearer insight into the failure mechanisms, compared to imaging of synthetic graphite electrodes, because MCMB particles have many upward-faced edge planes in the composite electrodes, on which the particles are cracked. Hence, cracking can be easily imaged during the cathodic polarization of the electrodes while surface reactions occur.[34] In the case of the synthetic graphite flakes, mostly the basal planes (the wider dimension of the particles) face upward, and hence cracking can be followed only indirectly, appearing as a lifting up of the particles' edges.

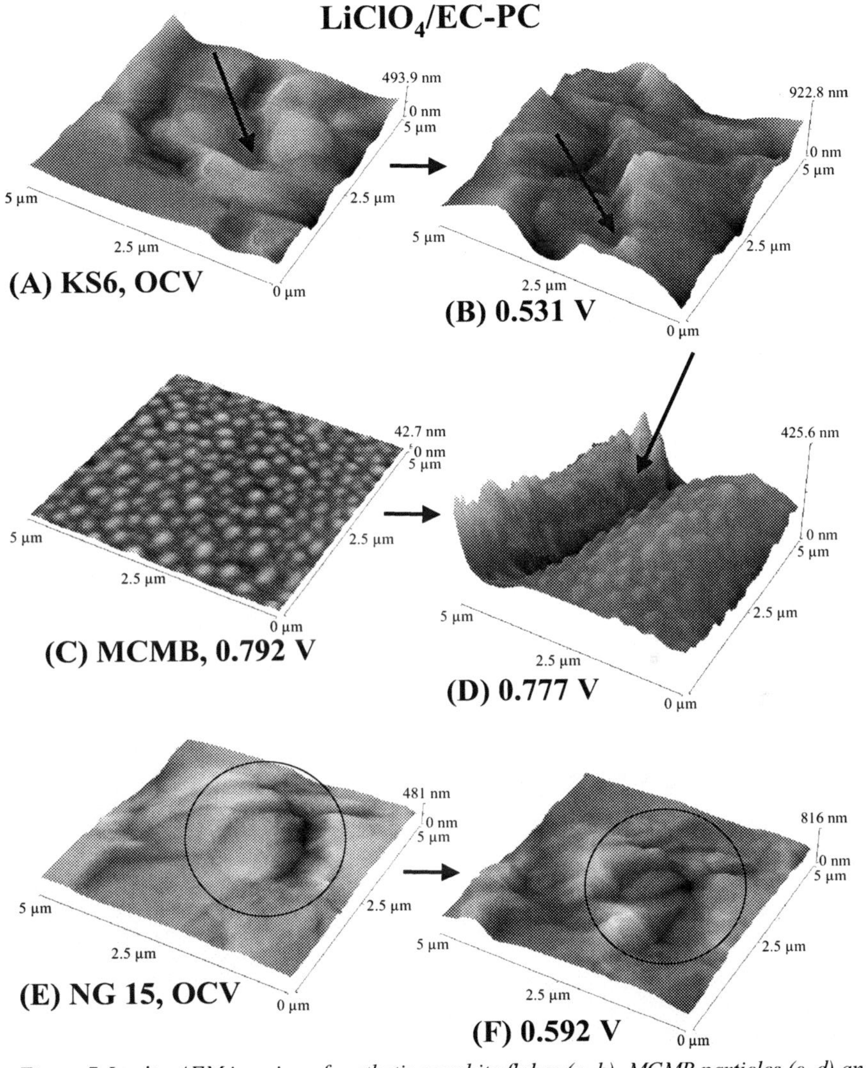

Figure 7. In-situ *AFM imaging of synthetic graphite flakes (a, b), MCMB particles (c, d) and natural graphite particles (e, f) during the first cathodic polarization of the electrodes in the probe solution ($LiClO_4$/EC-PC), measured at the indicated potentials vs. Li/Li^+. The arrows and circles point to the relevant morphological processes, as detailed in the text (see ref. 26).*

Finally, Figures 7e, 7f show AFM images of an electrode comprised of natural graphite flakes, measured *in situ* in the probe solution (EC-PC/$LiClO_4$) during the first galvanostatic polarization from OCV. The images reflect a relatively stable morphology with only minor changes, which should be attributed to the formation of surface films on top of the particles. While imaging natural graphite electrodes, one could hardly observe cracking of particles, as was observed with the synthetic flakes and the MCMB electrodes. It is suggested that electrodes comprising natural graphite flakes behave differently in PC-containing solutions because their edge planes are much smoother and contain fewer crevices, as compared to synthetic flakes or MCMB particles.

The possible detrimental build-up of pressure inside the graphite particles should depend critically on the quality of the electrode's passivation.

In cases where the solid reduction products of the solvent molecules form highly cohesive and adhesive surface films, the surface reactions are quickly blocked before further massive reduction of solution species (which also form the gas molecules) takes place. When passivation of the graphite is not reached quickly enough (as in the case of PC solutions), intensive surface reactions build up the internal pressure that cracks the particles and leads to their deactivation.

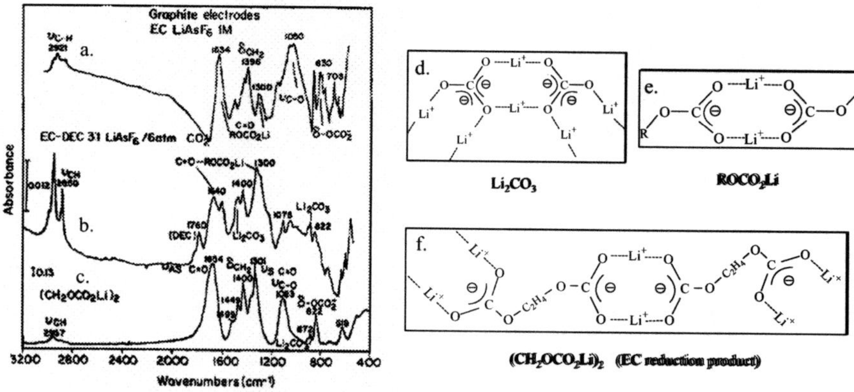

Figure 8. Typical FTIR spectra measured from graphite electrodes cycled in EC solutions at 25 °C (a, b, c) taken from ref. 21 and the relevant structures to which these spectra belong according to ref. 36 (d, e, f).

a. A graphite electrode after a first Li insertion-deinsertion in EC - $LiAsF_6$ 1 M.
b. Same as a, EC-DEC 3:1, 1 M $LiAsF_6$ solution, 6 atm CO_2.
c. FTIR spectrum of $(CH_2OCO_2Li)_2$ in KBr pellet.
d. The expected structure of Li_2CO_3 films.
e. The expected structure of Li_2CO_3 films.
f. The expected structure of $(CH_2OCO_2Li)_2$ films.

Figure 8 shows typical FTIR spectra measured from graphite electrodes cycled in different EC-based electrolyte solutions. The basic features belong to species such as $(CH_2OCO_2Li)_2$, as previously discussed in detail.[21] However, as was found by rigorous calculations,[36] these spectra belong to dimmers of these species, as also demonstrated in Figure 8.

The adhesion of the $ROCO_2Li$ species, formed by reduction of the alkyl carbonate solvents, to the carbon surfaces, as well as the cohesion of the $ROCO_2Li$ species, goes through the carbonate groups and the Li ions that span the negatively charged oxygen atoms[36] and the negatively charged carbon atoms (see Figure 8). EC and PC reduction forms $(CH_2OCO_2Li)_2$, and $CH_3CH(OCO_2Li)CH_2OCO_2Li$, respectively, which differ from each other in the methyl group. It is suggested that the $(CH_2OCO_2Li)_2$ species formed by EC reduction precipitate as adhesive and cohesive surface films more quickly and efficiently than PC reduction products, which have an extra methyl group. It is assumed that such a methyl group should interfere badly with both adhesion and cohesion of the Li dicarbonate species, due to steric hindrance (which does not exist with $(CH_2OCO_2Li)_2$).

From the above discussion, it is clear that the stabilization or failure of graphite electrodes depends on a delicate balance between passivation phenomena (due to the formation of highly cohesive and adhesive surface films) and a buildup of internal pressure due to the reduction of solution species inside crevices in the graphite particles. This delicate balance can be attenuated by both solution composition (EC-DMC vs. EC-PC or PC, etc.) and the morphology of the graphite particles (i.e. the structure of the edge planes and the presence of crevices).

4. SUMMARY

Understanding the failure mechanisms of graphite electrodes in Li insertion-deinsertion processes is an important issue in the field of Li batteries. In this respect, the failure of Li-graphite electrodes in PC solutions poses an interesting question, in light of the fact that in EC-based solutions lithiated graphite electrodes behave highly reversibly, even though EC and PC are so similar in their structure and properties.

In addition to the known cointercalation and exfoliation mechanism which provides a good explanation of the failure of graphite electrodes in ethereal solutions, this article suggests a mechanism for the failure of these electrodes in PC solutions, which is based on cracking of the particles, due to the build-up of internal pressure, that can be caused for instance, by the accumulation of gas inside the crevices in the edge planes of the particles during cathodic polarization of the electrodes. This mechanism is supported by electrochemical and morphological results. This paper compares three types of graphite electrodes; synthetic flakes, natural flakes and MCMB, which show different behavior in EC-PC/$LiClO_4$ (used as a probe electrolyte solution) in terms of irreversible capacity and stability. This difference is connected to the different morphologies of the graphite particles.

The different behavior of graphite electrodes in EC and PC solutions can be explained in light of their surface chemistry and the fact that in addition to surface species of the $ROCO_2Li$ (Li alkylene dicarbonates) type, reduction of both EC and PC form alkylene gas. When EC reduction dominates the surface chemistry, the $(CH_2OCO_2Li)_2$ thus formed can precipitate with good adhesion and cohesion to form highly passivating surface films. When the surface chemistry is dominated by PC reduction, the surface species that contain the methyl group are less cohesive and adhesive; thus, formation of passivating surface films requires an intensive reduction of solution species, and is accompanied by a massive propylene gas formation. Reduction of PC in crevices in the absence of a rapid formation of passivating surface films, may lead to an internal pressure buildup, which leads to the cracking and splitting of the particles and to an increase of the surface area, and hence, massive surface reactions take place and form surface films that electrically isolate most of the active mass. In cases of EC-PC mixtures, where EC reduction products also contribute to the formation of passivating surface films, electrodes comprised of natural graphite flakes whose edge planes are relatively "smooth" can behave reversibly with relatively low irreversible capacity loss. However, when the electrodes are comprised of synthetic flakes or MCMB whose edge planes have a high concentration of crevices, the passivation in an EC-PC based solution is not sufficient to avoid the buildup of internal pressure, which cracks the particles. Nevertheless, cracking of these particles in EC-PC based solutions is not massive enough to interfere adversely with the electrode integrity, nor to changes in the particles' orientation that lead to an electrical isolation of the active mass, as happens in pure PC solutions. Therefore, the electrodes can reach passivation in EC-PC solutions after pronounced consumption of irreversible charge and then lithium can be inserted and deinserted reversibly.

ACKNOWLEDGMENT

Partial support for this work was obtained by the BSF, Israel-USA binational foundation.

REFERENCES

1. Li-ion Batteries, Fundamentals and Performance; Wakihara, M.; Yamamoto, O., Eds.; Wiley-VCH: Weinheim (1998).
2. Dahn, J. R.; Sleight, A. K.; Shi, H.; Reimers, J. N.; Zhong, Q.; Way, B. M. *Electrochim. Acta*, **38**, 1179 (1993).

3. Aurbach, D.; Gofer, Y. in *Nonaqueous Electrochemistry*, D. Aurbach, Editor, Chap. 4, Marcel Decker, Inc., New York (1999).
4. Peled, E. in *Lithium Batteries*, J. P. Gabano, Editor, Chap. 3, Academic Press, Inc., New York (1983).
5. Aurbach, D.; Youngman Chusid, O.; Carmeli, Y.; Babai, M.; Y. Ein-Eli *J. PowerSources*, **43**, 47 (1993).
6. Aurbach, D.; Ein-Eli, Y.; Markovsky, B.; Carmeli, Y.; Yamin, H.; Luski, S. *Electrochem. Acta*, **39**, 2559 (1994).
7. Fong, R.; von Sacken, U.; Dahn, J. R. *J. Electrochem. Soc.*, **137**, 2009 (1990).
8. Naji, A.; Willmann, P.; Billaud, D. *Carbon*, **36**, 1347 (1998).
9. Yang, C. R.; Wang, Y. Y.; Wan, C. C. *J. Power Sources*, **72**, 66 (1998).
10. Winter, M.; Nova'k, P.; Monnier, A. *J. Electrochem. Soc.*, **145**, 428 (1998).
11. Winter, M.; Besenhard, J. O. in *Handbook of Battery Materials*, J. O. Besenhard, Editor, Part III, Chap. 5, Wiley-VCH Verlag GmbH, Germany (1999).
12. Rosoken, J. M.; Decker, F. *J. Electrochem. Soc.*, **143**, 2417 (1996).
13. Tran, T. D.; Feikert, J. H.; Pekala, R. W.; Kinoshita, K. *J. Appl. Electrochem.*, **26**, 1161 (1996).
14. Winter, M.; Besenhard, J. O.; Spahr, M. E.; Nova'k, P. *Adv. Mater.*, **10**, 725 (1998).
15. Dahn, C. H.; Kim, H. S.; Moon, S. I. *J. Power Sources*, **101**, 96 (2001).
16. Lampe-Onnerud, C.; Shi, J.; Onnerud, P.; Chamberlain, R.; Barnett, B. *J. PowerSources*, **97-98**, 133 (2001).
17. Oliver, J.; Winger, M. *J. Power Sources*, **97-98**, 151 (2001).
18. Joho, F.; Rykart, B.; Blome, A.; Novak, P.; Hilkelm, H.; Spahr, M. E. *J. PowerSources*, **97-98**, 78 (2001).
19. Zanem, D.; Antonni, A.; Pasquali, M. *J. Power Sources*, **97-98**, 146 (2001).
20. Zaghib, K.; Nadeau, G.; Kinoshita, K. *J. Power Sources*, **97-98**, 97 (2001).
21. Aurbach, D.; Markovsky, B.; Weissman, I.; Levi, E.; Ein-Eli, Y. *Electrochim. Acta*, **45**, 67 (1999).
22. Chung, G. C.; Kim, H. J.; Jun, S. H.; Choi, J. W.; Kim, M. H. *J.Electrochem. Soc.*, **147**, 4398 (2000).
23. Wang, Y.; Balbuena, P. B. *J. Phys. Chem.*, **A105**, 9972 (2001).
24. Aurbach, D.; Levi, M. D.; Levi, E.; Schechter, A. *J. Phys. Chem. B*, **101**, 2195 (1997).
25. Aurbach, D.; Teller, H.; Levi, E. *J. Electrochem. Soc.*, **149**, A1255 (2002).
26. Aurbach, D.; Koltypin, M.; Teller, H. *Langmuir*, **18**, 9000 (2002).
27. Aurbach, D.; Teller, H.; Koltypin M.; Levi, E. *J. PowerSources*, **119-121**, 2 (2003).
28. Cohen, Y.; Aurbach, D. *Rev. Sci. Instrum.*, *70*, 4668 (1999).
29. Aurbach, D.; Markovsky, B.; Levi, M. D.; Levi, E.; Schechter, A.; Moshkovich, M.; Cohen, Y. *J. Power Sources*, **81-82**, 95 (1999).
30. Peled, E.; Golodnitsky, D.; Pencier, J. in *Handbook of Battery Materials*, Besenhard, J. O. Editor, Part III, Chap. 6, Wiley-VCH Verlag GmbH, Germany (1999).
31. Aurbach, D.; Markovsky, B.; Schechter, A.; Ein-Eli Y.; Cohen, H. *J. Electrochem. Soc.*, **143**, 3809 (1996).
32. Koltypin, M.; Cohen, Y.S.; Cohen, Y.; Markovsky, B.; Aurbach, D. *Electrochem. Commun.*, **4**, 17 (2002).
33. Mortinent, A.; Le Gorrec, B.; Montella, C.; Yazami, R. *J. PowerSources*, **97-98**, 83 (2001).
34. Aurbach, D.; Moshkovich, M.; Gofer, Y. *J. Electrochem. Soc.*,**148**, E155 (2001).
35. Aurbach, D.; Gottlieb, H. *Electrochim. Acta*, **34**, 141 (1989).
36. Matsuta, S.; Asada, T.; Kitaura., K. *J. Electrochem. Soc.* **147**, 1695-1702 (2000).

NEW DEVELOPMENTS IN THE ADVANCED GRAPHITE FOR LITHIUM-ION BATTERIES

Francois-Xavier Henry, Igor V. Barsukov, Joseph E. Doninger, Scott Anderson, Peter R. Booth*, Peter L. Zaleski, Richard J. Girkant, David J. Derwin, Maritza A. Gallego, Tomás Huerta, and Gabriela Uribe

Superior Graphite, Peter R Carney Technology Center
4201 W. 36th Street, Chicago, IL 60632, USA, www.SuperiorGraphite.com

Abstract

In this paper, three generations of natural graphite are described in order to show the evolution in understanding of which physical and electrochemical requirements are needed to make usable graphite for practical lithium-ion batteries. Of the three alternative types of carbonaceous materials discussed in the paper, the optimum has been found to be natural graphite. Its behavior has been compared with the behavior of graphitic mesophase carbon microbeads and hard carbon materials. The comparison allowed summarizing key guidelines to design new generations of active materials. The capacity, irreversible capacity loss, rate capability, safety and packing density are some of the key characteristics, we concluded, are essential when developing new generations of high packing density and low surface area graphitic carbons.

1. INTRODUCTION

Due to its high energy density (3,860 mAh/g) and low voltage, lithium is the most attractive metal of the periodic table for battery application. Unfortunately lithium metal, and most of its alloys cannot be used in rechargeable batteries because of their poor cyclability. Therefore, lithium intercalation compounds and reversible alloys are among today's materials of choice for subject application. The most common active materials for the negative electrodes in lithium-ion battery applications are carbonaceous materials. The ability of graphitized carbonaceous materials to

* Corresponding author. E-mail: PBooth@SuperiorGraphite.com

I.V. Barsukov et al. (eds.), New Carbon Based Materials for Electrochemical Energy Storage Systems, 213–229.

be intercalated makes them excellent reservoirs of lithium [1]. Even with a maximum theoretical capacity of 372 mAh/g (stage I of the graphite intercalation compounds with lithium, LiC_6), which is an order of magnitude lower than the theoretical capacity of lithium metal, these materials can be efficiently cycled, when lithium metal cannot.

Synthetic carbonaceous materials are widely used in these applications. Several types of synthetic materials (e.g. graphitized mesophase carbon microbeads (MCMB), graphitized milled carbon fiber, and even, initially, hard carbons) became the materials of choice at the time of commercialization of first successful lithium-ion batteries in late 1980s. New trends, mainly driven by cost reduction and need for improved performance, currently shift focus towards application of natural graphite.

Besides lower cost, natural graphite has several advantages when compared to its synthetic counterparts. One of them is higher reversible capacity of natural graphite. The downside of natural graphite is that unfortunately, its almost perfect lamellar structure makes processing of graphite uneasy. Natural graphite is also more sensitive to co-intercalation of certain solvents, which is usually being associated with increase of the irreversible capacity and reduction of cell's cyclability [2].

The irreversible capacity results from formation of a surface-electrolyte interface (SEI) layer, and is believed to be caused by decomposition of the electrolyte on the surface of active material during few first charge cycles [3-5]. The values of irreversible capacity and the SEI are functions of the type of active material and the electrolyte. Also, the safety issue, which is believed to be associated with stability of SEI, has been identified as a major parameter in the equation [6-7]. The contribution of the negative electrode to the thermal runaway is believed to be related to the nature and also to the surface area of the active material [8-9].

Said subjects are being analyzed in this work. Also, the authors have attempted to show that in order to be suitable for lithium-ion applications, a carbon-based active material has to meet a complex number of physicochemical and electrochemical characteristics. A simple check of galvanostatic behavior, which is often used today to conclude about carbon's suitability for lithium-ion battery technology, is rarely enough for making an accurate assessment.

2. EXPERIMENTAL

2.1. Graphite Purification

Natural flake graphite has been initially upgraded by flotation technologies to reach a carbon level in the order of 95%C. Such precursor has been heat-treated at high temperature to bring the carbon content up to

99.95%C, and to eliminate potential battery poisons contained in graphite's ash. The sublimation of impurities is performed on an industrial scale in Superior Graphite's continuous thermal purification furnaces under inert gas. During the process of purification, the temperature, which flake is exposed to, reaches up to 2,800°C.

2.2. Graphite Processing

The purified flake is ground on various industrial mills in order to reduce size of the flakes and to synthesize application-specific morphologies.

In order to generate proper size distribution of graphite powder, the milling parameters are adjusted. Changes in particle size distributions for the end-products are controlled with laser particle size analyzers (example: Microtrac-X100, of Microtrac, Inc., Montgomeryville, PA, USA).

2.3. Physicochemical Characterization

The surface area of graphite is determined by nitrogen adsorption using a BET single point method (equipment is available from Quatachrome Instruments, Boynton Beach, FL, USA).

The packing density of graphite is determined by the tap density analysis. In this test, we are determining the volume of 100g of graphite powder after automatic tapping (Quatachrome Autotap model) of a graduated cylinder filled with a powdered sample, for 1,500 times in approximately 4 minutes. During tapping, the travel height of a graduated cylinder with graphite is 3.2 mm.

The Microtrac-X100 (a laser particle size analyzer) was used to determine the size distribution of the powder. The method is based on a liquid mode of operation, that is before laser scanning, the powder is added to a solution of deionized water, containing a surfactant.

The morphology of particles was assessed by synergistic analysis of packing density values, along with Scanning Electron Microscopy imaging. Most of the pictures of powders used in this paper were taken using JEOL's JSM-6320F instrument outsourced from Drexel University, Philadelphia, PA, USA.

Two methods were used to determine the purity of the final material. The ash content was obtained by calculation of a weight loss of 1 g size graphite sample, after its exposure to 900°C for 4 hours (Superior Graphite AIMS procedure FBQ0001). This test gives a good idea of the total amount of impurities. The nature of impurities and their specific concentrations as measured in ppm or ppb levels, was determined with an ICP inductively coupled plasma spectrometer (Model Jobin Yvon Ultima POXX/681, available from Horiba Group, Longjumeau, France) after leaching impurities out from graphite in a mixture of nitric and sulfuric concentrated acids.

Moisture has been determined by Karl Fisher moisture titration at 300°C (model Metrohm 756 KF Coulometer with 707 KF oven, available from Brinkmann Instruments, Inc. of Westbury, NY, USA).

2.4. Electrochemical Characterization

The electrochemical testing of active materials has been performed on a 24-channel battery cycler (model BT2000 Basic Charge, available from Arbin Instruments of College Station, TX, USA).

The electrodes were prepared by mixing graphite powder with the NMP-predissolved PVDF binder (grade KF#9306, available from KUREHA Advanced Materials Div., Tokyo, Japan) at the ratio 9/1 of C/PVDF, respectively. The viscosity of the slurry was adjusted at 5,000 cps by adding more solvent (NMP).

The slurry was pasted on a copper foil at slow motion (<2.54 cm/s) using an automatic draw down coater available from Paul N. Gardner Co. of Pompano Beach, FL, USA) and the electrode was dried for a day under vacuum at 120°C. After this drying step, the density of the coating was adjusted to ~1.45-1.55 g/cc through calendaring operation.

The carbon / Li metal half-cells were built inside the Ar-filled glove box (UniLab system available from MBraun Inc. of Stratham, NH, USA).

The design of the 2-electrode electrochemical cells used in this work is state of the art and is based on HS Test Cell available from Hohsen Corp. of Tokyo, Japan [10]. Unless stated otherwise, these cells were tested at 32°C at different cycling rates (from C/20 to C rate) under continuous current. The electrolyte used was EC:DMC (1:1), $LiPF_6$ (1M) made by Cheil Industries, South Korea.

3. RESULTS AND DISCUSSION

3.1. Electrochemical Prescreening

The capacity and profile of the galvanostatic cycling curves are among the first key characteristics of the active material.

The first galvanostatic charge/discharge at C/20 rate for a natural graphite (grade LBG1025 of Superior Graphite, Chicago, IL, USA), hard carbon (grade Carbotron of Kureha Co., Tokyo, Japan) and graphitized mesophase material (grade MCMB2528 of Osaka Gas Co., Tokyo, Japan) are presented in Figure 1.

The hard carbon has a totally different electrochemical behavior than the natural graphite and the MCMB2528. Close to half of its capacity during the intercalation and during the de-intercalation manifests itself at a potential above 0.4V versus lithium, when it is less than 10 percent of that for the graphitic active materials. The fact that there is no staging process, as observed on the curves for hard carbon, is correlated to the crystalline structure of the material. Hard carbons have an extremely limited lattice organization and the electrochemical curve without any plateaus shows this clearly [11-12]. In these disorganized materials, Lithium is believed to be adsorbed on the available sites.

This, sometimes, allows hard carbons to have higher capacity than the theoretical capacity of graphite (was not seen to be the case with a particular sample of commercial hard carbon). During the charge and discharge, the potential decreases or increases almost continuously. Two other highly graphitic materials show a classic staging process during cycling. Each plateau corresponds to a specific phase or stage [13-14].

During intercalation, the first plateau that appears at approximately 0.2V versus lithium corresponds to the stage IV. For this stage, four graphine layers separate two layers of intercalated lithium. Stages II and III can be observed before the first stage appears at a potential close to 90 mV. The theoretical capacity of purified graphite (372 mAh/g) is based on composition of stage I, (LiC_6).

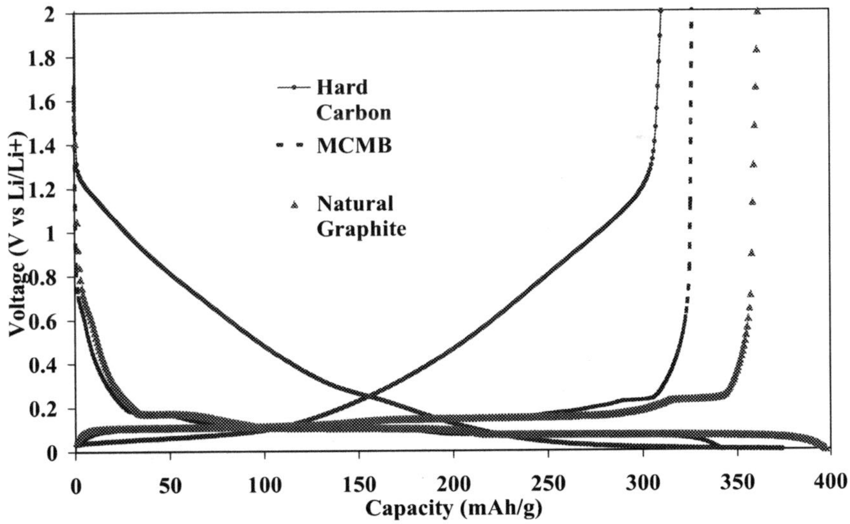

Figure 1. Initial Galvanostatic Charge/Discharge Curves of Three Types of Active Carbonaceous Materials at C/20 rate.

The electrochemical characteristics of these three materials are summarized in Table 1. The reversible capacity, the irreversible capacity loss and capacity below a certain voltage (0.5V in this case) were identified to be the key important electrochemical parameters at the time.

These results show that purified natural graphite has the highest capacity and the highest capacity below 0.5 V. The irreversible capacity is favorable in case of MCMB2528. With its low capacity and its higher irreversible capacity loss, the hard carbon doesn't seem to be an interesting candidate for conventional lithium-ion battery applications, as its application will result in reduced cell discharge voltages.

Table 1. Electrochemical characteristics at C/20 rate of a commercial hard carbon, MCMB 2528 and a natural graphite (LBG1025) in half cells with EC-DMC (1:1), $LiPF_6$ (1M).

Material / Performance Property	Hard Carbon	MCMB2528	Natural Graphite (LBG1025)
Reversible capacity (mAh/g)	*310*	*330*	*360*
Capacity < 0.5 V (mAh/g)	*206*	*320*	*354*
Irreversible Capacity (%)	*16.8*	*5*	*8.9*

From these data, the two types of materials that can be considered as suitable for lithium-ion applications are MCMB and natural graphite. MCMB are widely used at the moment. Data in Table 1 shows that natural graphite can be an attractive candidate for the application in question as well.

Five years ago, we were considering that these electrochemical data were enough to define if an active material would be a good candidate for lithium-ion batteries. Based on this understanding, Superior Graphite developed its "LBG" grades.

3.2. Natural Graphite LBG1025: Electrochemical Results

Figure 2 presents the results of galvanostatic cycling of LBG1025. The data is summarized in Table 2. Our two design guidelines were achieving low irreversible capacity and the highest possible capacity at high rate.

Knowing that for any pure natural graphite the capacity at low rate should be very close to the theoretical capacity, it did not become a surprise when the reversible capacity value of LBG1025 reached 365 mAh/g at C/20 rate; it was still as high as 349.5mAh/g at C/5 rate.

Due to issues of polarization in a specific cell design we use for testing, the cut-off potential at C/2 is fixed at – 40mV vs Li/Li^+. At this negative potential we never observed with LBG material any lithium plating.

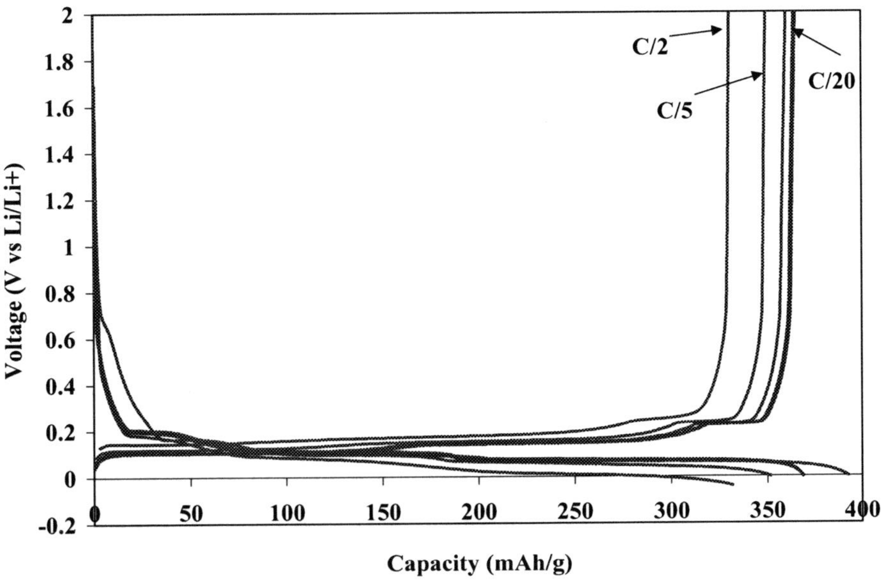

Figure 2. Galvanostatic charge/discharge curves of natural graphite LBG1025 ***at C/20, C/5 and C/2 rates. Electrolyte: EC-DMC (1:1) $LiPF_6$ (1M).***

Table 2. Electrochemical characteristics of natural graphite LBG1025 in half cells Electrolyte: EC-DMC (1:1) $LiPF_6$ (1M).

Material / **Performance Property**	**Charge capacity at 0V vs Li/Li^+, mAh/g**	**Discharge capacity (2V vs Li/Li^+), mAh/g**	**Irreversible capacity loss, %**
First cycle at C/20	*392.8*	*360.2*	*8.3%*
Second cycle at C/20	*364.2*	*364.2*	-
Third cycle at C/20	*368.7*	*365.2*	-
C/5	*351*	*349.5*	-
C/2	*287* *331 at (-0.04V)*	*330*	-

Under these conditions (intercalation from 2V to -0.040V vs Li/Li^+), the capacity during de-intercalation is 330mAh/g. At C/2 rate, the capacity at 0V during intercalation is 287mAh/g. This still represents 77% of the theoretical capacity value for graphite. LBG1025 clearly shows some good high rate performance and the irreversible capacity of 8.3% is well below the 10%, which we were targeting to achieve in our development goals.

3.3. Natural Graphite LBG1025: Physical Characteristics and Behavior in Lithium-ion Electrode Assemblies

The physical characteristics of the powder and the mechanical properties of the electrode made from these powders were seen to be among key important parameters. Some physical characteristics of the LBG1025 and its typical Scanning Electron Microscope image can be found in Table 3. The SEM shows a flaky, rounded edge smooth morphology.

Table 3. Physical characteristics and typical Scanning Electron Microscope image of purified natural graphite LBG1025.

Surface Area (m^2/g)		*<4.9*	
Tap Density (g/cc)		*0.47*	
Particle Size Distribution (μm)	D_{10}	*<9*	
	D_{50}	*19*	
	D_{90}	*<35*	

Mechanical properties of the electrode turned out to be a critical factor not clearly understood at the time of designing of LBG1025 graphite. Authors have learned that a battery cannot be cycled if the coating does not adhere properly to the current collector.

There are some standard adhesion tests used by the paint and coatings industry, which we have adopted for our work. The most common is the "Scotch tape" test, which is described by an ASTM D3359, method B.

In this test, a strip of a specific grade of a Scotch tape is applied to the coating whose surface has been cross-cut with a multi-teeth scratching tool, and then removed. The percentage of coating that remains on a substrate is assessed. Usually, less than 65% of coating left on a substrate suggests poor adhesion. The manual version of this test used in our investigations is somewhat operator dependent but it gives a good idea about the level of adhesion of graphite coating to the copper current collector.

Figure 3 presents results of adhesion for LBG1025/PVDF composites. As one can see, in the particular testing conditions, LBG1025 has a poor adhesion (100% of the coating has been removed by the Scotch tape even without a need to make cross-cut scratches on the surface of current collector).

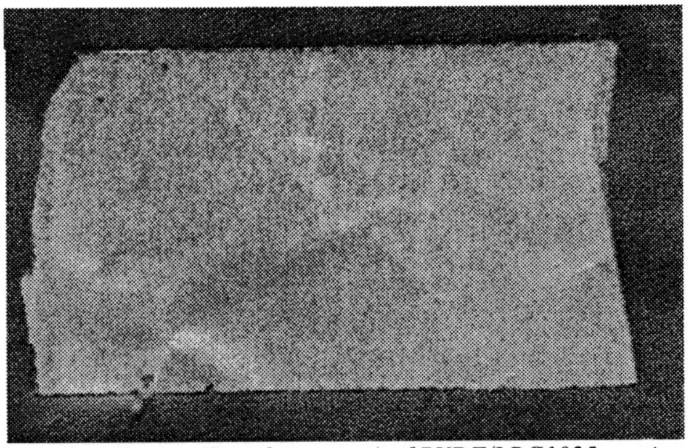

Figure 3. Adhesion (Scotch tape test) of PVDF/LBG1025 coating showing poor adhesion of graphite to copper current collector.

Analysis of results reported for natural graphite LBG1025 allowed us to draw a preliminary conclusion that regardless of excellent (near theoretical) electrochemical performance, this graphite, in its pure form, is not suitable for application in the negative electrodes of lithium-ion batteries. This is true at least in cases when LBG1025 is used as a sole active carbon ingredient of the anodes. Noteworthy mentioning: this grade of natural purified graphite has been found by the industry to be of efficient use as diluents for MCMB (up to 50% dilution) with primary goal of cost reduction of the negative electrode.

Partially disappointing results with LBG1025 helped us define new targets needed to be achieved in order to develop a successful natural graphite-based material for application in the lithium-ion batteries. New product design guidelines incorporated a combination of key physical characteristics (packing density, morphology) and the electrochemical behavior (capacity, irreversible capacity loss). Following these guidelines, Superior Graphite has developed the “SL” line of products, which is described in the following section.

3.4. The “SL” / “SLA” Product Line of Thermally Purified Natural Graphite: Electrochemical Results

Figure 4 presents typical results of galvanostatic cycling of purified natural graphite SLA1020 (formerly known as SL-20) from C/20 to C rates. The results have been summarized in Table 4.

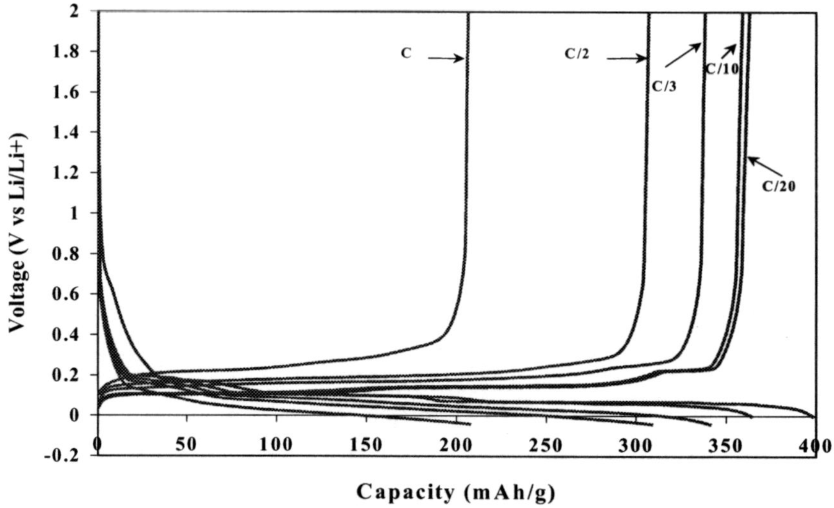

Figure 4. Galvanostatic charge/discharge curves of natural SLA1020 at C/20, C/10,C/3, C/2 and C rates. Electrolyte: EC-DMC (1:1) $LiPF_6$ (1M).

Table 4. Electrochemical characteristics purified natural graphite SLA1020 in half cells. Electrolyte: EC-DMC (1:1) $LiPF_6$ (1M).

Properties	Charge (intercalation) at 0V, mAh/g	De-intercalation at 2V, mAh/g	Irreversible capacity, %
First cycle at C/20	*398.1*	*362.7*	*8.8%*
C/10	*363.8*	*358.9*	-
C/3	*315 (341 at -0.04V)*	*338.4*	-
C/2	*257 (308 at -0.04V)*	*306.8*	-
C	*151 (207 at -0.04V)*	*206*	-

The capacity of SLA1020 at C/20 is 362mAh/g and the irreversible capacity loss is 8.8%. From C/20 to C/3 the profile of the galvanostatic cycling curve features the plateaus, which are specific to highly graphitic materials. At higher rates (C/2 and C), the curve is smooth; there are almost no plateaus and the capacity drops significantly at the C rate.

We believe that this type of smooth profile results from kinetic limitations. Still at C/2, the capacity with cutoff potential 0.0V vs Li/Li^+ is 257mAh/g. It drops to about 151mAh/g at C rate. When compared with

LBG1025 (discussed in the previous sections of this paper), the electrochemical performance of SLA1020 is slightly lower in terms of material's rate capability. This becomes noticeable at C/2 and C rates. From what we have reported previously, the electrochemical behavior should be considered as only one of several critical parameters.

3.5. Natural Graphite SLA1020: Physical Characteristics and Behavior in Lithium-ion Electrode Assemblies

Natural purified graphite SLA1020 has near similar particle size distribution than LBG1025, which has been discussed earlier. However, due to spheroidal morphology of SLA1020, its packing density exceeds 0.8 g/cc (this is almost a factor of 2 of the packing density of LBG1025).

Thus SLA1020 features a particular type of morphology, which we were targeting to manufacture in order to improve the packing density, the adhesion to the copper and the cycling behavior. The adhesion test of an electrode prepared with SLA1020 is presented by Figure 5.

The adhesion is noticeably improved over that with LBG1025. When doing the "Scotch tape" test, as described by an ASTM D3359, method B (discussed in detail earlier), only a fraction of the anode coating could be peeled off the copper substrate.

Table 5. Physical characteristics and typical Scanning Electron Microscope image of purified natural graphite SLA1020 (formerly known as SL-20).

Surface Area (m^2/g)		*< 5.5*
Tap Density (g/cc)		*>0.8*
Particle Size Distribution (µm)	*D_{10}*	*<11*
	D_{50}	*20*
	D_{90}	*<35*

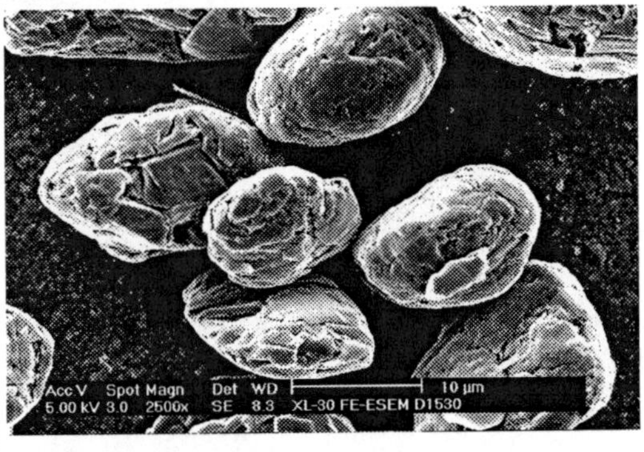

Authors believe that this shows an effect of particle morphology on the adhesion of the coating to the current collector. Having a spheroidal morphology has also the advantage to limit negative effects of specific orientation of the particles of graphite parallel to the current collector. The term of "specific orientation", is often associated with flaky morphologies, and is believed to be the reason for poor cyclability of flaky graphite that have a tendency to expend during the intercalation in a direction perpendicular to the current collector. This process is highly undesirable in practical batteries. It can be associated with loss of contact between copper current collector and active material.

The guidelines, which we have followed during the development of SLA1020 and similar types of graphite (those identified by Superior Graphite's product nomenclature as "SL" and "SLA" family of products) were based on an improved understanding of a matter that for a particular application in lithium-ion batteries it is not enough to have a material with good electrochemical performance; rather, material's packing and adhesion have to be improved accordingly.

As a consequence, a significant deal of positive response has been received with regards to SLA1020 graphite from a number of end-users. Application of this graphite was found to be acceptable by certain lithium-ion battery technologies, yet its use did not become overwhelming throughout the lithium-ion battery industry, suggesting there has been room left for further improvement of this grade.

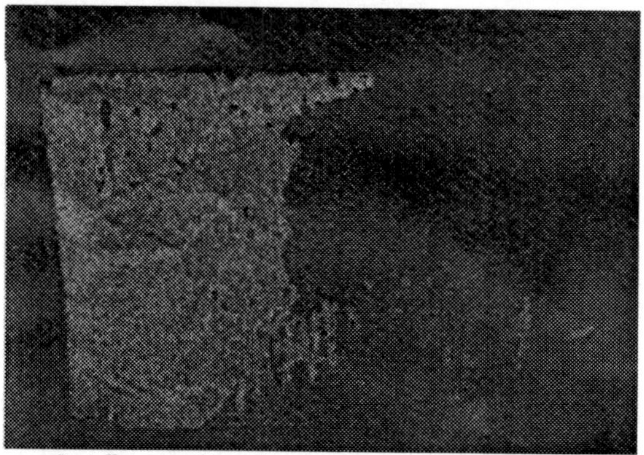

Figure 5. Adhesion (Scotch tape test) of PVDF/SLA1020 coating showing improved adhesion of graphite to copper current collector.

After having developed SLA1020 and two other grades of similar nature (one being finer, and another one being coarser), a new parameter missing in the specification had to be addressed. That is ability of graphite to withstand prolonged calendar life and long term cycling.

Novel design approaches to addressing these issues resulted in introduction of the "SLC" family of products, which is at the active stage of commercialization by Superior Graphite at the time of writing of this paper.

3.6. The SLC Family of Purified Natural Graphite

We have learned that contribution of the negative electrode of lithium-ion battery to the on-set temperature and kinetics of the thermal runaway reaction, as well as battery's ability to demonstrate increased calendar life and long term cycling performance depends on the type of active carbonaceous material, its morphology, and also on material's specific surface area [9, 15]. In particular, we have learned that "ideal" graphite that meets above design criteria should feature reduced surface area.

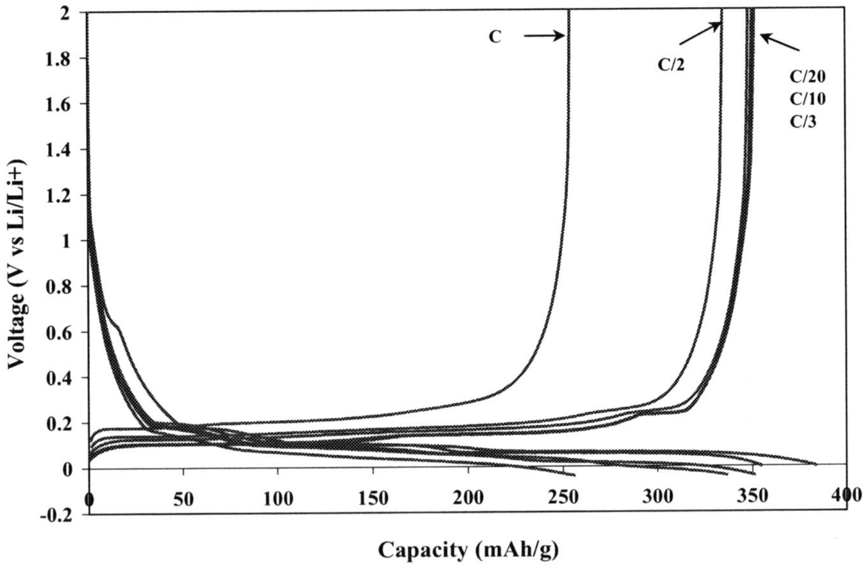

Figure 6. Galvanostatic charge/discharge curves of natural SLA1015 at C/20, C/10,C/3, C/2 and C rates. Electrolyte: EC-DMC (1:1) $LiPF_6$ (1M).

SLC1015 is what Superior Graphite has developed in order to provide lithium-ion battery market with reduced surface area graphite.

SLC1015 is a natural graphite-based product, which features spheroidal shapes of particles and a surface area of below 1.2m^2/g. Said significant reduction of material's specific surface area has been achieved through special surface treatment of natural graphite surface. This treatment is believed to be resulting in blocking cavities and cracks in precursor natural graphite particles, many of which are thought to be responsible for increased surface activity (increased surface area) of graphite.

3.7. Natural Graphite SLC1015: Electrochemical Results

The electrochemical galvanostatic charge-discharge curves at different rates are presented by Figure 6. The data is also summarized in Table 6. In our testing, the irreversible capacity loss of SLC1015 was seen to be in the range of 6 to 8.5% (function of electrolyte used). The capacity of SLC 1015 even at C/2 rate (continuous charge) is 282mAh/g at a potential of 0.0V vs Li/Li$^+$ reference electrode.

3.8. Natural Graphite SLC1015: Physical Characteristics and Behavior in Lithium-ion Electrode Assemblies

The proprietary surface treatment done on SLC1015 and other SLC grades available from Superior Graphite reduces dramatically the specific surface area of this group of carbon products.

Table 6. Electrochemical characteristics purified natural graphite SLA1020 in half cells. Electrolyte: EC-DMC (1:1) LiPF$_6$ (1M).

Properties	Charge (intercalation) at 0V, mAh/g	Discharge at 2V, mAh/g	Irreversible Capacity, %
First cycle at C/20	*383.5*	*351.8*	*8.3%*
C/10	*354.3*	*350.4*	-
C/3	*322.7 (351 @ -0.04V)*	*348.3*	-
C/2	*282.3 (336.4 @ -0.04V)*	*335.1*	-
C	*225.5 (255.7 @ -0.04V)*	*254*	-

Table 7. Physical characteristics and typical Scanning Electron Microscope image of surface treated purified natural graphite SLC1015.

Surface Area (m^2/g)		*<1.2*
Tap Density (g/cc)		*>0.85*
Particle Size Distribution (μm)	*D_{10}*	*<11*
	D_{50}	*25*
	D_{90}	*<35*

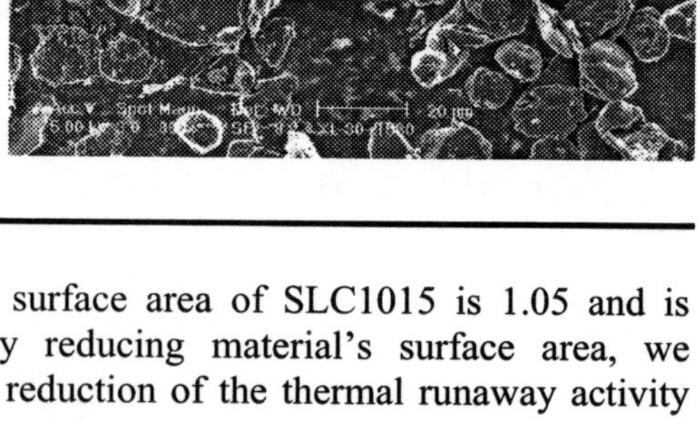

For instance, the typical surface area of SLC1015 is 1.05 and is statistically below 1.2 m^2/g. By reducing material's surface area, we expected to achieve a significant reduction of the thermal runaway activity of carbon (the later is believed to have correlation with improved abuse tolerance in the full cells and batteries). Also, extension of cyclability and calendar life of battery, we thought, could be achieved.

These expectations came true. Once developed, SLC1015 demonstrated increased packing density, reduced surface area and the preliminary results about its abuse tolerance performance are extremely encouraging.

A set of physical characteristics for the product under consideration is presented in Table 7. SLC1015's adhesion result is presented by Figure 7.

Figure 7. Adhesion (Scotch tape test) of PVDF/SLC1015 coating showing excellent adhesion of graphite to copper current collector.

The adhesion of SLC1015 is noticeably better than that with LBG1025 and also SLA1020, described earlier. Only a minute fraction of the SLC1015 based coating could be removed from the copper substrate when an adhesion check in accordance with the ASTM D3359, method B procedure has been performed. The interface "copper/coating" was seen to be much stronger than with any previously described generation of natural graphite.

Performance of the surface treated purified natural graphite SLC1015 and other similar grades in Superior Graphite's "SLC" product family has attracted a number of lithium-ion battery manufacturers. The number of satisfied end-users is steadily growing.

The above success of SLC product line indicates that design parameters taken as targets for development meet expectations of the lithium-ion battery application.

Besides "high energy density" batteries (lap top computers, cellular phone power sources, etc.), new material looks promising for the emerging "high power" segment of the battery market. For instance, authors would like to make a reference to the important work by developers at the Argonne National Laboratory, USA, in qualifying the SLC1015 graphite as a viable candidate for electrode active materials for the emerging application in hybrid electric vehicles. The SLC1015 graphite has been reported to be fully suitable for said application, as can be seen in [16, 17].

4. CONCLUSIONS

The comparison in between natural graphite and other carbonaceous materials has shown that natural graphite having sufficient purity and an optimal set of surface properties can be an outstanding candidate for lithium-ion battery applications.

In this work we have considered three generations of natural graphite. Using examples of these three product families, the authors have shown evolution in understanding of requirements necessary to achieve for an active material to be used at an industrial level. The progress achieved in the processing of natural graphite currently allow Superior Graphite to produce very low surface area graphite having high packing density. The packing density of regular SLC type materials is in between 0.80 to 1.1g/cc with surface area in between 0.85 to 1.5 m^2/g.

Given the complex process to produce mesophase carbon (graphitized microbeads and fibers), natural graphite can be very competitive in terms of its manufacturing costs [18]. The physical characteristics of certain SLC type materials are extremely close to the characteristics of state-of-the-art MCMB grades.

At the electrochemical performance level, these novel natural graphite-based materials surpass mesophase carbon's characteristics as related to cell/battery safety performance, low irreversible capacity loss, and good rate capability even at high current densities.

At a broader level, a battery is a complex product and the negative electrode is just one of its many components. The electrochemical behavior of graphite depends on a lot of parameters (such as, for instance, current collector, binder, electrolyte [19], special additives, application, etc.). LBG1025 is a good example to show that the electrochemical performance data is only one piece of the puzzle. Good initial galvanostatic performance of graphite is not enough for material to be suitable for lithium-ion battery application. Laboratory testing of initial performance of carbon is always necessary, but at the end, the only way to test the material is in real application, in full lithium-ion cells and batteries.

REFERENCES

1. Hossain, Sohrab. Rechargeable Lithium Batteries (Ambient Temperature). In Handbook of Batteries, David Linden ed., New York, N.Y., 36.1-36.77, 1995.

2. Winter M., Novak P. and. Monnier A. Graphites for Lithium-ion Cells: The Correlation of the First-Cycle Loss with the Brunauer-Emmett-Teller Area. J. Electrochem. Soc., 145, 428-435 (1998).
3. Lithium Batteries. E. Peled, J.P.Gabano, eds., Chap 3, 43-69 Academic Press, London (1983).
4. Naji A., Willmann P., and Billaud D. Electrochemical Intercalation of Lithium into Graphite: Influence of the Solvent Composition and the Nature of the Lithium Salt. Carbon, 36, 1347-1352 (1998).
5. Endo E., Tanaka K. and Seika K. Initial Reaction in the Reduction Decomposition of Electrolyte Solutions for Lithium Batteries, J. Electrochem. Soc., 147, 4029-4033 (2000).
6. Henriksen G. *Advanced Process Research: Overview/Advances./* Proceedings of the Annual Merit Review Meeting of the U.S. DOE ATD Program, Argonne, IL, (8/2001).
7. Cao F., Barsukov I.V., Bang H.J., Zaleski P. and Prakash J. Evaluation of Graphite Materials as Anodes for Lithium-Ion Batteries. J. Electrochem Soc. 147 (10), 3579-3583 (2000).
8. Barsukov I.V. Development of low-cost, novel carbonaceous materials for anodes in lithium-ion rechargeable batteries – Superior Graphite Co. Snapshots of CARAT (Cooperative Automotive Research for Advanced Technology) Projects. Publication of OAAT & OTTEE&RE, U.S. Department of Energy, 9/2001, 26-27.
9. Barsukov I.V. Development of low-cost, carbonaceous materials for anodes in lithium-ion batteries – Superior Graphite Co. Snapshots of CARAT (Cooperative Automotive Research for Advanced Technology) Projects. Publication of Office of FreedomCAR and Vehicle Technologies, EERE, U.S. Department of Energy, 5/2003, 22-23.
10. Battery R&D Equipment. Hohsen Corp. product catalog. Tokyo, Japan (2002), 1.
11. Zheng, L., Zhong Q., and Dahn J.R. High-Capacity Carbons Prepared from Phenolic Resin for Anodes of Lithium-Ion Batteries. J. Electrochem. Soc., 142, 211-214 (1995).
12. Dahn J.R., Zheng L., Liu Y. and Xue J.S. Mechanism for Lithium Insertion in Carbonaceous Materials, Science, 270, 590-593 (1990).
13. Billaud D., Henry F.X. and Willmann P. Electrochemical Synthesis of Binary Graphite-Lithium Intercalation Compounds. Mat. Res. Bull., 28, 477-483 (1993).
14. Billaud D., Henry F.X., Lelaurain M., and Willmann P. Revisited Structures of Dense and Dillute Stage II Lithium-Graphite intercalate Compounds. J. Phys. Chem Solids, 57, 775-781 (1996).
15. Joho F., Novak P., and Spahr M.E. Safety Aspects of Graphite Negative Electrode Materials for Lithium-Ion Batteries. J. Electrochem. Soc., 149, 1020-1024 (2002).
16. Amine K., Liu J., Belharouak I. and Kahaian A. Material screening from and with suppliers. In Chapt. Overview of cost reduction and Technology Transfer Efforts. Proc. Of the Advanced Technology Development Program, U.S. DOE FreedomCAR annual merit review meeting, June 24-25 (2003) Argonne National Laboratory, Argonne, IL, USA, P.8.
17. Liu J., Vissers D.R., Amine K., Barsukov I.V., Doninger J.E. Surface Treated Natural Graphite as Anode Material for High-Power Li-Ion Battery Applications. In New Carbon-Based Materials for Electrochemical Energy Storage Systems: Batteries, Fuel Cells and Supercapacitors. Barsukov I., Johnson C., Doninger J., Barsukov V. eds. NATO ARW series Volume. Springer (2005) - in this book.
18. Barsukov I., Henry F., Doninger J., Gallego M., Huerta T., Girkant R. and Derwin D. On the electrochemical performance of lithium-ion battery anodes based on natural graphite with various surface properties. ITE Letters on Batteries, New Technologies & Medicine, V.4, N.2 (2003), 163-166.
19. Kohs W, Santner H.J., Hofer F., Schröttner H., Doninger J., Barsukov I., Albering J.H., Möller K.-C., Besenhard J.O., and Winter M. A study on electrolyte interactions with graphite anodes exhibiting structures with various amounts of rhombohedral phase. J. Power Sources, 199-121, 528-538 (2003).

MECHANISMS OF REVERSIBLE AND IRREVERSIBLE INSERTION IN NANOSTRUCTURED CARBONS USED FOR Li-ION BATTERIES

F. Béguin[a*], F. Chevallier[a], M. Letellier[a], C. Vix[c], C. Clinard[a], J.N. Rouzaud[a], and E. Frackowiak[b]

[a]*CRMD, CNRS-Université, 1B Rue de la Férollerie, 45071 Orléans Cedex 02, France*
[b]*ICTE, Poznan University of Technology, ul. Piotrowo 3, 60-965 Poznan, Poland*
[c]*ICSI, CNRS, 15, rue Jean Starcky, 68057 Mulhouse, France*

1. INTRODUCTION

Taking into account the high-energy demand of electric vehicles, an important research effort is devoted for developing a new generation of power sources. In the case of lithium-ion batteries, the main attention is paid to designing materials with improved electrochemical performance. The negative electrode of commercial lithium-ion batteries is usually graphite, which reversible lithium intercalation capacity is limited to one lithium for six carbon atoms in standard conditions [1]. Due to their higher storage capacity, nanostructured carbons, such as hard carbons, seem to be promising materials in replacement of graphite [2]. However, they demonstrate a noticeable hysteresis and irreversibility, which can preclude a practical application [2,3]. Despite the huge number of publications on the subject within the last years, the causes of these two important drawbacks are not yet clearly elucidated. Hence, the understanding of the insertion-deinsertion mechanisms in nanostructured carbons is a key for improving the performance of this kind of materials.

In this paper, new highlights are proposed to interpret the reversible and irreversible capacities of nanostructured carbons in lithium batteries. A proportional dependence between the irreversible capacity and the active surface area (ASA) of carbon materials will be demonstrated, showing the ASA concept more universal than any of the other parameters which were previously considered. In-situ ^{7}Li NMR will be also presented as a means to

[*] Corresponding author. E-mail: beguin@cnrs-orleans.fr

I.V. Barsukov et al. (eds.), New Carbon Based Materials for Electrochemical Energy Storage Systems, 231–243.

follow stepwise the sequence of reversible lithium insertion, without all the disadvantages of the *ex-situ* methods.

2. RELATION BETWEEN THE IRREVERSIBLE CAPACITY AND THE ACTIVE SURFACE AREA

2.1. Literature Review

The irreversible capacity of lithium-ion batteries is essentially attributed to the electrolyte decomposition on the carbon electrode during the first reduction cycle, forming a passivation layer (also called Solid Electrolyte Interface, SEI). The main components of the SEI identified by infrared spectroscopy are Li_2CO_3, $ROCO_2Li$, etc. [4-6]. According to some publications, the extent of SEI formation is directly related to the surface area accessible for the electrolyte. Therefore it is often compared to the BET specific surface area, giving in some cases a linear relationship [7-9]. Since nanostructured carbons are generally characterised by a noticeable value of BET specific surface area, it could be a reasonable explanation for the important value of irreversible capacity encountered with these materials. However, Buiel et al. showed that this contribution to the irreversible capacity cannot be higher than 0.13 lithium for 6 carbon atoms, and some other parameters, for example the surface oxygenated functionality might be involved [10]. Furthermore, for carbon materials with a mesoporous character, such as carbon nanotubes or carbons from templates, the irreversible capacity is always very high [11], and its evolution is better related to the mesopore volume [12].

The influence of heteroatoms involved in surface functional groups, such as oxygen and sulfur, on the irreversible capacity has been demonstrated by Larcher [13]; higher the surface concentrations of groups, higher the contribution to the irreversible capacity. Upon increasing the highest treatment temperature (HTT) of the carbon precursor, the irreversible capacity decreases. Following by XPS the evolution of -C-Li or $-CO^- Li^+$ surface concentration *vs* HTT, it has been proposed that it is due to a decrease of the amount of free radicals or surface hydroxyl groups, respectively [14,15]. Similarly the drastic increase of irreversible capacity which occurs after ball milling of graphite in argon atmosphere is interpreted by a highly developed active surface, i.e. dangling bonds [16]. However, the involvement of free radicals to form C-Li bonds is quite hypothetic [17,18]. Nevertheless, there is no doubt that a part of lithium is irreversibly trapped by surface functional groups. The species which are usually adsorbed on a carbon surface, i.e. water, di-oxygen, etc. are also contributing to the irreversible capacity, by increasing the thickness of the SEI [19,20]. Finally, considering a series of carbons, a linear relationship has been found between the irreversible capacity and the micropore volume [21]; from *ex-situ* ^{7}Li NMR results, the authors suggested that metallic lithium is irreversibly trapped in the micropores [22].

Since claims attribute the irreversible capacity, at least partly, either to the surface functionality or to adsorbed species, it has been attempted to cure the surface of carbon materials. When the surface oxygenated groups are reduced by di-hydrogen, the irreversible capacity decreases and simultaneously the reversible capacity increases due to the introduction of C-H functionality on the carbon surface [23]. A positive effect of HTT at ca. 900-1000°C under vacuum has been observed, which could be due to surface groups elimination [24,25]. However, surprisingly, the performance does not improve after high temperature treatment under secondary vacuum [26]. Chemical vapour deposition of pyrocarbon allows the irreversible capacity of non-graphitisable carbons to be strongly depressed [10]. Since the specific surface area measured by nitrogen adsorption is not modified by this treatment, the observed decrease could be due to the elimination of surface groups or adsorbed species. By analogy, increasing the hydrophobic character of carbon by copper deposition on the surface allows to get a smaller value of irreversible capacity [27]. The irreversible capacity of graphite is also decreased after pyrolytic carbon deposition [28] or coke coating from the carbonisation of a pitch/resin blend [29] or coal tar pitch [30]. It has been suggested that the pyrolytic carbon coating prevents the solvent co-intercalation into graphite [28]. It also allows reducing the BET specific surface area and the O/C ratio [31]. Similarly after coke coating, there is less gas evolution during the first discharge, that it is attributed to a decrease of the exposed surface area of graphite [30].

To summarise, mainly two parameters seem to contribute to the irreversible capacity of nanostructured carbons: the SEI formation, which would depend on the BET specific surface area and the presence of species adsorbed or bonded to the surface [10]. However, our previous observation that the irreversible capacity is unchanged after annealing under secondary vacuum, disclaims a direct influence of oxygenated surface groups or adsorbed species [26]. Moreover, it is not worth to use the BET specific surface area, which is obtained from the amount of nitrogen physisorbed on the carbon surface in order to explain the SEI formation where high energies are involved. Therefore, we suggest relating the irreversible capacity to the number of active sites on the carbon surface, so-called active surface area (ASA) [26]. The active surface area of a carbon material corresponds to the cumulated surface area of the different types of defects present on the carbon surface (stacking faults, single and multiple vacancies, dislocations) [32]; these sites are responsible for the interactions with the adsorbed species. Since the surface functionality and the BET specific surface area are not two independent parameters (in general, higher the BET surface area, higher the number of surface groups), the ASA appears as a more universal concept.

In this paper, the effect of surface functionality will be studied by comparing the irreversible capacity of viscose carbon fibres outgassed at 900°C to that of the pristine material. Graphite ball-milled in different atmospheres is selected to demonstrate the irreversible capacity *vs* ASA relationship.

2.2. Experimental

The materials chosen for our investigations were natural graphite (UF2 from Kropfmühl <5µm) or a carbon cloth prepared by carbonisation of viscose under neutral atmosphere for 15 minutes, successively at 400, 700 and 1000°C. The carbon cloth is used as a self-standing electrode, without any binder or conductivity additive, which is an advantage for investigating only the electrochemical properties of the disordered carbon. The natural graphite specimens (1 g) were ball-milled with 7 g of stainless steel balls under controlled atmosphere using a Spex 8000 mill. For some experiments, the carbon samples were coated with pyrolytic carbon by chemical vapour decomposition of propylene (2.5 ml/mn) diluted in nitrogen (100 ml/mn) during 10 minutes at 900°C.

Nitrogen adsorption/desorption isotherms were determined at 77 K, using a Micromeritics ASAP 2010, after outgassing the samples at 150°C until the pressure reached 10^{-6} mbar. The Active Surface Area (ASA) was measured by the method of Walker et al.[33] which is based on di-oxygen chemisorption. The sample (a mass corresponding to a surface area of ca. 0.5 m^2) was first outgassed under dynamic vacuum (10^{-3} Pa) at 950°C for 5 hours to release the active sites. In a second step, oxygen is chemisorbed at 300°C during 15h (initial pressure: 66.5 Pa) leading to the formation of surface oxygenated complexes. After elimination of the remaining di-oxygen, the amount of oxygenated complexes is determined by mass spectrometry analysis of evolved CO and CO_2 by subsequent outgassing at 950°C.

For the electrochemical investigations, two electrode cells (Swagelok type) with a glass micro-fibre separator (Whatman® GF/D cat 1823070) were assembled in an argon atmosphere glove box. The working electrode was a disk of the carbon cloth or a plastic carbon pellet prepared according to the Bellcore technique [34], and metallic Li was used as counter and reference electrodes. The electrolyte was 1M $LiPF_6$ in a mixture of EC/DEC (1/1) (Merck). The cells were investigated by galvanostatic charge-discharge in the potential range from 2 V to 0 V *vs* Li, at C/20 rate (corresponding to a capacity of 372 mAh/g in 20 h), using a Mac Pile multi-channel potentiostat/galvanostat (Biologic, France).

2.3. Results and Discussion

As detailed in the literature background, it is often attempted to relate the irreversible capacity either with the BET specific surface area or with the presence of surface groups. Below, we will show two experiments, which disclaim these kinds of correlations.

Figure 1 presents the plot of the BET specific surface area *vs* the irreversible capacity measured for graphite samples milled in different atmospheres and sometimes post-treated by pyrolytic carbon deposition. The experimental values are quite scarce and, contrarily to several claims [7-9], there is not any linear dependence between these two parameters. It seems that the linearity would exist only for samples from the same family with comparable microtextures.

The effect of surface functionality has been investigated in a next experiment. Figure 2 shows the galvanostatic charge/discharge on the carbon cloth outgassed at 150°C under vacuum in order to eliminate the adsorbed water. As it is often the case for hard carbons, the irreversible capacity is quite high, $x_{irr} = 0.71$ (taken as x in the formula Li_xC_6). Taking into account the literature data [13], such a high value could be due to a developed oxygenated surface functionality. Therefore, for a second experiment, the surface groups have been eliminated by heat-treating the carbon cloth at 900°C under vacuum, condensing the vapours of the vacuum line in a liquid nitrogen trap. Figure 3 presents the galvanostatic curve for the resulting material, which has been transferred into the swagelok cell without any contact with the air atmosphere, in order to avoid any pollution. It is remarkable, that the electrochemical characteristics are even worst than for the as received specimen.

Hence, it is clear that the irreversible capacity is not directly related with the presence of surface oxygenated groups. In the conditions used for outgassing at 900°C, the number of active sites has not been modified [33]. Therefore, taking into account that the irreversible capacity is almost the same for the experiments shown in Figures 2 and 3, it might mean that only the active sites are responsible of the electrolyte decomposition, and consequently of the SEI development.

In order to verify this assumption, the Active Surface Area (ASA) of the various graphite samples has been expressed as a function of the irreversible capacity, and the result is presented in Figure 4. A perfect linear dependence is observed, which proves that the main contribution to the irreversible capacity is the decomposition of the electrolyte on the active sites during the first galvanostatic reduction. A similar trend has been found for the carbon cloth treated in different conditions [26]. Figure 4 shows also that all the samples (4, 5, 6, 7) after milling present the highest values of ASA, whatever the atmosphere for milling, due to the production of many edge planes [16].

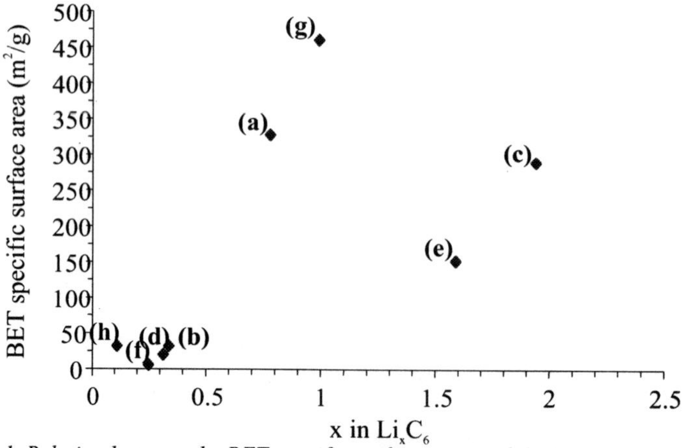

Figure 1. Relation between the BET specific surface area and the irreversible capacity x of graphite samples treated in different conditions. (a) 10 h in vacuum; (b) 10 h in vacuum + pyrolytic carbon deposition; (c) 10 h under H_2; (d) 10 h under H_2 + pyrolytic carbon deposition; (e) 10 h under O_2; (f) 10 h under O_2 + pyrolytic carbon deposition; (g) 20 h in vacuum; (h) 20 h in vacuum + pyrolytic carbon deposition.

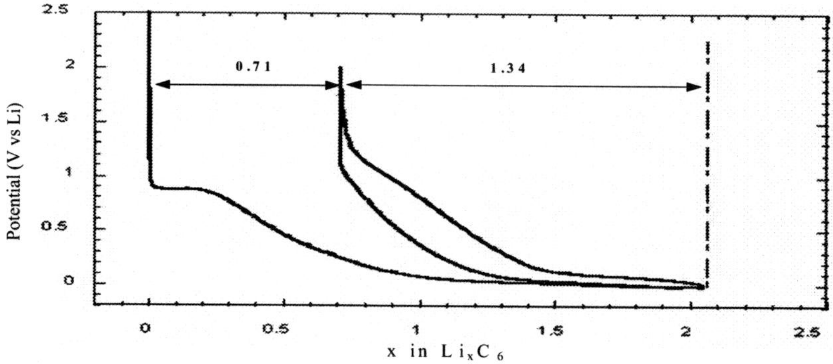

Figure 2. Galvanostatic charge/discharge of the as-received carbon cloth outgassed at 150°C under vacuum.

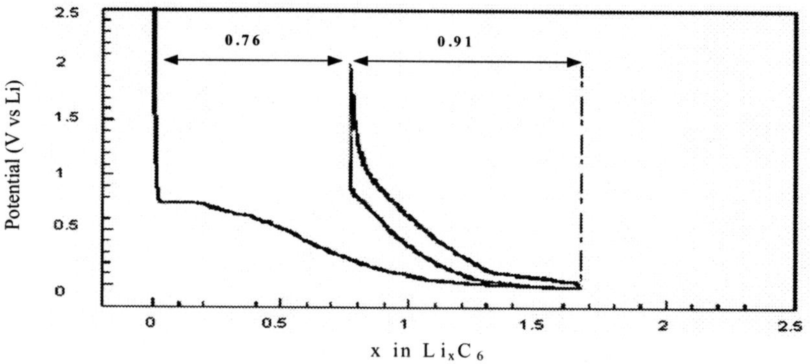

Figure 3. Galvanostatic charge/discharge of the carbon cloth outgassed 24 hours under vacuum at 900°C.

It is remarkable that the samples 1, 2, 3, 8 which have been coated in the same conditions by chemical vapour deposition, present comparable values of ASA and irreversible capacity (Figure 4). After coating, the active sites of milled graphite are blocked and only those of the pyrolytic carbon layers are effective in the electrolyte decomposition. Similarly, the decrease of irreversible capacity observed by other authors after pyrolytic carbon deposition [28] or coke coating [29,30] on graphite is due to a blockage of active sites. Since the active sites are rather located on the edge planes, the BET surface area, which theoretically expresses the total surface of the basal planes, cannot be a relevant parameter. Hence, the proportionality of irreversible capacity with the BET specific surface area will only be found with non-microporous solids of the same family which have a comparable density of active sites, for example a series of graphites [9].

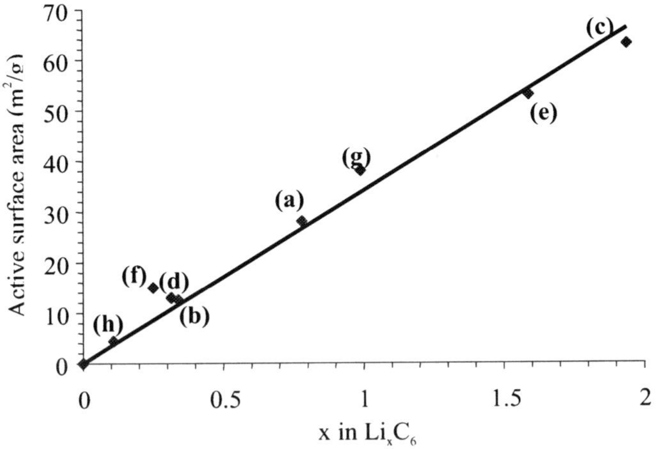

Figure 4. Relation between the BET specific surface area and the irreversible capacity x of graphite samples treated in different conditions. (a) 10 h in vacuum; (b) 10 h in vacuum + pyrolytic carbon deposition; (c) 10 h under H_2; (d) 10 h under H_2 + pyrolytic carbon deposition; (e) 10 h under O_2; (f) 10 h under O_2 + pyrolytic carbon deposition; (g) 20 h in vacuum; (h) 20 h in vacuum + pyrolytic carbon deposition.

3. *IN-SITU* ^{7}LI-NMR STUDY OF THE REVERSIBLE LITHIUM INSERTION MECHANISM

3.1. Introduction

Although some nanostructured carbons are able to store reversibly higher amounts of lithium than graphite [2], they generally demonstrate a high hysteresis (see for example Figure 2) that still precludes their use in lithium-ion batteries. In order to improve their electrochemical properties as electrode materials, a current effort is made to better understand the insertion-deinsertion mechanism. *Ex-situ* solid-state ^{7}Li-Nuclear Magnetic Resonance (^{7}Li-NMR) is one of the most useful methods which has been extensively applied for the determination of the states of lithium at various steps of reduction or oxidation of nanostructured carbons [35,36].

The principle of the measurements by *ex-situ* ^{7}Li NMR is as follows. During the galvanostatic charge/discharge of a lithium/carbon cell, the experiment is stopped at a selected potential. Then, the carbon material is extracted from the cell, eventually washed with a clean solvent, dried and ground before being put into a sealed tube for the NMR analysis. Using such a methodology, the equilibrium composition of the sample might be modified by washing or by moisture and oxygen from the atmosphere of the glove box during the different transfers. On the other hand, each value of potential requires the preparation of a new sample that limits the number of possible measurements if one wants to keep a reasonable experimental time. Recently, to overcome these drawbacks, we have performed *in-situ* ^{7}Li-NMR during the galvanostatic cycling of a composite carbon electrode in aN ultra-thin plastic lithium/carbon cell [37,38]. In these conditions, the carbon substrate is the same for all the spectra, eliminating possible uncertainty in the interpretations that might always exist when comparing the results obtained with different starting specimens and electrochemical cells.

In the following part of this paper, *in-situ* ^{7}Li-NMR is used to better know the state(s) of lithium reversibly stored in disordered carbons.

3.2. Experimental

For these experiments, we used the carbon cloth prepared by carbonisation of viscose up to 1000°C. It was additionally coated with a thin layer of pyrolytic carbon by CVD under propylene. The conditions for the sample elaboration were previously given in Part 2. As seen before, the advantage of pyrolytic carbon coating is to depress the amount of lithium irreversibly inserted, that should make easier the interpretation of the NMR data related with the reversible part of insertion. For the *in-situ* ^{7}Li NMR experiments, we used a supple ultra-thin plastic lithium/carbon cell built according to the conditions described in reference [38]. The liquid electrolyte was 1M $LiPF_6$ in a 1:1 volume ratio of ethylene carbonate EC and diethyl carbonate DEC (Merck). ^{7}Li-NMR was performed using a Bruker Avance DSX 360 spectrometer equipped with a broadband probe head. Once the cell to be investigated was positioned in the 10 mm coil, the copper current collectors were welded to wires connected to a Mac Pile II galvanostat (Biologic, France). The cell was cycled at a C/20 rate in the potential range from 2 V to 0 V *vs* Li. In these conditions, the acquisition time (8 min) was short compared to the time of a complete cycle (60 hours), therefore the cell cycling was not stopped during recording the NMR spectra.

High Resolution Transmission Electron Microscopy (HRTEM, Philips CM20, 200 kV) was applied to get structural and nanotextural information on the fibers, by imaging the profile of the aromatic carbon layers in the 002-lattice fringe mode. A carbon fiber coated with pyrolytic carbon was incorporated in epoxy resin and a transverse section obtained by ultramicrotomy was deposited on a holey carbon film. An in-house made image analysis procedure was used to get quantitative data on the composite.

3.3. Results and Discussion

The HRTEM observation of the cross section of a coated fiber showed that the core is constituted of aromatic layers highly misoriented, whereas they are preferentially oriented in parallel for the thin coating; pairs of stacked layers form mainly Basic Structural Units (BSUs) in which the average interlayer distance is smaller than between the aromatic layers in the bulk of the fiber. Since the nanotexture is more dense for the pyrolytic carbon than for the fiber itself, it acts as a barrier which prevents the diffusion of the large solvated lithium ions to the core of the fiber, allowing the passivation layer to be less developed after this treatment. Hence, the major amount of lithium inserted is involved in the reversible contribution; therefore this composite material is extremely interesting for the *in-situ* ^{7}Li NMR study of the reversible insertion.

Figure 5 shows the result of *in-situ* ^{7}Li NMR during galvanostatic cycling of the supple cell. Initially, the spectrum comprises one line at 0 ppm corresponding to Li^{+} from the dissolved salt and another line at 263 ppm due to metallic lithium from the counter electrode. Since, the latter line is completely out of the shift range noted for the new lines, it is not shown in Figure 5. During the electrochemical reduction step, two lines related with lithium insertion are observed. The minor one progressively shifts in the first cycles from 0 ppm to 18 ppm, where it stabilizes until the end of the reduction process. The position of the most important one progressively downfield shifts to reach values of ca. 110 ppm at full discharge (Figure 5). During the oxidation step, the reverse phenomenon is observed with some hysteresis, the two new lines shifting back to 0 ppm and vanishing completely when the potential reaches 2 V *vs* Li.

Lithium peaks at constant position are also found for graphite intercalation compounds (GICs), at 41 ppm for the 1st stage LiC_6 or the 2nd stage LiC_{12} compounds [39,40], and at 10 ppm for the 2nd stage LiC_{18} phase [41]. Hence, the peak found at 18 ppm for the carbon composite is attributed to lithium intercalation (with a charge transfer to carbon) in graphitic type intervals. HRTEM image analysis shows that such kinds of intervals are rather found in the pyrolytic carbon coating and to a lesser extent in the core of the fiber. The peak, which shifts up to 110 ppm is attributed to quasi-metallic bidimensional lithium clusters located in narrow nanopores formed by pairs of spaced and misoriented aromatic layers in the core of the fibers.

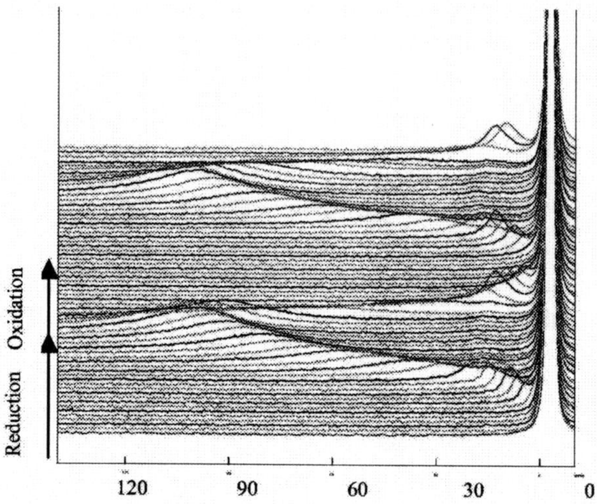

Figure 5. ^{7}*Li NMR spectra recorded during 3 cycles of reversible lithium insertion-deinsertion in the composite carbon electrode. The peak at 0 ppm is due to ionic lithium* ($Li^+PF_6^-$ *and passivation layer). The peak of lithium at 263 ppm is not shown.*

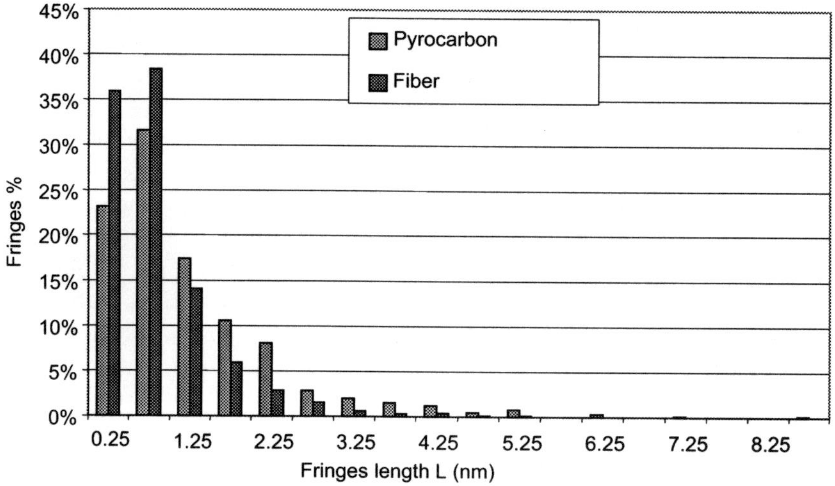

Figure 6. Histogram of the fringes length obtained by computer analysis of the HRTEM image of the fiber/pyrolytic carbon composite.

Table 1. Values of electron affinity for few aromatic hydrocarbons and for graphite.

	Electron affinity (eV)
C_6H_6	0
Naphtalene	1.1
Anthracene	1.8
Graphite	4.39

The histogram of the composite fringes lengths obtained by computer analysis of the HRTEM images is shown in Figure 6. Short fringes are more numerous in the fiber, whereas long fringes up to 6 nm essentially exist in the pyrolytic carbon coating. Then, it is interesting to consider the data reported in Table 1, which shows the values of electron affinity for some aromatic hydrocarbons and for graphite. From this table, we observe that higher the number of aromatic rings, from benzene to graphite, higher the electron affinity.

The high charge transfer in graphite is correlated with the lithium intercalation between the layers of infinite size. Hence, we can confirm that the Knight shift of 18 ppm corresponds to lithium intercalated between layers of large size, which are essentially located in the pyrolytic carbon coating, whereas the values up to 110 ppm are due to lithium close to layers of smaller size. It means that lithium becomes more and more "metallic" (i.e. less ionic) during the reduction process, occupying nanopores where it is close to layers of decreasing diameter.

4. CONCLUSION

In this paper, we presented new information, which should help in optimising disordered carbon materials for anodes of lithium-ion batteries. We clearly proved that the irreversible capacity is essentially due to the presence of active sites at the surface of carbon, which cause the electrolyte decomposition. A perfect linear relationship was shown between the irreversible capacity and the active surface area, i.e. the area corresponding to the sites located at the edge planes. It definitely proves that the BET specific surface area, which represents the surface area of the basal planes, is not a relevant parameter to explain the irreversible capacity, even if some papers showed some correlation with this parameter for rather low BET surface area carbons. The electrolyte may be decomposed by surface functional groups or by dangling bonds. Coating by a thin layer of pyrolytic carbon allows these sites to be efficiently blocked, without reducing the value of reversible capacity.

Regarding the reversible capacity of disordered carbons, we found that lithium is mainly in a quasi-metallic state, another minor part being in a state comparable to intercalation. If we consider that the lithium density in the intercalated domains is comparable to that of graphite, we can conclude that the enhanced reversible capacity of the disordered carbons is related with the quasi-metallic state of lithium. In this paper, we postulate that the electron density of the quasi-metallic lithium clusters is related with the size of the neighbour carbon layers. In future work, we plan to investigate different materials in order to determine if there is any relationship between the reversible capacity and the fringe length.

ACKNOWLEDGMENTS

This research was supported by the NATO Science for Peace programme (Project CARBON SfP 973849) and by a PhD grant from the Conseil Régional du Centre (France) to Frédéric Chevallier. Thanks are due to Luc Aymard for providing the samples of milled graphite and to Mathieu Morcrette for building the supply cells; both are from Université de Picardie, Amiens (France).

REFERENCES

1. Guérard D., Hérold A., Carbon 1975; 13: 337
2. Zheng T., Dahn J.R., Applications of carbon in Lithium-ion batteries, In Carbon materials for advanced technologies, Burchell T.D. Ed., Elsevier, Oxford, 1999, pp 341-388.
3. Frackowiak E., Béguin F., Carbon 2002; 40: 1775.
4. Naji A., Ghanbaja J., Humbert B., Willmann P., Billaud D., J. Power Sources 1996; 63: 33.
5. Aurbach D., Zaban A., Ein-Eli Y., Weissman I., Chusid O., Markovsky B., Levi M., Levi E., Schechter A., Granot E., J. Power Sources 1997; 68: 91.
6. Kanamura K., Tamura H., Shiraishi S., Takehara Z., J. Electroanal. Chem. 1995; 394: 49
7. Fong R., Von Sacken U., Dahn J.R., J. Electrochem. Soc. 1990; 137: 2009
8. Simon B., Flandrois S., Fevrier-Bouvier A., Biensan P., Mol. Cryst. Liq. Cryst. 1998; 310: 333
9. Winter M., Novak P., Monnier J., J. Electrochem. Soc. 1998; 145: 428.
10. Buiel E., Dahn J.R., J. Electrochem. Soc. 1998; 145: 1977.
11. Duclaux L., Frackowiak E., Béguin F., J. Power Sources 1999; 81-82: 323.
12. Béguin F., Méténier K., Chevallier F., Azaïs P., Pellenq R., Bonnamy S., Duclaux L., Rouzaud J.N., Frackowiak E., Mol. Cryst. Liq. Cryst. 2002; 386: 151.
13. Larcher D., Mudalige C., Gharghouri M., Dahn J.R., Electrochim. Acta 1999; 44: 4069.
14. Y. Matsumura, S. Wang, J. Mondori, J. Electrochem. Soc. 1995; 142: 2914.
15. Y. Matsumura, S. Wang, J. Mondori, Carbon 1995; 33: 1457.
16. Azaïs P., Duclaux L., Faugère A.M., Béguin F., Appl. Phys. Lett. 2002; 81: 775.
17. Y. Matsumura, S. Wang, Y. Nakagawa, C. Yamaguchi, Synth. Met. 1997; 85: 1411.
18. M. Winter, H. Buqa, B. Evers, T. Hodal, K.C. Moller, C. Reisinger, M.V. Santis Alvarez, I. Schneider, G.H. Wrodnigg, F.P. Netzer, R.I.R. Blyth, M.G. Ramsey, P.

Golob, F. Hofer, C. Grogger, W. Kern, R. Saf, J.O. Besenhard, Monatsh. Chem. 2001; 132: 473.
19. Xing W., Dahn J.R., J. Electrochem. Soc. 1997; 144: 1195.
20. Aurbach D., Schechter A., Electrochim. Acta 2001; 46: 2395.
21. Guérin K., Février-Bouvier A., Flandrois S., Simon B., Biensan P., Electrochim. Acta 2000; 45: 1607.
22. Guérin K., Ménétrier M., Février-Bouvier A., Flandrois S., Simon B., Biensan P., Solid State Ionics 2000; 127:187.
23. Lu W., Chung DDL, Carbon 2001; 39: 493.
24. M. Kikuchi, Y. Ikezawa, T. Takamura, J. Electroanal. Chem. 1995; 396: 451.
25. R. Alcantara, F.J. Fernandez-Madrigal, P. Lavela, J.L. Tirado, J.M. Jimenez-Mateos, C. Gomez de Salazar, R. Stoyanova, E. Zhecheva, Carbon 2000; 38: 1031.
26. Béguin F., Chevallier F., Vix C., Saadallah S., Rouzaud J.N., Frackowiak E., J. Phys. Chem. Solids 2004; 65: 211.
27. Y. Wu, C. Jiang, C. Wan, E. Tsuchida, Electrochem. Comm. 2000; 2: 626.
28. M. Yoshio, H. Wang, K. Fukuda, Y. Hara, Y. Adachi, J. Electrochem. Soc. 2000; 147: 1245.
29. Kuribayashi I., Yokoyama M., Yamashita M., J. Power Sources 1995; 54: 1.
30. Yoon S., Kim H., Oh S.M., J. Power Sources 2001; 94: 68.
31. Natajaran C., Fujimoto H., Tokumitsu K., Mabuchi A., Kasuh T., Carbon 2001; 39: 409.
32. Lahaye J., Dentzer J., Soulard P., Ehrburger P. Carbon gasification : The active site concept. In Fundamental Issues of Control of Carbon Gasification Reactivity, Lahaye J, Ehrburger P. Ed., Academic Publishers, London, 1991, p.143-158
33. Laine N.R., Vastola F.J., Walker P.L. Jr., J. Phys. Chem. 1963; 67: 2030.
34. Tarascon J.M., Gozdz A.S., Scmutz C., Shokoohi F., Warren P.C., Solid State Ionics 1996; 86-88: 49.
35. Gautier S., Leroux F., Frackowiak E., Faugère A.M., Rouzaud J.N., Béguin F., J. Phys. Chem. A 2001; 105: 5794.
36. Guérin K, Ménétrier M., Février-Bouvier A., Flandrois S., Simon B., Biensan P., Solid State Ionics 2000; 127: 187.
37. Letellier M., Chevallier F., Clinard C., Frackowiak E., Rouzaud J.N., Béguin F., J. Chem. Phys. 2003; 118: 6038.
38. Chevallier F., Letellier M., Morcrette M., Tarascon J.M., Frackowiak E., Rouzaud J.N., Béguin F., Electrochem. Solid State Lett.2003; 6: A225.
39. C. Conard J., Estrade H., Mat. Science Engineering 1997; 31: 173.
40. Imanishi N., Kumai K., Kokugan H., Takeda Y., Yamamoto O., Solid State Ionics 1998; 107: 135.
41. Tatsumi K., Zaghib K., Sawada Y., Abe H., Ohsaki T., in Rechargeable Lithium and Lithium-Ion Batteries, Megahed S., Barnett B.M., Xie L. Editors, PV 94-28, p. 97, The Electrochemical Society Proceedings Series, Pennington, NJ (1994).

SOME THERMODYNAMICS AND KINETICS ASPECTS OF THE GRAPHITE-LITHIUM NEGATIVE ELECTRODE FOR LITHIUM-ION BATTERIES

Rachid Yazami[1,2*], Audrey Martinent[1] and Yvan Reynier[1,2]

[1]*LEPMI, UMR CNRS-INPG 5631, 38041 St Martin d'Hères, France*
[2]*California Institute of Technology, Division of Engineering and Applied Science, Pasadena, California 91125, USA*

1. INTRODUCTION

The chemistry of lithium-ion batteries is based on the lithium-ion shuttling between the graphite negative electrode and the transition metal(s) oxide positive electrode. The overall reaction can be schematized as:

$$Li_xC_6 + Li_yMO_2 \leftrightarrow 6C + Li_{y+x}MO_2 \qquad (1)$$

Due to differences in the chemical potential of lithium in the Li_xC_6 and $Li_{y+x}MO_2$ materials, the operating voltage of the battery may be as high as 4V.

The lithiated graphite electrode has many attractive features that have made it the best choice for negative electrode in commercial batteries. These include high mass and volume capacity, low operating voltage, long cycle life and good safety. From the crystal chemistry point of view, lithiation of graphite undergoes a characteristic series of phase transitions called stages. An intercalation stage describes the 2D stacking sequence of the lithium layers between the graphene layers. The stage number is the number of periodically stacked graphene layers between two adjacent intercalated layers. Starting from pure graphite, successive stages are formed as the lithium concentration increases. One way to intercalate lithium into graphite is by electrochemical cathodic polarization of a graphite electrode in a Li organic liquid or solid-state polymer electrolyte cell. Based on in situ and ex-situ x-ray diffractometry, the stages that form as the lithium concentration increases are: dilute stage 1, stage 4, stage 3, liquid-type stage 2, stage 2 and stage 1 [1-4]. The composition induced stage transitions occur

* Corresponding author. E-mail: yazami@caltech.edu

I.V. Barsukov et al. (eds.), New Carbon Based Materials for Electrochemical Energy Storage Systems, 245–258.

at relatively well defined 'x' values around 0.13, 0.18, 0.23, 0.33 and 0.5 respectively. In dilute stage 1, lithium is randomly distributed between the graphene layers, which causes a slight shift of the (002) main peak of hexagonal graphite to lower diffraction angles. Liquid-like stage 2 has lithium in every other layer, with no long-range in plane ordering. This contrasts with the stage 2 compound of nominal composition $Li_{0.5}C_6$ (LiC_{12}) in which the lithium forms an ordered array with the characteristic hexal structure [5]. The same hexal structure occurs in stage 1 LiC_6 in which lithium occupies each Van der Waals gap between the graphene sheets.

Figure 1 shows typical lithiation/delithiation voltage profile curves obtained with graphite under a low galvanostatic cycling regime (i.e. C/100 rate) [6]. It shows plateaus and semi-plateaus corresponding to the stage transitions described above. A flat plateau is the signature of a two-phase system as in the case in the composition range of $0.5<x<1$. It has been well accepted that the plateau is related to a two-phase system that is a mixture of stage 2 LiC_{12} and stage 1 LiC_6 phases following the equilibrium reaction:

$$LiC_{12} + yLi \leftrightarrow 2y\ LiC_6 + (1-y)\ LiC_{12} \qquad (2)$$

According to equation 2, the amounts of each phase should vary linearly with intercalated lithium intercalated beyond LiC_{12}.

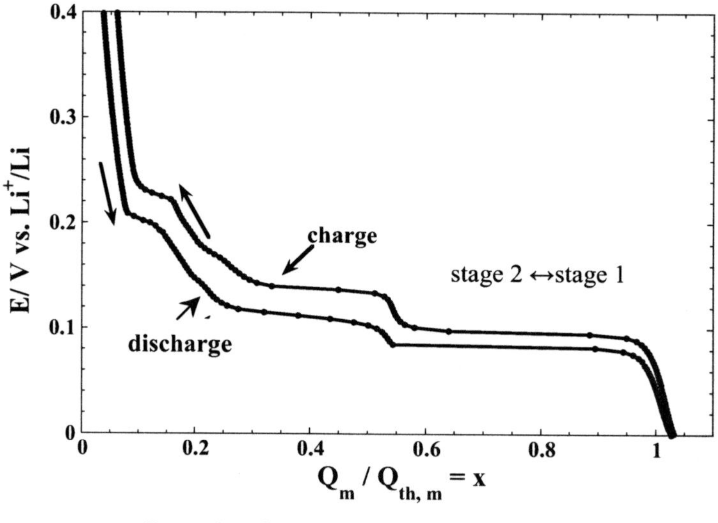

Figure 1. Voltage vs. composition curves during the graphite electrode lithiation and delithiation at C/100 rate.

Besides staging, the other important characteristic of the graphite negative electrode is the formation of the solid electrolyte interphase (SEI) during the first cycles [7, 8]. The SEI comes mainly from surface reactions between the electrolyte and the graphite electrode. The SEI plays an important role in the characteristics of the Li_xC_6 electrode including the irreversible capacity loss and the thermal and chemical stability that govern the electrode self discharge and cycle life [9, 10].

In this paper we present results from independent studies on the stage 2 to stage 1 transition area that show some unexpected features (anomalies). The results are obtained by electrochemical impedance spectroscopy (EIS), entropy measurements ($\Delta S(x)$) and in situ x-ray diffractometry (XRD). The aim is to understand the mechanism of stage transition dealing with the observed anomalies.

2. EXPERIMENTS AND RESULTS

2.1. The Kinetics (EIS) Anomaly

Coin-type half-cells with metallic lithium as the counter (and reference) electrode and natural graphite as the working electrode were used for the EIS study. The electrolyte consisted of a molar solution of $LiPF_6$ in equivolumic mixture of ethylene carbonate (EC) and dimethyl carbonate (DMC) [5]. The cell was cycled several times between 1.6V and 0V. Starting from the fully delithiated state at 1.6V, EIS spectra were taken every 200mV in the voltage window between 1.5 and 200mV and then every 50 mV between 200mV and 0mV. For the latter, EIS spectra were taken in both decreasing and increasing voltage sweeps.

Typical Nyquist plots of the Li /Li_xC_6 cell contain three frequency domains: a high frequency (HF) range covering the ~10^4-10 Hz, a medium frequency (MF) range between 10 and 1 Hz, and a low frequency range in the 10-10^{-2} Hz. Each frequency region may be associated with various physical chemical phenomena that were shown and discussed in the literature [6, 11-21]. The focus of this work is on the anomalous dimensional changes observed in the HF semi circle during the stage 2-stage 1 transition.

Figures 2a and 2b show a series Nyquist plots during delithiation (a) and re-lithiation (b) of graphite in the 0V -0.2V voltage window. The HF area of the plots consists of a semi circle of composition dependant radius R. Starting from ~0V where the electrode composition is close to LiC_6, R remains constant with increasing voltage. At a threshold voltage, R suddenly decreases and then remains almost constant until 200mV. When the voltage is decreased again from 200mV, the diameter does not vary much. It then increases suddenly to reach a new mostly constant size. If the intercalation - de-intercalation cycle is repeated again, a similar behavior is observed; that is the HF semi circle changes suddenly in size at well-defined voltages during lithium de-intercalation and re-intercalation. The reversible character of the HF semi-circle "breathing" suggests it should be associated with either surface' or bulk-related reversible electrode phenomena.

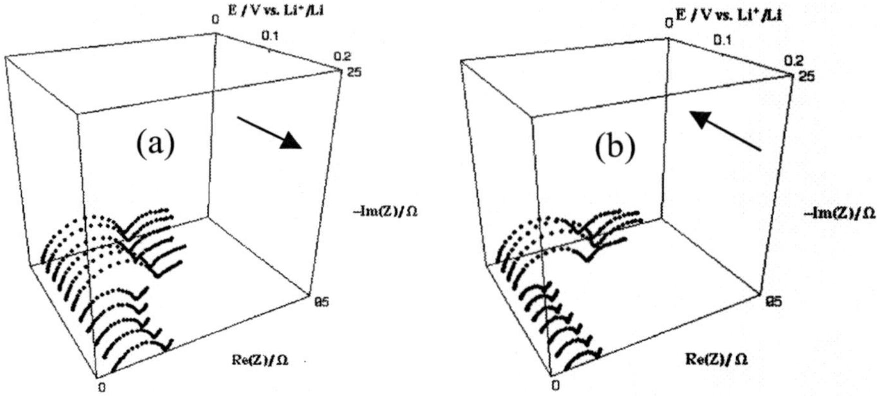

Figure 2. Evolution of the high frequency EIS semi-circle during delithiation (a) and lithiation (b) of the graphite electrode.

We found an equivalent electrical circuit that fits best the Li_xC_6 electrode behavior at high frequency. The circuit consists of a resistor R in parallel with a constant phase element (CPE). The latter is defined with a pseudo-capacitance Q and a parameter α with $0< \alpha <1$ [6]. The impedance of the CPE is expressed as:

$$Z_{CPE} = \frac{1}{Q(j\omega)^{\alpha}} \qquad (3)$$

where j is the complex unity number ($j^2 = -1$) and $\omega = 2\pi\nu$, ν = frequency. The impedance of the equivalent circuit is therefore:

$$Z_{circuit} = \frac{R}{1 + RQ(j\omega)^{\alpha}} \qquad (4)$$

where 2R is the diameter of the HF semi circle.

Figure 3 shows the voltage dependence of R during the delithiation (3a) and re-lithiation (3b) cycle. In the lithiation step (Fig. 3a), as the voltage decreases R remains mostly unchanged until it suddenly increases and reaches a new plateau. The corresponding voltage onset is at about 70mV vs. Li^+/Li. During de-intercalation, R varies a little, as the voltage is more negative, it then suddenly decreases to stabilize at a lower plateau. The onset voltage here is about 100mV. The onsets of increase and decrease in the R value occur at the same voltages of 70 and 100mV in the following cycles. It is interesting that 70 and 100mV match respectively the voltage plateaus during intercalation and de intercalation in the stage 2 to stage 1 phase transition showed in Figure 1.

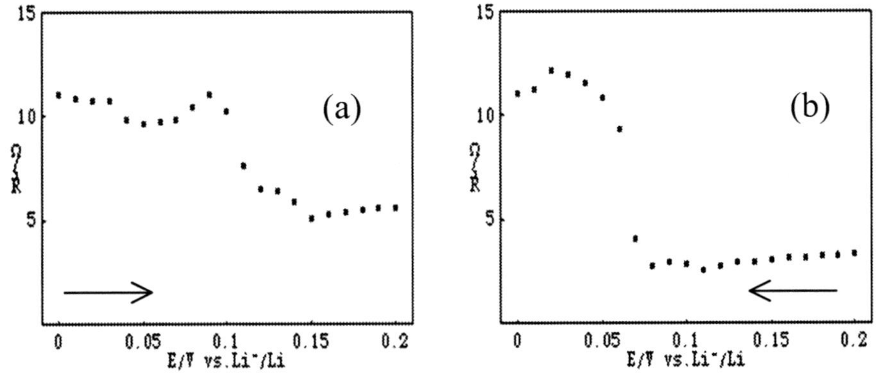

Figure 3. Voltage dependence of the HF semi circle radius during Li de-intercalation (a) and re-intercalation (b) into graphite.

2.2. The Crystal Structure 'Anomaly'

In-situ x-ray diffraction (XRD) was performed on a coin type cell with a 4x6 mm Kapton window coated with conductive thin copper layer. The graphite electrode was pressed against the Kapton window so as to be reached by the x-ray beam. After several lithiation/delithiation cycles under a C/10 rate between 1.5 and 0V, the cell was fully delithiated up to 1.5V. The cycle capacity achieved with the graphite electrode is about 360mAh/g. The cell was then re-lithiated under a slower rate of C/20. XRD patterns were taken for about five minutes every hour while the cell is under continuous discharge. As result the lithium composition x in Li_xC_6 was incremented by 0.05 between two successive XRD scans.

Figure 4 shows the evolution of the XRD spectra during the discharge (lithiation, Fig. 5a) and the charge (Fig. 5b). The diffraction angle area shown is around the strongest peaks (i.e. the 002 of graphite and stage 2 and the 001 of stage 1). Starting from fully delithiated graphite with 002 peak at around 26.6° (d_{002}=3.35Å, Cu K_α radiation) the peak shifted continuously to lower angles until it reached about 25.2° at x=0.33. This angle is the one of the 002 stage 2 compound.

The angular position of the peak then remained almost unchanged. No other peak appeared in the $0.5<x<0.7$ composition domain, where stage 1 is expected to form according to the stage 2 – stage 1 two-phase model. Such a model has been widely used in the literature to explain the flat voltage plateau in the $0.5<x<1$ composition range. It was only after $x>0.7$ that a new peak at $2\theta\sim24°$ could be noticeable. The latter is the 001 peak of stage 1 corresponding to $d_{001}=3.71$ Å. As x increased from 0.7 to 1, the 002 peak of stage 2 decreased in intensity while the 001 peak of stage 1 increased.

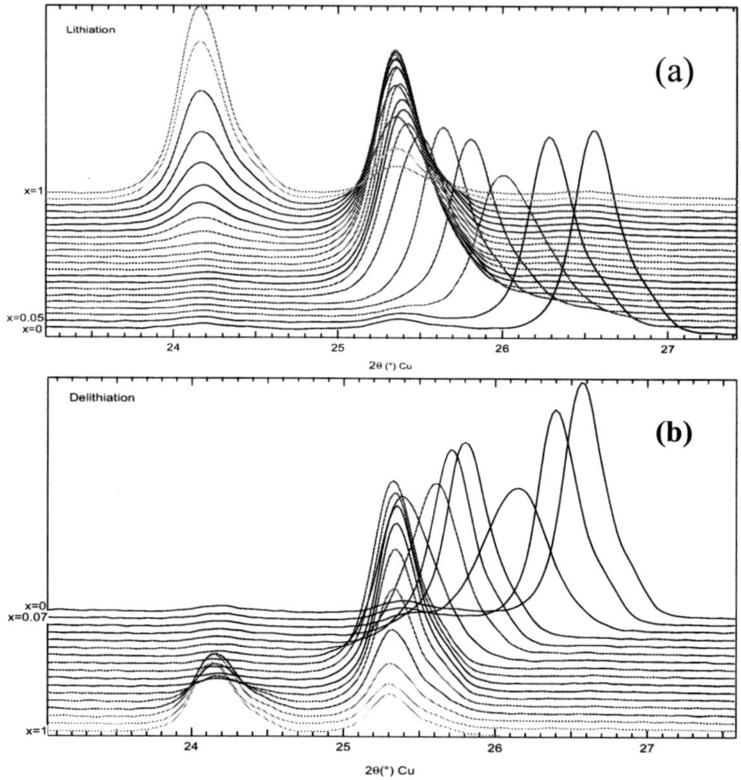

Figure 4. XRD (w/ Co K□) spectra during a lithiation (a) and de-lithiation (b) cycle of graphite under C/20 rate. The spectra were taken for 5 minutes after each hour charge or discharge.

During lithium de-intercalation the peaks intensities varied in opposite direction. However the decrease rate of the 001 peak is lower at the beginning of this process that its increase rate during intercalation. In order to illustrate this hysteretic behavior, we calculated the average interlayer spacing <d(x)> based on the integrated 00s peak intensities of stage s. for pure graphite we took the 002 peak intensity. Taking $\alpha(s)$ the fraction of stage s in the mixture, <d(x)> is given by:

$$< d(x) >= \sum_{s} \alpha_{s} \frac{3.71 + 3.36(s-1)}{s} \qquad (9)$$

$$\sum_{s} \alpha_{s} = 1$$

Figure 5 displays the composition dependence of <d(x)> during intercalation and de intercalation. This curve seems to feature four distinguishable domains.

1. $0<x<0.15$ where <d(x)> varies linearly with almost no hysteresis.
2. $0.16<x<0.33$: a small hysteresis appears with <d(x)> is slightly higher during intercalation than de-intercalation.
3. $0.33<x<0.5$: <d(x)> is almost constant around 3.51Å and
4. $0.5<x<1$: strong hysteresis where <d(x)> is lower during intercalation than during de-intercalation (arrows indicate the HF semi circle 'anomaly' (EIS)

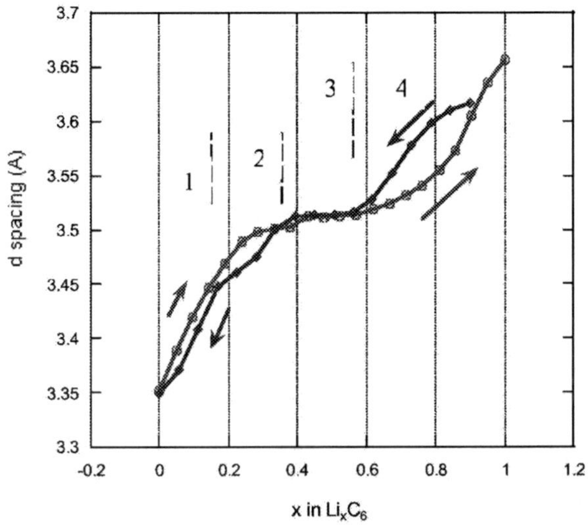

Figure 5. Composition dependence of the average graphene interlayer spacing during the lithium intercalation and de-intercalation.

2.3. The Entropy 'Anomaly'

We used the temperature dependence of the open circuit voltage (OCV) at constant composition x in Li_xC_6 to measure the entropy variation $\Delta S(x)$ expressed by:

$$\Delta S(x) = F\frac{\partial(OCV(x))}{\partial T}\Big|_x \qquad (5)$$

Figure 6 shows the entropy curve obtained during the lithiation process. Starting from a high and positive value of 60J/(mol.K) at x~0, $\Delta S(x)$

decreases steadily, changes in sign and reaches a local minimum of about -15J/(mol.K) at $0.4<x<0.5$. Then $\Delta S(x)$ suddenly increases at $x\sim0.5$ and takes values close to zero then it makes two plateaus before decreasing sharply in the $x=0.9-0.95$ composition range.

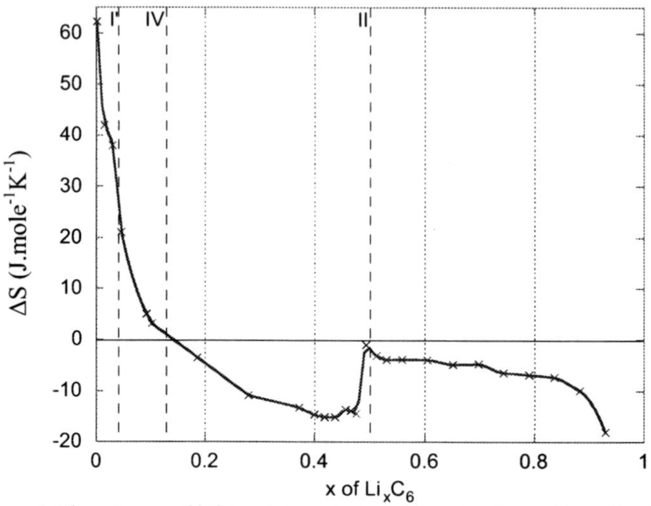

Figure 6. The entropy of lithium intercalation into natural graphite, after [22].

3. DISCUSSION

3.1. Kinetics and Strains in the SEI

We found three anomalies during the stage 2 to stage 1 phase transition. The kinetics anomaly observed by EIS in the high frequency area where the semi circle reversibly changes in size in the 70-100 mV voltage domain where the transition takes place. There is a slight voltage difference during intercalation and de-intercalation (~30mV) probably due to kinetics effects such as phase nucleation and growth.

The origin of the HF semi circle, which is present even at the initial state of the cell before any cathodic polarization of graphite, has been discussed in details in the literature [6, 11, 12, 21]. Several phenomenological models were proposed to deal with the HF semi-circle including the grains-substrate and grains/grains contact resistances within the electrode [12]. Electrosorption was also considered in relation with the voltage dependence of the HF region during the early stage of the cathodic polarization [6, 21].

The formation of the solid electrolyte interphase (SEI) on the surface of the graphite grains occurs at electrode potentials above 250mV vs. Li^+/Li.

The characteristic reduction semi plateau associated with the SEI formation appears in the 800-600mV region in EC-DMC/$LiPF_6$ electrolyte under C/10 discharge rate [23].

Depending on the graphite and electrode composition, the SEI formation typically accounts for 5 to 20% of the total electrode capacity loss. The SEI plays a key role in the Li_xC_6 electrode operation. It has been described as a heterogeneous multilayers structure [23–25]. The inner layer is believed to consist mainly of inorganic lithium compounds that result from the reductive decomposition of the lithium salt anion and solvent. Because of the high reticular energy of inorganic compounds, the inner layer is expected to prevent solvent co-intercalation since it acts as a molecular sieve with efficient lithium ion de-solvation. The upper SEI layer should consist of organic lithium salts coming from solvent decomposition whereas the outmost layer has more diffuse structure. It is likely made of oligomers and polymers, which result from ring opening of cyclic carbonates solvents.

As the graphite electrode reduction proceeds, the lithium ion transfers from the liquid electrolyte to the graphite surface through the successive SEI layers. Reduced lithium on the surface will intercalate between the graphene layers under a diffusion controlled mechanism. During the lithium intercalation, the average interlayer spacing $\langle d(x) \rangle$ increases between $x = 0$ and 0.33, then stabilizes between $x = 0.33$ and 0.5 as shown in Figure 5. Above $x = 0.5$ $\langle d(x) \rangle$ increases again although smoothly. The inner inorganic and rigid SEI layer should accompany the morphological changes of the active surface. Such changes should induce distortions of the SEI' inner layer that most likely affect its lithium ion transport characteristics in particular increasing its resistivity. We anticipate that the increase in the interfacial resistance due to distortions of the SEI should be at the origin of increase of the HF semi-circle size during the stage 2 to stage 1 transition. During the reverse stage 1 to stage 2 transition, $\langle d(x) \rangle$ decreases and re-stabilizes below $x = 0.5$ at the same value than during intercalation. This should result in strain relaxation and the recovery of the ion conductivity of the SEI internal layer. Since stage 2 to stage 1 transition is overall reversible, the HF semi circle will decrease to almost its original value.

3.2. Possible Origins of the Entropy Anomaly

We attributed the sudden increase in entropy at $x \sim 0.5$ where stage 1 should begin to form to combined contributions from configurational and vibrational entropy [22]. Configurational entropy goes with in-plane or the out-of-plane lithium vacancy 'disorder'. This translates into either the formation of lower Li density layers or the presence of Li stacking faults. Indeed a possible mechanism of the stage 1 phase formation may involve a 'dilute lithium layer' (noted dil.-Li) in similar way than in dilute stage 1 that forms at early stages of the graphite lithiation. Such a compound would have alternating 'normal' Li layer (Li) with the hexal structure and a dilute lithium layer following the sequence (Li)-G-(dil.-Li)-G. Figure 7 illustrates the structure of such pseudo stage 1 compound. Another source of disorder is the formation of "fractionary" stages such as stages 3/2 and 4/3 [26]. The latter have the following stacking: (Li)-G-(Li)-G-G and (Li)-G-(Li)-G-G-(Li)-G respectively. Figure 7 illustrate the fractionary stages as possible path during the formation of stage 1, the other path involves the pseudo stage 1 formation. A combination of fractionary stages may also be considered to deal with the lithium layers stacking disorder during the early formation of stage 1. The most likely fractionary stages are the $\frac{n+1}{n}$; (n>1) stages, which consists of a periodic stacking of n lithium layers separated by n+1 carbon layers. Is such a phase the periodicity along the c axis is:

$$c = 3.35 + n\,3.71\ (\text{Å}) \quad (6)$$

The strongest x-ray diffraction line is then the (00 n+1) line, with corresponding reticular distance $d_{(00\,n+1)}$ of:

$$d_{(00n+1)} = \frac{3.35 + 3.71\,n}{n+1} \quad (7)$$

and Bragg diffraction angle 2θ of:

$$2\theta = 2\text{Arc}\sin\frac{(n+1)\lambda}{2(3.35 + 3.71\,n)} \quad (8)$$

The Bragg diffraction angle 2θ is a decreasing function of n and varies between 2θ~25.2° and 2θ~24° corresponding to the pure stage 2 (n=1) and pure stage 1 (n=∞) respectively. Since we didn't observe any continuous shift of the 002 peak of stage 2 towards lower angles in the 0.5<x<~0.7 composition domain and no 001 peak of stage 1 compound were detected in that domain, we suggest the formation of superdense stage 2 compound. As a matter of fact the average interlayer spacing in superdense stage 2 compound should be very close to that in normal stage 2. The difference lies in the in-plane density of the lithium atoms as illustrated in Figure 8-a.

The hysteresis in the average interlayer spacing in the 0.5<x<1 composition range during intercalation and de-intercalation in Figure 5 suggests that different crystallographic paths are followed during intercalation and de-intercalation.

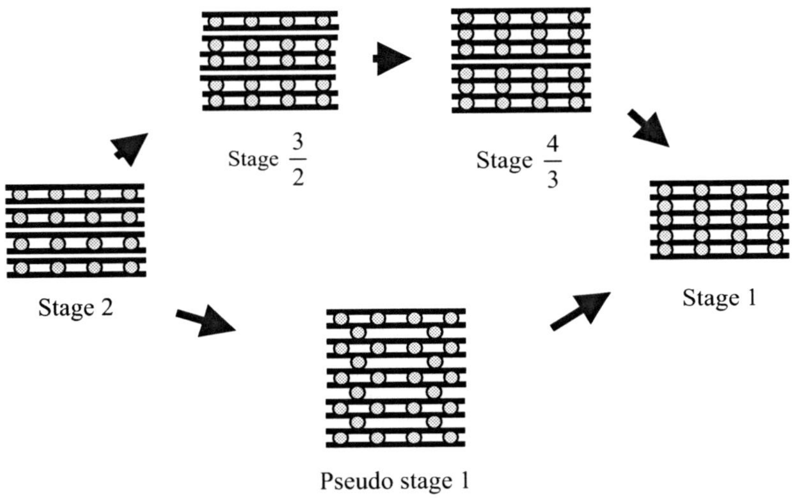

Figure 7. Possible paths of intermediary phases between stage 2 and stage 1.

As in the latter the interlayer spacing is higher than in the former, it is very likely that starting from LiC_6, lithium is de-intercalated from each equivalent layer, which preserves the stage 1 stacking sequence as sketched in Figure 8-b. This implies that at the early stage of de-intercalation, lithium layer have cation vacancies, which may be a mirror phase to dilute stage 1 formed at early steps of intercalation. Li vacancies would then form a dilute stage 1 compound keeping almost constant the average interlayer spacing.

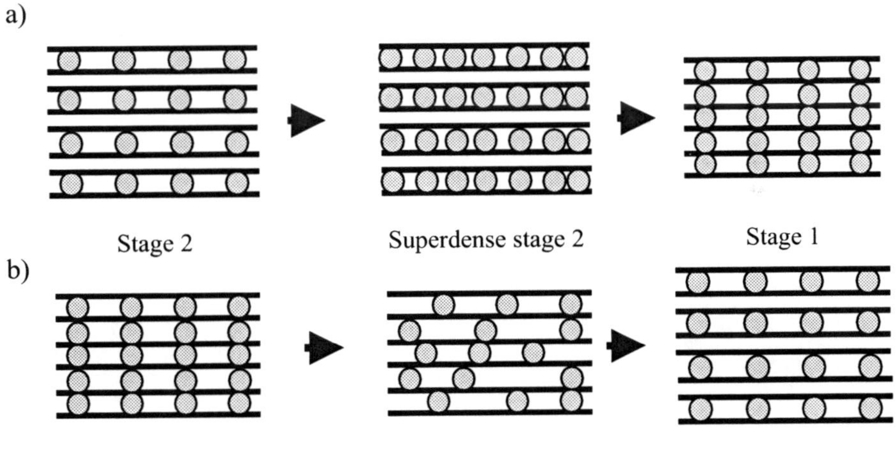

Figure 8. Schematic representation of a) the stage 2 → stage 1 transition through superdense stage 2 compound such as $Li_{1.5}C_{12}$ (or $Li_{0.75}C_6$) and, b) stage 1 → stage 2 transition involving dilute stage 1 compound.

In super dense phases, the average Li-Li distance within the layer is lower than 4.26 Å of the hexal structure of normal stage 2 and stage 1 compounds. In bcc lithium, the Li-Li distance is 3.04 Å higher than 2.46 Å in stage 1 superdense LiC_n phases (n<6) shown in Figure 9. The latter phases cannot be formed by electrochemical intercalation under normal pressure and temperature conditions [27-29]. Rather they are typically obtained under high pressure/high temperature synthesis conditions [30] or by ball milling [31]. Stage 1 superdense LiC_2 decomposes to intermediary and "kinetically" more stable stage 1 phases such as Li_7C_{24} and Li_7C_{24} [32]. The ultimate phase of decomposition is LiC_6. All superdense phases described in the literature have the lithium sub-lattice commensurate with that of graphite, with short Li-Li distances. The latter are at origin of metastability as the electrostatic Li-Li repulsion becomes dominant. Should superdense stage 2 form as intermediary phase between stage 2 and stage 1, its stabilization may go through a change in the Li in-plane ordering. This may involve a non-commensurate Li sub-lattice having a Li-Li distance lying between 2.46 Å and 4.26 Å of LiC_2 and LiC_6 respectively. More work is necessary to confirm the formation of superdense stage 2 and eventually the type of lithium in-plane organization.

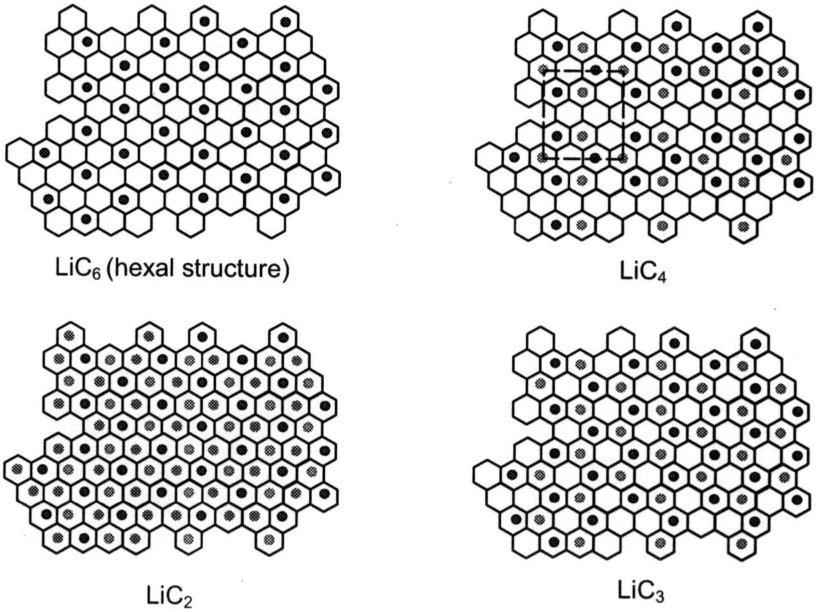

Figure 9. In-plane organization of the LiC_n (n≤6) phases including some superdense phases.

The vibrational entropy is related to a disorder originated from the Li vibrational motion about the equilibrium position at the center between two adjacent carbon hexagons. Such a motion can be decomposed in two terms: a vibration along the c axis and a vibration within the medium plane between the graphene layers [22]. Each vibration mode is defined by its Debye temperature, θ and $\theta_{//}$ for movement perpendicular and parallel to the graphene plane respectively. The vibrational entropy during the stage 2 to stage 1 transition is given by:

$$\Delta S_{vib}=R(\ln(\theta_{//1}/\theta_{//2}) + 2\ln(\theta_{1}/\theta_{2})) \quad (8)$$

The Debye temperatures of stages two and one were determined by inelastic neutron scattering measurements [33]. The total entropy variation using equation 8 is in the order of about 2 J/(mol.K). Although smaller in value, such variation accounts for 10-15% of the total entropy and should not be neglected. We are currently carrying on calculations of the vibrational entropy from the phonon density of states in Li_xC_6 phases.

4. CONCLUSION

Independent EIS, XRD and entropy measurements show some anomalies at the stage 2 ↔ stage 1 phase transition. It seems there are converging evidences that for compositions between $0.5<x<1$ in Li_xC_6 the compound doesn't consist of a linear mixture of stage 2 and stage 1 as suggested by equation 2. Rather intermediary phases may form such as superdense stage 2 and Li vacancy containing stage 1. Fractionary stages are also possible, which may increase the configurational entropy at transition. We also showed that the crystallographic paths may be different during the transition one way the other. These paths involve intermediary states with close in energy. Phase transition kinetics should account for the differences.

We suggested a mechanism involving distortion of the inner (inorganic) layer of the SEI to deal with the "breathing" of the high frequency semi circle in the Nyquist plots. Such a distortion should be induced by a large (>5%) volume change during the stage 2 ↔ stage 1 transition. More studies are needed to confirm some of the models we proposed including theoretical calculations and in-situ NMR and diffraction studies.

ACKNOWLEDGEMENTS

The authors thank Prof. Brent Fultz from Caltech for fruitful discussions and Prof. Claude Montella from LEPMI-Grenoble for his important input in the EIS part of this paper. They acknowledge the financial supports from Ato-Fina S.A. and CNRS both from France and the US-DOE through Basic Energy Sciences Grant DE-FG03-00ER15035.

REFERENCES

1. R. Yazami, P. Touzain, J. Power Sources 9 (1983) 365.
2. J.R. Dahn, Physical Review B. 44 (1991) 9170.
3. D. Billaud, F.-X. Henry, P. Willman, J. Power Sources 54(1995)207.
4. D. Aurbach, B. Markovsky, I. Weissman, E. Levi, Y. Ein-Eli, Electrochim. Acta 45 (1999) 67.
5. D. Guérard and A. Hérold, Carbon 13 (1975) 337.
6. A. Martinent Ph. D. Thesis, Univ. Grenoble (2001)
7. R. Fong, U. Von Sacken, J.R. Dahn, J. Electrochem. Soc. 137 (1990) 2009.
8. R. Yazami, D. Guérard, J. Power Sources 43-44 (1992) 39.
9. R. Yazami, Electrochemica Acta 45 (1999) 87.
10. R. Yazami and Y. F. Reynier, Electrochemica Acta 47 (2002) 1217.
11. E. Barsoukov, J.H. Kim, C. Oh Yoon, H. Lee, J. Electrochem. Soc. 145/8, 1998, 2711.
12. Y.C. Chang, H.J. Sohn, J. Electrochem. Soc. 147/1, 2000, 50.
13. A. Funabiki, M. Inaba, Z. Ogumi, J. Power Sources 68, 1997, 227.
14. M.D. Levi, D. Aurbach, J. Phys. Chem. B 101, 1997, 4641.
15. E. Barsoukov, J.H. Kim, C. Oh Yoon, H. Lee, J. Electrochem. Soc. 145/8, 1998, 2711.
16. P. Liu and H. Wu, J. Power Sources 56, 1995, 81.
17. B. Markovsky, M.D. Levi, D. Aurbach, Electrochim. Acta, 43/16-17, 1998, 2287.
18. A. Funabiki, M. Inaba, Z. Ogumi, S. Yuasa, J. Otsuji, A. Tasaka, J. Electrochem. Soc., 145/1, 1998, 172.
19. D. Aurbach, M.D. Levi, E. Levi, A. Schechter, J. Phys. Chem. B, 101, (1997) 2195.
20. T. Piao, S.M. Park, C.H. Doh, S.I. Moon, J. Electrochem. Soc. 146/8, (1999) 2794.
21. M. Holzapfel, A. Martinent, F. Alloin, B.Le Gorrec, R.Yazami, and C. Montella, J. Electroanal.Chem. 546: 41-50 APR 10 2003
22. Y. Reynier, R. Yazami, B. Fultz, J. Power Sources 119 (2003) 850-855.
23. S. Genies, Ph. D. Thesis, Univ. Grenoble (1998).
24. R. Yazami, Proc. Int. Semin. Prim. Sec. Batt. Deerfield Beach, Florida, 1997
25. E. Peled, D. B. Tow, A. Merson, A. Gladkich, L. Burstein, D. Golodnitsky, J. Power Sources 97-8 (2001) 52.
26. C. D. Fuerst, J. E. Fischer, J. D. Axe, J. B. Hastings and D. B. McWhan, Phys. Rev. Lett., 50 (1983) 357.
27. R. Yazami, A. Cherigui, V. A. Nalimova and D. Guérard in Proc. Of the 182nd Meeting of the Electrochemical Society, 1992 - Toronto, Canada, p. 17.
28. C. Bindra, V. A. Nalimova, D. E. Sklowsky, Z. Benes and J. Fischer, J. Electrochem Soc. 145 (1998) 2377.
29. R. Tossini, R. Janot, F. Nobili, D. Guèrard and R. Marassi, Electrochem. Acta 48 (2003) 1419.
30. K. N. Semenenko, V. V. Adveev, V. Z. Mordkovich, Doklady An. SSSR 271 (1983) 1402 and V. A. Nalimova, V. V. Adveev and K. N. Semenenko, High Pressure Res. 6 (1990) 11.
31. D. Guérard and R. Janot, J. Phys. Chem. Sol. 65 (2004) 147.
32. J. Conard, V. A. Nalimova and D. Guérard, Mol. Cryst. Liq. Cryst. 245 (1994) 25.
33. R. Moreh, N. Shnieg and H. Zabel, Phys. Rev. B 44 (1991) 13111.

CHARACTERIZATION OF ANODES BASED ON VARIOUS CARBONACEOUS MATERIALS FOR APPLICATION IN LITHIUM-ION CELLS

Alexandr N. Kozhevnikov, Yuri A. Podalinski, Olga R. Yakovleva, Vladimir S. Kotlyar*, Mikhail E. Petropavlovski, Victor G. Smirnov, and Vladimir V. Dzhurzha

Accumulator Research and Design Institute "Istochnik"
10 Dahl street, St.-Petersburg 197376, Russian Federation

Abstract

Testing a number of carbon materials was carried out with the purpose to choose one or several for application as a reagent of the anode for Lithium-Ion cells for their experimental and batch production. The testing consists of cycling of an electrode of the material being examined against Lithium metal electrode, or $LiCoO_2$ electrode. The basic control parameters were: specific weight capacity, irreversible capacity of the first cycle, capacity degradation degree during cycling. The best parameters among the tested anodic samples have the materials based on natural graphite "Formula BT SLC 1115" and "Formula BT SLA 1020" by Superior Graphite Co. (USA), and graphite SPG (Russian Federation).

Keywords

Lithium-Ion cell, graphite, specific capacity, irreversible capacity.

1. INTRODUCTION

The characteristics of a carbon material used as active reagent of the negative electrode (anode) of a Lithium-Ion cell considerably influence the power characteristics of the cell as a whole. Thus, the major parameters are: the values of specific capacity per unit weight and volume, and also the

* Corresponding author. E-mail: vkotlyar@nwgsm.ru

I.V. Barsukov et al. (eds.), New Carbon Based Materials for Electrochemical Energy Storage Systems, 259–268.

value of irreversible capacity. The last parameter determines the amount of superfluous active material of the positive electrode, necessary for formation of Solid Electrolyte Interface (SEI) of the anode on the first cycle of life of the cell [1]. SEI is determined as a surface film of solid electrolyte, permeable to diffusion by the ion of Lithium (Li^+). SEI is formed as a result of interaction of the active surface of the carbon anode with the electrolyte. Presence and the properties of SEI on the surface of the carbon anode determines such characteristics of a Lithium-Ion cell, as retentivity of the charge, degree of reversibility of the capacity in a "charge-discharge" cycle and the rate of capacity fade during long cycling of the cell. It is accepted, that ionic conductivity of SEI is caused by presence of insoluble Lithium carbonates in the film, during formation of which a part of Li^+, passing into the electrolyte from the lithiated reagent of the cathode, is being spent [1].

In other words, SEI is formed at the expense of loss of a part of capacity of the cathode (the ways of preliminary chemical processing of a carbon material with the purpose of formation of SEI are not considered in this paper). Taking into account, that the cathode reagent is the heaviest and most expensive component of a Lithium-Ion cell, it is desirable, that anode reagent, having the largest possible specific capacity, would require the least quantity of Li^+ for formation of SEI. In turn, the characteristics of a carbon material depend on the set of factors (the nature of a material, the technology of its processing, its particle size distribution, true surface area, etc.). From the point of view of practical application, the characteristic of a carbon material can be most completely and most precisely evaluated by carrying out the "charge-discharge" cycles. The models of electrode of the material under examination were cycled in the model of a "half-element" against a Lithium electrode, or in the model of a cell against an electrode of lithiated metal oxide, or other cathode capable to reversibly intercalate Li^+. In this work, an attempt to evaluate the characteristics of a set of carbon materials by the approach mentioned above was carried out. These materials are: pyro-carbon; nano-tubes; "expanded" graphite; spectrally pure graphite; the material based on natural graphite developed by RIECP (the above are produced in Russia); the materials based on natural graphite "Formula BT SLA 1020" and "Formula BT SLC 1115" by Superior Graphite Co. (USA).

2. TECHNIQUE

The electrodes for examination were produced by smearing of suspension of active materials in the solution of binding substance - polyvinylidenefluoride (PVDF) - on the metal substrate by means of applicator "Doctor Blade" by Hohsen Corp. N-methyl-2-pyrrolidone was used as the solvent for the binding substance.

For manufacturing of negative electrodes, suspensions containing 45wt% powder of the carbon material being examined, 5wt% PVDF, and 50wt% of the solvent. Copper foil with thickness of 0,02 mm was applied as substrate.

For manufacturing of positive electrodes, pastes with the following ratio of the ingredients were applied: Lithium cobaltate by Merck or by "Baltiyskaya Manufaktura" (Russia) - 42,5wt%, conductive additive (acethylene soot) - 3,5wt%, PVDF – 4wt%, solvent - the balance. Aluminium foil with the thickness of 0,02 mm was used as a current collector.

Removal of the solvent was carried out by exposure of intermediate products of electrodes at the temperature 130-150°C within 30 - 40 minutes. Then the intermediate products were rolled for giving the required density to the electrode layer; finished electrodes were cut from the obtained tape by means of the cliché.

In the case of using the metallic Lithium electrode, it was cut out of Lithium foil with the thickness of 0,2 mm.

As the electrolyte, 1M solution of Lithium hexafluorophosphate in the mixture of ethylene carbonate and diethyl carbonate (1M $LiPF_6$ EC:DEC=1:1 - electrolyte LP-40 by Merck) was used.

As the separator, microporous polypropylene film (PORP by NPO UFIM, Russia) with the thickness of 25 μm was used.

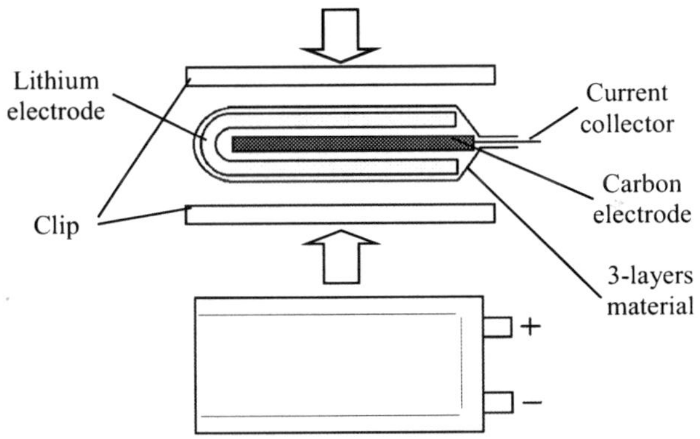

Figure 1. Test cell in "soft" pouch.

On the Figures 1 and 2, the schematical drawings of cells for “charge-discharge” cycles are shown. On the Fig. 1, the cell in the "soft" pouch, formed by welding of three-layer material (lavsan-aluminium-polyethylene) on polyethylene layer, is shown.

During the cycling, the cell was compressed by a clip with the purpose to provide dense assembly of the electrodes. By the Figure 2, the cell made with application of parts of "coin" element CR-2016 is presented. The capsule and the top of the cell are made of Nickel-plated steel, the sealing lining is of polypropylene. The models of half-elements Li-C and the cells $LiCoO_2$-C for testing were made both in the first and in the second variants of cells.

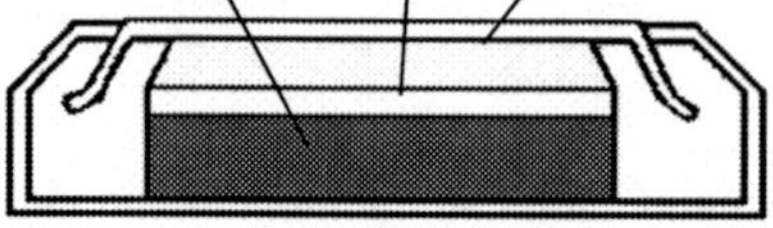

Figure 2. CR2016 test cell.

All assembly operations were carried out in tight camera in the atmosphere of dry air with the dew point of -40°C. Before the assembly, the parts of the capsule and the separator were dehydrated under vacuum at 50°C within 24 hours; the electrodes being tested were exposed in vacuum furnace at 150°C within 24 hours.

Charge and discharge current densities were in the range of 0,15-1,0 mA/cm^2, the final charge voltage of the model of the cell was 4,18-4,20 V.

During the cycling of the half-elements on the first charge (intercalation of Li^+), the minimal current density was chosen, thus the theoretical capacity was given to the electrode (calculated from the weight of reagent and in view on formation of LiC_6). The discharge (deintercalation of Li^+) was stopped upon achieving voltage of 1V.

Irreversible capacity was evaluated as the ratio between the difference of the values of discharge capacity on 3-4th and 1st cycles and the fixed capacity (3-4th cycles).

3. RESULTS OF EXPERIMENTS

3.1. Nano-Tubes

The samples of nano-tubes designated as NTC and NTA were given for tests by research-and-production firm "Astrin" (Russian Federation). Due to limited quantity of the material, the tests had rough character. During the tests it was revealed, that specific capacity per unit weight of nano-tubes is comparable to those of other carbon materials. During 50 (for NTC) and 76 (for NTA) cycles “charge-discharge”, the tendency to good stability of discharge capacity was revealed. The tests were stopped on spending of metal Lithium electrode. Good retentivity of charge in charged state during storage within 50 day was also revealed. The loss of capacity was no more than 7%. The irreversible capacity was 30±10%.

3.2. Pyro-Carbon

The samples of pyro-carbon (PC) were synthesized by the Central Research Institute of Materials (Russian Federation). [2]

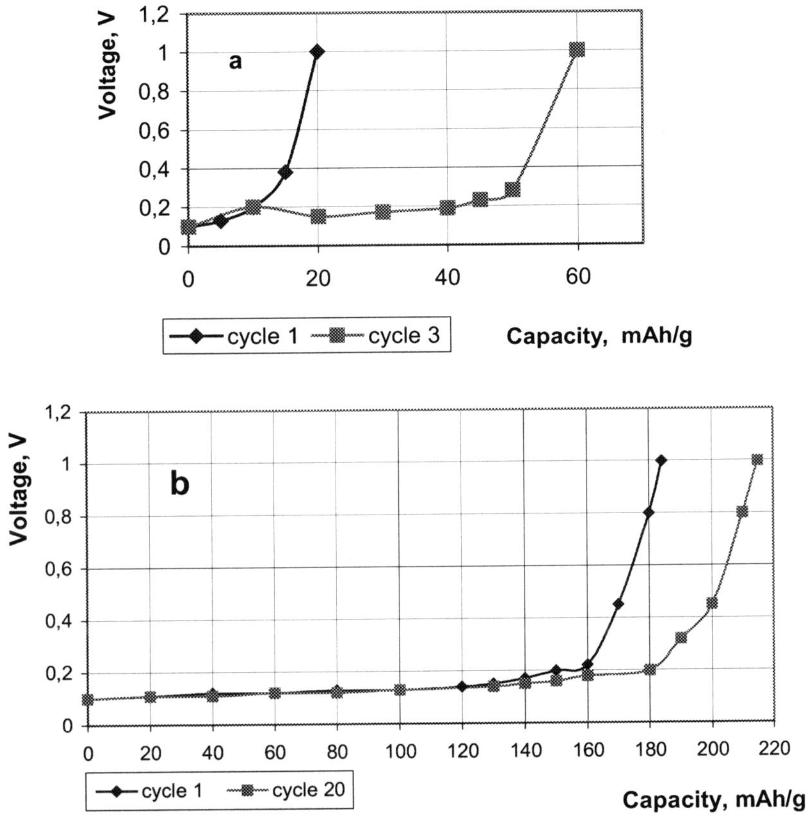

Figure 3. Discharge curves of PC anodes based on ND (a) and NC (b).

The raw material for the synthesis was methane. Powder of Nickel carbonyl (NC) or powder of nano-diamond (ND) was the catalyst. Attempts to synthesize pyro-carbon on copper powder were not successful. Powder with the composition 70%PC, 30%NC, and also the set of powders with various ratios of PC and ND were tested. Anodes made of the powder 70PC30NC showed satisfactory cycle behavior and had specific capacity 180 mA·h/($g_{\text{of powder}}$) (260 mA·h/($g_{\text{of carbon}}$)) (Fig. 3a). The anodes made of powder xPCyND, irrespective of the components ratio, had specific capacity no more than 70 mA·h/(g $_{\text{of powder}}$) (Fig. 3b). The results obtained show the activating influence of Nickel on the processes of synthesis of pyro-carbon and intercalation of Li^+ on it. It should be noted also, that the anodes of powder 70PC30NC in propylene carbonate electrolyte showed considerably lower values of capacity, than in electrolyte LP-40 (Figure. 4).

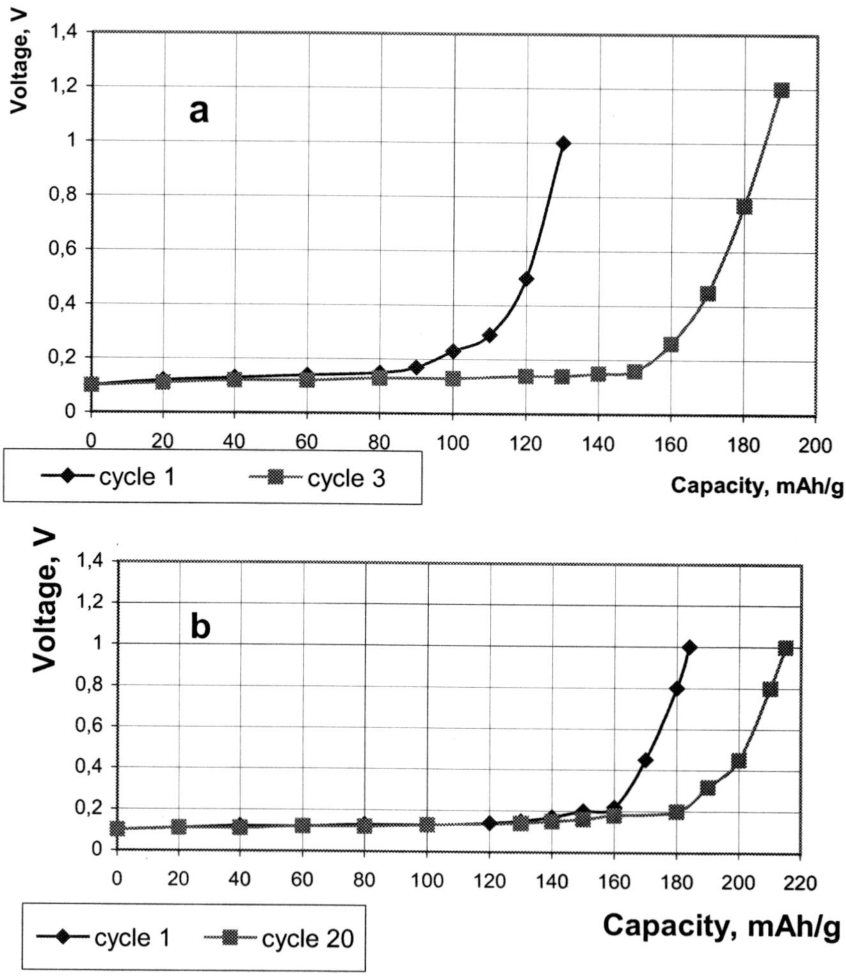

Figure 4. Discharge curves of PC anodes in
a) 1 M $LiPF_6$ in propilene carbonate + DME
b) 1 M $LiPF_6$ in ethelene carbonate + diethylcarbonate.

3.3. Expanded Graphite

The samples of expanded (thermally expanded) graphite (TEG) were provided by the Central Research Institute of Materials (Russian Federation). TEG is a powder of light grey color with low bulk density and extremely developed true surface; it can be easily pressed in a flexible plate ("cardboard"). The anodes made of TEG, had the capacity about 200 mA·h/g (Figure 5). However, in spite of the fact that the anodes were made of pressed powder, they had volumetric capacity in 3-4 times smaller, than other tested materials. The discharge curve is shown on Figure 5.

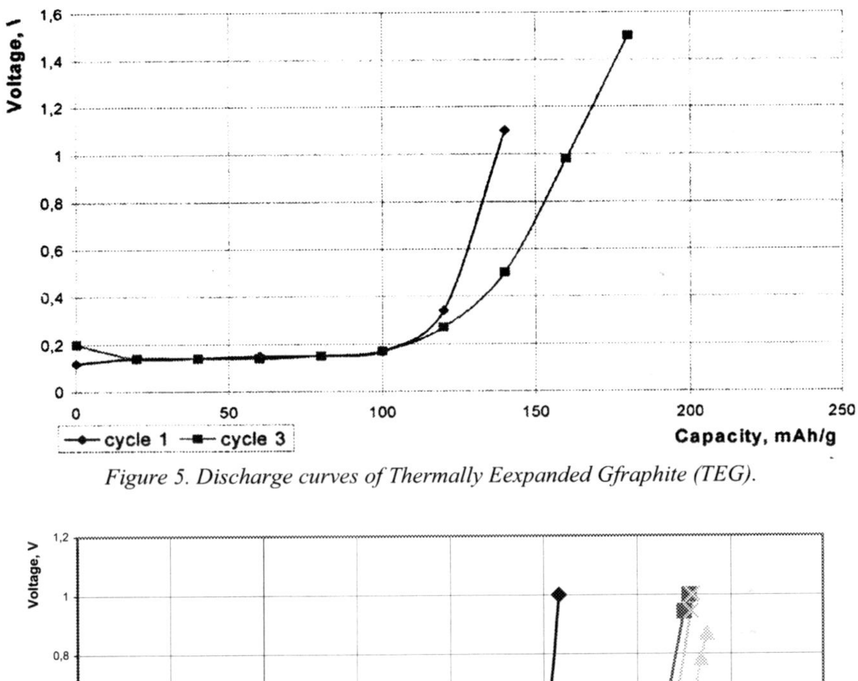

Figure 5. Discharge curves of Thermally Eexpanded Gfraphite (TEG).

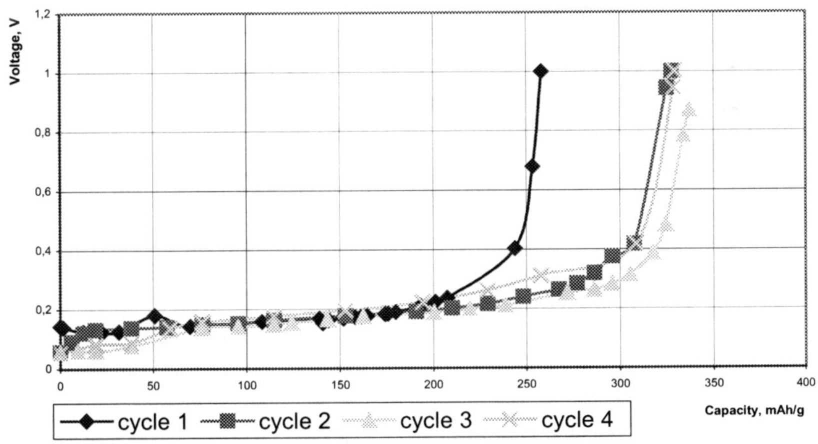

Figure 6. Discharge curves of the anodes based on RIECP carbon.

3.4. Carbonaceous Material RIECP

The material was developed by the Research Institute of Electrocarbon Products (RIECP, Russian Federation) on the basis of natural graphite specially for application in Lithium-Ion cells. The samples were given with symbolic numbers, without certificates. The anodes, made of RIECP powder, had the capacity about 300 mA·h/g. The typical dicharge characteristic is submitted on Fig. 6. Irreversible capacity is about 16%. Concerning the cycle behavior, this material is not as stable as other ones tested (Fig. 10).

3.5. SPG Graphite

SPG is a spectrally pure graphite. It is a powder containing rather large fractions (up to 500 microns). This material was not intended for application in Lithium-Ion cells and has no technological properties necessary for manufacturing of a thin electrode. The fraction up to 100 microns was selected from it, and this allowed making the required electrodes.

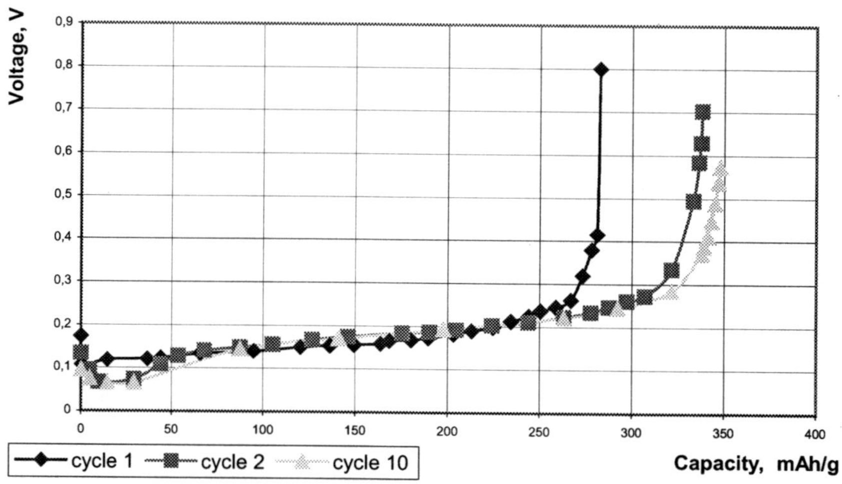

Figure 7. Discharge curves of C (SPG).

On Figure 7, the typical discharge curves of SPG electrode, obtained with use of half-element, are shown, and on Figure 8 discharge curves of the model cells are presented. As it is visible from the diagrams, on SPG electrodes the value of specific discharge capacity reached 340 mA·h/g, the stability of discharge capacity was observed during 40 cycles (Fig. 10), the irreversible capacity loss was not more than 20%. The tests are being continued.

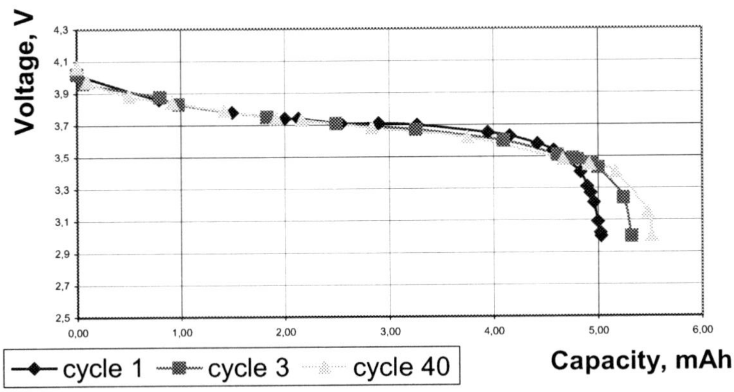

Figure 8. $LiCoO_2$ – C(SPG) discharge curves.

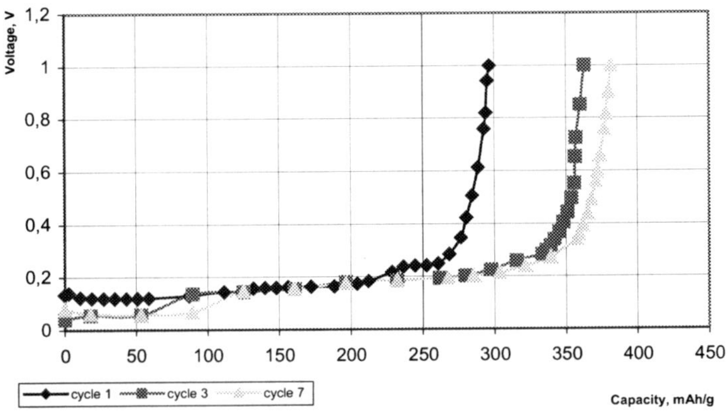

Figure 9. Discharge curves of anode Formula BT SLA 1115.

3.6. Performance of Thermally Purified Spheroidal Natural Flake Graphite SLA 1020 and Surface Coated Natural Graphite SLC1115

Carbonaceous materials provided by Superior Graphite Co. (USA) were developed on the basis of natural graphite specifically for application in Lithium-Ion cells. Samples had designations of "FormulaBT SLA1020" and "FormulaBT SLA1115". The anodes made out of these materials, during the tests in half-elements, have shown high and comparable values of specific capacity (up to 350 mA·h/g and up to 360 mA·h/g, respectively). Figure 9 shows these results. Both grades demonstrated good stability under cycling (Fig.10). The tests for stability under cycling will be continued on model cells with the purpose to prevent the possible influence of the limited reversibility of metal Lithium electrode on the results of tests. The irreversible capacity loss on both materials was no more than 20%.

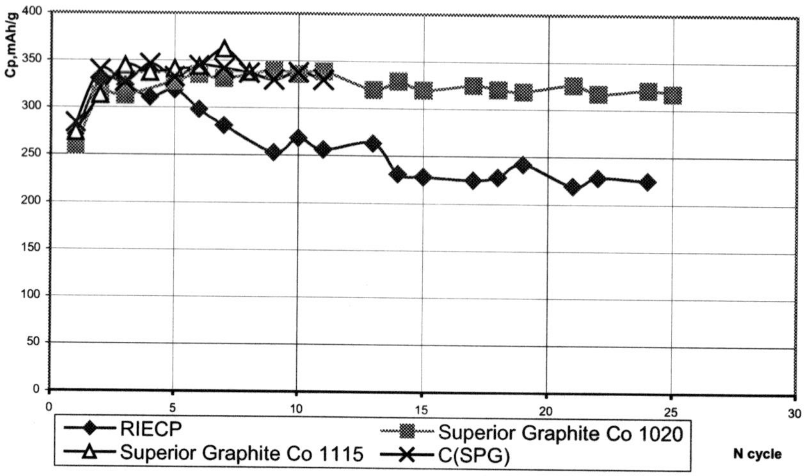

Figure 10. Capacity fade of anodes based on different materials.

4. CONCLUSIONS

In summary, we wish to note, that the results reported provide only initial characterization of the materials tested not to the full. The further profound and extended research and tests in scale of full-sized cell, first of all concerning to long cycling, is necessary. Nevertheless, in our opinion, when choosing an anode material for Lithium-Ion cells being developed, it is necessary to prefer the carbon materials based on natural graphite - "Formula BT SLA 1020" and "Formula BT SLA 1115" by Superior Graphite Co. (USA). Spectrally pure graphite (Russia) in its modified form, also can be recommended for application. Besides, the carbon material RIECP (Russia) remains a promising candidate for application in question.

ACKNOWLEDGEMENTS

The authors would like to thank all companies, which presented the samples for tests; Dr. Igor Barsukov (Superior Graphite Co.) for his efforts in providing of our participation in the conference; U. S. Civilian Research & Development Foundation (CRDF) for financial support of our trip.

REFERENCES

1. Peled E. Lithium Stability and Film Formation in Organic and Inorganic Electrolyte for Lithium Battery Systems.// in Gabano J.-P., Ed. Lithium Battery. London etc. Academic Press, 1983. Pp. 43-72.
2. Gordeev S. K., Grechinskaya A. V., Zhdanov V. V., and Krasnobrizhiy A. V. Patent of Russian Federation № 2122589.

A CARBON COMPOSITE FOR THE NEGATIVE ELECTRODE OF LI-ION BATTERIES

A.V. Churikov*, N.A. Gridina, N.V. Churikova

Chemistry Department, Saratov State University, Astrakhanskaya 83, 410012 Saratov, Russian Federation

Abstract

Composite electrodes made of two carbon components were evaluated experimentally as anodes for Li-ion batteries. The electrochemical activity of these electrodes in the reaction of reversible lithium intercalation from/to a solution of $LiPF_6$ in ethyl carbonate and diethyl carbonate was studied. Compositions of the electrode material promising for the usage in Li-ion batteries were found.

1. INTRODUCTION

As it is known, graphite and layered carbon materials capable of reversible lithium intercalation, which form Li-C compounds of various compositions are widely used as the active substance of the Li-ion battery anodes [1-10]. When carbon is used as a matrix for lithium, the structural characteristics of a specific material are of importance. Though the literature devoted to correlations between the intercalation depth and various structural parameters of a material is rather inconsistent, there is, however, a certain common point of view. From the microscopic standpoint, the optimal carbon material for the anode of Li-ion batteries should be neither perfect crystalline nor completely amorphous. It may contain centers of crystalline graphite (a mesophase) distributed over an amorphous carbon matrix. According to Dahn et al. [10-13], the suitable carbon materials fall into one of three classes, namely, graphitic materials, hydrogen-containing materials, and single-layer hard carbons.

There are several classifications of carbon materials, given in accordance with their different properties. Hard carbon is meant as a carbon

* Corresponding author. E-mail: ChurikovAV@info.sgu.ru

I.V. Barsukov et al. (eds.), New Carbon Based Materials for Electrochemical Energy Storage Systems, 269–276.

material whose structure is insensitive to heat treatment. Its precursors usually contain heteroatoms, mainly oxygen. That is why the hard carbon structure comprises crosslinking groups to prevent the formation of graphite stacks and their coalescence at heat treatment [9-13]. The disordered structure of soft carbons changes with high-temperature treatment towards the perfect graphite one and approaches it at about 3000°C. These materials are composed of more or less uniformly oriented crystallites whose sizes and ordering degree increase with graphitization heat treatment [9].

Graphites exhibit an optimal discharge curve with a low-voltage plateau and a good intercalation capacity (300 – 370 mAh/g or $x = 0.8–1.0$ in the theoretical formula Li_xC_6). Many of them allow long cycle life and a weak hysteresis of the charge-discharge curves to be achieved. However, graphites were noted to be less stable at cycling [8,14]. Their degradation is usually caused by exfoliation of the structure by coarse penetrating particles, which leads to electrode failure. Besides, graphites were found to be unable to provide high discharge rates because of slow lithium diffusion [14].

Partially graphitized cokes produced by means of thermal decomposition of organic raw materials, polymers and graphitized carbon fibers show a good performance [1,2,6-17]. The properties of carbon materials are often improved due to large amounts of dopants (H, O, S, N, P, Si, etc.) [9,18].

Therefore, there is a wide spectrum of carbon materials suitable for the usage in Li-ion batteries; the choice of a specific one determined by many factors. According to Ref. 7, the percentage of various carbon materials used in commercial Li-ion batteries was as follows: graphites – 43 %, hard carbons – 52 %, soft carbons – 5 %.

The interlayer distance d_{002} and the crystallite sizes L_a and L_C are the principal microstructural characteristics of carbon. d_{002} = 0.3354 nm for the perfect graphite structure while disordered carbon materials have noticeably larger d_{002}'s due to the graphene planes being incompletely parallel to each other. At the early stage of graphitation (about 1200°C) the crystallites achieve about 5 nm in both directions with d_{002} being about 0.344 nm. An increase of the treatment temperature improves the structure and raises the crystallinity degree of soft carbon. Simultaneously, the shape of the discharge curve (lithium deintercalation) gets better; it becomes more and more horizontal with a sharp potential elevation at complete lithium extraction, the discharge plateau approaching the potential of pure lithium. Therefore, a certain correlation between the electrochemical and crystallographic properties exists and both depend on the perfection degree of the carbon structure [9].

As to the intercalation capacity, its relation to the carbon crystalline structure is ambiguous. The reversible capacity is reported to go through a minimum at d_{002} = 0.343 nm. The L_C dependence of capacity is reported to be of a similar shape (a minimum near $L_C \approx 5$ nm, an increase when

$L_C \approx 1$ nm or L_C>20 nm). Highly graphitic materials (d_{002} within 0.335 – 0.338 nm and L_C>30 nm) show some increase of the capacity when the lowest limit of d_{002} is achieved and L_C increases. The specific reversible capacity vs crystallite size or vs treatment temperature diagrams are often published. They look as smooth curves with a minimum within medium temperatures or medium crystallite sizes [7,11,17].

On the one hand, such a behavior of capacity is explained by great importance of the perfection degree of the structure of graphitic materials at lithium intercalation. According to numerous papers, the ability of carbon to reversibly absorb significant amounts of lithium enhances with the graphization degree (graphitization is meant as the carbon structure getting more ordered and the areas with perfect graphite structure expanding). On the other hand, the capacity of heavily disordered carbon materials prepared by low-temperature (500-1000°C) pyrolysis of organic compounds, on the contrary, rises with disordering due to involvement of extra mechanisms of lithium absorption, besides interlayer insertion. The carbon residue after primary carbonization at 500-600°C has almost the same structure. The primary crystallite stacks are comprised only of 2–4 almost parallel to each other graphite planes of 10–20 aromatic rings each [9]. Some of those disordered substances exhibit supercapacity (i.e. x>1, up to 1 Ah/g) at lithium absorption [3,7-9,11-14]. The nature of the mechanisms of excessive lithium storage in low-temperature carbons is different. In particular, a part of lithium atoms is assumed to be in the interlayer space while the other part is bounded immediately to carbon atoms with hydrogen atoms involved [19-21], or the fixed excessive lithium atoms are located on the edges and breakages of the graphite structure [22]. The lithium immediately bounded to a carbon atom is fixed much stronger than that in the interlayer space of graphite, thus the bounded Li splits out at a more positive potential. The uncertainty of the nearest environment in a disordered structure leads to an almost linear shape of the discharge curve with no noticeable plateau observed. Thermal treatment of such materials causes a sharp drop of their capacity with its minimum within 1500-2000°C, and then, when approaching 3000°C, the capacity increases due to structure becoming more graphitic. There are some demerits of such materials, namely, instability of the structure at lithium insertion/extraction, a large irreversible capacity for the first charging, and a strong hysteresis of the charge-discharge curves [10-13].

Though a common viewpoint has been produced as to the structure-property relationship after intense studies of this problem, there remain many contradicting examples.

The replacement of an individual carbon material by a mixture (composite) in order to improve the electrode's performance seems promising [23]. Discharging at low potentials (0–0.25 V vs Li/Li^+) has been proven to be provided by the graphite-structure component while it is the non-graphitic one, which works above 0.25 V [4,9]. The present work deals with mixtures of two sorts of carbon. Substances are rather different in structure, e.g. non-graphitic carbon and crystalline graphite (graphite and low-temperature coke) should be taken as source materials for making a composite. The composite prepared may have a rather negative discharge plateau like graphite with a long-term cycling life of the electrode to be provided by amorphous carbon. Using such an approach, we expected the mixed electrodes to combine the advantages of both carbon materials and to compensate their demerits.

The goal of the present work was to examine binary carbon + graphite mixtures in reversible electrochemical intercalation of lithium from a nonaqueous solution. The following carbon materials different in structure were chosen: soft carbon prepared by means of pyrolysis of bitumus oil and thermally expanded graphite (TEG). Thermally expanded graphites have a larger d_{002} distance, and lithium insertion distorts their crystal lattice weaker in comparison with natural and artificial graphites.

2. EXPERIMENTAL

The thermally expanded graphite was used without any preliminary treatment; the carbon was synthesized by carbonation of the raw material in an argon flow at 500°C with subsequent thermal treatment. Carbon residues strongly differing in the crystallinity degree were obtained as depends on the temperature which ranged from 500°C to 1300°C. The identity of each material was checked by X-ray spectroscopy.

The working electrodes were prepared on a nickel support with polytetrafluoroethylene in N-methylpyrrolidone as a binder. All the operations of assembling and hermetization of cells were carried out in a glove box. Electrochemical measurements were made in mockups of the elements with lithium counter electrode and reference electrode filled with a 1 M $LiPF_6$ solution in a mixed solvent (ethylene carbonate (EC) + diethyl carbonate (DEC), both from Merck). Charge-discharge characteristics were registered on a series of samples at 35°C. Cycling was performed with a current density 0.2 mA/cm^2, charging/discharging were advanced to potential $E = 0$ and 3 V (relative to the Li/Li^+ electrode), respectively.

3. RESULTS AND DISCUSSION

The source carbon materials show a significant electrochemical activity for lithium intercalation though the reversible capacity is relatively low and tends to reduce with cycling. For the thermally expanded graphite the reversible capacity was about 220 mAh/g and 290 mAh/g at discharge to 0.5 and 3 V, respectively.

The mesocarbons prepared by bitumus oil pyrolysis at low temperatures (500 and 600°C) turned out to be practically incapable of reverse functioning as a lithium intercalating material. The highest values of the specific capacity (200 mAh/g) were obtained for the carbons synthesized within 700–1100°C. They also show a tendency of gradual reduction of performance characteristics with cycling. Increasing the pyrolysis temperature up to 1300°C led to a decrease in the specific capacity. A positive effect of the treatment temperature on the shape of the discharge curve was noted. The substances obtained by low-temperature pyrolysis have a practically linear discharge curve. As the pyrolysis temperature rises, the discharge curve undergoes flattening, i.e. the greater part of the discharge capacity is released by the electrode at low potentials. These observations are consistant with the literature data for graphitized carbon materials [7-14].

Thus, the electrochemical properties of the individual carbon materials are not so high as to enable their commercial usage in Li-ion batteries. In order to improve the performance, we started making composite materials from two individual carbon ingredients. Figure 1 shows a typical result of electrochemical tests of an electrode made of a blend of graphite and soft carbon treated at 1100°C (C1100) in comparison with the discharge curves of the individual constituents.

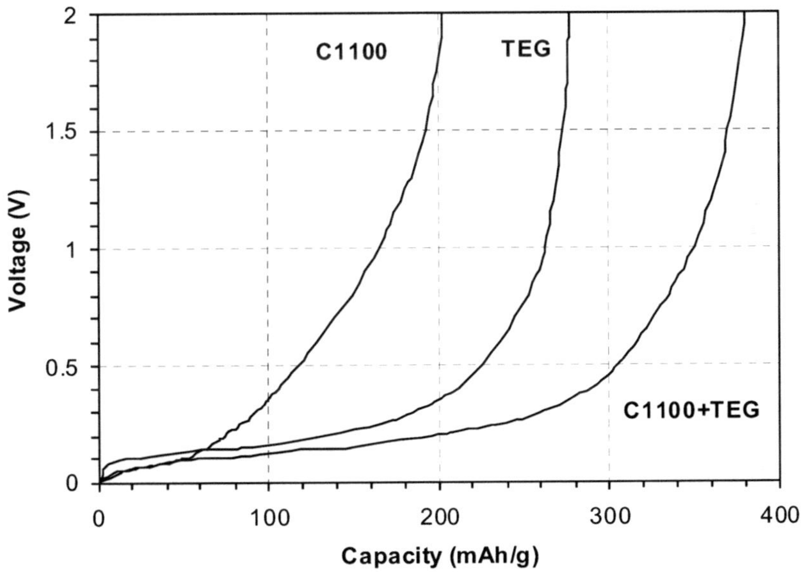

Figure 1. Discharge curves (10th cycle) of the electrodes made of the individual carbon materials and the 1:1 composite. The current density 0.2 mA/cm².

A positive effect of blending was observed in each case, i.e. the cycling capacity of the mixed electrode was significantly higher than that of the electrodes made of the constituents instead of the arithmetic mean which one might expect. At the same time, the composite carbon+graphite electrode preserved its discharge plateau near 0.1 V, similar to pure graphite. The C1100+TRG electrodes have exhibited the highest and most stable characteristics of all those under study.

Experiments aimed at determining the optimal ratio of the components in the composite were conducted. The maximum capacity corresponds to a 1:1 wt ratio while the 7:3 and 3:7 electrodes show a somewhat lower capacity.

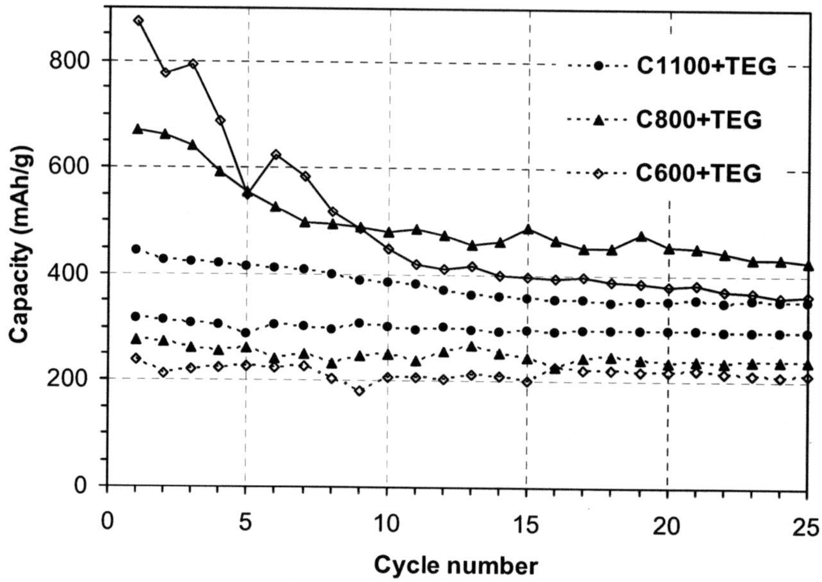

Figure 2. Discharge capacity of the electrodes of different composition versus the number of cycles. The cycling is within 0 – 3 V (solid lines) and 0 – 0.5 V (dashed lines). The current density is 0.2 mA/cm².

Some data on the cycling stability of the composite electrodes are presented in Figure 2 as the discharge capacity vs the number of cycles. Practically, of the most importance is the behavior of the negative electrode within 0 – 0.5 V where the major part of the capacity stock is realized. Thus, we used two modes of cycling, namely, soft discharge to 0.5 V in order to prolong the cycle life, and deep discharge to 3 V in order to evaluate the maximum capacity released by the material. The C1100+TRG electrodes have exhibited the best performance, the cycling efficiency approaching 100 %. The average capacity at 0.5 V discharge was 300 and 265 mAh/g for the 1:1 and 3:7 (7:3) carbon : graphite ratio, respectively. This index was lower for C800+TRG (240 mAh/g at a 1:1 wt ratio) and much lower for

C600+TRG (200 mAh/g at a 1:1 wt ratio). Thus, decreasing the treatment temperature worsens the capacity characteristics of the composite electrode as well as those of the individual components.

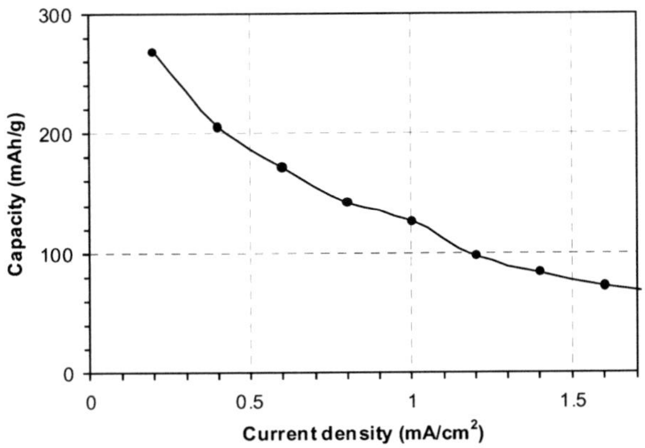

Figure 3. Dependence of the discharge capacity of the C1100+TRG electrodes on the charge-discharge current density.

However, deep discharge is a different matter. As Figure 2 demonstrates, values of the discharge capacity over 400 mAh/g were registered for the C1100+TRG electrodes at discharge to 3 V, which are more than the theoretical limit of 372 mAh/g at x=1. A supercapacity considerably exceeding the theoretical limit was detected at deep discharge of the composites with the low-temperature carbon. The electrodes containing the carbon treated at 600(C show the initial values of the discharge capacity up to 900 mAh/g. These results for the carbon not subjected to high-temperature treatment are quite comparable to the literature data on the supercapacity of disordered carbon structures formed by low-temperature pyrolysis of organic substances [3,7-9,11-14]. This supercapacity disappears at potentials above 0.5 V, and then, when E<0.5 V, the discharge capacity, on the contrary, falls with the treatment temperature, which is also in good agreement with the literature data on the role of graphitic and non-graphitic components in lithium absorption.

Attention should be paid to the steadiness of the average discharge capacity in cycling to 0.5 V and the gradual decrease of this value in the course of deep cycling. The electrodes with low-temperature C600 exhibit rather fast reduction of the discharge capacity while those with C1100 and C800 reduce this value slower. These facts are in accord with some literature data on the instability of disordered carbon structures at cycling [10-13].

The influence of the charge-discharge current density on the specific capacity properties of the electrodes is depicted in Fig. 3. One can conclude that even at a 0.2 mA/cm^2 current density the potentialities of the material are not developed fully. Further increasing the cycling current density up to

1.5 mA/cm2 led to a threefold reduction of the capacity taken up. However, the usage of a combined mode of battery charging (the galvanostatic mode in combination with the potentiostatic one) instead of constant current charging allows to avoid deterioration of the electrode under high current loads.

4. CONCLUSION

The electrochemical activity of thermally expanded graphite and soft carbon produced by means of bitum oil pyrolysis was studied in reversible lithium intercalation from a 1 M solution of $LiPF_6$ in the EC+DEC mixture. The effect of mixing different-type carbon materials for improvement of the performance is proposed to be employed. The cycling capacity of the composite electrode was found to be higher than that of the electrodes made of the constituents. The electrodes made of the 1:1 blend exhibit the maximum discharge capacity. The highest discharge characteristics were found for the thermally expanded graphite + soft carbon (treated at 1100°C) system, namely, a capacity of 300 and 400 mAh/g at discharge to 0.5 and 3 V, respectively; the Coulombic efficiency approaching 100%.

REFERENCES

1. Scrosati B. J. Electrochem. Soc. 1992. V.139. P.2776.
2. Abraham K.M. Electrochim. Acta. 1993. V.38. P.1233.
3. Armstrong A.R., Bruce P.G. Nature. 1995. V.373. P.598.
4. Fujimoto H., Mabuchi A., Tokumitsu K., Kasuh T. Carbon. 2000. V.38. P.871.
5. Levi M.D., Aurbach D. Electrochim. Acta. 1999. V.45. P.167.
6. Noel M., Santhanam R. J. Power Sources. 1998. V.72. P.53.
7. Endo M., Kim C., Nishimura K., Fujino T., Miyashita K. Carbon. 2000. V.38. P.183.
8. Broussely M., Biensan P., Simon B. Electrochim. Acta. 1999. V.45. P.3.
9. Flandrois S., Simon B. Carbon. 1999. V.37. P.165.
10. Dahn J.R., Zheng T., Liu Y., Xue J.S. Science. 1995. V.270. P.590.
11. Buiel E., Dahn J.R. Electrochim. Acta. 1999. V.45. P.121.
12. Xing W., Xue J.S., Zheng T., Gibaud A., Dahn J.R. J. Electrochem. Soc. 1996. V.143. P.3482.
13. Xing W., Xue J.S., Dahn J.R. J. Electrochem. Soc. 1996. V.143. P.3046.
14. Shi H. J. Power Sources. 1998. V.75. P.64.
15. Mabuchi A., Fujimoto H., Tokumitsu K., Kasuh T. J. Electochem. Soc. 1995. V.142. P.3049.
16. Takami N., Satoh A., Ohsaki T., Kanda M. J. Electrochem Soc. 1998. V.145. P.478.
17. Takami N., Satoh A., Ohsaki T., Kanda M. Electrochim. Acta. 1997. V.42. P.2537.
18. Larcher D., Midalige C., Gharghouri M., Dahn J.R. Electrochim. Acta. 1999. V.44. P.4069.
19. Peled E., Eshkenazi V., Rosenberg Y. J. Power Sources. 1998. V.76. P.153.
20. Wang Z, Huang X., Xue R., Chen L. Carbon. 1999. V.37. P.685.
21. Zheng T., McKinnon W.R., Dahn J.R. J. Electrochem. Soc. 1996. V.143. P.2137.
22. Wu Y.-P., Wan C.-R., Jiang C.-Y., Fang S.-B., Jiang Y.-Y. Carbon. 1999. V.37. P.1901.
23. US Patent 5,677,082. Compacted carbon for electrochemical cells (Prior. Oct.14, 1997).

ELECTROCHEMICAL INTERCALATION OF PF_6^- AND BF_4^- INTO SINGLE-WALLED CARBON NANOTUBES

Rachid Yazami[1], Irina V. Goncharova[2*] and Vladimir N. Plakhotnik[2]

[1]CNRS-Grenoble, PB 75, 38401 Martin d'Heres, France and
California Institute of Technology, Pasadena, California 91125, USA
[2]Department of Chemistry and Engineering Ecology,
Dniepropetrovsk National University of Railway Transport, Dniepropetrovsk 49010, Ukraine

Abstract

Thin films single-walled carbon nanotubes (SWNTs) are used as the positive electrode in lithium half-cells with $LiPF_6$ and $LiBF_4$ in liquid organic electrolytes. Upon driving the potential to a high 4.5V, oxidation peaks appear in the cyclic voltammograms, suggesting the anions may intercalate into the SWNT triangular structure. By analyzing the shape of voltammogram changes upon cycling, we found that the anion intercalation into SWNT is partly reversible.

Keywords

Single-walled carbon nanotubes; SWNT; anion; intercalation.

1. INTRODUCTION

Since their first discovery by Iijima in 1991 [1], carbon nanotubes have attracted a great deal of interest due to their very exciting properties. Their structure is characterized by cylindrically shaped enclosed graphene layers that can form co-axially stacked multi-wall nanotubes (MWNTs) or single-walled nanotubes (SWNTs). Like in graphite, carbon atoms are strongly bonded to each other in the curved honeycomb network but have much weaker Van der Waals-type interaction with carbons belonging to

* Corresponding author. E-mail: gonch_irina@mail.ru

I.V. Barsukov et al. (eds.), New Carbon Based Materials for Electrochemical Energy Storage Systems, 277–282.

adjacent nanotubes. Such a weak interaction is believed to be at the origin of the bundle structure of carbon nanotubes. Intercalation compounds have been reported in MWNTs and SWNTs, with intercalated species occupying the Van der Waals gaps. Most reported intercalation compounds were concerned with alkali metals [2, 3]. So far, there have been no reports on anion intercalation although covalent compounds especially with fluorine [4] were successfully prepared. In this paper we have performed some preliminary experiments on the anion electrochemical intercalation into SWNTs with the following reaction equation:

$$nC + A^- \rightarrow C_nA + e^- \qquad (A^- = PF_6^- \text{ or } BF_4^-) \qquad (1)$$

2. EXPERIMENTAL

SWNT obtained by pulsed laser ablation of graphite [5], were shaped into a thin film (buckypaper) of approximately 50 μm in thickness. XRD (Co K□radiation) and TEM (high resolution imaging and electron diffraction) were used to characterize their microstructure. Discs of 6 mm in diameter were cut from the film and dried in vacuum at 200° C for 12 hours before their use as positive electrodes in coin-type (2016) cells. Cells contained metallic lithium as the negative pole and a polypropylene microporous separator soaked in EC:DMC-$LiPF_6$ or EC:DMC-$LiBF_4$ molar solutions as electrolyte. The cells were then cycled between initial open-circuit voltage and 4.5V, and then between 4.5V and 2V by cyclic voltammetry technique under 70 μV/s voltage sweeping rate.

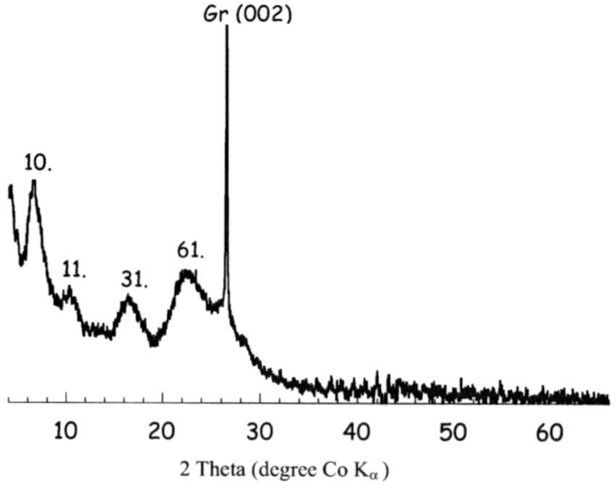

Figure 1. XRD pattern of purified SWNT.

3. RESULTS AND DISCUSSION

3.1. Structural Characterization

Characterization of the initial SWNTs included XRD and HR-TEM imaging and diffraction. Fig. 1 shows the XRD chart and Fig. 2 the HR-TEM image of the starting SWNTs. The low angle XRD diffraction peaks are characteristic of the bundle structure, with characteristic hk lines. The 'a' crystal parameter, the triangular structure is about 1.7 nm. Graphite impurity (<5%) is clearly evidenced by the (002) characteristic peak. The bundle structure is visible in the HR-TEM image in Fig. 2a. Fig. 2b shows typical diffraction rings of the triangular structure with corresponding indexation.

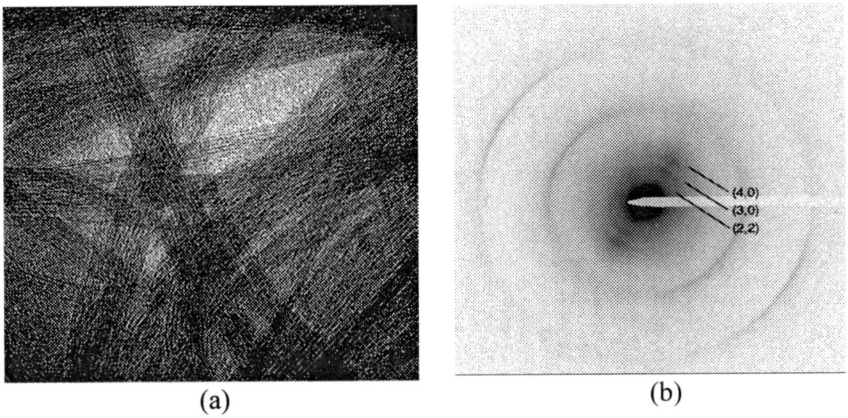

Figure 2. HR-TEM of SWNTs. Insert show the diffraction rings.

3.2. Electrochemical Characterization

Figure 3 and Figure 4 show the first cyclic voltammogram (CV) obtained with PF_6^- and BF_4^-, respectively. Arrows in the figure indicate the direction of the voltage sweep. During the first voltage sweep, starting from ca. 1.1 V vs. Li^+/Li, the oxidation current remains very low until a threshold value of about 4.15V is reached, then it increases sharply. When the sweep is reversed after 4.5V, a strong oxidation peak appears at about 4.4V followed by several other oxidation peaks especially around 3.5V. The current is then reversed and a reduction peak appears in the 2.5-2.2V area.

In the following cycles, the CV pattern does not show anymore the 4.15V anodic potential threshold, indicating a lowering of the anion intercalation overpotential.

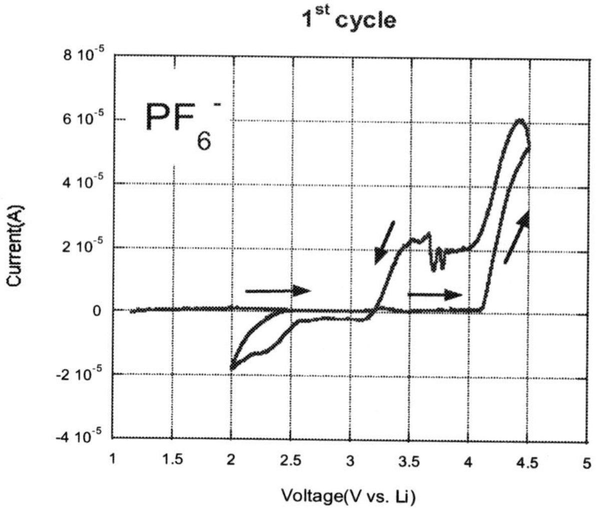

Figure 3. First CV (70 mV/s) of Li/EC:DMC-LiPF$_6$/SWNT cell.

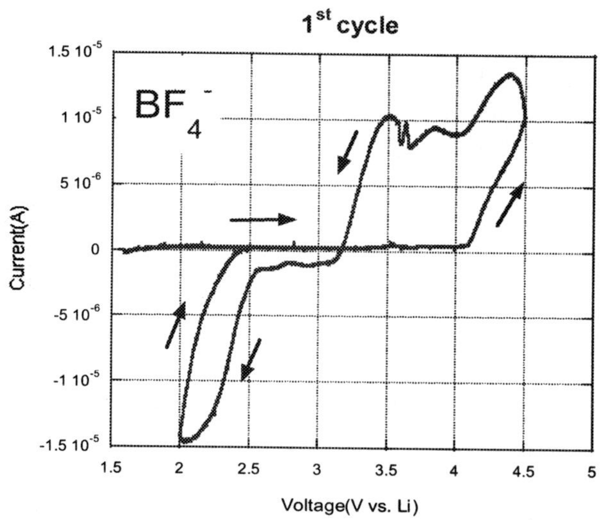

Figure 4. First CV (70 mV/s) of Li/EC:DMC-LiBF$_4$/SWNT cell.

New anodic and cathodic peaks appear which indicates some degree of reversibility of the anion intercalation. Figure 5 and Figure 6 show the second cyclic voltammogram (CV) obtained with PF_6^- and BF_4^- respectively. However, the anodic part remains dominant.

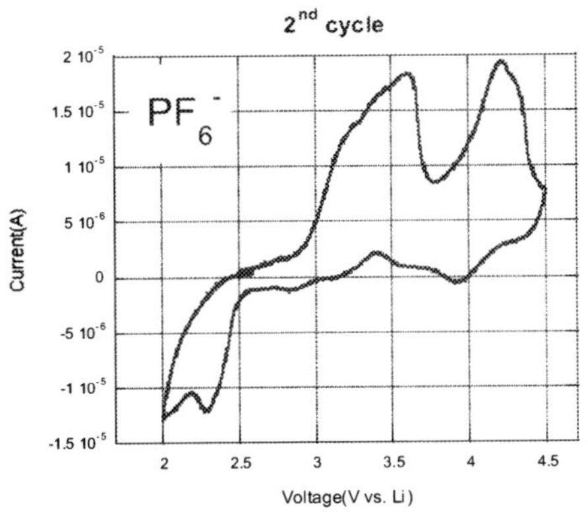

Figure 5. Second CV (70 mV/s) of Li/EC:DMC-LiPF$_6$/SWNT cell.

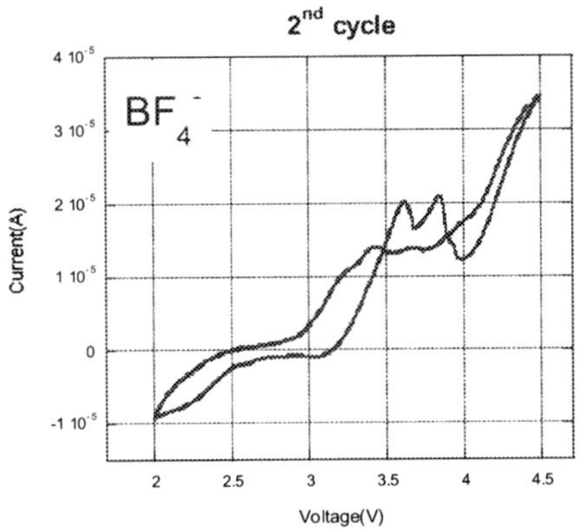

Figure 6. Second CV (70 mV/s) of Li/EC:DMC-LiBF$_4$/SWNT cell.

The appearance of oxidation peaks in the CV is a direct evidence of PF_6^- and BF_4^- anions incorporation within the SWNT structure. The macromolecule C_n should then bear a positive charge as a result of subsequent oxidation following. In the case of PF_6^- the anodic oxidation is:

$$nC + PF_6^- \rightarrow (C_n^+, PF_6^-) + e^- \qquad (2)$$

It is very likely that PF_6^- occupies the Van de Waals gap between the SWNTs in a similar way than in graphite intercalation compounds [6] and in anion-doped fullerenes [7]. Because of lower surface area between SWNTs, the stability of the (C_n^+, PF_6^-) ionic compound should be lower than in flat graphite layers. Therefore, during the electrochemical intercalation a chemical de-intercalation (decomposition) may take place, which explains the low faradaic yield of the anodic intercalation.

The observation of a high threshold voltage during the first intercalation suggests high activation energy to form the C_n^+ macrocation. Since alkali metals do intercalate into SWNTs with associated formation of macroanions C_n^-, SWNTs are amphoteric in character. However it seems that the macroanion form is more stable and reversible as reported for the lithium intercalated SWNTs.

4. CONCLUSION

We have shown that BF_4^- and PF_6^- intercalate into SWNTs in a similar way than in graphite or in fullerenes (C60 and C70 molecules). The intercalation (equation 1) takes place in different steps, which evolve with the cycle number. We suppose that in the first cycle it needs to break the tube-tube interaction (separation) and to transfer the electron. After separation is made, it becomes easier to intercalate and deintercalate the anions, although the reaction is not 100 % reversible. The technological output of different cycles upon anion intercalation into SWNTs is obvious, knowing the need to improve the energy density and the power density of existing lithium-ion batteries. We believe that increasing the energy and power density will make the nanotube-based lithium batteries even more attractive.

ACKNOWLEDGEMENTS

The authors give thanks to Dr. Channing Ahn from California Institute of Technology for providing the SWNTs sample and for allowing us to show his TEM results. Thanks also to Dr. Dayal Meshri from Advance Research Chemicals (Catoosa, OK, USA) for providing high purity $LiPF_6$ and $LiBF_4$ for this investigation.

REFERENCES

1. Iijima S. Nature 1991; 354: 56-58.
2. Thess A., Lee R., Nikolaev P., Dai H., Petit P., Robert J., Xu C., Lee Y. H., Kim S. G., Rinzler A. G., Colbert D. T., Scuseria G. E., Tomarek D., Fischer J. E. and Smalley R. E. Science 1996; 273: 483-487.
3. Zhou O., Fleming R. M., Murphy D. M., Chen C. H., Haddon R. C., Ramirez A. P. and Glarum S. H. Science 1994; 263: 1744-1747.
4. Yazami R., Gabrisch H., Fultz B. J. Chem. Phys. 2001; 115: 10585-10588.
5. Ajayan P. M. and Ebessen T. W. Rep. Prog. Phys. 1997; 60: 1025-1062.
6. Yazami R, Touzain P. Solid State Ionics 1983; 9-10: 489-494.
7. Yazami R, Cherigui N. Mol. Cryst. Liqu. Cryst. A-1994; 244: 209-214.

SURFACE TREATED NATURAL GRAPHITE AS ANODE MATERIAL FOR HIGH-POWER LI-ION BATTERY APPLICATIONS

J. Liu[1*], D. R. Vissers[1], K. Amine[1], I.V. Barsukov[2], J.E. Doninger[3]

[1]*Electrochemical Technology Program, Chemical Engineering Division Argonne National Laboratory, 9700 South Cass Ave., Argonne, IL 60439, USA*
[2]*Superior Graphite Co., 4201 West 36th Street, Chicago, IL 60632, USA*
[3]*Dontech Global, Inc. 427 E. Deerpath Road, Lake Forest, IL. 60045, USA*

Abstract

High power application of Li-ion battery in hybrid electrical vehicles requires low cost and safe cell materials. Among the various carbon anode materials used in lithium-ion batteries, natural graphite shows the most promise with advantages in performance and cost. However, natural graphite is not compatible with propylene carbonate (PC)–based electrolytes, which have a lower melting point and improved safety characteristics. The problem with it is that the molecules of propylene carbonate intercalate with Li^+ into graphite, and that frequently leads to the exfoliation of the graphite matrix.

In this work, we improved the performance of the natural graphite by surface modification. In this effort, two types of natural graphite were evaluated for high power applications in Hybrid Electrical Vehicles. One was a round edge type natural graphite SL-20, and the second graphite sample was surface treated with an amorphous carbon coating (version, SLC1015), both materials are available from Superior Graphite of Chicago.

Test cells using the above graphite materials were evaluated in PC-based electrolytes. For this work, the hybrid pulse power characterization (HPPC) test was performed on cells with the different graphite anodes and PC-based electrolytes to evaluate their high power capabilities. These electrochemical experiments indicate that cells containing the surface-modified natural graphite can meet the power requirement set by the FreedomCar partnership for the hybrid vehicle applications.

Keywords

Natural graphite, carbons, surface modification, coating, lithium-ion cells, high power, hybrid electric vehicles (HEV).

* Corresponding author. E-mail: LiuJ@cmt.anl.gov

I.V. Barsukov et al. (eds.), New Carbon Based Materials for Electrochemical Energy Storage Systems, 283–292.

1. INTRODUCTION

Carbon materials are commonly used as the negative electrode (anode) in lithium-ion cells. Among the various carbon materials, natural graphite appears to be the most promising candidate because it has a high capacity (372 Ah/kg, with a stoichiometry of LiC_6), a potential near lithium, a relatively flat potential profile, good cycle-ability, lower cost and reasonable safety. Thus Natural graphite is among the most attractive negative electrode materials for lithium-ion batteries [1-3]. Moreover, natural graphite was also successfully investigated in high power lithium-ion batteries for hybrid electrical vehicle propulsion [4].

But it is known that natural graphite has a problem in some of the more promising electrolyte solvents, like those containing propylene carbonate (PC). PC has several advantages over other organic carbonates as a Li-ion battery electrolyte solvent component, e.g., lower cost, lower volatility and higher flash point and better low temperature performance. However, when using PC in Li-ion battery, the solvated PC lithium ion will co-intercalated into the graphite layers and cause subsequent exfoliation and failure of graphite electrode [1].

The performance of natural graphite in PC-containing electrolyte can be improved by its surface modification. An amorphous carbon coating was one method to prevent graphite exfoliation which was based on a concept that a core-shell-structured carbon composite may be applied to the graphite anode material in Li-ion batteries by Kuribayashi et al.[5]. In this effort, two types of natural graphite were evaluated for high power application. One was a round edge type natural graphite SL-20, and other sample was surface treated (carbon coated) version, SLC1015. Both materials are the grades available from Superior Graphite Co. of Chicago IL.

It is known that the electrolyte solvent may be decomposed during the first lithium ions intercalation (or formation cycle) into graphite to form a solid electrolyte interphase (SEI) film on the graphite negative electrode surface. It is this SEI film (or layer) that largely determines the performance of graphite as a negative electrode in Li-ion batteries. Aurbach *et al.*, postulates that the initial surface reaction and SEI film controls the nature of further reactions [6]. Clearly, the electrode performance of natural graphite depends on the nature of solvent and the graphite surface modifications which may inhabit the solvated lithium ions from insertion into the graphite structure.

One of the promising techniques to modify the surface of natural graphite is soft carbon or an amorphous carbon coating. Mitsui Mining Company used the thermal vapor decomposition (TVD) treatment [2] to obtain modified natural graphite and Superior Graphite Co. uses a new technique with lower cost to form a thin film carbon coating on the natural graphite materials. It is therefore anticipated that the modified natural graphite will have stable performance in PC based electrolytes and will reduce the cost for both the negative electrode and electrolyte in lithium ion battery systems.

The objective of the present study is to evaluate the performance of surface coated natural graphite in PC based electrolyte under high power test conditions and to investigate the effect and impact of surface coating the natural graphite in Li-ion batteries for high power applications and their safety. Our goal is to select the best natural graphite as an advanced negative electrode material for Li-ion battery that are safe, low cost, high power, possess excellent calendar life and PC based electrolyte tolerant. From our results, modified natural graphite in Li-ion batteries has shown excellent performance and meets the requirements for high power applications in hybrid electrical vehicles (HEV) [7].

2. EXPERIMENTAL

The graphite materials used in this study include: SL-20 which is round edge natural graphite, and SLC-1015 which is the surface coated natural graphite from the type of untreated precursor SL-20, and both samples were obtained from Superior Graphite Co.

The anodes of these two graphite samples were fabricated from a slurried mixture which contains 92 wt% of active graphite powder and 8wt% polyvinylidene difluoride (PVDF) polymer binder (Kureha #9130) and using 1-methyl-2-pyrrolidinone (NMP) (Aldrich, >99%) as the solvent. After getting the homogenous slurry, the electrode laminates were coated on Cu current collector foil using a doctor blade in the laboratory-made laminate-coater. The laminates were then dried first at 75°C in air for 3 hrs and then the final heat treatment was carried out in a vacuum oven at 75°C for 10 hrs. Finally, the laminates were calendared to about 35% porosity in a dry room.

Coin-type cells (size CR2032, Hohsen Co.) with a lithium foil counter electrode (FMC Corporation) and a polypropylene separator (Celgard 3501) were assembled inside a helium-filled glovebox (<5ppm, H_2O and O_2). Laboratory made cell fixtures were also used for the HPPC test at room temperature.

The electrolyte used in the lithium cell studies was typically 1.2M LiPF6 in ethylene carbonate (EC): propylene carbonate (PC): methyl ethyl carbonate (MEC) in a 3:3:4 mixture. The cells were cycled at room temperature using Maccor Series 4000 control unit in a galvanostatic mode under a constant current density of 0.1 to 1 mA/cm^2.

Postoperative investigations of the anode material were carried out using scanning electron microscope (SEM, JEOL model JSM-6400 at 20 kV) to view the morphology of graphite particles.

To evaluate the high power cell performance, hybrid pulse power characteristic (HPPC) tests [8, 9] were carried out at Argonne National Laboratory using their own designed cyclers that are operated by VAX station computers to control the experiments. ANL designed cell fixtures were also used for cell tests. In these cell studies, $LiNi_{0.8}Co_{0.15}Al_{0.05}O_2$ was used as a positive electrode and 1.2M $LiPF_6$, EC:PC:MEC (3:3:4) was used as electrolyte.

Safety studies of the graphite anode samples were performed using a Perkin-Elmer Differential Scanning Calorimeter (DSC, model Pyris 1) instrument. The temperature scanning rate was 10°C/min over a temperature range of 50 to 375°C.

3. RESULTS AND DISCUSSION

3.1. Morphology of Materials

SEM representative images of the surface treated natural graphite SLC-1015 and its untreated precursor SL-20 are shown by Figure 1. The graphite particles with the rounded edges having less active sites tend to limit the reaction on its surfaces and thus improve its cycling performance and safety.

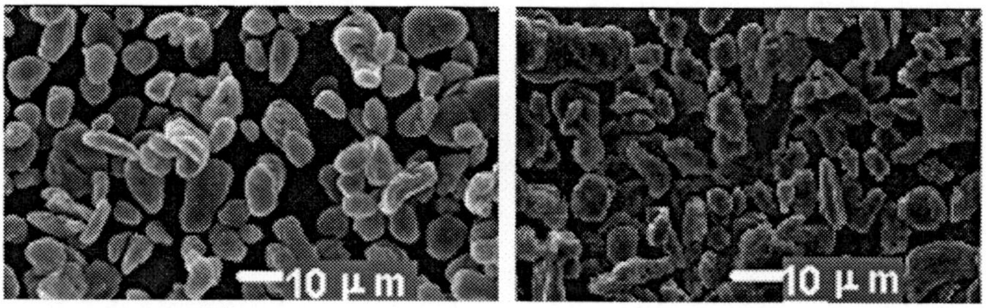

(a) (b)

Figure 1. SEM images of (a) untreated natural graphite precursor SL-20 and (b) surface treated natural graphite with round edge shape (SLC-1015).

SEM images indicate that both graphite samples have very similar morphologies. Negative electrode laminates with active materials of SL-20 and SLC-1015 were then prepared using similar compositions. These laminates were then used to prepare negative electrodes that were inserted into Li-ion cells having similar cathodes and electrolyte materials.

3.2. Capacity of Graphite Anodes

The initial charge/discharge capacity curves of SL-20 and SLC-1015 electrodes vs. Lithium counter electrodes are present in Fig.2 and 3 at the C/5 rate. The carbon surface-coated material, SLC-1015 in 1.2M $LiPF_6$, EC/PC/MEC electrolyte has a higher discharge capacity (335mAh/g in comparison to 310mAh/g for SL-20 uncoated material. These results indicate that this material is more stable in PC-based electrolyte than the uncoated material, SL-20. Since the electrolyte contains 30 wt% PC, it is interesting that the uncoated round edged SL-20 material showed no apparent exfoliation during the initial cycle. This is very different from the natural graphite with flake type morphology, where the graphite exfoliation takes place immediately with much reactivity in PC-based electrolytes.

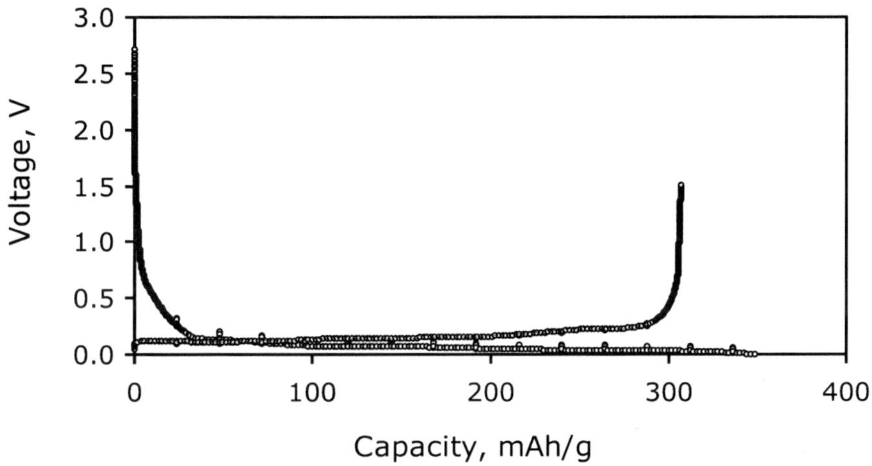

Figure 2. Initial charge/discharge curve of graphite sample SL-20, in 1.2M LiPF6, EC/PC/MEC (3:3:4) electrolyte. Test cell was cycled at C/5 rate.

In Figure 3, the discharge capacity of surface modified graphite sample is about 335mAh/g in 1.2M $LiPF_6$, EC/PC/MEC electrolyte for SLC-1015 graphite. This is about 8% higher than un-coated graphite precursor SL-20. The reason for this may be due to the better morphology of the powder and less capacity loss in PC-based electrolyte during cycling. For this reason we focus on the surface modified natural graphite SLC-1015 in the following experiments.

To verify the rate capability of the SLC-1015 material, cell with lithium counter electrodes were cycled at the C/10 and C/2 rates, respectively, as shown in Figure 4. The practical discharge capacities for the surface treated material, SLC1015 are superior to those obtained for the untreated graphite SL-20. As indicated in Fig.4, SLC1015 shows reversible capacity values of about 349mAh/g at the C/10 rate and 320mAh/g at C/2 rate.

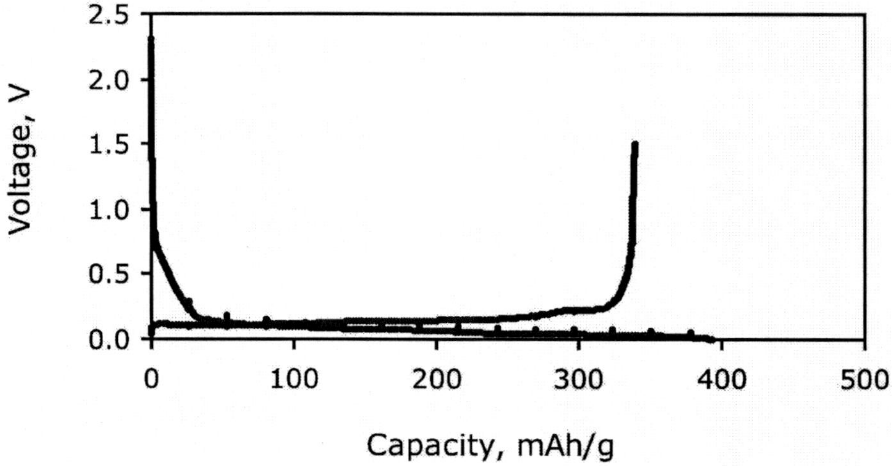

Figure 3. Initial charge/discharge curve of graphite sample SLC-1015, in 1.2M LiPF6, EC/PC/MEC (3:3:4) electrolyte. Test cell was cycled at C/5 rate.

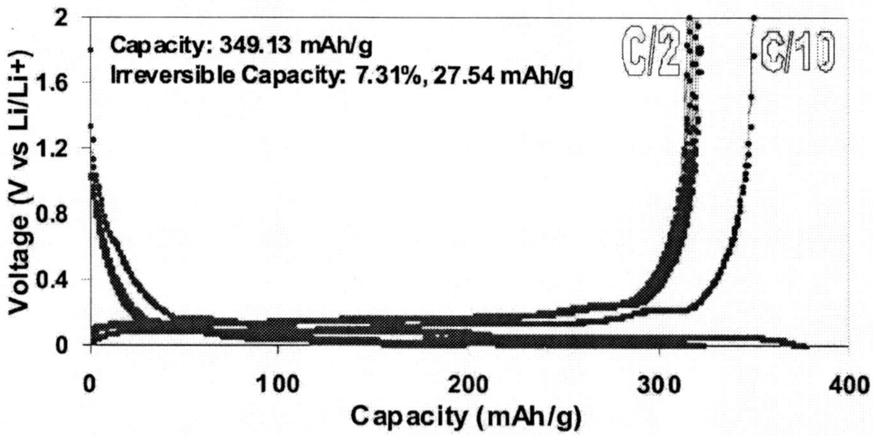

Figure 4. Charge/discharge curves of the surface treated natural graphite SLC1015 vs. Li/Li^+ in 1M $LiPF_6$, EC:DMC (50:50 wt%).

In the above study, SLC-1015 showed a low irreversible capacity when the cell was cycled at the C/10 rate, e.g., 7.3% of its capacity, that is a very good performance.

3.3. High Power Studies

Characterization (HPPC) test of full cells were performed with graphite anodes in accordance with the procedure described in the PNGV test manual [8, 9]. Figure 5 and 6 show the typical HPPC test results. This cell had $LiNi_{0.8}Co_{0.15}Al_{0.05}O_2$ as the positive active material and carbon-coated natural graphite SLC1015 as the negative active material.

Figure 5 shows the area specific impedance (ASI) for the HPPC 18-s discharge and 2-s charge pulses as a function of depth of discharge (DOD). The ASIs for both charge and discharge are very low over a wide range of DOD. These values are lower than the ASIs, needed to meet the FreedomCar performance goal, which is 35 ohm-cm^2 for 18-s discharge and 25 ohm-cm^2 for 2-s charge pulses. The cell pulse-power capability for both charge and discharge was thus found to meet the FreedomCar power requirements, especially in the cell "sweet spot" of 30-70% DOD.

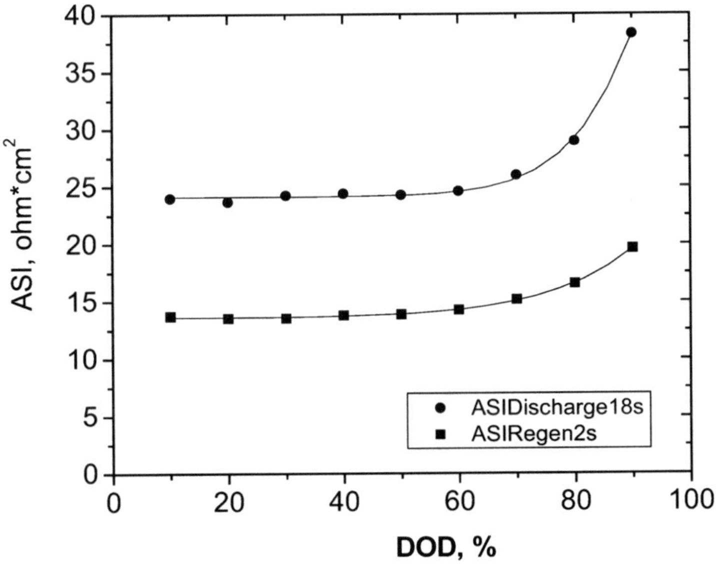

Figure 5. ASI vs. DOD for cell using surface treated natural graphite SLC1015 as anode in 1.2M $LiPF_6$, EC/PC/MEC (30:30:40 wt%). Pulse discharge at 10C rate.

The power capability of the cell against DOD was plotted in Figure 6. The performance of such cells and test results thereof indicate that cells containing the surface-modified natural graphite meet and exceed the power requirements set by the FreedomCar partnership for the hybrid electric vehicle applications.

In comparison to SLC-1015 graphite, one cell was tested with 92wt% SL-20 vs. $LiNi_{0.80}Co_{0.15}Al_{0.05}0_2$ in 1.2M LiPF6, EC/PC/MEC (3/3/4) by the same procedure and conditions described earlier. The ASI vs. DOD plot is shown in Figure 7.

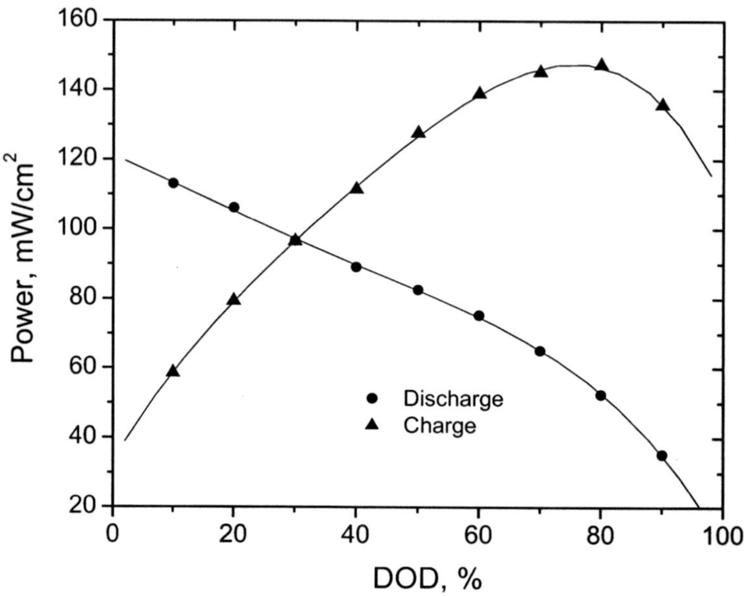

Figure 6. Power capability of cell using surface treated natural graphite SLC1015 as anode, and,$LiNi_{0.80}Co_{0.15}Al_{0.05i}$ cathode in 1.2M $LiPF_6$, EC/PC/MEC (30:30:40 wt%).

The ASI of cell using SL-20 as negative electrode is relatively higher compare to SLC-1015 in Fig. 5. But this value also meets the power requirements of the FreedomCar program for HEV application.

3.4. Safety Study

To evaluate the safety property of SLC-1015 graphite, DSC was used to study the thermal behavior of fully charged anode samples in electrolyte as a function of increasing temperature. Results are plotted in Figure 8.

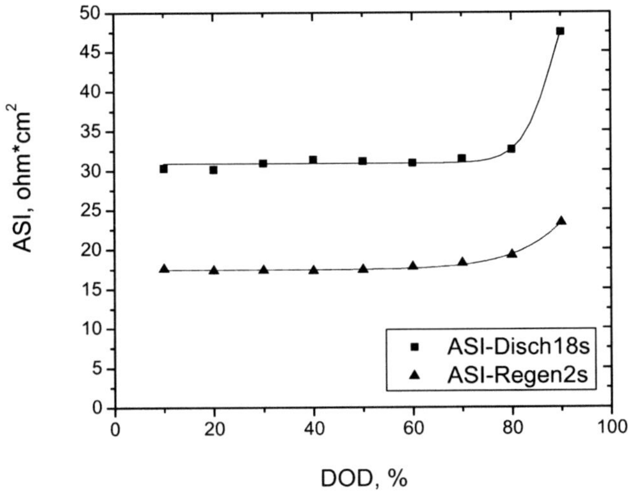

Figure 7. ASI vs. DOD for cell using surface treated natural graphite SL-20 as anode and $LiNi_{0.80}Co_{0.15}Al_{0.05}$ cathode in 1.2M $LiPF_6$, EC/PC/MEC (30:30:40 wt%). Pulse discharge at 10C rate.

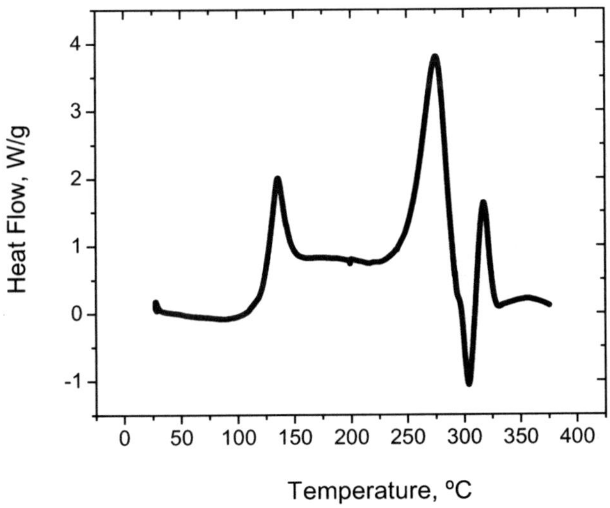

Figure 8. DSC of graphite sample (SLC-1015) at 100% SOC with electrolyte: 1.2M $LiPF_6$, EC/PC/MEC (30:30:40 wt%).

The natural graphite at 100% SOC (state of charge) was tested with electrolyte using DSC and it was found that the heat generation is small. The first peak at 125°C was mainly due to SEI layer breakdown and until 250°C the main reaction of the lithiated carbon is the reaction with electrolyte that results in gaseous products. This second peak indicates that the heat generation was largely limited by using PC-based electrolyte and this may improve the safety of Li-ion batteries for high power application, e.g., HEV.

4. CONCLUSIONS

In conclusion, the surface modified natural graphite has good performance in PC based electrolyte and also meets the power requirements for hybrid electrical vehicle applications. Surface carbon coated natural graphite SLC1015 is a very promising material in high power Li-ion batteries with lower cost, reasonable safety, and low irreversible capacity.

ACKNOWLEDGEMENT

Authors representing ANL acknowledge the financial support of the U.S. Department of Energy, FreedomCAR & Vehicle Technologies Program, under Contract No. W-31-109-Eng-38. Also, all authors are very grateful for the continued support of their DOE sponsor, Dr. Tien Duong.

REFERENCES

1. A. N. Dey, B. P. Sullivan, J. Electrochem. Soc., 117, 222 (1970).
2. M. Yoshio, H. Wang, K. Fukuda, Y. Hara and Y. Adachi, J. Electrochem. Soc., 147 (4), 1245 (2000).
3. H. Wang and M. Yoshio, J. Power Sources, 93, 123 (2001).
4. J. Liu, A. Kahaian, I. Belharouak, S. Kang, S. Oliver, G. Henriksen and K. Amine. "Screening Report on Cell Materials for High-Power Li-ion HEV Batteries", ANL-03/16, DOE report, April, 2003.
5. I. Kuribayashi, M. Yokoyama, and M. Yamashita, J. Power Sources, 54, 1 (1995).
6. D. Aurbach, M. D. Levi, E. Levi, A. Schechter, J. Phys. Chem. B, 101, 2195 (1997).
7. K. Amine and J. Liu, ITE Letters, Vol.1, No.1, B39 (2000).
8. "PNGV Battery Test Manual", Revision 1, Idaho National Engineering Laboratory, Department of Energy, DOE/ID-10597 (May, 1998).
9. PNGV Test Plan For Advanced Technology Development Gen 2 Lithium Ion Cells, EHV-TP-121, Revision 6, October 5, 2001.

CHAPTER 4:

EMERGING METAL/CARBON COMPOSITE ANODES FOR NEXT GENERATION LITHIUM-ION BATTERIES

Chapter 4: Subject Overview

The following chapter contains a collection of six papers specifically dedicated to the topic of metal/graphite composites as candidate active materials for the negative electrodes of the lithium-ion batteries of the near future. Editors believe this chapter to be a very first attempt made in the worldwide electrochemical literature to group metal/graphite composite lithium-ion battery developers into a stand-alone section of a book.

Recently the metal-carbon composites are being widely viewed as the most probable short-term solution to obtain a significant improvement of the classic carbon-based anode materials for lithium-ion batteries. There appears to be a significant driving force to commercialize these materials. It is based on a fact that fierce market competition already resulted in creation of lithium-ion batteries with near theoretical performance for a given cell size. Thus, editors estimate that it is going to be a matter of no more than one to two years before the practical limit of approximately 2,500 mAh* for the 18650-size lithium-ion cell is reached. The ever-growing power demand will require new electrode chemistries with higher specific capacities, and, likely, these materials shall be metal/graphite composites for the anode portion of the lithium-ion cells. The recent announcement by Sony of new commercially viable 'Nexelion' cells that contain a Co-Sn-C anode signals a trend and expands the types of future Li-ion cells that are envisioned.

In the opening paper of this chapter, a collaborative partnership of authors from Ukraine and USA (Prof. V. Barsukov and Prof. J. Doninger) formulate key theoretical principles for designing anode materials of new generation lithium-ion batteries. The authors propose to coat graphite with nano-dispersed Si or Sn, then imbed these composites in some highly stable shell, which would suppress volumetric changes of Si and Sn upon reversible formation of an alloy with lithium. The paper also goes into the theoretical grounds of the role of expanded graphite and conducting polymers for use in various battery and supercapacitor systems.

In the next paper by Y. Illin *et al.*, capabilities of Sn anodes are considered as a possible alternative to carbon. Thin films of Sn were deposited onto current collector in vacuum, and tested in the coin cells. Authors were able to obtain reversible alloying reaction, which stabilized at 100 mAh/g between cycle number 100 and 400. The stability of Sn and its characteristics upon cycling was seen to be a function of the current collector material. The best results were achieved with non-copper-based substrates.

The third paper in this chapter is a product of successful international collaboration, which came into being through assistance of

* This capacity limit for a given cell size has been determined as a result of computer simulation by GP Battery Technologies, as referenced from: J. Fan and D. Magnuson, in "Rechargeable Lithium Batteries" K.M. Abraham, E.S. Takeuchi, M. Doyle eds., PV 2000-21, the Electrochemical Proceedings Series, Pennington, NJ (2000).

NATO Science for Peace, NATO Advanced Research Workshop, and US Civilian Research and Development Foundation's programs. Novel metal/graphite composite materials proposed for development by Ukraine's Kiev National University of Technologies and Design were produced and further upgraded by Superior Graphite in the U.S. In the next stage, development effort linked more organizations in the U.S., one of which is an industry leading specialty lithium-ion battery manufacturer, Lithion, Inc. Even though the project has not yet finished, materials being developed look promising, as judged by their testing in the full prismatic lithium-ion cells. The paper is on carbon coated Si, and coated Si- and Al-modified graphite.

A paper by L. Reiter *et al.* is a joint contribution from two academic institutions and a battery technology development organization in Ukraine. It introduces carbon/graphite materials modified by d-metal complexes via reactions of pyrolysis of precursor carbons with various organics. The approach taken appears to be promising for both gas-diffusion electrodes and anodes of lithium-ion batteries. While non-graphitic materials used by the authors as precursors, did not appear to have benefited from grafting, graphites were seen to show improved high rate performance in the half-cells with metallic lithium as a counter electrode. Editors would like to comment that potential issues of high cost and stability of performance characteristics of graphites grafted by d-metals in the practical lithium-ion battery environments may need to be further investigated by the authors.

In the paper by Dr. L. Matzui *et al.*, authors seek to develop advanced composite materials based on natural graphite and nano-scale metals, which are known to reversibly form alloys with lithium. Authors submitted interesting results on plating of nano-sized particles of metal onto graphite surface. Of a little reservation to editors is the choice of carbon precursor, which is bulky expanded graphite. Expanded graphite's packing density properties typically make it unsuitable for application as anode active material in the lithium-ion batteries. The paper is lacking electrochemical performance data; hence it is difficult to assess how stable the composites are in practical lithium-ion cell environments.

The last paper of this chapter (C. Johnson *et al.*) summarizes important investigative work by researchers at the Argonne National Laboratory in the U.S. on the subject of improvements of graphitic anodes for high power lithium-ion batteries. The anode carbons were modified by Ni, Cu or carbon coatings, and then tested, among others, in PC-containing electrolytes. Authors conclude that Cu coating of graphite improves capacity of electrodes to a greater extent in PC than Ni coating; CVD carbon coating on graphite was also seen to also improve anode performance versus that based on uncoated graphite. The authors hypothesize that Cu may act as a better de-solvation catalyst than Ni when stripping the PC coordination sphere around the Li cation, thus leading to improved electrochemical performance.

ON THE THEORETICAL PREREQUISITES FOR APPLICATION OF NOVEL MATERIALS IN PROMISING ENERGY SYSTEMS

Viacheslav Z. Barsukov[1*] and Joseph E. Doninger[2]

[1]*Electrochemical Power Engineering & Chemistry Department, Kiev National University of Technology and Design, 2 Nemirovich-Danchenko str., Kiev 02001, Ukraine;*
[2]*Superior Graphite Co, 10 South Riverside Plaza, Chicago, 60606,IL / Dontech Global, Inc. 427 E. Deerpath Road, Lake Forest, IL 60045, USA*

Abstract

Some theoretical prerequisites for application of modified and expanded graphites, Si- and Sn-based composites and alloys, electro-conducting polymers as active materials, catalysts and electro-conductive additives for lithium - ion batteries, metal-air batteries and electrochemical capacitors are considered. The models and the main concepts of battery-related use for such materials are proposed.

Keywords

Modified graphite; expanded graphite; Si- and Sn- composites; alloys; conducting polymers; lithium-ion batteries; metal-air batteries; electrochemical capacitors.

1. INTRODUCTION

Lithium-ion batteries, high-energy lithium and metal-air batteries, electrochemical capacitors have been changing radically, in recent years, the structure of the market of power sources.

Constant technological improvements dictated by ever-increasing demands from the market, require the development of new materials. This process involves extensive basic electrochemical research into the nature and

* Corresponding author. E-mail: chemi@mail.vtv.kiev.ua

I.V. Barsukov et al. (eds.), New Carbon Based Materials for Electrochemical Energy Storage Systems, 297–307.

mechanisms of electrode processes, quantum-chemical and physicochemical modeling, a new look at the development and testing of battery materials.

The aim of this challenging task is to be always coming up with technological solutions that allow reaching and surpassing theoretically calculated energy density output limits for a particular size of electrochemical devices. This paper seeks to introduce theoretical concepts, which became founding stones for some of the new materials, which our groups have jointly proposed and developed for the markets of anode active materials in lithium-ion batteries, conductivity enhancement applications in various types of power sources, and conducting polymer matrixes for the emerging electrochemical supercapacitors.

2. RESULTS AND DISCUSSION

2.1. "Classic" Graphite and Modified Graphite Materials

Special types of natural and/or synthetic graphite are commonly used as negative active materials (AM) for lithium-ion batteries (LIB). After replacing their more expensive predecessors, mesophase carbon micro beads (MCMB), they have become state-of-the art active materials in recent years, and, therefore, we refer to them as "classic" lithium-ion grade graphite. These AM feature stabile capacity of Q^{-}_{c}=365-370 (mA·h/g) under prolonged cycling conditions due to their high purity and special structure of particles; also they demonstrate rather low irreversible capacity loss during the first SEI formation cycles ($\Delta \leq$ 10%). An illustrative result showing stable cycling of purified round-edge natural graphite is shown by Figure 1.

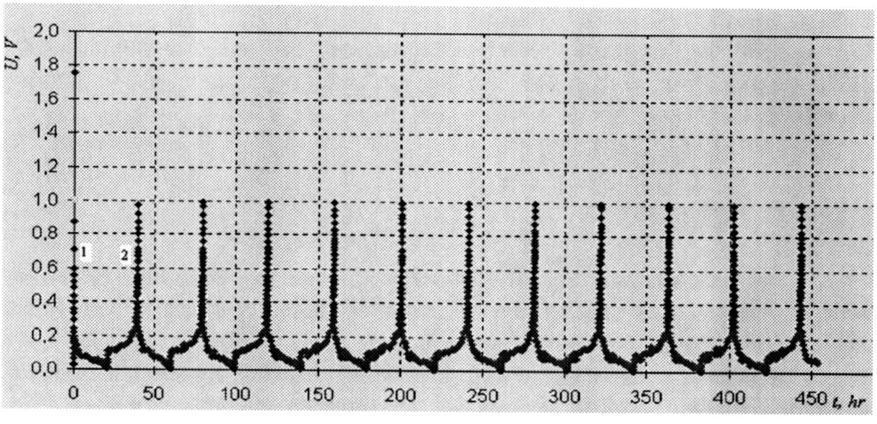

Figure 1. Typical galvanostatic charge (1) – discharge (2) curves of the lithium-ion battery grade graphite, SL-20 (Superior Graphite Co., USA), as tested at C/20 rate in 2016 coin cells having Li metal foil as counter electrode and electrolyte: EC:DMC + 1M $LiPF_6$.

The specific capacity obtained in such Ams, actually corresponds to near theoretical limit of 372 mA·h/g, as calculated on a basis of classical LiC_6 stoichiometry). Further increase of capacity is possible only via switching to new or modified materials, composites or alloys, which are capable for reversible and stable intercalation of lithium.

It is often proposed in the literature to use SnO_2, Si, carbon materials with disordered structure, and many similar materials as an alternative to graphite (see, for example [1-4]). Our studies show that indeed it is possible to achieve very high capacity with such active materials in the first cycle (Q^-=700-800 mA·h/g and higher), however this capacity decreases quite rapidly during cycling. Moreover, the irreversible loss of capacity in the first cycle is very high and is often over 50%, which brings to naught their advantages.

Different investigations of the mechanisms of capacity degradation during cycling show that one of the main reasons of this degradation is significant change (by a factor of 3-4) in AM volume on the intercalation-deintercalation of lithium, followed by AM destruction [2-3].

Therefore, a possible theoretical concept of using such materials in LIB's is based on the development of various composites (alloys), in which the volume variations of the electroactive constituents of AM during cycling can be compensated by the elastic properties of the electrically conductive matrix (Figure 2).

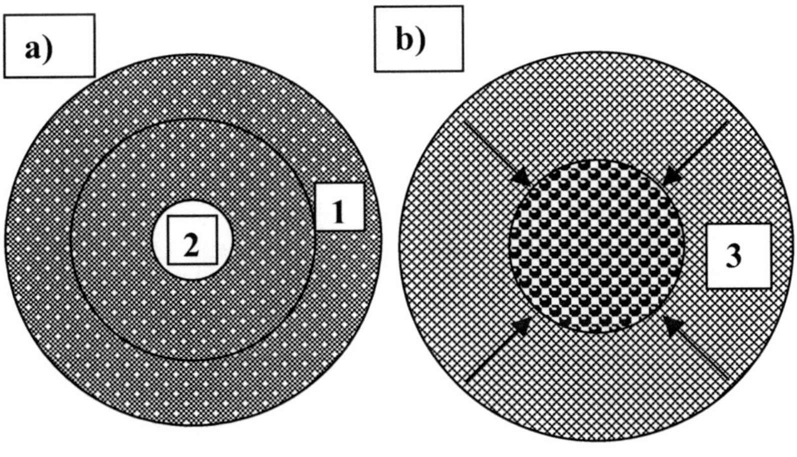

Figure 2. The concept of using additives of Si (or Sn) in the negative electrode: a) the beginning of charge; b) the end of charge. 1 and 3 – elastic porous graphite (or alloy) matrix with Li^+ ions; the particle of Si (or Sn) before (2) and after Li^+ ions intercalations.

The summary of our many try-n-error experiments suggests that for realization of concept proposed by Figure 2, it is necessary to enable the following conditions:

1) Furnish uniform & highly dispersed nano-sized particles of Si or Sn;
2) The amount of Si or Sn has to be small (usually on the order of 3-5wt%);
3) The bulk electrode has to have very good elastic properties, which can be created by proper optimization of conductive matrix.

In accordance with this concept, the experimental modified rounded-edge graphite grade was synthesized with 3wt% of the Si coating on LBG-25 graphite (the later being a commercial product of Superior Graphite Co. (SGC), a U.S. industrial carbon manufacturer). While reversible capacity of LBG-25 is approximately 370 mAh/g, its Si-coated version (grade SO # 3-43-18), demonstrated a noticeably improved discharge capacity of 412 mA·h/g. The above value of capacity is higher than theoretical capacity for pure graphite (372 mAh/g), which leads to the average utilization coefficient $\mu \cong 111\%$ (Figure 3, Table 1) for this product.

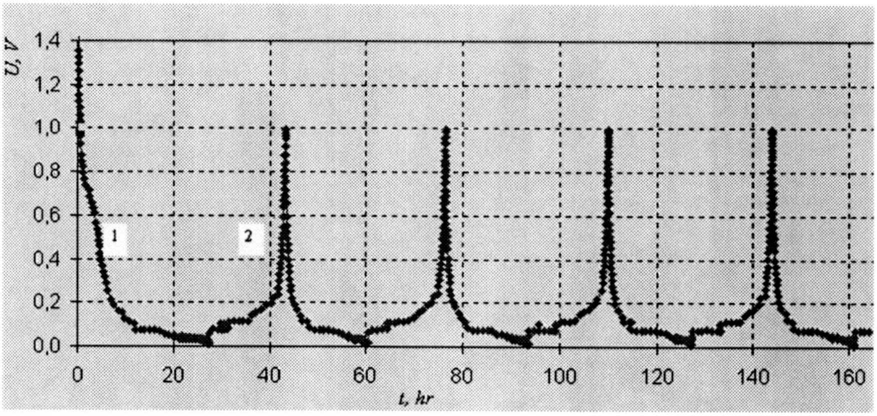

Figure 3. Galvanostatic charge (1) – discharge (2) curves of SO# 3-43-18 (LBG-25 + 3% Si) graphite at C/20 rate in 2016 coin cell. Electrolyte: EC:DMC + 1M $LiPF_6$. Counter electrode – Li foil.

It is interesting to understand the reasons for exceeding the theoretically possible capacity limitations with graphite SO # 3-43-18.

It is known about the existence of lithium silicide, Li_6Si_2, which is close to intermetallic compounds, and also that silicon is capable to form with lithium different alloys. We have calculated the theoretical specific capacity of such possible compounds, as well as pure silicon (Table 2). It is possible to explain from the Table 2 the noticeable increase of capacity for graphite electrodes (11%) even at the small content of Si (3wt%). We can suppose that some of such compounds (Li_XSi_Y) with high capacity may form during electrochemical cycling, and in doing so will give a noticeable boost to the total capacity of graphite. Creation of uniform highly dispersed network of particles of Si in an elastic graphite matrix leads to a relatively stable dependence of capacity vs. Si loading on graphite surface.

Table 1. Results of initial galvanostatic cycling of experimental 2016 coin cells with different types of graphite from SGC at C/20 rate ($\Delta U = 0.01$ - $1.0V$). Electrolyte – LP-30 from Merck (EC:DMC + 1M LiPF6). Counter electrode – Li foil.

Graphite grade and its main properties	**Average discharge capacity* (mA·h/g)**	**Utilization coefficient μ (%)**	**1st cycle loss (%)**
LBG-73 ($D_{50} = 43$ μm; $S = 2.85$ m^2/g)**	368	98.9	7.83
LBG-25 ($D_{50} = 18$ μm; $S = 4.9$ m^2/g)**	370	99.5	6.77
SO # 3-43-18 (LBG-25 + 3% Si)	412	111	21.4

Notes for the Table:
* *Results for an average sample in a series of 3-5 coin cells, during the first 4-5 cycles.*
***According to the Product Information Bulletins of SGC Purified Battery Grade Graphite.*

Table 2. Calculated theoretical specific capacity of some silicon compounds and pure Si.

Compounds	Li_6Si_2	$Li_{12}Si_7$	Li_7Si_3	$Li_{13}Si_4$	$Li_{22}Si_5$	Si
q_T, mA·h/g	1,641	1,149	1,411	1,716	2,005	4,000

Similar effect of relatively stable cycling on the level of discharge capacity of 500 mA·h/g can be achieved, for example, using **tin-based alloys** with different metals (please, see the detailed investigation of such alloys in this book [5]). These metals perform probably the functions of elastic matrix in such alloys and gave possibility to compensate for the volumetric changes of Sn.

Thus, the experiments carried out suggest that using the theoretical concept proposed by Figure 2, it is possible to create negative electrodes that would feature reversible capacity exceeding 400 mA·h/g.

2.2. Thermally Expanded Graphite

One of possible ways of improving the specific characteristics of LIB electrodes is the use of thermally expanded graphites (TEG) as electrically conductive additives of positive active mass. The efficiency of TEG's use may be illustrated by analyzing a simple theoretical model proposed by us: "AM – electrically conductive additive" (Figure 4).

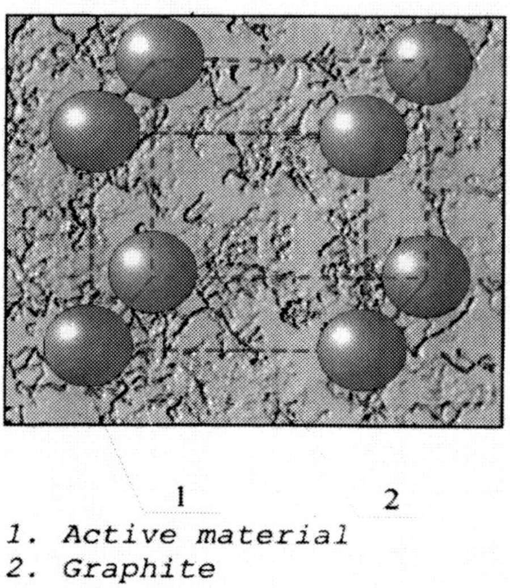

Figure 5. A model for composite porous electrode in a cubic close packing approximation.

The main idea of the model is that in order for the electrically conductive additive to effectively fulfill its functions, it must form a closed cluster (skeleton of the interconnected carbon particles, which is the conducting pass in electrode matrix). Once the sufficient conductive network was formed, further considerable increase of additive content is not needed, as it leads to decrease in the percentage of the electrochemically active constituent in the electrode.

On the basis of an analysis of this model in a cubic closed packing approximation, it may be shown that the necessary additive content per unit AM mass, $\boldsymbol{M}$, can be described by the following relation:

$$M = \frac{M_{AD}}{M_{AM}} = \frac{\rho_{AD}}{\rho_{AM}} \left[\frac{6}{\pi} \left(\frac{t_{AD}}{D_{AM}} + 1 \right)^{3} - 1 \right], \quad (1)$$

where M_{AD} is a content of the electrically conductive additive; M_{AM} is a content of the electrochemically active material; ρ_{AD}, ρ_{AM} are the *effective* densities of the additive and AM respectively; D_{AM} is an average size of AM spherical agglomerates; t_{AD} is an average size of additive interlayer (the average distance between the spheres).

It is clear from the equation (1) that the optimal specific capacity of such two-component composite porous electrode would be realized when

$$M \Rightarrow M_{min}. \qquad (2)$$

In this case a porous electrode has a maximal possible amount of the electrochemically active material and the minimal amount of the conductive additive, which is necessary for formation of closed cluster (the touch of the spheres at the Figure 5) and a good conductivity of electrode mass.

The ratio (2) can be realized when

$$(t_{AD} / D_{AM})\,\pi\pi << 1. \qquad (3)$$

In this case

$$M_{min} = \left(\frac{6}{\pi} - 1\right) \times \frac{\rho_{AD}}{\rho_{AM}}. \qquad (4)$$

It follows from the relation obtained that the minimum electrically conductive additive content is directly proportional to the effective density of the additive. By "effective density" we understand the density of the material under real conditions of making the electrode (with allowance for the actual molding (rolling) pressure, humidity, temperature, etc). In this respect, TEG has unique advantages over all existing types of additives. The density of this material in free state (bulk density) is 0.05 g/cm^3, which is about one-fourth of that for the ordinary graphite and one-fifteenth to one-twentieth of that for the metal powders (e.g. nickel, copper powders, etc.).

Particularly, let us make comparison of densities for common and expanded graphite in a free state:

$$\frac{\rho_{GR}^{com}}{\rho_{GR}^{exp}} \cong 4;$$

For this extreme case

$$\frac{M_{min}^{com}}{M_{min}^{exp}} = \frac{\rho_{GR}^{com}}{\rho_{GR}^{exp}} \cong 4. \qquad (5)$$

Thus, using TEG under the same conditions, the content of graphite in the different types of batteries could be reduced as far as by the factor of 4 with corresponding raise of the capacity and voltage of batteries. At this point of time, we have few examples of experimental evidence of such phenomena.

Application of TEG creates prerequisites for a considerable reduction of the electrically conductive additive's weight content in the electrode (provided, the choice of technological operations in making electrodes, which do not reduce considerably ρ_{GR}^{exp}, is correct). The concept is applicable to batteries with non-aqueous (earlier considered example of a lithium-ion battery system), as well as to aqueous electrochemical systems.

Besides the conductive additive, TEG may sometimes be a very effective catalyst support, for example, in the catalytic active composite with conducting polymers for the new air-metal batteries, which we proposed [6].

When used in the aqueous electrolytes, TEG has to provide high stability towards oxidation during the entire battery life. For achieving this target, it is often necessary to modify of TEG surface, for example, by Boron [7].

2.3. Conductive Polymers

The relatively high electronic conductivity of conducting polymers is connected with the presence in polymers of internal system of poly-π-conjugated bonds. This is easy to understand on the example of a simplest type of conducting polymer (polyacetylene) in the main and activated states (Figure 6).

(a)

(b)

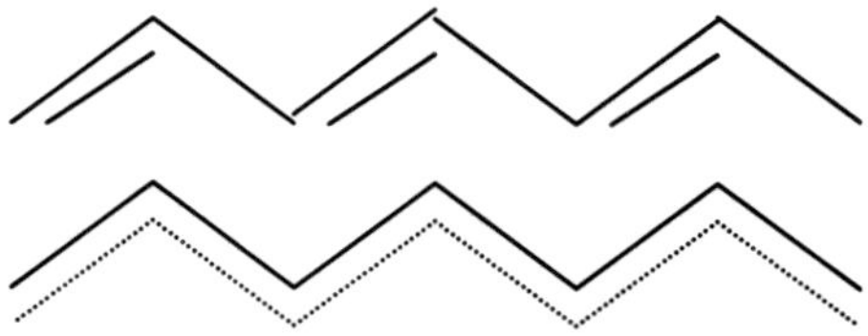

Figure 6. The main (a) and the activated (b) state of polyacetylene.

As our quantum-chemical calculations show, similar transformation and delocalization of bonds takes place in the conductive forms of some other types of CPs (polyaniline, polypyrolle, etc.). Delocalization of chemical bonds after activation leads to appearance of an electronic conductivity in such types of conducting polymers and creates prerequisites for their application as electrode materials of electrochemical power sources. Such activation can be stimulated by intercalation of ions, applying the potential, and by use of some other low energetic factors.

The conducting polymers show a significant non-faradaic component of the electrochemical mechanism. The essential differences of faradaic and non-faradaic systems in equilibrium behavior, trends of galvanostatic charge - discharge curves and cyclic voltammograms have been shown, and criteria for the identification of these mechanisms are proposed [8].

For example, the investigations of the current-generating mechanism for the polyaniline (PANI) electrode have shown that at least within the main range of potential ΔEn the "capacitor" model of ion electrosorption/ desorption in well conducting emeraldine salt phase is more preferable. Nevertheless, the possibilities of redox processes at the limits and beyond this range of potentials ΔEn should be taken into account. At the same time, these processes can lead to the fast formation of thin insulation passive layers of new poorly conducting phases (leucoemeraldine salt, leucoemeraldine base, etc.) near the current collector (Figure 7). The formation of such phases even in small amounts rapidly inhibits and discontinues the electrochemical process.

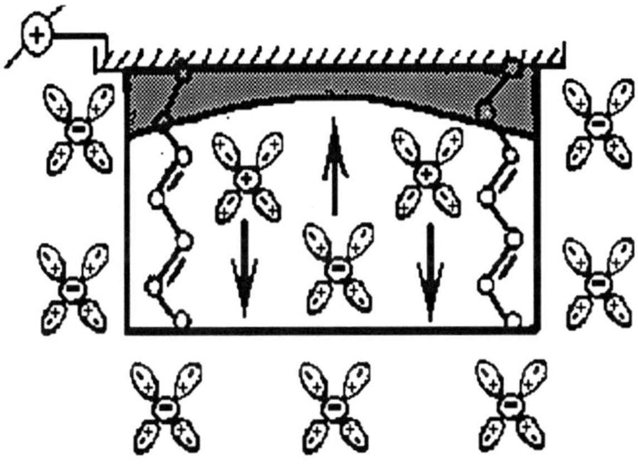

Figure 7. Schematic representation of processes of electromigration / diffusion of ions and formation of insulation passive layer on the boundary current collector / conductive polymer.

As it logically follows from the considered mechanism and the existing experimental data, PANI-type conductive polymers can certainly be used to improve the characteristics of *electrochemical capacitors*. According to our estimations, a specific capacity of electrodes with conductive polymers could be increased by as much as to 3-5 times over that of the classical carbon electrodes.

The detailed research and development on the promising hybrid system for electrochemical capacitors with conducting polymers is described in paper [9]. Besides the main electrochemical process (Figure 7), the side reaction of oxygen reduction takes place at the surface of PANI and some other conducting polymers. The above phenomena has found a practical application for development of air-metal battery mockups with low costs PANI/Expanded Graphite composite catalysts [6].

In order to study the possible reasons and mechanisms of the catalytic activity of conducting polymers (CP), the electronic structure of some molecular CPs clusters and its adsorption complexes with oxygen were modeled [6]. In the CP-O_2^* complex, the CP surface is an electron density donor. For example, in the case of PANI, the bond orders in adsorbed O_2^* molecules decrease by about 30%, and the bond lengths L increase by about 24%. Thus, the adsorbed O_2^* molecules have a fairly high degree of activation and can readily interact with the protons.

Therefore, quantum-chemical analysis confirms the mechanism of O_2 electroreduction and gives possibility to understand the reasons of PANI catalytic activity.

3. CONCLUSION

1. We have proposed a theoretical model, and have experimentally proven that it is possible to increase the capacity of negative graphite electrodes for lithium-ion batteries using the additives of Si and Sn applied as coatings on graphite. A negative active mass has to contain a composite (or alloy) of highly dispersed (nano-sized) particles of the above additives. To complement the efficient operation of Si- or Sn-coated graphite electrode, it is necessary to create a very elastic conductive matrix in the electrode. This conductive matrix will compensate the volumetric changes of Si or Sn. Such concept of development of the modified graphites can ensure a relatively stable cycling at high level of specific capacity of definitely more than 400 mA·h/g.

2. Thermally expanded graphite is, in principle, one of the most efficient conductive additives to various battery active materials, first of all due to its very low effective density $\rho_{GR}^{exp} \cong 0.5 \frac{g}{cm^3}$; high enough conductivity and stability towards oxidation. For realization of these key advantages, TEG has to be of an optimum particle size (function of active material's particle size) to ensure a ratio $t_{AD} / D_{AM} \cdot \pi\pi << 1$. Under such condition the content of graphite in the different batteries could be reduced noticeably and changed for the corresponding amount of active material.

3. The conducting polymers show a significant non-faradaic component of the electrochemical mechanism inside the main range of potentials ΔEn. Nevertheless, the possibilities of redox processes at the limits and beyond this range of potentials should be taken into account. At the same time, these processes can lead to rapid formation of thin insulating

passive layers of new poorly conducting phases. As it follows from the mechanism and the existing experimental data, PANI-type conductive polymers can certainly be used in order to improve the characteristics of electrochemical capacitors. Besides, these materials demonstrate good perspectives for application in air-metal batteries and fuel cells due to pronounced catalytic activity of PANI/TEG composites toward the reaction of oxygen reduction.

ACKNOWLEDGEMENTS

Authors would like to gratefully acknowledge Superior Graphite Co's Dr. Igor V. Barsukov for his valuable consultations on the subject, as well as for development of customized samples of graphite/Si used in our research. We also wish to acknowledge KNUTD's Ph.D. students: V.V. Khomenko and E.A. Illin, Ph.Ds. N.A. Chmilenko, V.V. Kryukov, E.A. Kryukova, A.S. Katashinskii and T.A. Motronyuk for various experimental examinations and discussions, which were useful in creation of support evidence for the above theoretical models. One of the authors, V.B. wishes to thank the Scientific Committee of NATO Science for Peace Program for partial financial support of this work under the umbrella of "Carbon" SfP 973849 project.

REFERENCES

1. Idota Y., Kubota T., Matsufuji A., Maekawa Y., Miyasaka T. Tin-Based Amorphous Oxide: A High-Capacity Lithium-Ion-Storage Material. Science 1997; 276:1395-97.
2. S. Bourderau, T. Brousse, D.M. Schleich. Amorphous silicon as a possible anode material for Li-ion batteries Journal of Power Sources 1999; 81–82:233–36.
3. Yang J., Takeda Y., Li Q., Imanishi N., Yamamoto O. Lithium insertion into Sn- and SnSb-based composite electrodes in solid polymer electrolytes. J. Power Sources 2000; 90:64-69.
4. H. Buqa, P. Golob, M. Winter, J. O. Besenhard. Modified carbons for improved anodes in lithium ion cells. J. Power Sources 2001; 97-98: 122-25.
5. Ye.O. Illin, V.Z. Barsukov and V.S. Tverdokhleb. Capabilities of tin thin films as negative materials for lithium-ion batteries. This book 2004.
6. V.G. Khomenko, V.Z. Barsukov, A.S. Katashinskii and T.I. Motronyuk. New concept for the metal-air batteries using composites: conducting polymers/ expanded graphite. This book.
7. T. Motronyuk, I. Barsukov, V. Barsukov, E. Frackowiak and F. Béguin. New types of graphite stable or the prolonged electrochemical oxidation. Ext. Abstract of the 4th Int. Symp. on New Materials for Electrochemical Systems, Montreal, Canada, July 9-13, 2001; 116-17.
8. V.Z. Barsukov, V.G. Khomenko, S.V. Chivikov, I.V. Barsukov and T.I. Motronyuk. On the faradaic and non-faradaic mechanisms of electrochemical process in conducting polymers. Electrochim. Acta, 2001; 46: 4083 – 94.
9. V. Khomenko, E. Frackowiak, V. Barsukov, F. Béguin. Development of supercapacitors based on conducting polymers. This book 2005.

CAPABILITIES OF THIN TIN FILMS AS NEGATIVE ELECTRODE ACTIVE MATERIALS FOR LITHIUM-ION BATTERIES

Yevgen O. Illin, Viacheslav Z. Barsukov*, and Viktor S. Tverdokhleb

Electrochemical Power Engineering & Chemistry Department, Kiev National University of Technology and Design, 2 NemIrovich-Danchenko str., Kiev 02001, Ukraine

Abstract

The capabilities of thin tin films and tin-based alloys with different metals as active materials for lithium - ion battery negative electrodes are considered. Electrochemical characteristics of such films at different substrates and mechanisms of their functioning are discussed.

Keywords

Tin, tin-based alloys, lithium-ion batteries, negative active material

1. INTRODUCTION

In recent years lithium - ion power sources have occupied one of the first places among other modern energy storage systems. Their functioning is based on the possibility of reversible intercalation of lithium ions in active materials (AM). Substances of tin are investigated often as negative materials for lithium-ion batteries.

It is known that in the process of electrochemical cycling of tin oxide-based thin films versus lithium metal there are at least two reactions in standard alkylcarbonate electrolyte based on lithium salts [1]. The first of them leads to formation of metallic tin on the first cycle:

$$SnO_2 + 4Li \rightarrow Sn + 2Li_2O. \qquad (1)$$

* Corresponding author. E-mail: chemi@mail.kar.net

I.V. Barsukov et al. (eds.), New Carbon Based Materials for Electrochemical Energy Storage Systems, 309–316.

This reaction is irreversible. In the next cycles tin metal forms lithium-tin alloys with different stoicheometry:

$$m\text{Sn} + n\text{Li} \leftrightarrow \text{Li}_n\text{Sn}_m. \quad (2)$$

There are several Li-Sn intermetallics [2]. Their structure and theoretical capacities differ in the wide range (Table 1).

Table 1. Li-Sn alloys and their theoretical capacity.

Alloy formula	Li/Sn atomic ratio	Theoretical specific capacity, mA·h/g
Li_2Sn_5	0.4	88
Li_2Sn_3	0.7	145
Li_3Sn_2	1.5	311
Li_2Sn	2.0	405
Li_7Sn_3	2.3	465
$Li_{13}Sn_5$	2.6	589
Li_7Sn_2	3.5	656
$Li_{22}Sn_5$	4.4	790
Sn	0.0	994

Thus, in the process of Li^+ reaction with tin, theoretical capacity of Li-Sn alloys can reach up to 790 mA·h/g. Theoretical capacity of pure Sn is 994 mA·h/g (7234 mA·h/cm^3). Formation of such alloys occurs in the range of potentials from 0 to 0.8V versus lithium (Table 2) or in the temperature window 0-800^{0}C. Charge-discharge curves have an inclined form.

Table 2. Formation potentials of different Li_ySn alloys, [3].

Stoichiometry	Potential of alloy formation, V (versus Li)
$0.40 < y < 0.70$	0.660
$0.70 < y < 2.33$	0.530
$2.33 < y < 2.63$	0.485
$2.63 < y < 3.50$	0.420
$3.50 < y < 4.40$	0.380

The reactions (1) and (2) were investigated from the electrochemical point of view (see, for example [4-7]). The particles of tin formed in the first cycle, were embedded in the lithium-oxide matrix. Existence of such matrix causes high cycleability of electrode. Therefore along with graphite usage of tin oxides as negative materials was started. They take part in cycling with some high discharge capacity loss from 600-450 mA·h/g during more than 100 cycles [8].

Earlier it was considered, that metallic tin by itself couldn't be used even as thin films because of the big volumetric changes during lithium intercalation-deintercalation [9]. In the formation of $Li_{4.4}Sn$ the volume of alloy makes 360 % of initial volume of tin [10]. In Sn metal each tin atom occupies $2,7\cdot10^{-2}$ nm^3. $Li_{22}Sn_5$ has a volume of 0,485 nm^3, corresponding to $9,7\cdot10^{-2}$ nm^3 per Sn atom [11]. In the process of tin thermal deposition on different metal surfaces it is mainly possible to obtain β-modification as so-called grain tin. It has a tetragonal structure similar to diamond, but compressed in the direction of **c** axis. β - modification has a high density and conductivity [12]. Under certain deposition conditions clusters of grain tin have rather small size. It is possible to assume that intercalation of some lithium ions will not result in the fast destruction of electrode. If layers of tin are alternated with layers of electrochemically inert substance with ionic conductivity, such electrode can provide high cycleability.

Reference [13] discusses the capabilities of thin-film tin anodes for lithium - ion batteries. Tin was electrodeposited from acid water solutions on a copper foil. Thickness of coating did not exceed 2 mμ. For studying thermal influence, samples were annealed under vacuum at 200^0C during 24 h. Galvanostatic tests were carried out in standard electrolyte in the range of potentials 0 - 2 V versus lithium foil electrode. It is interesting that all samples have shown the theoretical capacity of tin on the first charge. But further in the next cycles samples, which were not annealed failed discharge capacity down to 200 mA·h/g. Annealed samples have a discharge capacity 900 mA·h/g during more than 10 first cycles. Formation of interface consisting of Cu_6Sn_5 and Cu_3Sn systems in the second case is rather probable [14]. Moreover, these two phases can coexist in wide temperature window and appear almost simultaneously. Semi-successful attempts of tin-containing alloys deposition on various matrixes, for example on graphite, are reported [15]. High cycleability gave materials made by electroplating natural graphite with the systems $Sn_{65}Sb_{18}Cu_{17}$ and $Sn_{62}Sb_{21}Cu_{17}$. During the cycling versus lithium-cobalt oxide in standard alkylcarbonate electrolyte, materials saved more than 50% of discharge capacity after 34 cycles. Interesting results were obtained on material, made by co-sputtering of tin (45-60 mol. %) and molybdenum on a copper foil [16]. Thickness of film is 4-10 mμ. Crystalline structure consists of alternating tin and molybdenum layers with thickness of 1 mμ or less. These layers were separated by $Mo_{1-x}Sn_x$ film alloy. Discharge capacity in 18650 elements reaches 350 mA·h/g, and during more than 200 cycles exceeds 200 mA·h/g.

2. EXPERIMENTAL AND DISCUSSION

Subject of our investigations became thin films of tin deposited under high vacuum on different current collectors.

All samples were galvanostatically tested in coin cell 2016 semi-elements and special Teflon T-type testing cells in the potential window 0.03 – 0.80 V versus lithium foil in different modes from C/2 to C/40. We used standard electrolyte LP 71(Merck).

For example, we deposited β-tin on a copper current collector similar to Sanyo Electric Co [14]. It is interesting that as-deposited samples, obtained without annealing, have near 200 mA·h/g of specific discharge capacity (Figure 1) in spite of smooth coating with hundred nμ-size grain tin structure. This fact directly shows necessity of bronze interface between active material and copper current collector independently of coating methods. Good cycleability of all samples even in semi-elements versus lithium observed for as-deposited tin on copper current collectors without bronze intermediate layer. Very small hysteresis between charge and discharge capacity presents both in case of unwoven polypropylene separator (Mogilev, Belarus) and Celgard separator. Gradual decreasing of charge and discharge capacity changes on reversible regularity after four hundred cycles due to electrode structure reorganization.

We made attempts to deposit 1 mμ layer of β-tin film on 20 mμ tungsten foil. This foil was previously etched in nitric acid. We showed by XRD the absence of Sn-O and Sn-Mo bonds (Figure 2). Charge-discharge curves in all cases have sloping shape typical of tin (Figure 3).

Some hysteresis between charge and discharge capacity is caused obviously by collateral processes and growth of dendrites in cells in the case of usual separators usage. Growth of dendrites occurs even on three-layer separators (Figure 4). It is known that tin is active material with drastic inclination to dendrite-making. Because of the use of special separator materials (Celgard, Teklon, etc.) we were able to increase discharge capacity; also, decrease hysteresis between charge and discharge capacity, and achieved certain stability upon cycling. Application of modern lithium-ion grade separators does increase capacity output characteristics and improves it towards having higher stability, thus falling off of discharge capacity is not rapid (being the case for traditional separator materials), but rather gradual (Figure 5). This discharge capacity fade became essential after 10 cycles during cycling versus lithium foil.

The reason of slow charge/discharge capacity reduction is probably gradual loss of contact between the active particles and current collector. Nevertheless, in the case of copper current collector usage we observed even smaller increase of discharge capacity after 400th cycle (Figure 1).

In spite of higher capacity of tin electrode on tungsten current collector in comparison with tin electrode on copper current collector, specific capacities for unit of electrode mass have the reverse correlation (Table 3).

Nevertheless, such parameters are quite acceptable for testing in mockups of lithium-ion batteries.

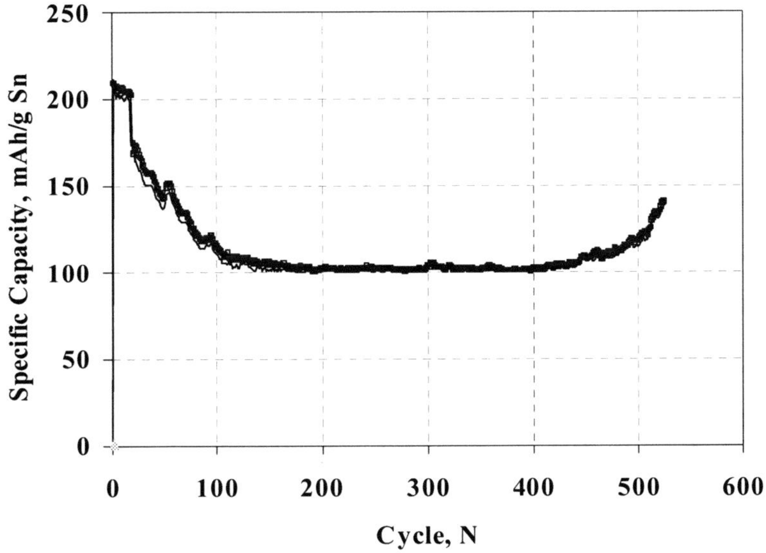

Figure 1. Specific charge (thick line) and discharge (thin line) capacity of tin electrode without annealing. Bronze interface is absent. Current collector – copper. Testing mode C/5. Counter electrode - lithium foil. Separators – 2 layers of unwoven polypropylene (Mogilev, Belarus).

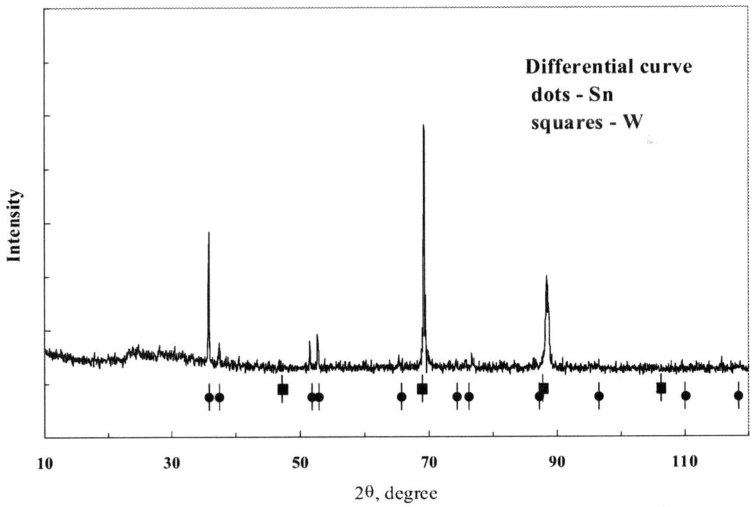

Figure 2. XRD pattern of tin coated tungsten.

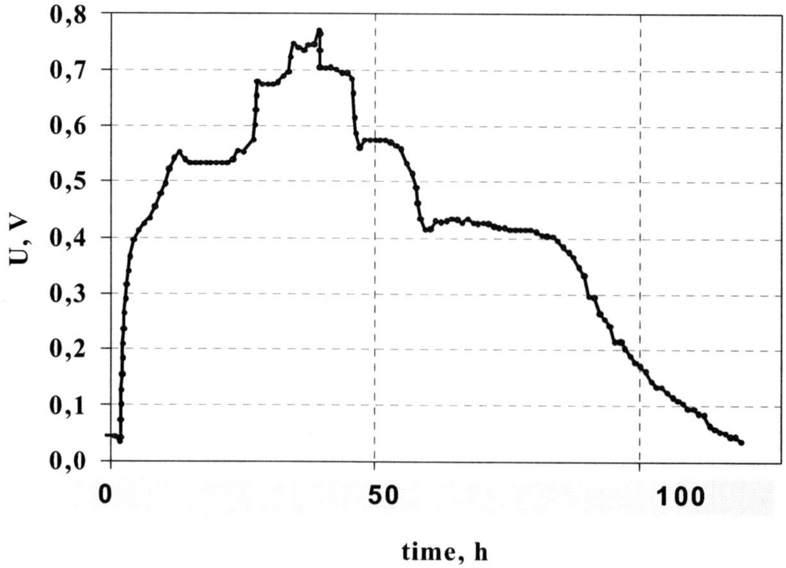

Figure 3. Charge-discharge curves of tin on tungsten current collector.

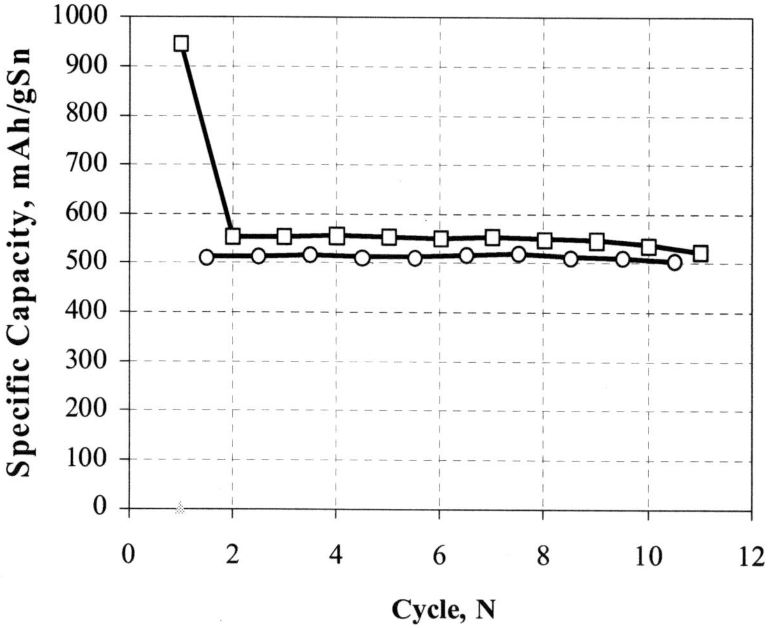

Figure 4. Specific charge (squares) and discharge (dots) capacity of tin electrode. Current collector – tungsten. Testing mode C/5. Counter electrode - lithium foil. Separator – 2-3 layers of unwoven polypropylene (Mogilev, Belarus).

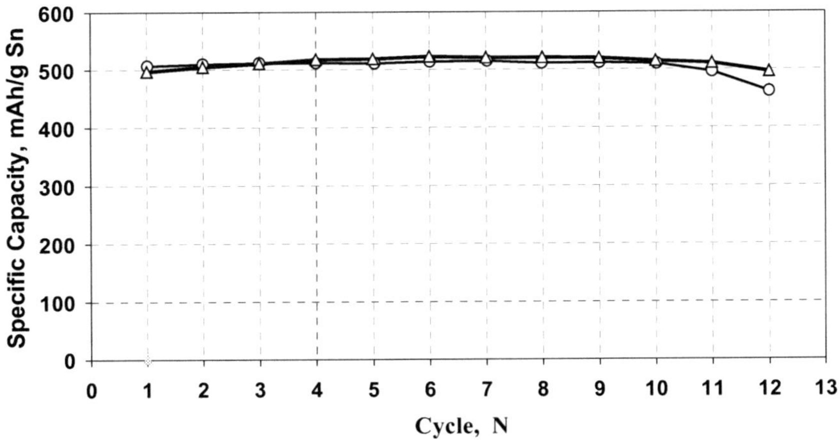

Figure 5. Specific discharge capacity of tin electrode without annealing. Current collector – tungsten. Testing mode C/5. Counter electrode - lithium foil. Separators: 1 layer of unwoven polypropylene (Mogilev, Belarus) + 1 layer of Celgard – triangles; 2 layers of unwoven polypropylene (Mogilev, Belarus) – dots.

Table 3. Specific characteristics of active material and electrodes.

Current collector	**Maximum discharge capacity on the unit of active material mass, mA·h/g**	**Maximum discharge capacity on the unit of electrode square, μkA·h/cm^2**	**Maximum discharge capacity on the unit of electrode mass, mA·h/g**
W	520	758	70
Cu	200	291	95

3. CONCLUSIONS

Nanostructured tin, tin films, alloys and composites can be considered as promising negative materials for lithium-ion energy storage systems. Tin based materials have a good cycleability, high charge and discharge capacity in wide range of cycling modes (C/2 - C/40). Tin-coated tungsten electrodes have a higher capacity than tin-coated copper electrodes, but larger hysteresis between charge and discharge curves and quicker capacity fade. Tin-coated copper electrodes have better cycleability than tin-coated tungsten electrodes. Tin-coated copper electrodes must be annealed for bronze interface creation.

ACKNOWLEDGMENT

Authors would like to express their gratitude to Dr. Kryukov V.V. (Kiev National University of Technologies and Design) for creation of original electrochemical test cells; Dr. Chmilemko M.A. for the help in assembly of standard size 2016 elements (Interdepartmental Division of Electrochemical Power Energetic and Engineering, NAS of Ukraine), Dr. Matzui L. U. for her assistance with XRD investigations (Kiev National T. Schevchenko University).

Also, authors would like to acknowledge NATO Science for Peace Program for financial support of this work through "Carbon" 973849 project.

REFERENCES

1. Y. Idota, M. Mishima, M. Miyaki, T. Kubota, T. Miyasaka, Eur.Pat.Appl., 651450 A1 94116643.1. 1997.
2. Moffatt, William G. *The Handbook of binary Phase Diagrams*. New York: Genium Publisching Corp., Schnectady, 1990.
3. Elidrissi MML., Corrector JI., Tirado JL., C. Perez V. SnHP04: a promising precursor for active material as negative electrode in Li-ion cells Electrochimica Acta 2001;47:489-93.
4. Wang J., Raistrick ID., Huggins RA., Behavior of some Binary Lithium Alloys as Negative Electrodes in Organic Solvent-Based Electrolytes. J. Electrochem. Soc. 1986; 133:457-60.
5. Anani A., Crouch-Baker S., Huggins RA. Kinetic and Thermodynamic Properties of Several Binary Lithium Alloy Negative Electrode Materials at Ambient Temperature. J. Electrochem. Soc. 1987; 134:3098-101.
6. Huggins RA., Polyphase Alloys as Rechargeable Electrodes in Advanced Battery Systems. J. Power Sources 1988; 22:341-50.
7. Huggins RA., Materials Science Principles Related to Alloys of Potential Use in Rechargeable Lithium Cells. J. Power Sources 1989; 26:109-20.
8. Idota Y., Kubota T., Matsufuji A., Maekawa Y., Miyasaka T. Tin-Based Amorphous Oxide: A High-Capacity Lithium-Ion-Storage Material. Science 1997; 276:1395-97.
9. Yang J., Winter M., Besenhard JO. Small particle size multiphase Li-alloy anodes for lithium-ion batteries. Solid State Ionics 1996; 90:281-87.
10. Yang J., Takeda Y., Li Q., Imanishi N., Yamamoto O. Lithium insertion into Sn- and SnSbx-based composite electrodes in solid polymer electrolytes. Journal of Power Sources 2000; 90:64-69.
11. Brousse T., Retoux R., Herterich U., Schleich D.M. Thin-Film Crystalline SnO_2-Lithium Electrodes. J. of Electrochemical Society, 1998;145:1-4.
12. Remy, Heinrich, *Lehrbuch Der Anorganischen Chemie*. B. 1, Leipzig: Academische Verlagsgesellschaft Geest und Portig K.-G., 1960.
13. Tamura N., Ohshita R., Fujimoto M., Fujitani S., Kamino M., Yonezu I. Study of some Li Alloy for Lithium Secondary Battery. Proceedings of Joint (ECS & ISE) International Meeting; 2001 2-7 September; San Francisco, 2:Abstract No. 251, 2001.
14. Hansen, Max, *Constitution of Binary Alloys*, New York: McGraw-Hill, 1958.
15. Ulis A., Rosenberg Y., Burstein L., Peled E. Degradation of Carbon-Supported Tin Alloys in $LiPF_6$ EC:DEC. Proceedings of Joint (ECS & ISE) International Meeting; 2001 2-7 September; San Francisco, 2: Abstract No. 225, 2001.
17. Turner R.L., Amik B., Krause L.J., Christensen L., Dahn J.R. Electrochemical Characteristics of Sn-Mo Anodes for Lithium Batteries. Proceedings of Joint (ECS & ISE) International Meeting; 2001 2-7 September; San Francisco, 2: Abstract No. 257, 2001.

COMPOSITE ANODE MATERIALS FOR HIGH ENERGY DENSITY LITHIUM-ION BATTERIES

Joseph S. Gnanaraj[1], Malgorzata K. Gulbinska[2], Joseph F. DiCarlo[2*], Igor V. Barsukov[3], Nancy Holt[3], Viacheslav Z. Barsukov[4], Joseph E. Doninger[5*]

[1]*Worcester Polytechnic Institute, Chemical Engineering Department, 100 Institute Road, Worcester, MA 01609 USA;*
[2]*Lithion, Inc., 82 Mechanic Street, Pawcatuck, CT 06379, USA;*
[3]*Superior Graphite, 10 S. Riverside Plaza, Chicago, IL 60606, USA;*
[4]*Kiev National University of Technology and Design, Electrochemical Power Engineering & Chemistry Department, 2, Nemirovich-Danchenko street, Kiev, 252011, Ukraine;*
[5]*Dontech Global, Inc. 427 E. Deerpath Road, Lake Forest, IL 60045, USA*

Abstract

Novel composite anode materials were prepared for use in lithium-ion batteries. The composite material formation involved coating of powdered substrates (graphite-, or silicon-based), thus modifying their surface properties and cycling performance. Applying an additional surface coating resulted in the production of composites with enhanced stability in the practical lithium-ion battery environments. Coin cell cycling, as well as full 7Ah prismatic cell cycling, and impedance spectroscopy methods were used to evaluate the electrochemical properties of coated materials.

Keywords

Lithium-ion batteries; high energy density; purified natural graphite; carbon coated silicon; graphite/metal composites; chemical vapor deposition.

1. INTRODUCTION

Carbonaceous materials with varying degree of graphitic order are the most common commercial anodes in secondary lithium-ion batteries. Among carbon-based materials, natural graphite is the most promising anode

[*] Corresponding authors. E-mails: jdoninger@dontechglobal.com and jdicarlo@lithion.com

I.V. Barsukov et al. (eds.), New Carbon Based Materials for Electrochemical Energy Storage Systems, 317–331.

material because of the high theoretical lithium intercalation specific capacity of 372 mAh/g, flat voltage potential profile, low irreversible capacity, and low cost [1]. However, graphite electrodes typically fail in a number of commonly used electrolyte solutions, such as propylene carbonate (PC). The use of PC based electrolyte in the Li-ion battery would improve the rate performance, the safety of the battery, its low temperature performance and, eventually, costs. In addition, the rapidly increasing market for rechargeable lithium-ion batteries (portable computers, cellular phones, a variety of military applications) calls for batteries with much greater capacities (>500 mAh/g) than the capacity values inherent to carbonaceous materials.

Silicon-based materials are listed among the promising novel anode materials. One of the most important advantages of silicon-based anodes is their exceptionally high theoretical capacity (c.a. 4000 mAh/g). This impressive theoretical capacity value is related to the high Li/Si ratio in lithium silicides (e.g. $Li_{4.4}Si$) formed *in-situ* in silicon-based anodes [2,3]. Unfortunately, tremendous volume changes in silicon are associated with the formation of lithium silicides with a high Li/Si ratio. These volume changes cause rapid mechanical degradation of the anode that eventually leads to the loss of electrical contact between the current collector and the silicon-based anode material [4]. Thus, the excellent initial capacities (over 1000 mAh/g) offered by silicon-based anodes, fade dramatically during the subsequent cycles. The addition of carbon to silicon powder greatly improves its cycleability and prevents capacity fading that is observed with pure silicon anodes [5, 6, 7].

In the second part of this study, purified natural graphite substrates (PNG) were coated with Si and other metals, followed by an additional coating with a protective layer of disordered carbon, in order to prevent the electrolyte reduction and to increase the cycling stability of PNG. The use of such composite carbonaceous materials comprising a "core" of graphite, a second layer of an alloying metal, and a protective "shell" of disordered carbon combines the favorable lithium storage properties of graphite, reversible alloy formation reaction with lithium, with the solvent co-intercalation retarding properties of non-graphitic carbons. The PNG surface modification has been purposively narrowed down to silicon and aluminum. These coatings were deposited onto the graphite surface to provide sites for lithium intercalation and to form the electronic conducting layer in the composite anode.

Fluidized-bed Chemical Vapor Deposition (FBCVD) was a method of choice in forming both carbon, and silicon coatings on graphite and silicon powders, respectively [8, 9]. The major advantage of fluidized bed CVD method over other coating processes is the ability to deposit uniform coating with the controlled thickness onto large quantities of substrate particles (silicon and graphite powders, respectively). In the typical fluidized bed CVD reactor, a non-reactive gas flows through a bed of particles, at a carefully selected velocity. There is a drag exerted on the solid particles by the flowing gas. However, when the gas velocity increases beyond a certain value, the total drag on the particles equals the weight of the bed, and the particles begin to lift and fluidize. The gaseous precursor is then introduced into the reaction chamber in order to chemically react with the substrate particles and to form the uniform coating layer.

Initially, cycling in the coin cells, and later, in full prismatic cells with rated capacity of 7 Ah were used in our investigations. Also, advanced impedance spectroscopy methods were used to evaluate the electrochemical properties of coated materials.

2. EXPERIMENTAL

The coin half-cells were assembled using metallic lithium discs as the positive electrodes. The negative electrodes were made up of composite anode materials films spread onto the copper foil current collectors. Coated powders were provided by Superior Graphite Inc. (Chicago, IL, USA). The carbon and silicon coatings on graphite particles were done by fluidized-bed chemical vapor deposition (FBCVD) method. Anodes were prepared by mixing approximately 1g of natural graphite (PNG), coated with: (a) carbon, (b) silicon (5.7 wt%), and (c) aluminum, respectively. Kynar (PVdF), Super P carbon black, and NMP solvent were also added. The anodes were made such that there was 87% anode material mixed with 10% Kynar, and 3% Super P carbon. The amount of NMP used to make the slurry was adjusted such that the slurry consisted of 47 wt% solids. The slurry was then transferred onto the flat glass surface of hand coater. Anode slurry was coated onto copper foil. The coating gap controlled the thickness of the slurry being applied to the foil, which determined the weight loading. The coated foil was dried in a vacuum oven for at least 12 hours at 110°C. The coated foil was then blanked to the desired disc size. The surface area of the electrode was 1.25 cm^2. The two-electrode coin cells were assembled in a glove box, as cited elsewhere [9]. The following electrolytes and/or their mixtures were used in coin half-cells: 1M $LiPF_6$ in mixture a of either EC:DEC:DMC (1:1:1, v/v/v) or PC:DEC:DMC (1:1:1, v/v/v). The electrochemical measurements included galvanostatic charge-discharge tests on the graphite electrodes. Freshly prepared natural graphite electrodes had an open circuit potential of *ca.* 3.0 V (vs. Li/Li^+). Galvanostatic cycling tests were performed at 0.2, 0.4 mA/cm^2 current densities.

The full lithium-ion prismatic cells with a rated capacity of 7Ah were assembled and evaluated in the Lithion, Inc. battery test facility. The PNG-based anode materials from Superior Graphite Co. were coated onto copper foil and blanked into anode components for assembly of the so-called

Lithion's "low rate" 7Ah cells. The cells were fabricated with Setella™ E20 separator, 1M $LiPF_6$ in EC:DMC:DEC (1:1:1) electrolyte and standard cathode components. The charge capacity of each cell is designated at 9 Ah during the C/20 formation cycle. This value is based on a cathode specific capacity of 190 mAh/g at the C/20 rate. The 8 Ah designation for each cell during the initial two C/10 charge cycles is based on a specific cathode capacity value of 170 mAh/g. After a preliminary evaluation of the three formation cycles, the cells were sealed and cycled for additional 7 cycles to stabilize data for the C/10 rate. A 72 h stand test was performed on cells after completion of the ten formation cycles.

A Maccor multichannel system (Model 2000) tester and an Arbin Instruments Inc., computerized multichannel battery tester were used for galvanostatic cycling. For PITT measurements, a computerized Solartron 1286 electrochemical potentiostat interface driven by Corrware (Scribner Associates, Inc. Southern Pines, NC) was used. Impedance spectra were measured using the Solartron 1250 Frequency Response Analyzer (FRA) driven by a Pentium-Pro PC. The amplitude of the ac voltage was 5 mV and in the frequency range of 65 kHz – 5 mHz in automatic sweep mode from high to low frequency with 72 points. The cells were aged by 10-15 cycles of galvanostatic cycling at the current density of 0.2 mA/cm^2 and at appropriate equilibration potentials in open circuit conditions. Unless stated otherwise, all the tests were performed at a constant temperature of 20°C in an incubator.

Scanning Electron Microscopy images of powders used in this paper were taken using JEOL's JSM-6320F instrument at the Illinois Institute of Technology and at Drexel University, Philadelphia, PA, USA.

3. RESULTS AND DISCUSSION

3.1. Cycling of Silicon

Volume changes in silicon during lithium-ion battery cycling are believed to be related to the formation of lithium-silicon alloys with a high content of lithium (*e.g.*, $Li_{4.4}Si$). The volume changes lead to anode pulverization that eventually causes the loss of electrical contact between the current collector and the silicon-based anode material. Thus, the initial capacities of over 1000 mAh/g offered by silicon-based anodes, fade during the subsequent cycles (Figure 1). The data generated in our laboratory show that the carbon coating on silicon powder prevents capacity fading that is observed with pure (uncoated) silicon anodes.

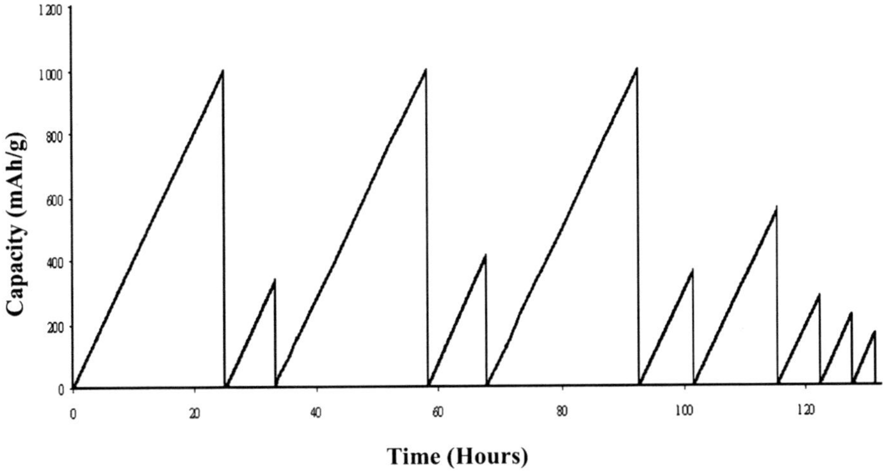

Figure 1. Capacity vs. time plot for uncoated silicon anode material.

3.2. Carbon / Silicon Composites

The ball milled silicon powder was ground in an agate mortar in order to break up any agglomerates formed during storage and/or transportation of silicon powder. About 0.5 g of sample was loaded into the fluidized bed CVD reactor. The temperature was raised to 950°C. Upon reaching 950°C argon carrier gas was passed through a bubbler filled with liquid toluene at a flow rate of 100 mL/min. After 30 min the carrier gas flow was stopped and the reactor was allowed to cool down.

The cycling properties of selected carbon-coated silicon powders were evaluated. The carbon-coated silicon powdered samples were cycled in the half-cells against the lithium metal anodes. In the preliminary coin cell test of the silicon powders, capacity cutoff was limited to 500 mAh/g in the voltage range of 0.01 to 1.3V. Galvanostatic cycling tests were performed at 0.2, 0.4 mA/cm^2 current densities. The charge-to-discharge efficiency was nearly 100% for the 27 cycles where discharge capacity was 500 mAh/g (20 initial cycles are shown in Figure 2). The coin cell cycling results showed an improved electrochemical performance of the carbon-coated silicon powders (Figure 2). However, we learned that optimization of the FBCVD process is required to achieve uniform carbon coating thickness on silicon powders in large quantities (>100 g).

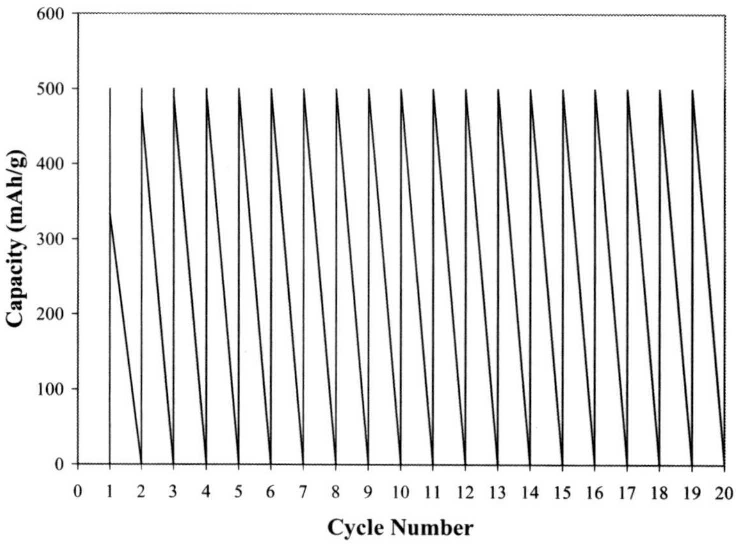

Figure 2. Capacity vs. time plot for cell containing carbon-coated silicon powder material.

3.3. Graphite / Metal Composites

In order to be able to work with scaled up quantities of the higher capacity material, it was decided to continue the work described in section 3.2 and coat Si onto carbon attempting to combine high capacity of Si and stability of performance characteristics of graphite substrates.

Two types of graphite precursors were used in this study as substrates for deposition by the alloying metals. The origin of the precursors was the same - purified crystalline flake graphite (commercial grade 2900G, being manufactured by Superior Graphite Co. of Chicago, IL, USA). This material is a 300+μm particle size flake of high purity (99.95%C), as achieved by the process of thermal purification [10]. The high resolution scanning electron micrograph (SEM) of such graphite is shown by Figure 3(a). The 2900G graphite has been converted through steps of grinding and classification into the 15 μm sized platelet (Figure 3b) and spheroidal (Figure 3c) structures.

Both carbon materials were tested for their initial electrochemical performance in the 2-electrode electrochemical cells with Li metal as a counter electrode. Our findings have shown that with both types of carbon materials, achieving near theoretical reversible capacity upon Li^+ deintercalation was possible. Thus, in a typical half cell environment (a CR2016 type coin cell with graphite and Li metal electrodes, a 1M LiPF6,

EC:DMC (1:1) electrolyte), at the formation rate of C/15, these carbons achieved the reversible capacity in between 350 - 360 mAh/g [10].

On the next stage, the two carbon precursors were coated with metals, which are known to form reversible alloys with lithium. It is noteworthy mentioning that the attempts to start using the process of reversible formation of the alloys between Li and certain metals have a long legacy and date back to the earlier works aimed at prevention of dendrite formation in the lithium metal – based electrochemical cells [11]. From the many metals, which form reversible alloys with Li, we have chosen to work with those, which are known to form Li alloys at the potentials, which coincide with or are as close as possible to those, at which the reversible intercalation of Li^+ into crystalline flake graphite takes place. Such approach will enable maintaining the same discharge voltage in the full lithium-ion cells as in those, based on conventional graphitic anodes.

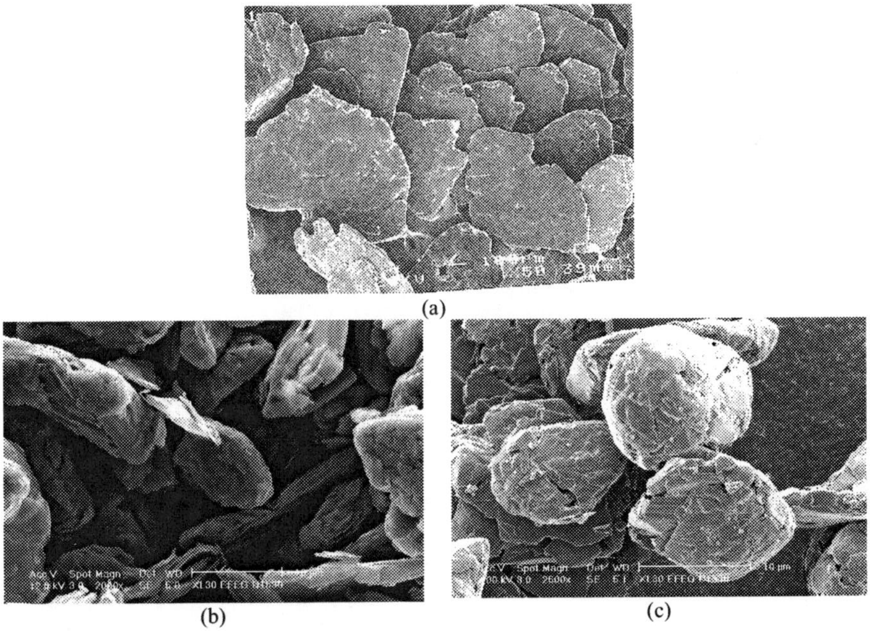

Figure 3. Scanning Electron Microscope Images of thermally purified natural crystalline flake graphite 2900G (a), and its ground versions displaying platelet (b) and spheroidal (c) morphologies.

The choice of metals has been narrowed down to Si and Al for this group of experiments. The approximate potentials of the reversible transformation (subject of polarization) of C into LiC_6 is taking place in the range of 0.6 – 0.01 V (vs Li/Li^+); while for conversion of Si into $Li_{22}C_5$ it is in the range of 0.3 – 0.05 V (vs Li/Li^+), and it occurs at approximately 0.27 V (vs Li/Li^+) for reaction of conversion of Al into LiAl [12, 13].

For illustrative purposes, the process of deposition of Si onto graphite is being used as an example. The 15 μm natural graphite precursors were introduced into the industrial size chemical vapor deposition reactor, where a thermal decomposition of silane (SiH_4) into the silicon and hydrogen was taking place under inert gas in accordance with the equation (1):

$$SiH_4 \Rightarrow Si + 2H_2. \quad (1)$$

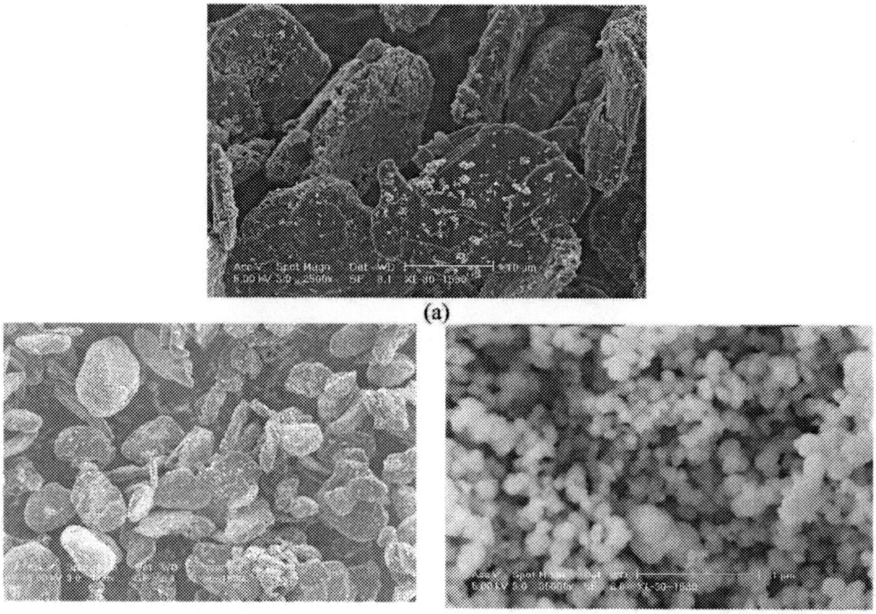

(a) (b) (c)

Figure 4. Less uniform Si coating on plate-like graphite (a), rather uniform Si coating on spheroidal graphite (b) and a magnified view of Si particulates on spheroidal graphite (c).

Silane decomposes to its elements at above 400°C. Process (1) is known as direct thermal decomposition, and produces either amorphous or polycrystalline Si (function of reaction temperature and other processing parameters), and is commonly used, for instance, in the solar cell industry to reduce silane to silicon.

The process was seen to be highly sensitive to impurities [14], among which are oxygen and water groups with basic pH. Their presence even in minute quantities may result in occurrence of the side reactions (2) and (3), respectively:

$$SiH_{4(g)} + 2O_{2(g)} \rightarrow SiO_{2(s)} + 2H_2O; \quad (2)$$

$$SiH_{4(g)} + (2+n)H_2O_{(g)} \rightarrow SiO_2 \cdot nH_2O_{(s)} + 4H_{2(g)}. \quad (3)$$

It was determined that plate-like morphologies are less desirable substrates for metal deposition, as compared with spheroidal substrates in terms of uniformity of metal deposition. This observation is consistent with classical teachings about metal deposition and corresponding crystal growth on foreign substrates [15]. In accordance with these teachings, metals get deposited on the most active sites of the substrate, rather than uniformly, on its entire surface. The most active sites are those, which feature cavities, cracks, pores, defects of crystallites and other un-uniformities. It is evident from the SEMs shown by Figures 3(b) and 3(c) that spheroidal graphite precursor had many more active sites, which allowed for more uniform distribution of the Si (please see a non-uniform and a rather uniform distribution of Si on two different substrates, as shown by Figures 4(a), and 4(b), respectively).

It is noteworthy that while the first droplets of metal deposit are being held to the substrate graphite, most likely, by particle-to-particle electrostatic forces, subsequent layers of metal tend to grow where the energy is lowest, e.g. on top and around the first layers thus creating a thermodynamical prerequisite for making a non-uniform coating. The smooth parts of the surface remain passive and are hard to coat (please see uncoated particles in Figure 4(a)).

Therefore, due to considerations of uniformity of coating, in further tests it was decided to focus on the spheroidal metal-graphite composites.

A departure from traditional approaches of depositing a significant amount of metal onto carbon precursor differentiated our experiments from the state of the art literature. Through several trial and error experiments, it was possible to formulate several key principles, which should constitute a design of the metal/carbon composites. Among those are:

(1) the amount of the alloying metal on carbon has to be kept at below 10wt%, otherwise it will be difficult to cope with large volume changes of metal upon alloying – de-alloying reaction;
(2) the particles of alloying metal have to be as small as possible, preferably nanosized or submicron (the reason is the same – suppression of the volume changes, as well as enhancement of electrostatic adhesion between the core graphite and the metal particle, which typically goes up as metal particles become smaller in size);
(3) the alloying metal has to be uniformly coated onto graphite substrate (a 15 μm base in our tests), and then it has to be additionally coated with a stabilizing coating (in order to suppress metal's volumetric expansion in the direction normal to the direction of core 15 μm graphite base).

Figure 4(c) depicts an image of the Si-coated spheroidal natural graphite substrate at higher magnification, as derived from Figure 4(b). The size of the substrate graphite particle is estimated to be about 15 μm, while the size of deposited Si particles appears to be in the range of 0.1-0.35 μm.

The comparative 2016 coin cell data of the uncoated vs. Si-coated spheroidized natural graphite precursor vs Li metal counter electrode (Figure 5) reveals a remarkable effect of metal addition onto the reversible capacity of anode. Thus, the starting material's reversible capacity increased from approximately 350 mAh/g to over 510 mAh/g (data taken at 0.8V vs Li/Li^{+}), which is 1.37 times higher than the theoretical value of the reversible capacity for graphite.

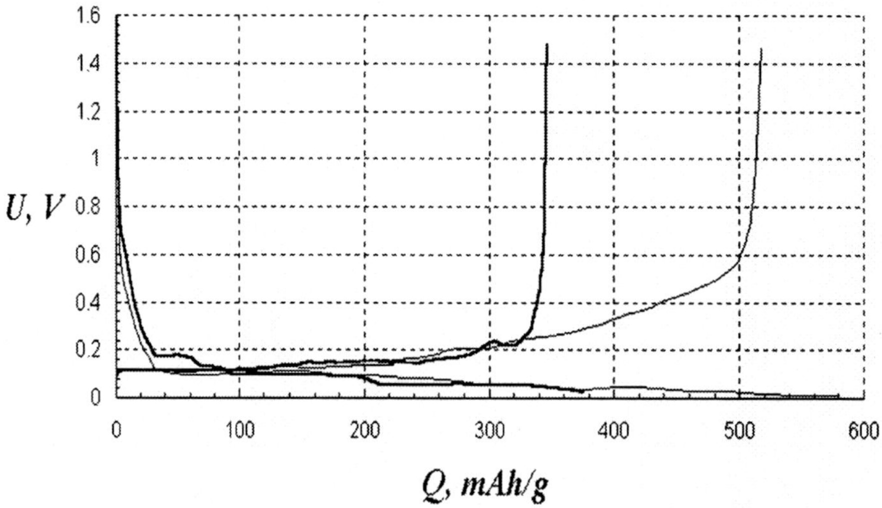

Figure 5. First charge and discharge of CR2016 coin type half-cells with anodes based on the 15 μm, spheroidized natural graphite (bold line) and the similar graphite modified by 8wt% Si at the C/15 rate.

Unfortunately, later findings resulted in an observation that composite materials lost their capacity upon further cycling unless the outer surface of particles was stabilized with an additional passivating coating, whose nature is described in a greater detail in [16] for the example of Superior Graphite's SLC product line. This outer stabilizing shell is made out of rigid disordered carbon material and uniformly coats each graphite particle. In the preliminary coin cell investigation, the outer shell was seen to be suppressing the volumetric changes in metal, thus improving stability of materials with cycling. It was observed that the surface shell was proven to be more efficient for the case of Al rather than for Si coatings.

Two most promising types of anode materials developed in this study were characterized by Lithion, Inc., a specialty lithium-ion battery manufacturer of Pawcatuck, CT, USA. The work has been carried out in standard Lithion's 7 Ah prismatic cells. The anode materials were: spheroidized purified natural graphite coated with Si additionally coated with a stabilizing carbon coating (sample is referred to as PNG-Si-2C), as well as spheroidized purified natural graphite coated with Al and also with a stabilizing carbon coating (sample is referred to as PNG-Al-2C). Performance of these cells has been compared with that of the control MCMB anodes.

Figure 6 compares the charge and discharge capacities vs. cycle number during the formation cycle. Si-2C coated graphite shows the highest irreversible capacity while MCMB has the minimum. Al-2C coating on graphite shows a small positive impact on the irreversible capacity but at the cost of capacity.

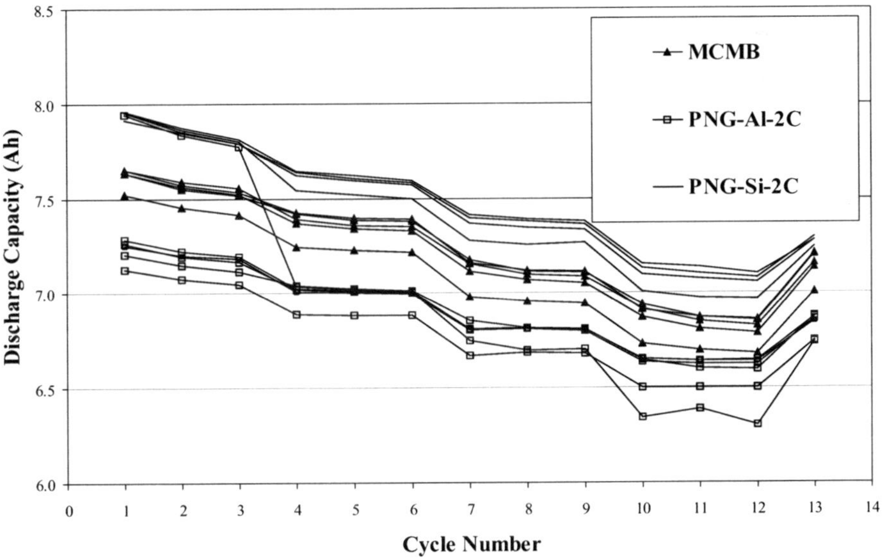

Figure 6. Prismatic 7 Ah battery; discharge capacity vs. cycle number plot.

Figure 7 shows the charge/discharge capacities vs. cycle number at different temperatures. It is interesting to note that the capacity at room temperature is regained after each low temperature cycling. Although Si-2C coating shows better capacity than that of Al-2C at low temperature, the capacity fading is large on long term cycling. At the room temperature, although Al-2C coating shows lower capacity, it was seen to be stable on cycling while Si-2C coating lost the capacity dramatically.

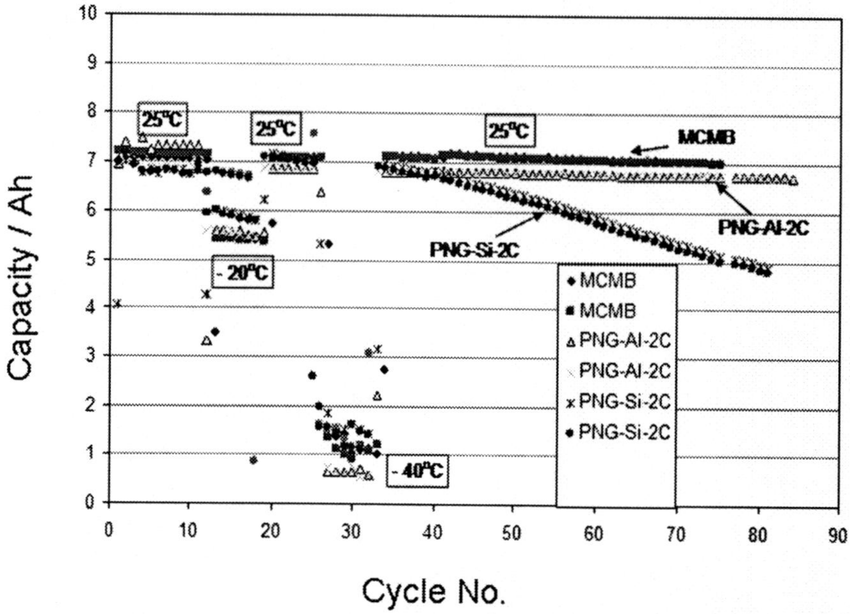

Figure 7. Prismatic 7Ah battery. Typical charge-discharge capacities vs. cycle number. Battery was cycled at different temperatures at C/4.

Figure 8 shows a family of impedance spectra for three types of carbon in full 7 Ah batteries. Four cells with each type of carbon were measured for impedance. The results appear to be very interesting. The original impedance data appears to be shifted on the real axis but when it was corrected it matches one another. Most of the times in the literature, it is advised to take high frequency intercept and consider it for the impedance value. This data provides valuable insight in to the real situation. The dc resistance value calculated from current interrupt measurements (done separately and not reported here) correlates well with the impedance' total resistance (extrapolation to <5mHz). The semicircle at high frequency side represents the SEI layer and shows that the metal coating actually increases the SEI resistance – a phenomena, which deserves a separate investigation.

Figure 9 shows the 72-hour charge stand test. The cells were charged to 4.1 V and stored at room temperature, followed by measuring their voltage change over the period of 72 hours. A known drawback of non-coated natural graphites is their inability to hold voltage on this test. It was pleasantly surprising to see the metal-coated natural graphite cells to hold charge quite well (on average better than MCMB control).

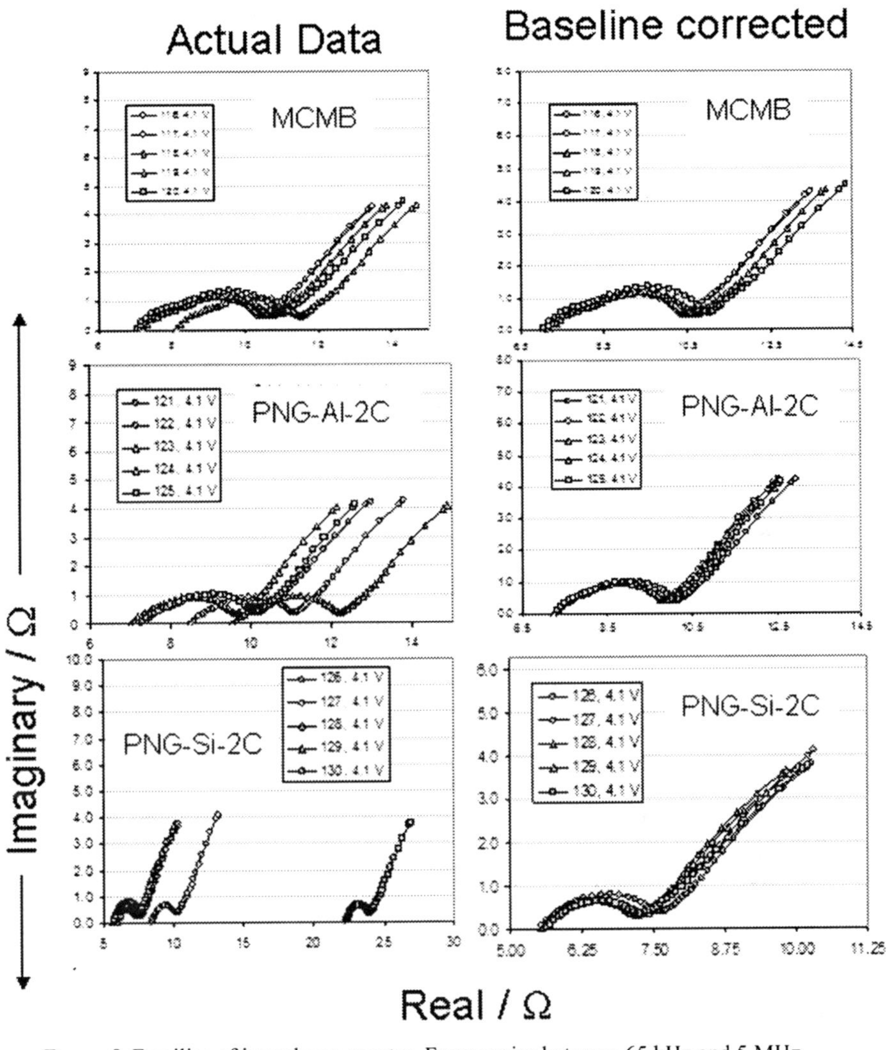

Figure 8. Families of impedance spectra. Frequencies between 65 kHz and 5 MHz.

4. CONCLUSIONS

Carbon coated Si-based anode materials were successfully synthesized by chemical vapor deposition methods. Coated materials performed well during the coin cell testing. Future work on this section involves assembly of 7Ah cells and the optimization of the FBCVD process parameters to achieve more uniform carbon coatings.

Stable cycling was achieved in the full 7Ah cells with a composite of spherical natural graphite coated with Al and then stabilized with a rigid carbon coating of the disordered nature. Further investigation is needed to fully understand the effect of rigid carbon shell on the electrochemical performance of graphite-based composite materials.

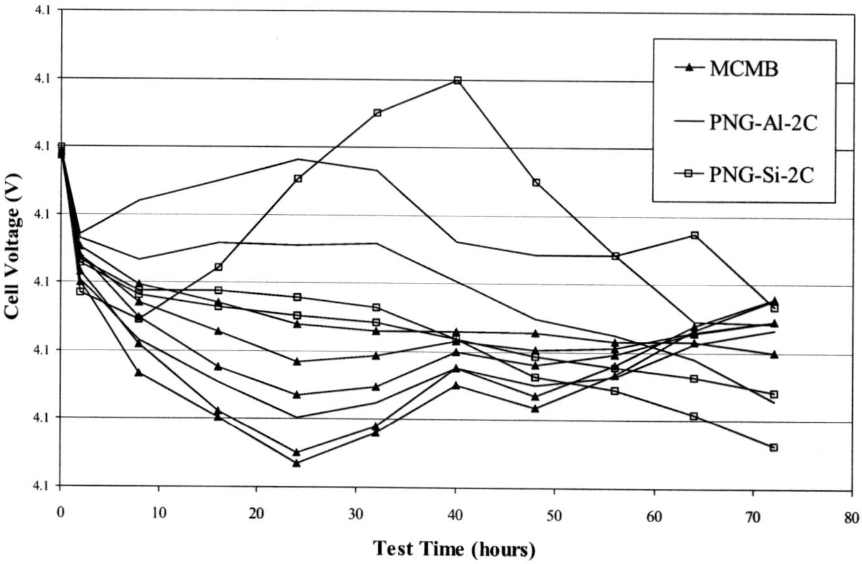

Figure 9. Prismatic 7Ah batteries. 72-hour charge stand test.

ACKNOWLEDGEMENTS

Authors representing Lithion, Inc., Superior Graphite Co. and Worcester Polytechnic Institute would like to thank the Army Research Laboratory for its partial financial support of this work through contract number DAAD17-01-C-0044.

Prof. V. Barsukov acknowledges financial support by the NATO Science for Peace project "SfP 973849", entitled: "Carbons as materials for the Electrochemical Storage of Energy". Him and Prof. J. Doninger wish to thank NATO's Dr. F. Pedrazzini (Program Director – Advanced Research Workshops), Dr. Chris De Wispelaere (Program Director - Science for Peace), and Chris Hilton (Program Manager – Travel Grants Program at the U.S. Civilian Research and Development Foundation, Arlington, VA, USA), whose efforts and coordination enabled bringing together parties – authors of this applied battery work.

REFERENCES

1. M. Wakihara, Materials Science and Engineering, R33, 109 (2001).
2. W. J. Weydanz, M. Wohlfahrt-Mehrens, R. A. Huggins, J. Power Sources, 81, 237 (1999).
3. S. Bourdeau, T. Brousse, D. M. Schleich, J. Power Sources, 81, 233 (1999).
4. H. Li, X. Huang, L. Chen, Z. Wu, Y. Liang, Electrochemical and Solid State Letters, 2, 11 (1999).
5. M. Yoshio, H. Wang, K. Fukuda, Y. Hara, Y. Adachi, J. Electrochem. Soc., 147, 1245 (2000).
6. C. S. Wang, G. T. Wu, X. B. Zhang, Z. F. Qi, W. Z. Li, J. Electrochem. Soc., 145, 2751 (1998).
7. P. C. Eklund, J. M. Holden, and R. A. Jishi, Carbon, 33, 7 (1995).
8. M. Gulbinska, F. S. Galasso, S. L. Suib, S. Iaconetti, P.G. Russell, J. F. DiCarlo, Mat. Res. Soc. Symp. Proc., 2003, 756, EE6.14.1- EE6.14.1.6.
9. S. Iaconetti, P. Russell, J. F. DiCarlo, M. Gulbinska, S. L. Suib, J. S. Gnanaraj, Proceedings of the 41st Power Sources Conference, 76-78, (2004).
10. I. Barsukov, Battery Power Products and Technology, 4, **9** (2000), 30.
11. Kabanov, B.N. and Chekavtsev, A.V. Electrochemistry of the Li-Al alloy and problems of creation of new power sources; in "Itohy Nauki i Tekhniki", Electrochimia eds., "VINITI" Publishing House, Moscow, Russia, 21, 140-176 (1984) – in Russian.
12. Kedrinskiy I.A., Yakovlev, V.G. Li-Ion Accumulators. Publishing House "IPK Platina", Krasnoyarsk, Russia, 124 (2002) – in Russian.
13. Winter, M. Besenhard, J.O. Electrochemical lithiation of Tin and tin-based intermetallics and composites. Electrochim. Acta, 45, 1-2, 31-50 (1999).
14. Cotton, F.A., Wilkinson, G., Murillo, C.A., and Bochmann, M. Advanced Inorganic Chemistry, Sixth Ed., Wiley, (1999).
15. Applied Electrochemistry, Kudryavtsev N.T. Ed., Second Ed., "Khimia" Publishing House, Moscow, Russia, 334-336 (1975) – in Russian.
16. F. Henry, I. Barsukov, J. Doninger, S. Anderson, P. Booth, P. Zaleski, R. Girkant, D. Derwin, M. Gallego, T. Huerta and G. Uribe. New Developments in the Advanced Graphite for Lithium-Ion Batteries – in this book.

ELECTROCHEMICAL ACTIVITY OF CARBONS MODIFIED BY d-METAL COMPLEXES WITH ETHANOLAMINES

L. G. Reiter[1], V. A. Potaskalov[1], A. A. Andriiko[1], V. S. Kublanovsky[2], M. A. Chmilenko[3], Yu. K. Pirskiy[2], V. I. Lisin[3], and S. M. Chmilenko[3]

[1]*National Technical University of Ukraine "Kyiv Polytechnic Institute", Chemical Technology Faculty, 37 Prospect Peremogy, 03057, Kyiv, Ukraine;*
[2]*Institute of General and Inorganic Chemistry, Ukrainian National Academy of Science, 32/34 Prospect Paladina, 04164, Kyiv, Ukraine*
[3]*Department of Electrochemical Power Engineering, Ukrainian National Academy of Science, 38-A Boulevard Vernadskogo, 04164, Kyiv, Ukraine*

1. INTRODUCTION

The organometallic complexes with d-metals are considered as promising electrocatalysts for oxygen electroreduction in air-metal electrochemical cells. Obviously, the first idea was to employ the catalytic mechanism of the oxygen reduction with porphyrin-like metal complexes [1] found in living beings (Figure 1).

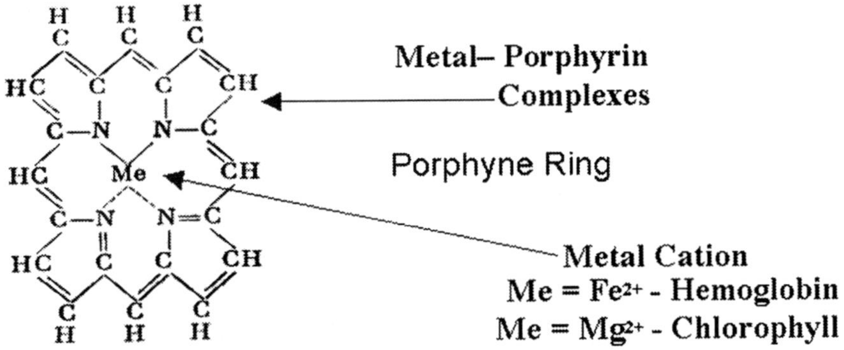

Figure1. Fragment of the structure of natural catalysts, N_4-metal complexes.

Numerous N_4 complexes of Co(II) [2-5], Fe(II) [6,7], Fe(III) [8], Cu (II) [9,10], and Mn(II) [11] were studied as electrocatalysts for modifying of the electrode surface for O_2 reduction. It was found that, varying ligands and

I.V. Barsukov et al. (eds.), New Carbon Based Materials for Electrochemical Energy Storage Systems, 333–344.

metal atoms, as well as by applying thermal treatment, one could obtain the catalysts with the parameters appropriate for metal-air batteries and electrochemical generators. Unfortunately, the N_4 complexes with porphyrin-like ligands are too expensive for mass production battery technology.

It is known that some spinel-structured 3d-metal oxides are good catalysts for many processes involving electron transfer [12]. However, their low conductivity does not allow for the direct use in the electrode of the battery, and grafting them onto the carbon matrix is also very difficult technical problem. It was found recently that this problem could be solved indirectly, creating the "spinel" catalytic centers on the surface of carbon by means of adsorption of some 3d-metal complexes on the graphite surface followed by subsequent pyrolysis at certain temperatures [13,14].

In this work we have studied the preparation of electrocatalysts on the graphite matrix using tri-nuclear complexes of 3d-metals with aminoalcohol ligands. Tri-nuclear complexes, $2[Co(Etm)_3]*Me(NO_3)_2$, where Etm = ethanolamine, Me = Zn^{2+}, Cu^{2+}, Ni^{2+}, Co^{2+}, were investigated.

As it was shown in [15-18], such compounds, with bridging atoms of deprotonated amino alcohol, are formed when aminoalcoholate complexes of metal (III) react with bivalent 3d-metal ions. Many representatives of these compounds were synthesized in crystalline state; the polynuclear compounds were also found to form in aqueous and methanol solutions. The structure of 2Co(III) – Ni(II) tri-nuclear complex, according to [17,18], is shown below in Figure 2.

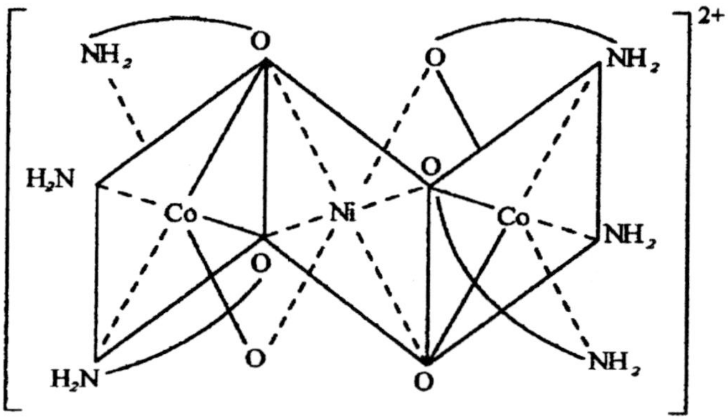

Figure 2. Structure of tri-nuclear complex ion $[Ni(Co(Etm)_3)_2]^{2+}$.

Being thermally decomposed onto the surface of carbon, this complex is expected to form very small catalytically active $NiCo_2O_4$ spinel centers. Thus, we have studied the catalytic activity of the products of pyrolysis at different temperatures toward two electrochemical reactions – reduction of oxygen in alkaline electrolyte and intercalation of lithium into carbons in aprotic electrolyte of Li-ion battery. To our knowledge, the catalytic effect of the metal complexes in the second reaction was not yet considered in the literature.

2. EXPERIMENTAL

Syntheses of the tri-nuclear complexes were performed in water-methanol solution according to the reaction

$$2\,f_{ac}\text{- Co(Etm)}_3 + \text{Me(NO}_3)_2 \rightarrow [\text{Me(Co(Etm)}_3)_2](\text{NO}_3)_2. \quad (1)$$

The f_{ac}- isomer of trice-aminoethanol complex of cobalt (III), with 3 coplanar oxygen atoms, was prepared according to the method [16].

The electrocatalysts for oxygen reduction were prepared as follows. These complex compounds were inoculated onto the carbon (AG-3, BET area near 800 m^2/g) by means of adsorption from dimethylformamide solutions. The portion of complex compound weighed so as to achieve 3% of Co content was mixed with the carbon, then 5 ml of dimethylformamide per 1 g of the carbon were added and the mixture was cured at room temperature for 24 hours. Series of samples obtained were thermally treated (pyrolyzed), and the resulting grafted carbons were tested as electrode materials in the reaction of molecular oxygen reduction.

Electrochemical performance of the modified carbons was examined by "floating" gas diffusion electrode [19] in an electrochemical cell with separated cathode and anode chambers at room temperature with PI-50-1.1 - potentiostat in 1N KOH aqueous solution.

The partial oxygen pressure was checked with an OP-210/3 microanalyser [20] before and after the bubbling of oxygen through the electrolyte. The potential was measured against an Ag/AgCl reference electrode.

Further on, the Co-Ni complexes were used for modification of Hohsen Carbon type (10-10) and Hohsen Graphite type (10-28) anode materials for Li-ion batteries applying similar procedure. These anode materials were tested in 2016 size lithium coin cells with a configuration *Li/electrolyte; (LP-30)/(modified anode material)*. The coin cells were assembled by standard technology in dry atmosphere of a glove box and then cycled with PC-governed galvanostatic equipment in the voltage window of 10 mV to 1.5 V.

3. RESULTS AND DISCUSSION

3.1. Electrochemical Properties of Modified Carbons in the Reaction of Oxygen Reduction

It was established that carbons grafted with pyrolyzed Co(III)-Me complexes increase its electrocatalytic activity, and the temperature of pyrolysis is very important. This can be seen from the plots in Figure 3. The maximum catalytic effect is achieved for the samples annealed near 600°C, decreasing at higher temperatures and almost vanishing at 800°C (Figure 4).

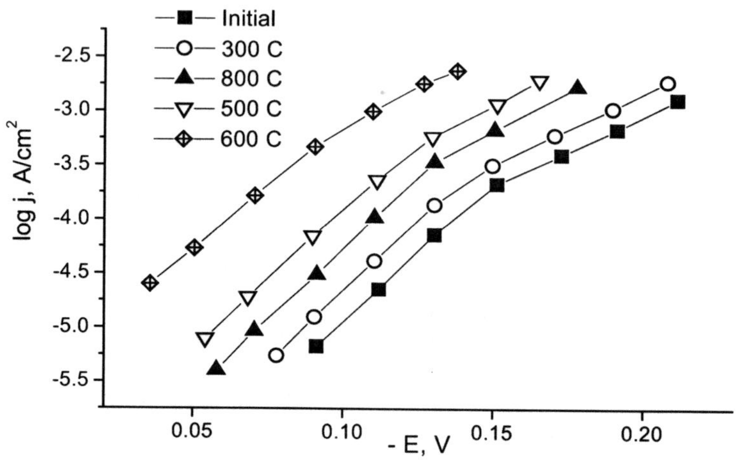

Figure 3. Current density versus electrode polarization (Ag/AgCl ref. electrode) for oxygen reduction in 1M KOH at 20 °C on Co-Ni modified AG-3 carbon.

Such regularities indicate that some active catalytic centers are formed at 600°C and destroyed at higher temperatures. As follows from the formula (Figure 1), the initial structure in the course of pyrolysis on the surface of graphite could rearrange itself into spinel-like groups bonded to the surface via remaining nitrogen atoms, $MeCo_2O_y$ –(N)– (C_x), as it was already presumed in ref. [13]. This hypothetic mechanism is in agreement with DTA data reported in [14] – though the individual polynuclear complex compounds exothermally decompose totally at 220-240°C, they become much more thermally stable being adsorbed onto the graphite matrix.

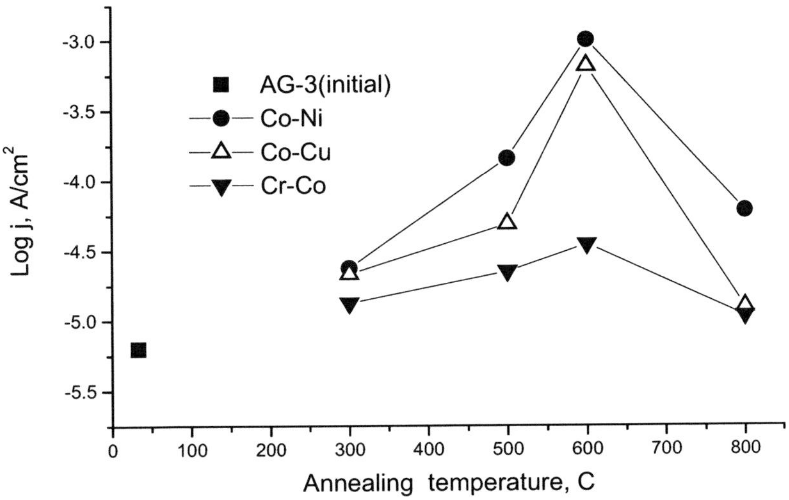

Figure 4. Logarithm current density at –0.1 V vs. the temperature of pyrolysis for trinuclear complexes on the surface of AG-3 carbon.

As is can be seen from Fig. 4, the better electrocataysts for oxygen reduction are derived from Co-Ni tri-nuclear complex, $[Ni(Co(Etm)_3)_2](NO_3)_2$.

It also consists with the "spinel" hypothesis for the structure of active catalytic sites: the spinel compound $NiCo_2O_4$ is well known for its catalytic activity in the reaction of oxygen reduction [21].

3.2. Electrochemical Properties of Modified Carbons in the Reaction of Lithium Intercalation

The electrochemical intercalation of Li was studied for carbon electrodes modified by the 2Co-Ni complex, which showed the best effect in the reaction of oxygen electroreduction. Galvanostatic charge-discharge technique (PC governed automatic bench) in 2016 coin type cells was used for this purpose.

The electrode active masses were prepared by mixing the active material with 10 wt.% of polyvynilidene fluoride slurred in 1-methyl-pyrrolidone solvent. The actual mass was then pasted onto one side of pre-cleaned Co foil, dried for 4 h at 100°C, pressed and the disks of 15,6 mm diameter were cut and placed into the cell's coins. Large excess of Li metal (foil) was used as a counter electrode. 1M $LiPF_6$ solution in mixture of ethylene carbonate (50 vol.%) and methyl carbonate was used as an electrolyte (Merck product LP30).

3.2.1. Carbon-Type Material

Figure 5 shows galvanostatic charge-discharge voltage profiles of the initial non-modified Hohsen Carbon-Type material. This is a typical behavior of non-graphite carbon, with sloped voltage-time curves, more than 20% first cycle irreversible capacity loss and good cycle retention with moderate reversible capacity (180-200 mAh/g depending on the current).

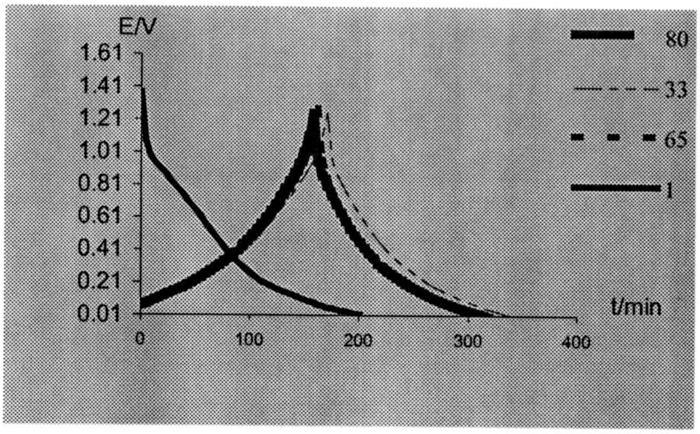

Figure 5. Charge-discharge curves of Hohsen Carbon-Type Material in the course of long-term cycling.

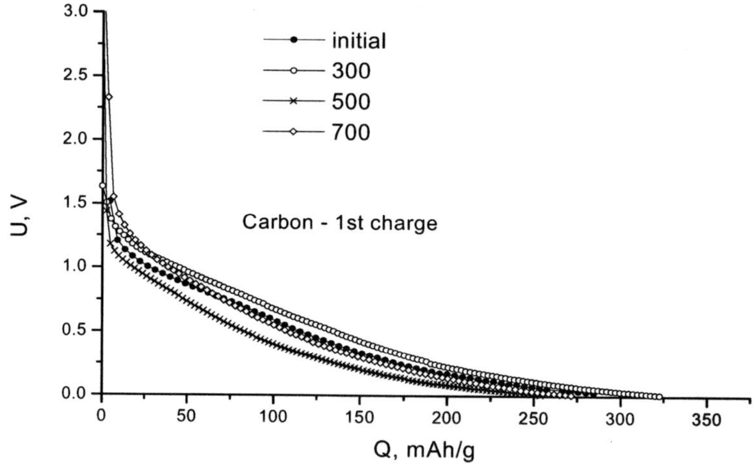

Figure 6. The first charge curves of Hohsen Carbon-Type material modified with Co-Ni complex. Annealing temperatures are shown in the figure. Rate ~ 0.1 C (10 hours).

The modification with Co-Ni complex pyrolysed at different temperatures affects the shape of first charge curve as shown in Figure 6. The sample-300 (annealed at 300°C) increases the charge capacity while sample-500 decreases, sample-700 remaining almost unchanged. However, no gain in the discharge capacity is observed for all but sample-500 modified materials (Figure 7). The sample-500 increases its first discharge capacity by 5%, and, more remarkably, decreases its first cycle capacity loss from 26 to 9%. However, these advantageous features of the modified material disappear in course of long-term cycling of the cells (Figure 8).

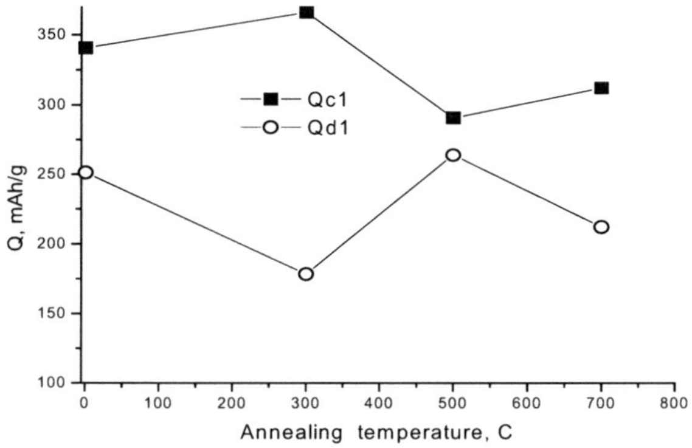

Figure 7. Specific capacity during the first charge (solid symbol) and discharge of modified carbon-type materials.

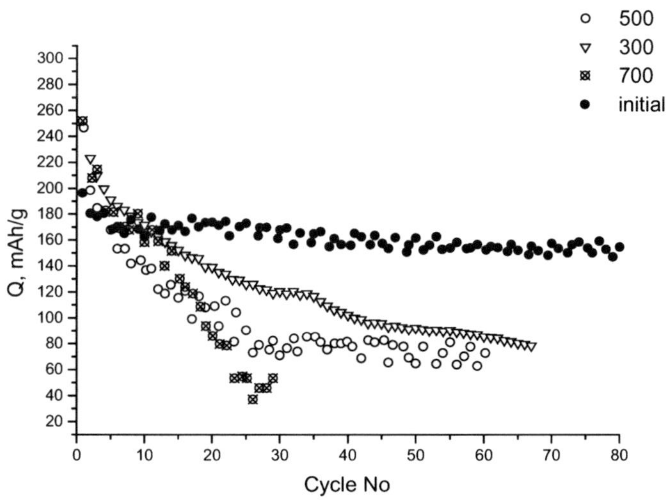

Figure 8. Behavior of modified carbon-type materials in the course of galvanostatic cycling, current load 30 mA/g.

As it is seen from the data of Figure 8, all modified materials have poor cycling performance; their reversible capacities fade faster than the one of initial non-modified material, and become lower after the first 8-10 charge-discharge cycles. Thus, we can conclude that no positive effect is achieved by means of modification of the Carbon-Type material with bimetal tri-nuclear complex of Co(III)-Ni(II).

3.2.2. Graphite-Type Material

Figure 9 shows voltage profiles of the initial and modified Graphite-type materials at first charge (A) and subsequent discharge-charge (B) processes.

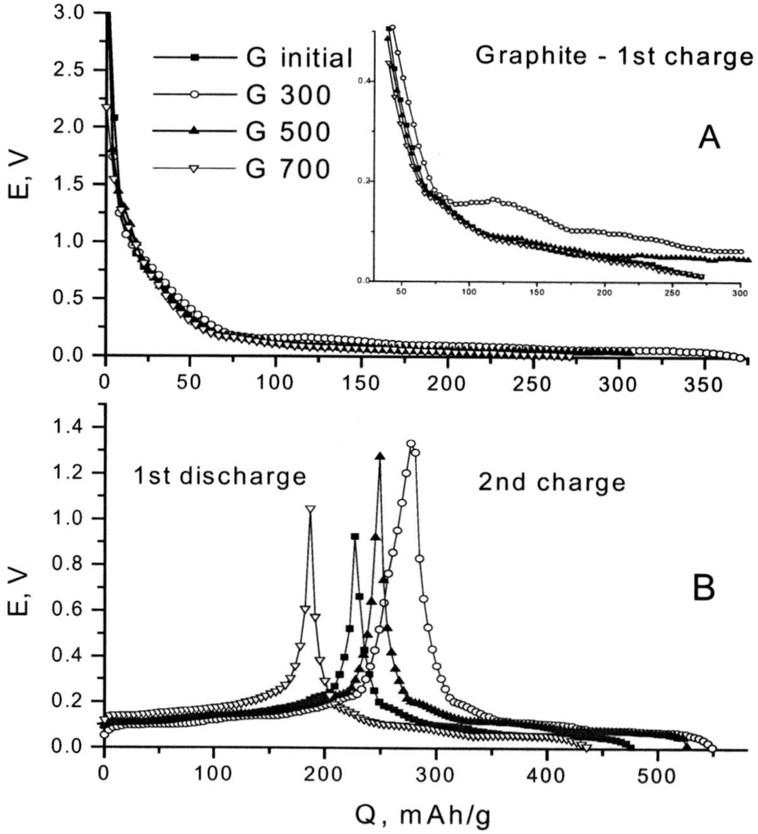

Figure 9. Voltage profiles of Hohsen Graphite-Type Material modified with the Co-Ni complex: A – first charge, part of it is enlarged in the insert; B – next discharge and charge half-cycles.

The higher first charge capacity and small plateaus at 0.16 and 0.10 V on the curves of "low-temperature" sample G300 (annealed at 300°C, Figure 9A) are probably the indication of the reduction of remaining metal complexes in this sample. This process is irreversible, because the plateaus do not appear in the discharge and next charge curves. The curves for other samples are typical for graphite electrode materials of Li-ion batteries.

The irreversible loss of capacity at first charge is an adherent property of each carbon anode material for Li-ion battery since it provides the charge necessary to buildup the Solid Electrolyte Interface (SEI), which is necessary for reversible operation of the battery. As Figure 10A shows, this quantity is larger for modified materials than for initial one (8%), being at maximum for G300 (26%) and descending to 10% for G700, which is close to the value of the initial material (8%).

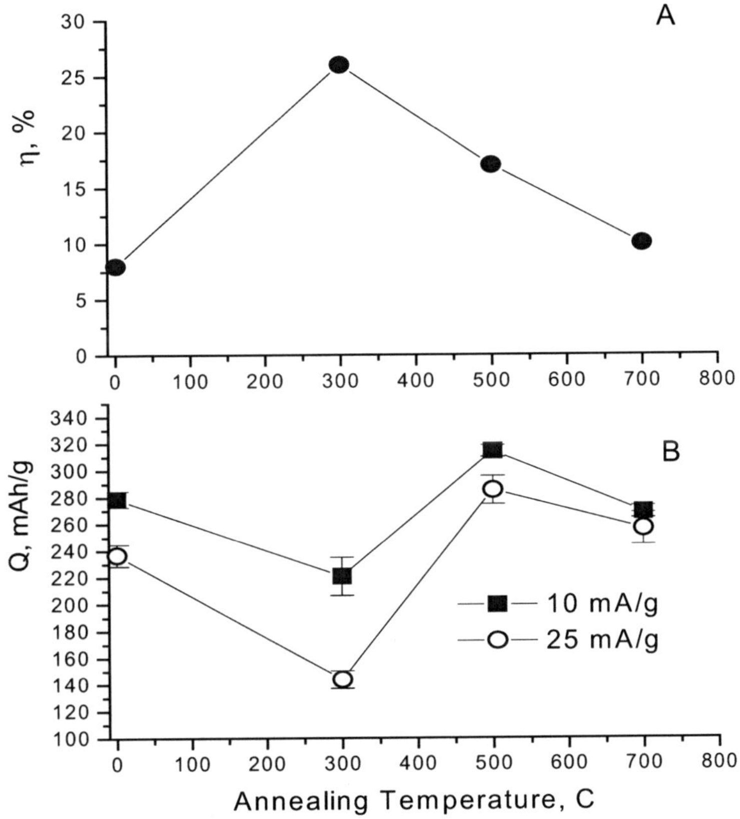

Figure 10. Irreversible capacity losses (A) and reversible capacity at 10-th cycle (B) for two specific current values of modified Graphite-type materials annealed at different temperatures.

The reversible capacity over the first few cycles (Fig. 10B) differs depending on the temperature of pyrolysis. In this respect the initial and "high-temperature" G700 sample are almost the same; the capacity is lower for "low-temperature" G300 sample and higher for the intermediate G500 sample. Thus, grafting of the Graphite-type anode materials with pyrolyzed at 500°C Co-Ni complexes enhance their electrochemical performance. The effect is the greater the higher is the charge-discharge current. According to the data in Fig. 10B, the capacity gain is 13% at specific current 10 mA/g and 21% at 25 mA/g.

The next important property of an electrode material is cycle retention in course of long-time operation. Figure 11 provides such data for the modified graphite materials.

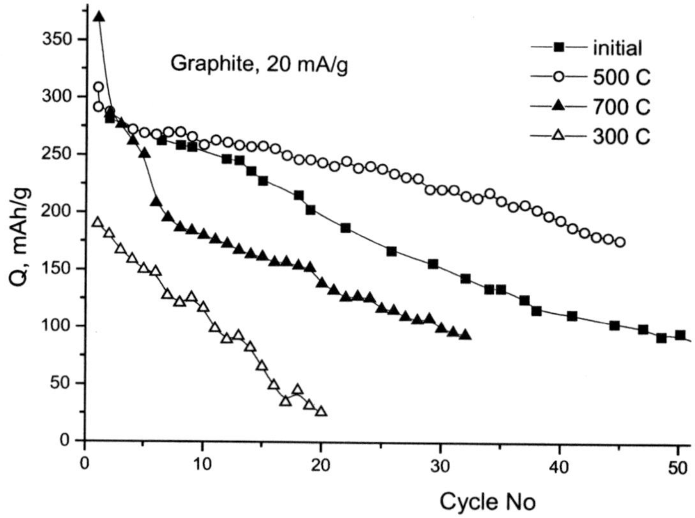

Figure 11. Capacity fading of initial and modified Graphite-Type materials in course of continuous cycling.

The reversible capacity of the modified G500 material is seen to be constantly higher than that of the initial graphite while the performance of other modified samples is worse.

Thus, unlike in the case of Carbon-type samples, modification of the Graphite-type anode materials with pyrolyzed at 500°C tri-nuclear $[Me(Co(Etm)_3)_2](NO_3)_2$ complexes enhance their electrochemical performance, especially at higher charge-discharge rates. The inner mechanism of such effect still remains unclear. We suggest the formation of nanosize cataliticaly active spinel-like groups, which promote the charge transfer across electrolyte/graphite interface. However, why this mechanism does not work in case of non-graphitized carbon material? We do not know the answer yet. Additional research is necessary in order to clear out the details and optimize the performance of the grafted anode materials for its implementation in future battery technologies.

4. CONCLUSIONS

1. Modification of carbon materials by tri-nuclear complexes of 3d-metals with ethanolamine ligands increases the catalytic activity with regard to the electrochemical reaction of oxygen reduction. The Co-Ni complex is most active in this reaction if pyrolyzed at 600°C.

2. Modification of Graphite-type materials by the Co-Ni complex pyrolyzed at 500°C improves significantly the performance of these material in the anode of Li battery. The inner mechanism of this effect is still unclear. We suggest the formation of nanosized spinel structures which facilitate the charge transfer across SEI.

3. No positive effect of the modification was found for Carbon-type (non-graphite) materials.

ACKNOWLEDGEMENTS

Part of this work was funded by the State Foundation of Fundamental Research of Ukraine, grant # F7/217-2001. The author A. A. Andriiko gratefully acknowledges the support by Civilian Research and Development Foundation (travel grant TGP-1056), which enabled presentation of this work at the NATO CARWC.

REFERENCES

1. Tarasevich M.R., Radyushkina K.A. Catalysis and Electrocatalysis with Metalporphyrins (in Russian). M.: Nauka, 1982, 168p.
2. Shi C., Anson F.C. Inorg. Chem. 1995; 34:4554.
3. D'Souza F., Hsieh Y.Y., Deviprasad R.G. J. Electroanal. Chem. 1997; 426:17.
4. Bhugun I., Anson F.C. J. Electroanal. Chem. 1997; 430:155.
5. Shamsipur M., Salimi A., Haddadzadeh H., Mousavi M.F. J. Electroanal. Chem. 2001; 517:37.
6. Forster P.A., Kuwana T. Inorg. Chem. 1983; 22:699.
7. Ozer D., Harth R., Mor U., Bettelheim A. J. Electroanal. Chem. 1989; 266:109.
8. Zhang J., Anson F.C. J. Electroanal. Chem. 1993; 353:265.
9. Zhang J., Anson F.C. J. Electroanal. Chem. 1992; 341:323.
10. Zhang J., Anson F.C. J. Electroanal. Chem. 1993; 348:81.
11. Bettelheim A., Ozer D., Parash R. J. Chem. Soc. Faraday Trans.1 1983; 79:1555.
12. Razina N.F. Oxide Electrodes in Aqueous Solutions (in Russian). Alma-Ata: Nauka, 1982, 160 p.
13. Kublanovsky V.S., Pirsky Yu.K. Russian J. Applied Chem. 2001; 27:1116.

14. Pirsky Yu.K., Levchuk Ya.N., Reiter L.G., Kublanovsky V.S. Ukrainian Chemical Journal. 2003; 69(3-4):77.
15. Gerasenkova A.N., Udovenko V.V., Russian Journ. Inorg. Chem. 1972; 18:2446.
16. Udovenko V.V.,Stepanenko O.N., Eroshok B.G. Russian J. Inorg. Chem. 1974; 19:2455.
17. Stepanenko O.N., Rejter L.G. Ukrainian Chemical Journal. 1992; 58(12):1047.
18. Stepanenko O.N., Potaskalov V.A., Trachevskiy V.V. Coordination Chemistry (in Russian). 2001; 27(1):57.
19. Steinberg G.V., Kukushkina I.A., Bagotskiy V.S., Tarasievich M.R., Electrochemistry (Russian), 1979; 15:527.
20. Kublanovsky V.S., Pirsky Yu.K., Yakimenko N.G. Russ. J. Phys. Chem. 1997; 71(1):54.
21. Efremov B.N., Tarasievich M.R. "Electrocatalysis and Electrocatalytic Processes" (in Russian). In *Sbornik Nauchnykh Trudov (Collection of Sci. Transactions),* Kiev: Naukova Dumka, 1986; 44-71.

METAL-GRAPHITE COMPOSITS AS MATERIALS FOR ELECTRODES OF LITHIUM-ION BATTERIES

Ludmyla Matzui*, Mykhailo Semen'ko, Mykol Babich, Leonid Kapitanchuk

Kyiv Taras Shevchenko National University, Dept. of Physics, 64, Volodymyrska Street, Kiev, Ukraine.

1. INTRODUCTION

Carbonaceous materials such as graphite, hard carbon, pyrolytic carbon and nanotubes have been extensively investigated as anode materials for Li-ion batteries. Among many types of carbon anodes that have been reported, several types of carbon materials exhibit superior performance from the point of view of reversible capacity and cyclic efficiency of the lithium-ion batteries. These carbon and graphite materials possess different microstructure, texture, crystallinity and morphology. Two typical types of carbon materials, namely, highly ordered graphite treated at the temperature of about 3000^0C and non-graphitized carbons treated at low temperatures of about 1000^0C have been used in commercial batteries. The precursor materials include cokes, polymers, fibers etc. Besides, the behavior and the mechanism of lithium ions penetration into different kinds of carbon and graphite host have been extensively studied both experimentally and theoretically. As for the highly ordered graphite, the charge/discharge reaction is based on Li^+ intercalation and deintercalation, and theoretical lithium storage capacity of graphite anode for lithium-ion secondary battery has been considered to be 372 mAh/g [1,2]. On the other hand, disordered carbons, reportedly show reversible capacity from 400 to 900 mAh/g in the voltage interval of Li/C cells ranging between 0 and 2,5V [3]. Though the theoretical models of this phenomenon are still under extensive discussion, there are several suggestions to explain the high capacity of disordered carbon anodes [4]:

1. lithium occupation of the nearest neighbor sites;
2. multilayered lithium deposition at the surface of crystallites;
3. extra lithium storage at the edge of crystallites;

* Corresponding author. E-mail: matzui@univ.kiev.ua

I.V. Barsukov et al. (eds.), New Carbon Based Materials for Electrochemical Energy Storage Systems, 345–356.

4. lithium ion clusters in nanoscaled cavities;
5. lithium binding in the vicinity of hydrogen atoms.

On the other hand, it is well known now that such metals as Al, Sn, Pb as well as Si are able to store Li via alloy formation though electrochemical means resulting in substantially higher capacities [5-8]. For example, Si or metallic Sn can electrochemically alloy with Li up to about 4,4 Li per Sn, which yields a maximum theoretical capacity of 990 mAh/g [5,7]. However, the cycling capacity of these materials is limited by the considerable volume changes during lithium introduction and by the brittleness of intermetallic lithium phase. In order to improve their structural stability, the superfine or even nanoscaled intermetallic compounds have been developed and investigated. In general, superfine or nanoscaled materials are expected to possess lower specific volume variation and, hence, better cyclability in electrochemical intercalation and extraction of lithium ions. For example, it was found that ultrafine Sn-SnSb particles with d<300nm could improve the electrochemical performance of the tin based alloys [9]. Yet superfine or nanoscaled particles are predisposed to particle aggregation. The resultant micro-sized particles are formed during lithation and de-lithiation reactions. One of the possible ways to solve these problems consists in the creation of the composite materials. The performed activities on the creation of metal-graphite composite materials by CVD deposition on mesoscale carbon or by mechanical mixing showed this approach to be prospective [10].

Therefore, taking into account the listed above, it is necessary to develop metal-graphite composite materials, which could provide the conditions for the realization of all mentioned lithium storage mechanisms with maximal efficiency. The main goal of this work was to synthesize graphite based composite materials containing superfine or nanoscaled Sn, SnO_2 and Si particles. To reach this goal it was necessary to work-off the methods of metal-graphite nanocomposite materials production and to determine the influence of the type of source graphite material and synthesis conditions on the phase composition, dispersity and morphology of the synthesized composite materials.

2. EXPERIMENTAL PROCEDURES

The natural crystalline flake graphite from Zavalie field in Ukraine's Kirovograd Province has been subjected to thermochemical treatment and the resultant product, thermoexfoliated graphite, has been used as carbon material for CM creation. This material was selected because of following considerations:

1. Thermoexfoliated graphite (TEG) is a material obtained by deep thermochemical treatment of natural graphite (intercalation by sulfuric acid with subsequent heat treatment in a regime of thermal shock). Practically, it does not differ from the source natural graphite by crystallographic parameters, namely, by the value of d_{002}. According to the X-ray diffraction data the interlayer distance d_n along c axis is equal to $d_n = (3.36 \pm 0.001)\,\mathring{A}$, for TEG samples used in this work. That is in a fairly good agreement with that for pure graphite. This result proves the absence of turbostratification of graphite layers and allows to consider the lithium penetration due to intercalation being occurred effectively.

2. TEG macrostructure differs from that of natural graphite: it possesses abnormally high porosity and highly developed active surface (40-50 m^2/g) (Figure 1). The performed thermochemical treatment leads to an essential exfoliation of graphite matrix with a formation of cellular structure. The thickness of cell's walls is equal to 20-25 nm. The surface of cell's walls contains a lot of macrocracks, outcrops of crystallites, etc. The thermochemical re-treatment was applied to enhance TEG dispersivity.

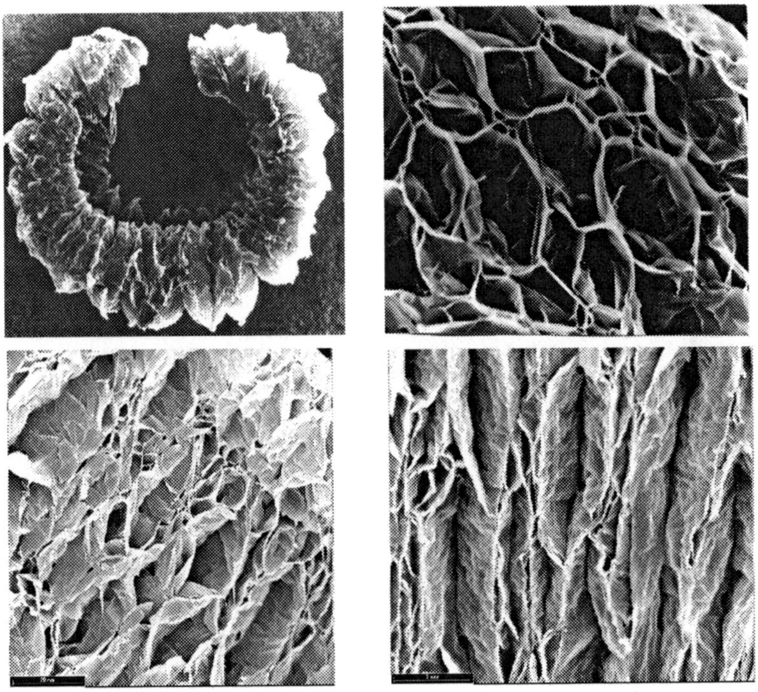

Figure 1. Parts of the surface of source TEG (a, b) and TEG oxidized by sulfuric acid and subjected to thermal shock at 800⁰C (b); part of the surface of TEG particle oxidized by nitric acid and subjected to thermal shock at 800⁰C (d).

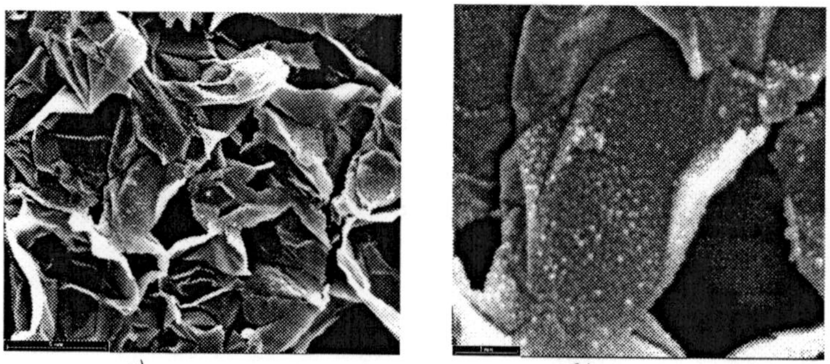

a) b)

Figure 2. Part of TEG particle with Si deposited in vacuum: a), b) are the different regions of TEG particle (Si content ~ 7wt%).

TEG structure refinement has distinctly observed in electron microscopy studies of the oxidized TEG powders subjected to the repeated thermal shock. In this case the size of TEG macropores was equal to 1.5-2 μm that is essentially lower that for source TEG. Figure 2 presents SEM images of the source TEG particle (a) and TEG particle oxidized by sulfuric acid and re-exfoliated at 800^0C (b).

The surface of the edges of graphite macroplanes became sheet-like and hilly due to the intensive exhausting of the oxidant during the thermal shock. Moreover, the morphology of TEG particle surface could be modified by the selection of the appropriate intercalant. For example, when the nitric acid is used instead of sulfuric, the essential refinement of TEG macrostructure is observed and TEG particles possess more fine structure (it is clearly seen from Figure 3): mean size of the macropores is equal to 1-3 μm, the thickness of graphite walls (verges) – 0.1-0.2 μm. Besides, the edges of graphite particles contain the outcrops of dislocations etc. However, according to the X-ray studies TEG treatment by HNO_3 oxidant does not result in (002) graphite reflection broadening as compared with the feed graphite. This is the evidence of the possible additional refinement of graphite macrostructure. The analysis performed, indicates that TEG possesses all the necessary features to realize lithium storage both through intercalation and the mechanism, which is dominant in disordered carbon materials, i.e. lithium occupation of the nearest neighbor sites, multilayered lithium deposit on the surface, and at the edge of crystallites, lithium ion clusters in nanoscaled cavities, etc.

Besides, the anomalous elastic-plastic characteristics of TEG are the important factors for application as a support. Due to these characteristics TEG-based matrix could accumulate the possible volume changes of metallic component at their cyclic changes during charge/discharge process.

Phase composition, distribution character of the modifier particles on the surface of graphite particles has been analyzed using the X-ray studies, the Auger spectroscopy, electron microscopy and secondary ion mass spectrometry. The investigations have been performed using the following equipment: DRON-4-07 X-ray diffractometer (filtered CoK_{α} radiation), LAS-2000 Auger spectrometer (Riber, France), JSM-840 electron microscope (Japan).

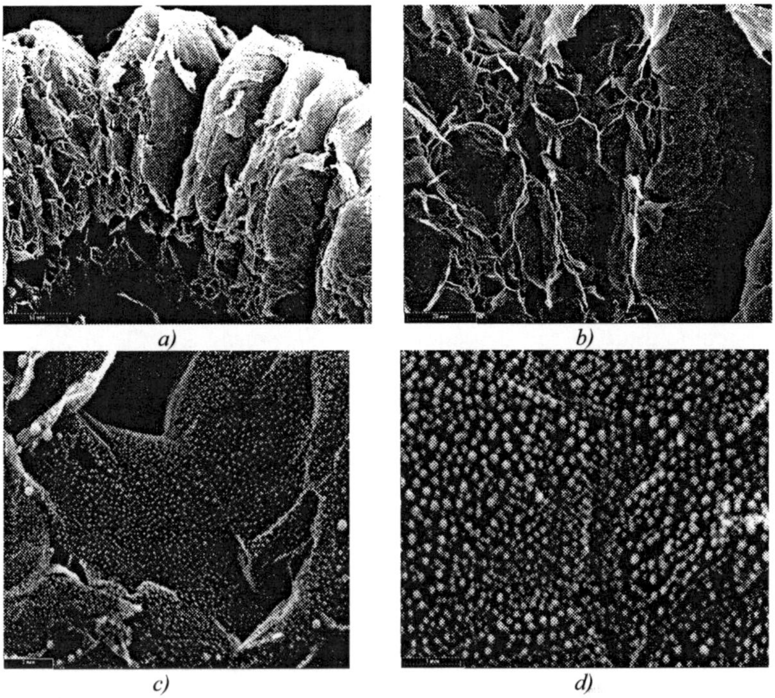

Figure 3. Part of TEG particle with Sn deposited in vacuum: a, b, c, d are the different regions of particle and different image magnification (Sn content ~ 7wt%).

The following procedures of modifying coating deposition on the surface of graphite suppor particles have been applied to produce TEG-CMs:

- the method of thermal vacuum deposition;
- the method of metal salt chemical plating from water-salt solution with subsequent thermal or liquid-phase reduction of salt to metal or metal oxide.

The method of thermal vacuum deposition has been used to obtain tin and silicon deposit on TEG surface. The main difficulty at metal component deposition on graphite support is to obtain uniform metal coating on the surface of disperse particles. The system of uninterrupted mixing of TEG powder during the material evaporation has been created. Its principle of operation is as follows: a contained holding TEG is an enclosed metal chamber, at the bottom of which there is gauze having a mesh of 100 ×100 μm in size. This container was suspended in electromagnetic vibrator and placed under the evaporator. The whole system was placed into the working chamber of VUP-5 vacuum station. Shaking frequency at deposition was equal to 15-20Hz. As a result, the particles change their position in the flow of metal vapor that allows relatively uniform deposition of material on graphite particles. Here the vacuum exceeded $5 \cdot 10^{-5}$ P.

The second method, i.e. electroless plating, has been used also to produce graphite - Sn, SnO_2 composite materials. The chemical deposition of metals on the graphite supports is performed through two principal steps.

The first step includes the impregnation of the oxidized graphite material by the water-salt solution of different concentration under the certain temperature-temporal regimes. The diffusion of salt solution into the pores of graphite material takes place at impregnation. Temperature and time are the main factors determining this process. Temperature increase accelerates the diffusion of salt solution, however, it should not be higher than the temperature of salt decomposition. With respect to the microporous structure of the set of the used graphite supports (e.g., TEG) the impregnation should be of a long duration. UZDN-A ultrasonic disperser has been applied in some cases to decrease the impregnation time and to enhance the impregnation efficiency. Thermal treatment of the carbon material impregnated by salt in the air was used to obtain graphite-tin dioxide CM.

The second processing step consisted of salt decomposition with the subsequent reduction to pure metal. The method of chemical deposition of metal salts from the water salt solution with the subsequent reduction to pure metal by liquid phase reducer has been applied to prepare graphite-tin CMs. In this case tin chloride was used for impregnation and potassium tetrahydroborate was used as liquid phase reducer.

This method of nanodispersed metal production has the advantages as compared with the previous one because metal reduction is performed at sufficiently low temperatures (20-200^{0}C). That is why the diffusion of atoms and migration of the formed metal particles is suppressed, no metal particles agglomeration occurred.

3. RESULTS AND DISCUSSION

The following materials were synthesized using the listed methods:

1. TEG-Si, TEG-Sn and dispersed natural graphite-Sn have been prepared by thermal vacuum deposition under continuos mixing.

SEM studies of the obtained TEG-Si powders have shown almost uniform distribution of silicon particles on TEG surface. These metal coated particles were found to be of the spherical shape, mostly similar in size

(50 nm). The absence of silicon particle segregation proves the absence of silicon diffusion on TEG surface upon preparation process. The absence of silicon reflections at the X-ray diffraction patterns could be the evidence of the amorphous structure of Si.

Some other situation is realized in a case of TEG-tin CMs. Electron microscopy studies of the obtained TEG-Sn powders revealed the uniform coverage of TEG surface by tin particles. Tin particles are of spherical shape and their sizes are about 40-80 nm, i.e. somewhat higher than in a case of silicon particles. Low scatter of particle sizes is observed as in a case of TEG-silicon system. However, as it is clearly seen from the data of the X-ray structure analysis (Figure 4) tin particles deposited on the surface of graphite support are in crystalline state. The distinct and narrow tin reflections at the X-ray diffraction pattern evidence this fact.

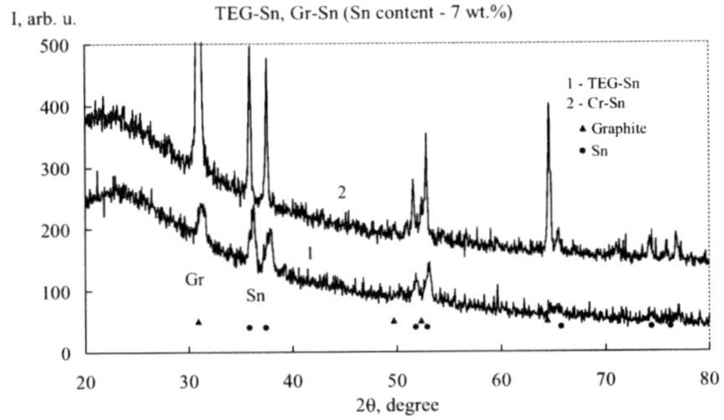

Figure 4. X-ray diagrams for TEG and disperse graphite with Sn.

A different situation is realized when the dispersed graphite is used as a graphite support. In this case tin particles distribution on graphite surface is extremely non-uniform (Figure 5). An essential scatter of particle sizes is observed. Along with small particles of 30 nm in size, which cover separate regions of graphite surface forming almost uniform single layer big particles of 150 – 200 nm in size are observed, which represent the aggregations of tin particles. According to the X-ray structure analysis the coherent domain size is almost 2 times higher in these materials as compare to TEG-tin CMs.

Thus, the method of thermal vacuum deposition in a regime of continuos mixing permits to obtain graphite – silicon and graphite – tin CMs containing modifier component about 10wt%. The particles 40-80 nm in size are almost uniformly distributed on graphite surface. However, certain disadvantages are inherent to any vacuum deposition method. It is difficult to control the content of the deposited component, to produce the materials of high modifier content and the efficiency of such methods is quite low.

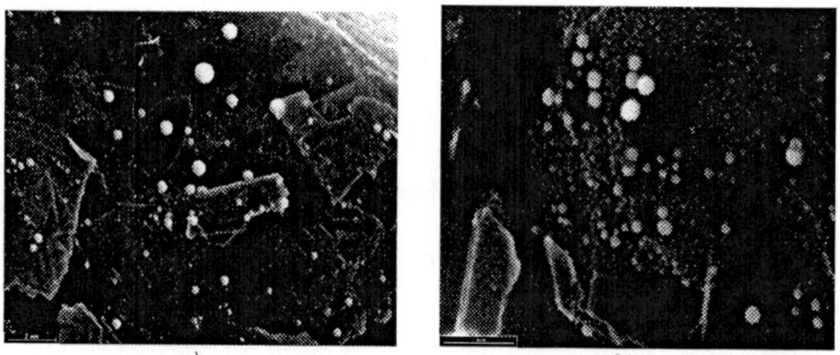

a) b)

Figure 5. Part of disperse graphite particle with Sn deposited in vacuum: a and b are the different regions and image magnification of particle (Sn content ~ 7wt%).

2. Our studies on graphite – transition metal systems [11] have shown the methods of chemical deposition of nano-scaled metal particles on the surface of graphite supporter to be powerful technique for production of CM with well dispersed, nanoscaled homogeneously distributed modifier component.

Graphite – tin dioxide CMs have been obtained by the method of chemical deposition of tin nitrates from the water-salt solutions with subsequent thermal decomposition of salt. According to the X-ray phase analysis (Figure 6) the impregnation of both TEG and graphite by water salt solution with the subsequent salt decomposition on air leads to the formation of tin dioxide on the surface of graphite matrix. Moreover, in spite of the sufficiently high content (60wt%) of modifier the Bragg's reflections corresponding to SnO_2 are wide and low intensive that could evidences the amorphous or disperse state SnO_2.

The character of the diffraction pattern remains almost invariable upon the increase of temperature of treatment and using both TEG and dispersed graphite as a support. Figure 7 displays the character of SnO_2 distribution on the surface of graphite matrix. As it is clearly seen SnO_2 forms sufficiently big segregations 0,1-0,2 μm in size nonuniformly distributed on the surface of graphite matrix. These data are in contrast to the results on production of graphite – transition metal oxide CMs where the modifier oxides were distributed on the surface of graphite supporter in a form of continuos layer of particles 30 – 50 nm in size. That is why thermal reduction of tin oxides in a hydrogen flow should be considered unsuitable for graphite-tin CMs production. So, the method of liquid phase reduction of metal cations in water medium by KBH_4 have been used to produce such CMs.

TEG-Sn CMs have been prepared according to the following scheme:

1. The impregnation of the oxidized TEG by $SnCl_2 \cdot 6H_2O$ water salt solution of certain concentration at 293 K for 20 hrs.

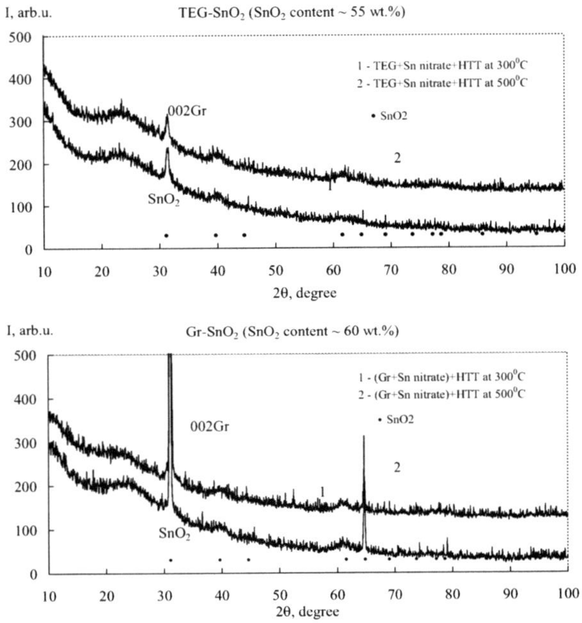

Figure 6. X-ray diagrams for TEG with SnO_2 and for dispersed graphite with SnO_2.

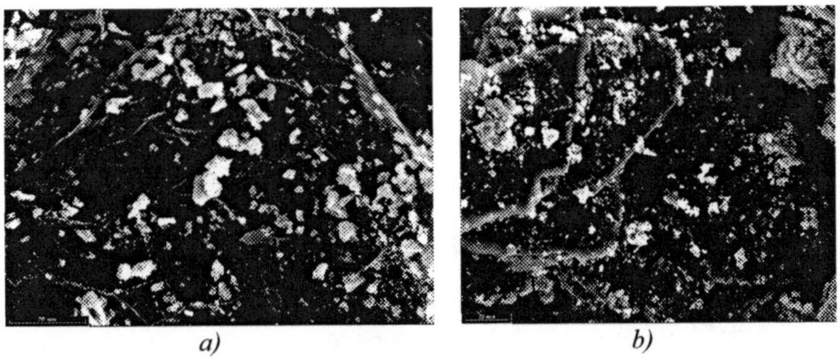

Figure 7. Part of TEG (a) and disperse graphite (b) particles with deposited SnO_2 (SnO_2 content ~ 55wt%).

2. Sn cations reduction by the solution of potassium tetrahydroborate.

The X-ray phase analysis of the CMs obtained by these methods (Fig. 8) showed that the modifier coating is pure tin. Moreover, the coherent domain size of tin particles in this case is almost the same as for the materials obtained by thermal vacuum deposition. Besides, it is necessary to note that tin content in CMs produced by this method is 6 - 7 times higher than that for CMs produced by thermal vacuum deposition method. The tendency to enlargement of coherent domain size is revealed when disperse graphite is used as a supporter instead of TEG.

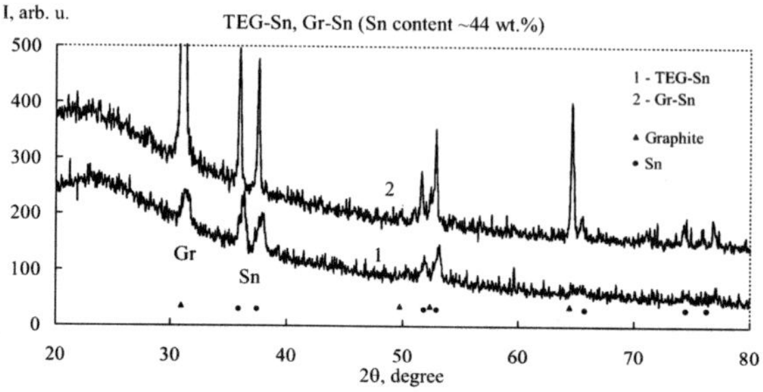

Figure 8. X-ray diagrams for TEG Graphite with Sn.

Figure 9 is a group of SEM images illustrating the character of metal distribution on the surface of graphite particles. It is seen that an essential dispersion of TEG particles upon chemical treatment is observed. The size of graphite particles after metal deposition is equal to 200-400 μm. The metal layer is sufficiently uniform and consists of the particles 40 nm in size. However, a lot of particles' agglomerations 80-140 nm in size is also observed. The nonuniformity of particles' distribution and their sizes in CMs on the base of disperse graphite is more substantial than in a case of CMs on the base of TEG (Fig. 10). But even in this case the size of agglomeration does not exceed 150-180 nm in spite of the high content of the deposited component. Thus, the method of liquid phase reductions of metal cations on TEG and dispersed graphite permits to produce graphite – Sn CMs of uniform phase composition. The metal in these CMs is distributed in a form of nanoscaled formations on the surface of the graphite matrix.

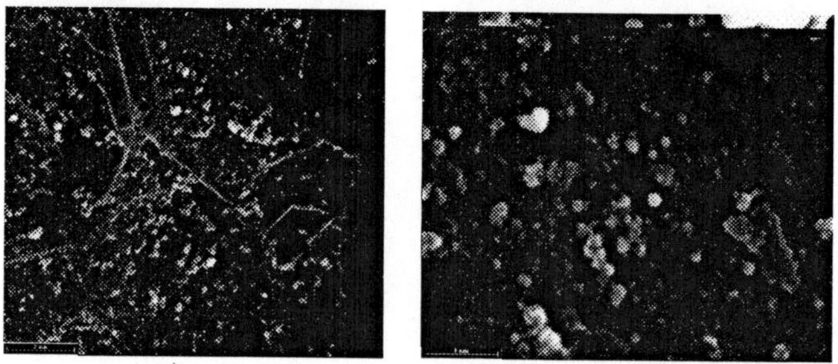

a) *b)*

Figure 9. Part of TEG particle with deposited Sn: a, b are the different regions of particle and different image magnification (Sn content ~ 44 w.t%).

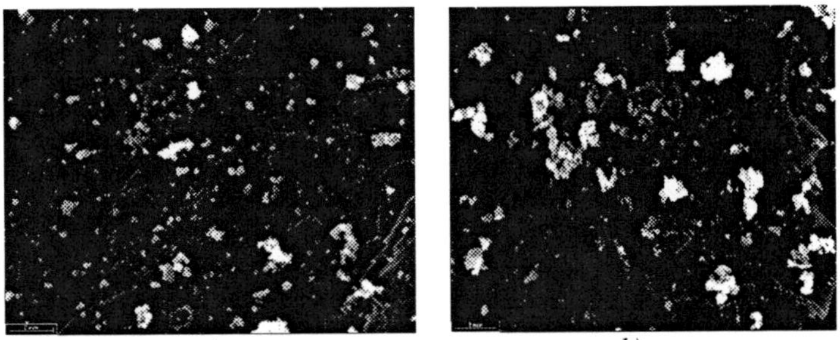

a) b)

Figure 10. Part of disperse graphite particle with deposited Sn: a, b, are the different regions of particle and different image magnification (Sn content ~ 44wt%).

4. CONCLUSSION

1. Graphite-tin and graphite–silicon nanocomposite materials have been obtained by the method of thermal vacuum deposition under continuous agitation, and also by the method of electroless plating of salts from water solutions with the subsequent liquid phase reduction.2. The size of modifier particles (Sn and Si) deposited on TEG was shown to be equal to 60-80nm and was uniformly distributed on TEG surface up to 60 wt.% of modifier; the migration of particles is essentially suppressed. This is in contrast to the case of disperse graphite being used as a supporter. 3. The application of TEG in the production of nanocomposite materials is preferable as compare to the other forms of graphite. Taking into account its morphology, the character of modifier particle distribution and their size, TEG-based CMs could be considered as prospective materials for the construction of graphite–anodes of lithium-ion batteries.

REFERENCES

1. Dahn J.R, Sleigh A.K., Shi H., Way B.M., Rweydans W., Reimers J.N., Zhong Q., Vonken U. In: *Lithium batteries, new materials and perspectives,* G. Pistoria, ed. New York and Amsterdam: Elsevier, 1993.
2. Endo M., Kim C., Nishimura T., Fujino T., Miyashita K. Recent development of carbon materials for Li ion batteries. Carbon, 2000; 38: 183-97.
3. Wang Z., Huang X., Xue R., Chen L. A new possible mechanism of lithium insertion and extraction in low-temperature pyrolitic carbon electrode. Carbon 1999; 37: 685-92.
4. Xiang H.Q., Fang S.B., Yjiang Y. A model for lithium in high–capacity carbons with large hysteresis. Carbon 1999; 37: 709-11.
5. Li N.C., Martin C.R. A high-rate, high–capacity, nanostructured Sn-based anode prepared using sol-gel template synthesis. J Electrochem. Soc.2001; 148(2): A164-70.

6. Mao O., Dunlap R.A., Danh J.R. Mechanically alloyed Sn-Fe(-C) powder as anode materials for lithium–ion batteries. I. Sn_2-Fe-C system. J Electrochem. Soc. 1999; 146(2): 405-513.
7. Crosnier O., Brousse T., Devaux X., Fragnaud P., Schleich D.M. New anode systems for lithium ion cells. J.Power Sourc.2001; 94: 169-74.
8. Moon H.S., Ji K.S., Ryoo K.C., Lee Y.K., Chung S., Park J.W. Characteristics of silicon-doped tin oxide thin films as anode materials for microbatteries. J. Cer. Proc. Res. 2002; 3: 34-7.
9. Chen W.X., Lee J.Y., Liu Z. The nanocomposites of carbon nanotube with Sb and $SnSb_{0.5}$ as Li –ion battery anodes. Carbon 2003; 41: 956-66.
10. Wang C.S., Wu G.T., Zhang X.B., Qi Z.F., Li W.Z. Lithium insertion in Carbon-Silicon composite materials produced by mechanical milling. J. Electrochem. Soc. 1998; 145: 2751-8.
11. Vovchenko L.L., Matzui L.Yu., Brusilovets A.I.S ynthesis of graphite-metal composite materials by salt.

ELECTROCHEMICAL PERFORMANCE OF Ni/Cu-METALLIZED & CARBON-COATED GRAPHITES FOR LITHIUM BATTERIES

Christopher S. Johnson[1*], Kevin Lauzze[1], Nick Kanakaris[1], Arthur Kahaian[1], Michael M. Thackeray[1], Khalil Amine[1], Giselle Sandí-Tapia[2], Stephen A. Hackney[3], and Robert O. Rigney[4]

[1] *Argonne National Laboratory, Chemical Engineering Division, Electrochemical Technology Program, 9700 S. Cass Ave., Argonne, IL, 60439 USA*
[2] *Argonne National Laboratory, Chemistry Division, 9700 S. Cass Ave., Argonne, IL, 60439 USA*
[3] *Department of Materials and Metallurgical Engineering, Michigan Technological University, Houghton, MI 49931 USA*
[4] *Fluid Air Inc., 2550 White Oak Circle, Aurora, IL 60504 USA*

Abstract

Synthetic and natural graphites were metallized with Cu or Ni using a fluidized bed coating and annealing process. With this method, crystalline nanometer-sized metal islands (~ 50 nm diameter) were deposited onto the surface of graphite. Post-metallization, the graphite materials were cycled in lithium coin cells to determine their electrochemical properties in a propylene carbonate (PC) solvent based electrolyte. Better cycling performance (316 mAh/g active graphite, 20th cycle; 26% irreversible capacity) was obtained with the Cu-coated graphites versus Ni-coated or uncoated types (247 mAh/g, 20th cycle; 62% irreversible capacity) in a 30% PC blend electrolyte at 50°C. The Cu nanosized deposits help to mediate charge transfer into the graphite particle via improved particle to particle contacts (electrical connectivity) and may contribute to improve kinetics of lithium ion insertion into the graphene layers by assisted Li-Cu inter-diffusion. Further, Cu may act as a de-solvation catalyst to remove PC-coordinated molecules from Li cations allowing Li insertion into grapheme layers and overall improved electrochemical performance. Carbon-coated graphite electrodes prepared by a vapor deposition process, show superior performance versus the uncoated graphite in a PC electrolyte system. Coated graphites are under investigation as possible new anodes for high-power lithium-ion battery applications.

---

* Corresponding author. E-mail: johnsoncs@cmt.anl.gov

I.V. Barsukov et al. (eds.), New Carbon Based Materials for Electrochemical Energy Storage Systems, 357–376.

Keywords

copper metallization, graphite, lithium batteries, metal coatings, fluidized bed, carbon

1. INTRODUCTION

Carbonaceous materials such as coke and graphite are the negative electrodes or anodes of choice in various commercial lithium-ion batteries. While intermetallic materials such as η'-Cu_6Sn_5 [1], InSb [2], and Cu_2Sb [3] have shown excellent promise as an alternative high volumetric capacity anode (e.g. InSb; ~ 1710 mAh/cm^3) in advanced lithium cells, they are still exploratory and remain as a long-term research objective [4]. Graphite electrodes are the state-of-the-art, and potential new developments to improve graphite and other carbonaceous materials in lithium batteries may have an immediate impact in the technology [5].

Recently there has been an interest to coat or modify the surface of graphite materials such that these electrodes may operate more effectively in advanced electrolyte systems containing propylene carbonate (PC) solvent [6]. PC, a cyclic di-ester, is a desirable solvent in these electrolytes because it has favorable low temperature properties such as increasing the solvation of the $LiPF_6$ salt as well as lowering the viscosity of the electrolyte. Improvement in these properties results in better electrolyte conductivity at lower temperatures. However, since Li cations become highly solvated in PC, it becomes difficult to strip the solvent coordination sphere during electrochemical reaction and as a consequence, the co-intercalation of PC between graphite layers during the first lithiation reaction occurs [7]. Recent studies have confirmed the production of propylene gas from a side-reaction breakdown of the intercalated PC molecule [8]. The formation of gas is detrimental to the local integrity of the graphite, subsequently leading to graphene layer exfoliation, and mechanical destruction and failure of the electrode [8, 9].

A control of the graphite electrode-electrolyte interface in a PC solvent electrolyte is therefore a key to obtain modest electrochemical performance in lithium cells [10]. Apart from electrolyte additives, an approach is to modify the SEI (solid electrolyte interphase) film to create a lithium-permeable but protective coating at the graphite edge surface to attempt to physically block co-intercalation of PC solvent, yet allow diffusion of Li through the coating to the graphite layers. Various approaches for coating graphite have been previously demonstrated using carbon [11-14], Cu [6, 15], Ni [16], Al [17], Sn [18], Ag [19-20], as well as controlled oxidation of the graphite surface [21].

This research is focused on ways to improve the electrochemical performance of graphite materials in lithium batteries. We have attempted to modify the graphite surface via metallization in order to minimize PC solvent co-intercalation. In a secondary effect, we may also expect to increase the conductivity of the graphite with metallization, which, in turn, could improve its electrochemical performance by lowering its impedance. In this paper, two laboratory scale processes for modifying the surface of synthetic or natural graphite are described. One new process was designed to produce thin metal deposits or, preferably, a coating by a type of fluidized bed process while the other was used to prepare a thin amorphous carbon coating on the graphite surface. These methods for coating electrode materials are general and many variations are possible. A few examples are highlighted in this paper. These metallized or coated graphite materials were tested as electrodes in lithium coin cells in order to determine their electrochemical properties in PC based electrolytes.

2. EXPERIMENTAL

2.1. Materials

Synthetic graphite powder (MAG-10) was used as received from Hitachi Chemical (Japan). Natural graphite (SL20; rounded edge type) was provided by Superior Graphite Co. (Chicago, IL). Other chemicals used were ACS reagent grade (99+% purity) and were purchased from Aldrich Chemical Company. Copper formate dihyrate and ($C_2H_2CuO_4$ •2 H_2O;) nickel formate dihydrate ($C_2H_2NiO_4$•2 H_2O), and formic acid were used in the fluid bed coating process. Propylene gas (99.99%) was supplied by AGA Gas Inc. Purified water (milli-Q; Barnsted Nanopore) was used to make an aqueous solution of the formate salts in 2% formic acid that were used as the spray solutions in the fluid bed reactor.

2.2. Characterization

Powder X-ray diffraction patterns of the samples were collected on a Siemens D5000 powder diffractometer with Cu K_α radiation between 10° and 80° 2θ at a scan rate of 0.6° 2θ/min. High-resolution TEM images of the 4.8% Cu-metallized MAG-10 graphite powder were collected on a JEOL-JEM 4000FX-1 transmission electron microscope under an accelerating voltage of 200 keV. Samples were prepared for the electron microscope by a standard procedure described elsewhere [22]. The level of metallization and dispersion was checked by analyzing the powders by energy dispersive analysis of X-rays (EDAX) and EDAX elemental mapping on a JEOL 6400 scanning electron microscope (SEM). Powder samples were spread onto a double-sided carbon conductive tape. The samples were then loaded in the SEM for EDAX analysis (Noran Voyager II System). In addition to EDAX, the coated samples were heated in air on a microbalance to 900°C in order to burn off the graphite. A computer-controlled simultaneous TG/DTA EXSTAR 6300 unit (Seiko Instruments, Inc.) was used for thermogravimetric analysis. The metal to graphite content was calculated from the weight loss and formation of the metal-oxide, CuO or NiO.

Electrodes were fabricated by intimately mixing 92 wt% of the appropriate graphite powder, and 8 wt% polyvinylidene difluoride (PVDF) negative electrode binder (Kynar, Elf-Atochem), in a 1-methyl-2-pyrrolidinone (NMP) solvent (Aldrich, 99+%). The mixed slurry was cast onto a copper foil current collector and dried at 75°C for 3-5 h. The electrode laminates were dried further at 70°C under vacuum overnight. Disk electrodes with an area of 1.6 cm^2 were punched from the laminates. The electrodes were evaluated in coin-type cells (size 2032, Hohsen Corporation) with lithium foil (FMC Corporation) as the counter electrode, a polypropylene separator (Celgard 3501 grade), and an electrolyte of 30% propylene carbonate (PC), 30% ethylene carbonate (EC), 40% ethyl-methyl carbonate (EMC) (by volume), with 1.2 *M* $LiPF_6$ (Tomiyama). Cells were constructed inside a dry room (< 2 % relative humidity). Comparative conditions of cycling was done at room temperature or 50°C at a constant current density of 0.2 mA/cm^2 between 1.5 and 0 V (Maccor Series 4000 cycler). Additional cycling experiments were done in a non-PC electrolyte solution consisting of 1 *M* $LiPF_6$ in a 1:1 EC:DEC (50:50; by volume) solvent mixture (Merck), and a concentrated PC electrolyte of 70% PC, 20% EC, 10% EMC (by volume) and 1.2 *M* $LiPF_6$ (Tomiyama) at room temperature or 50°C.

2.3. Fluid Bed Reactor – Cu and Ni Metallization

A benchtop fluid bed reactor (Model 2 Fluid Air fluid bed, Fluid Air Inc., Aurora, IL, U.S.A; www.fluidairinc.com) was used to spray coat the graphite particles with an aqueous solution of copper or nickel formate dihydrate. The coating was done at 40-50°C using heated, filtered room air. The weight percent aqueous solution (Ni or Cu formate salt) to graphite was typically 10-35% going in the fluid bed reactor. Final percentages of deposition were calculated as a weight percent after final firing of the samples or from calculations of weight loss from thermal gravimetric analysis (TGA) experiments used to burn away the graphite. In this paper we present data for a range of samples that had 3.0 to 11.7wt% Cu-metallized and another that was 4.6wt% Ni-metallized graphite. The as-deposited graphite was subsequently annealed in 4%H_2 (helium balance) gas flow at

325 - 400°C for 10-14 h in an alumina tray inside a Lindberg tube furnace under controlled gas flow. On some occasions, and as a control experiment, pure argon gas was employed during the annealing step. Resultant powders were used in electrochemical (coin cell cycling) studies.

2.4. Quartz Reactor - Carbon Coating

A quartz reactor was used for deposition of a carbon-coating onto the synthetic graphite MAG-10. In this experiment, the graphite sample was loaded on an alumina tray then fed into a quartz tube reactor. After purging the reactor with nitrogen gas for 3 hours, a propylene gas in a balance of nitrogen was premixed to a final percentage of 75% propylene in nitrogen then flowed into the reactor that was heated to 700°C and held for 3 h [23]. The EXSTAR 6300 TG/DTA was used to measure the carbon-coating content in the graphite sample [24]. Samples were heated at 10°C/min to 900°C in a flow of dry air (30 cc/min). Resulting weight percent of carbon on graphite was calculated as 22% +/- 0.5%.

3. RESULTS & DISCUSSION

3.1. Cu and Ni-metallized Graphite

The graphite particles were coated in a specially modified fluid bed reactor that provides two distinct zones of air-suspension. A partition or tube can be used to physically divide these zones and lessen particle-to-particle contact resulting in agglomeration. The inner zone incorporates a high velocity airflow that pneumatically accelerates particles past a spraying nozzle. Spray droplets impact upon, spread out over, and evaporate from the particles. The coated particles spout out and drop down to a bed of particles in the outer zone. The airflow velocity is much lower in the outer zone. The airflow is just enough to keep the particles moving back into the inner zone. This cyclic process can be continued until the desired level of coating is applied. Overall, this process, also known as Wurster coating, is a relatively simple, rapid and inexpensive method for powder coating, and has been used for some time in the pharmaceutical industry. In addition, as larger batch runs become necessary, the process can easily be adapted to proportionally larger-sized commercialized fluidized bed systems.

The final third step is a mild heat-treatment of the as-deposited graphite that is necessary to form the final metallized graphite. The graphite is heat-treated in 4%H_2 (helium balance) at fairly low temperatures of 325°C. Under these conditions, the copper and nickel formates form metallic

Cu or Ni directly during the heat-treatment step without the formation of unreduced side-products of CuO or NiO.

Figure 1 shows the XRD patterns of the Cu-metallized graphite before (Fig. 1a) and after heat-treatment (Fig. 1b and 1c). The many peaks seen in 1a. are those of the Cu-salt film present on the surface of the graphite prior to heat-treatment. Diffraction from Cu metal crystallites in these samples after heat-treatment were clearly observed in the patterns in 1b and 1c. The Cu diffraction lines intensities are proportional to the metal content present in the graphite. X-ray diffraction revealed the resultant particles were pure copper crystalline with a face-centered cubic (fcc) structure. The graphene layers and their *d*-spacing is unperturbed (less than 0.2% change) by the mild overnight heat-treatment process carried out in regeneration gas (4% H_2 in balance of helium) for metallized graphites shown in 1b and 1c.

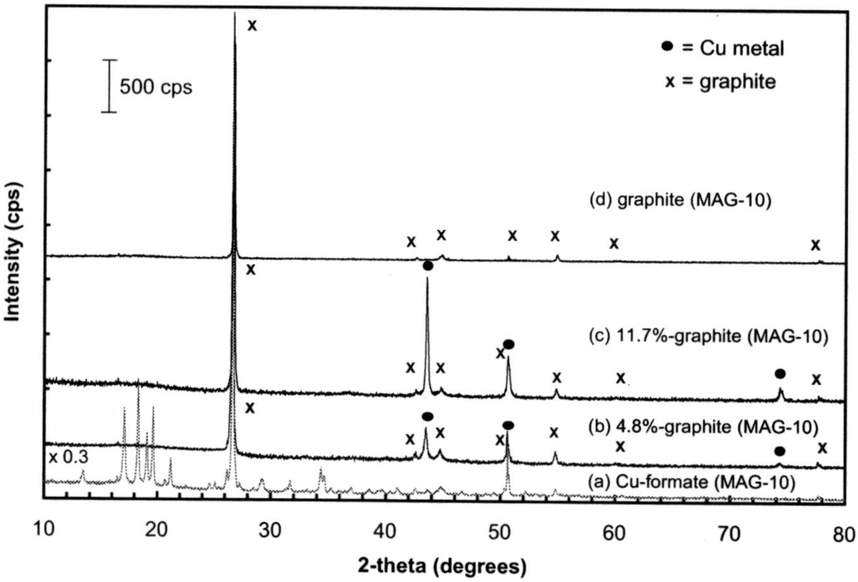

Figure 1. XRD of (a) Cu-formate coated graphite, (b) 4.8% Cu-graphite, (c) 11.7% Cu-graphite and (d) control (pristine) graphite.

Similar to copper, the nickel formate is also converted to Ni metal when heated in 4%H_2 in a balance of helium gas. This is demonstrated in the XRD pattern in Fig. 2a where a sample of nickel formate (in the absence of graphite) was heat-treated at 400°C for 10 h. One can clearly see that Ni metal is present. An interesting comparison may be made if the nickel formate sample is instead heated only in argon gas flow for 10 h at 400°C. In this case (Fig. 2b), both NiO and Ni are formed. Thus, it is important to heat-treat in a more reducing atmosphere to generate fully reduced metal.

The XRD for the sample of Ni-metallized on graphite (4.6wt%) is provided in Fig. 2c, however the strongest peaks of Ni metal are overlapping with those peak reflections from the MAG-10 graphite. But, the X-ray diffraction in Fig. 2a revealed the Ni particles were pure nickel crystalline with a face-centered cubic (fcc) structure.

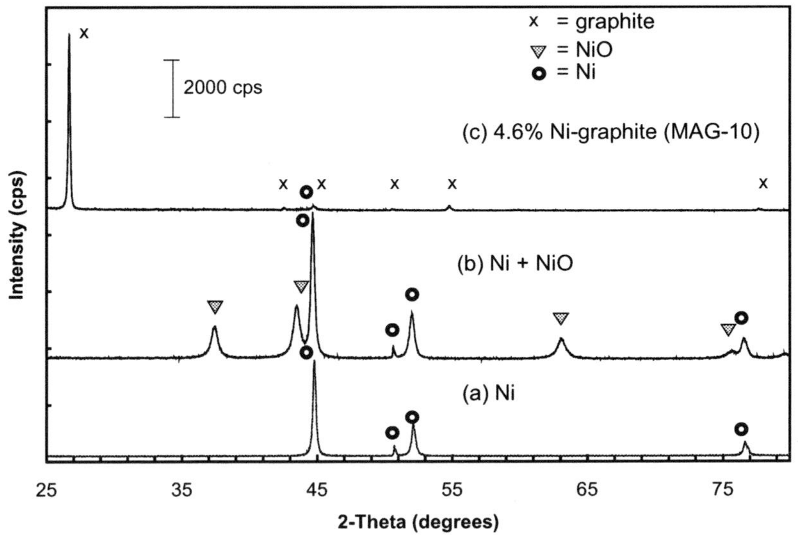

Figure 2. XRD of (a) metallic Ni from Ni-formate heated at 400 °C in 4%H_2, (b) metallic Ni plus NiO from Ni-formate heated at 400 °C in argon, and (c) 4.6% Ni-graphite.

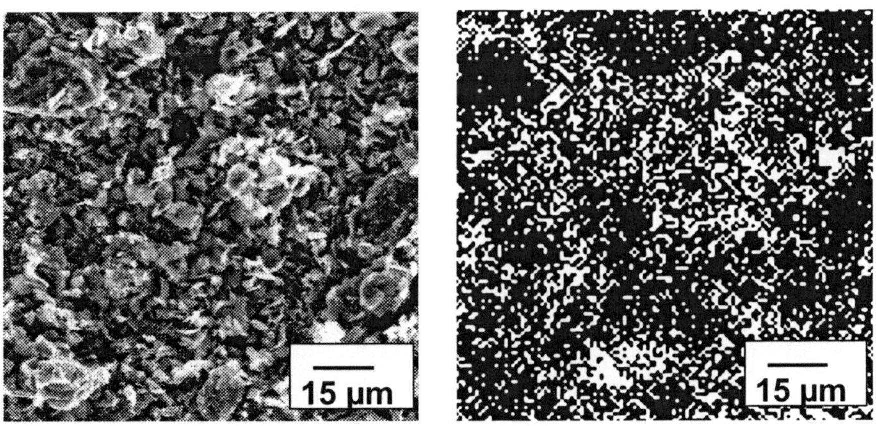

(a) (b)

Figure 3. (a) SEM micrograph and (b) EDAX Cu elemental mapping of 11.7 wt% Cu-graphite material (synthetic graphite MAG-10). In (b) the EDAX map was taken on the image in (a), and the lighter areas represent Cu regions.

The SEM micrograph and the corresponding EDAX elemental mapping for the 11.7 wt.% Cu-graphite (MAG-10) sample are shown respectively in Figures 3a and 3b. At the micron resolution of this micrograph, it is evident that the Cu is homogenously distributed on the surface of the graphite and the coverage is quite good. Thus, the fluid bed Wurster coating method was effective in laying down in even distribution of metal and may therefore provide a new alternative method for the coating of powdered battery materials.

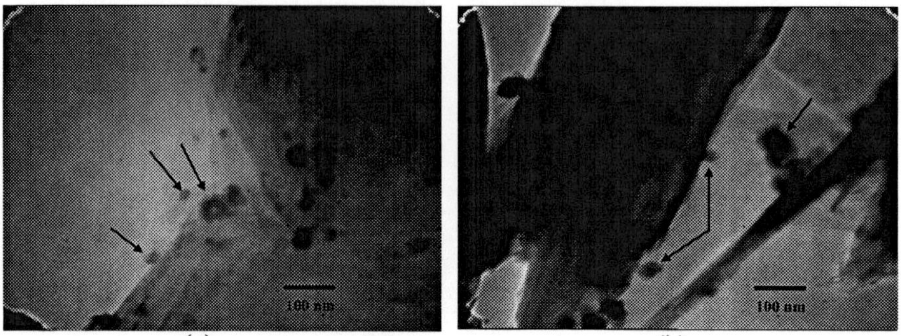

(a) (b)

Figure 4. HRTEM images (a) and (b) of 4.8 wt% Cu-graphite material (MAG-10); arrows indicate deposits of nano-Cu.

To gain a deeper understanding of the morphological nature of the Cu on graphite, the 4.8wt.% Cu-graphite (MAG-10) was examined by HRTEM analysis. In Figs. 4a and 4b, two TEM images are shown for the sample. In Figure 4a, 20-50 nm-sized crystallites of Cu are observed (marked by arrows) that are roughly spherical in shape. Figure 4b also shows another area of the sample, and in this case a slightly larger agglomeration of 100 nm in size appears along an edge of the graphite. Through an analysis of these figures and from careful electron diffraction imaging of the basal plane and edge plane of the graphite (not shown here) it appears that the nano-Cu does not preferentially migrate to the graphite edge plane (i.e. intercalate pathway) during the coating and heat-treatment step. Thus, a mechanism where there would be direct physical blockage of co-intercalation of solvent molecules during the first lithium insertion reaction may be limited in this case. The differences in electrochemical performance of the Cu-metallized graphite vs. un-metallized may include factors such as the extent and nature of the SEI or the kinetics of de-solvation that may be affected by the presence of the Cu.

Previous studies focusing on deposition of Cu [6, 15] or Ni [16] coatings and deposits on natural and synthetic graphics relied on an electroless plating method [25]. In this method many different chemical solutions are used in multiple steps to treat the graphite surface, such as chromic acid plus sulfuric acid, then tin chloride, and palladium chloride treating baths. Finally a copper electroless plating bath is used for the deposition. In our work, the fluid bed Wurster coating process was implemented and it appears more suitable for metallizing graphite versus that of the electroless method, which is not easy to scaleup and cumbersome and time-consuming, and also as compared to the metal-carbonyl process, which uses toxic gases such as tetracarbonyl nickel ($Ni(CO)_4$) during operations [26].

3.2. Carbon-Coated Graphite

Carbon-coating is an effective way to improve the performance of electrode materials for lithium batteries, particularly with graphites [11-14]. It is also known to aid in the surface conductivity for $LiFePO_4$ as a cathode material [27]. There are many ways to coat powders with carbon, but in this study, we have chosen to decompose a hydrocarbon vapor of propylene in a nitrogen carrier gas at a moderate temperature of 700 °C. Criteria for using this process include a material that is stable at this temperature and under a reducing environment.

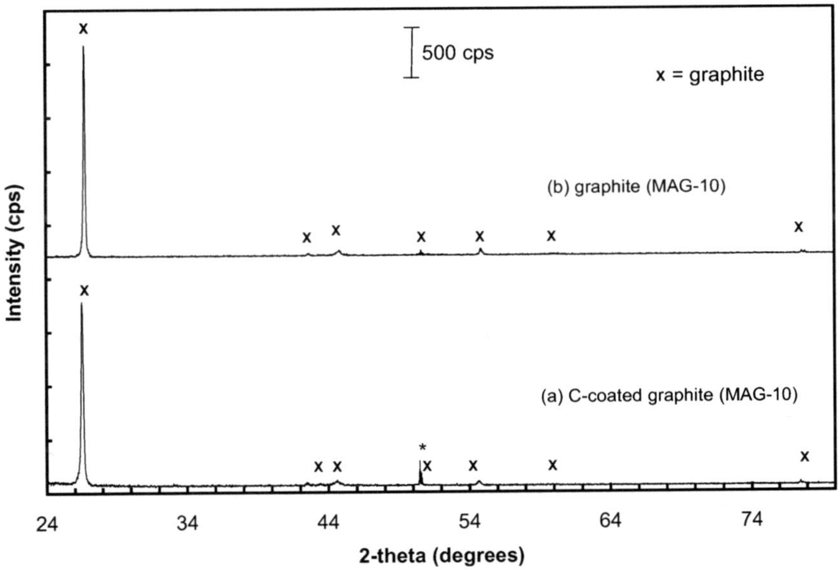

Figure 5. XRD of (a) 22 wt. % C-coated graphite (MAG-10) and (b) pure (pristine) MAG-10 graphite.

In our labs, we have successfully carbon-coated $Li_4Ti_5O_{12}$, $LiFePO_4$ and graphite powders with this method [28]. There was no change in the

MAG-10 XRD pattern before and after carbon-coating (22 wt%) in Fig. 5a-b which suggests that the carbon-coating is amorphous. All of these materials are quite thermally stable in a reductive gas environment which allows for propylene gas to be reactively cracked at the surface of these powders resulting in an amorphous carbon film. As suggested by Sandí et al. [23], the propylene gas after reaction tends to leaves a fairly homogenous and hydrogen-free film.

3.3. Electrochemical Processes

The MAG-10 was chosen as a candidate anode for high-power battery applications. It is flaky synthetic graphite that was provided by Hitachi. Electrochemical performance of this graphite appears satisfactory from high-power pulse (HPP) lab cell testing and modeling studies conducted in our laboratory [29]. As such, it has been studied as a "Generation -2" anode material within the Advanced Technology Program (ATD) at Argonne under the guidance and funding of the U.S. Department of Energy (DOE). However, eventual applications of this graphite directly in commercial hybrid vehicle batteries will necessitate further improvement in the material for operation at low temperatures possibly with a PC blended electrolyte. Further, the conductivity of the graphite needs to be increased to improve the rate performance at low temperature. It is towards these goals that many of the experiments done herein were conducted with MAG-10 graphite material to demonstrate improvements. However, future generations of advanced anode graphite materials, may employ using low cost natural graphite with film-forming additives or coated and/or metallized modifications of natural graphite without electrolyte additives.

An important physical property, which is distinctive for Cu as a metal is that it allows for Li diffusion [30], while Ni metal does not support Li solubility and is Li impervious. Further, the $Li_{0.22}Cu$ alloy is a known phase [31] at high and low temperatures, while no Li-Ni phases are known to exist [32]. Thus, for a lithium battery environment, if Cu coverage is predominate at the graphite edges, then the diffusion of Li through the metal would be allowed, unlike for Ni. For this reason, we concentrated our efforts in electrochemical testing of Cu-metallized graphite materials. Indeed, cycling tests conducted on Ni-metallized MAG-10 graphite did not demonstrate any improvements over the baseline uncoated samples, unlike that for Cu. Further, the Ni did not improve the performance of the graphite in a 70% PC electrolyte system.

The voltage curve collected at 50°C and in 30% PC electrolyte for the 11.7wt.% Cu-graphite (MAG-10) first full 2 cycles are shown in Figure 6 (top traces) together with the pristine (uncoated) MAG-10 graphite (bottom traces).

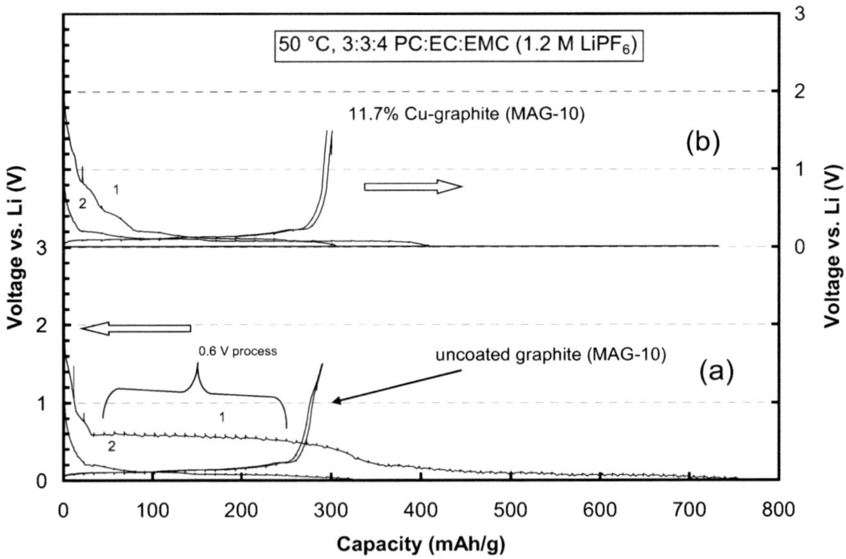

Figure 6. Voltage profiles of lithium cells with (a) uncoated MAG-10 graphite and (b) 11.7wt% Cu-graphite (MAG-10) electrodes in 30% PC blended electrolyte and at 50°C.

The vertical ticks in the voltage curves are due to current interruptions conducted every 20 minutes during the cycling to check the cell impedance. Note that the capacity values in Figure 6 have not been corrected for the inactive Cu in the graphite. There is a large first cycle irreversible capacity loss of 61.7% for the pristine sample, whereas the 11.7wt.% Cu-graphite shows only a 26.4% loss. Most of the irreversible capacity stems from the solvent co-intercalation plateau at about 0.6 V in the uncoated graphite [6, 10]. Note how the 11.7 wt.% Cu-graphite does not show as pronounced plateau. Inspection of the curve shows that approximately 30 mAh/g or ca. 10-12% total capacity originates from slight irreversible co-intercalation of solvent at a voltage range of 0.7 to 0.55 V in this profile. Thereafter the coulombic efficiency and cyclability of the 11.7wt.% Cu-graphite is excellent as shown in Fig. 7 out to 20 cycles (283 mAh/g (uncorrected for Cu weight)). This data was far better than that seen for the un-metallized (pristine) graphite sample in 30% PC blend electrolyte. From observation of the long plateau at ~0.6 V in Figure 6 and the cycling instability or higher fade in Figure 7, one may compare the two materials and see the improvement. This result implies that the performance is tied to a partial blockage of PC solvent molecule co-intercalation into the graphite on the first lithiation cycle in the Cu-metallized graphite electrode. As a control experiment, and as expected, the MAG-10 graphite itself (no coating) cycles normally in a non-PC electrolyte, such as LP-40 at 50°C (Figure 8).

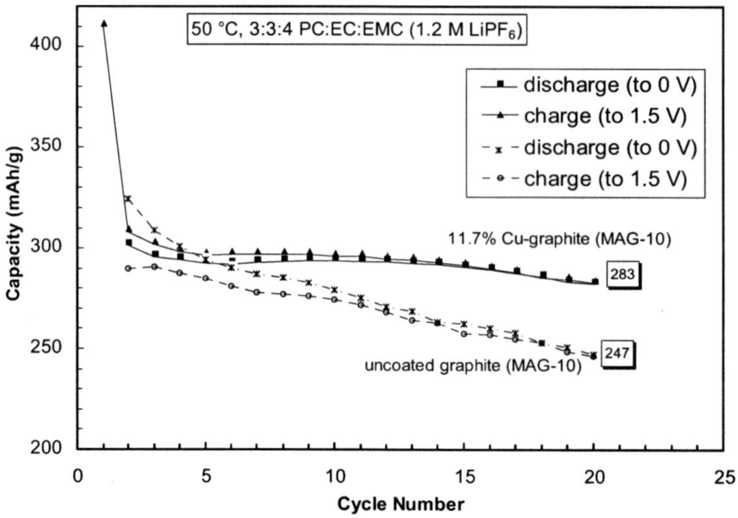

Figure 7. Capacity versus cycle number plots of the Li/graphite and Li/11.7%Cu-graphite cells at 50 °C in 30% PC blended electrolyte.

Similar results were found with additional tests on sister cells of 11.7 wt.% Cu-graphite in 30 % PC electrolyte. However, it was not possible for the 11.7 wt.% Cu-graphite material to cycle in the 70% PC blend electrolyte. Other lower wt.% contents of Cu or Ni-graphite in this high percentage PC also were unable to cycle. In general it appears to be quite difficult for graphite and modified graphite materials to cycle whatsoever in a 70% blend of PC (70% PC, 20% EC, 10% EMC and 1.2 *M* $LiPF_6$) as an electrolyte. More work is needed in optimizing the surface properties of natural graphites in order to cycle more reasonably in concentrated PC-based electrolytes.

One may expect the rate capability to be quite good in the Cu-metallized graphite sample, particularly since the particle-particle contact points and electrical conductivity may be improved. In Figures 8 and 10, the cycling stability and rate capability are given. This experiment was conducted in the non-PC based electrolyte. After 44 cycles the 11.7wt.% Cu-graphite delivers a specific capacity of 330 mAh/g of active graphite on discharge at a constant current C/5 rate at 50 °C in a lithium cell in LP-40 (1:1 EC:DEC, 1 *M* $LiPF_6$). After a constant current (0 V to 1.5 V) rate study up to 2C (2 mA/cm^2) during the next 8 cycles, the cell was returned to its initial rate of C/5 to see if the material's electrochemical performance was adversely affected. In Figure 8, the specific capacity of the cell returned back to a steady 327 mAh/g of active graphite. In Fig. 10, the % loss in capacity is plotted as a function of the C rate. At the 2C rate (2 mA/cm^2), the material delivered 60% of its C/5 rate baseline capacity. Further, the curve is somewhat linear implying that the material may support even higher rates such as 4C and still provide 20% capacity or about 60-70 mAh/g.

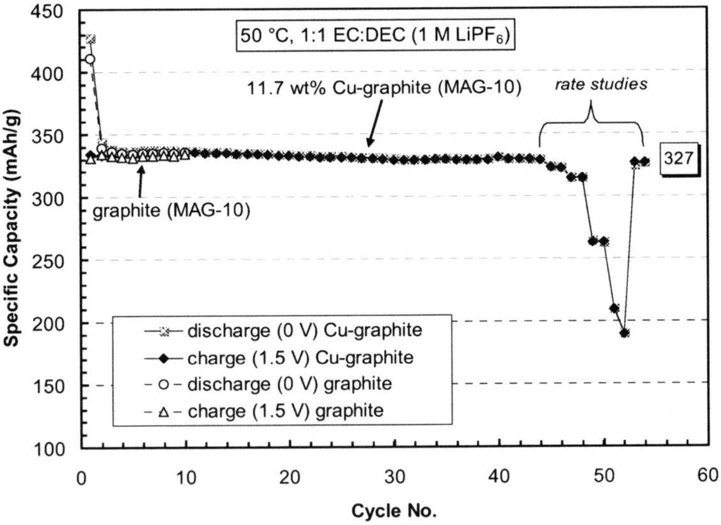

Figure 8. Specific capacity versus cycle number plots of the Li/graphite and Li/11.7% Cu-graphite cells at 50 °C in 1:1 EC:DEC (1 M $LiPF_6$; LP-40).

The current interruption technique was used to ascertain the impedance of the graphite materials. In Figure 9, a typical discharge curve (after first cycle break-in) is plotted for the 11.7%-Cu-graphite. Note the steps in the voltage profile indicative of graphite-staging reactions. The Fig. 9. inset is the expanded plot of a typical current interrupt (30 seconds) for the electrode together with one from the uncoated non-metallized (control) graphite cell. The cell's depth-of-discharge in this analyzed region was about 50-60% DOD and a cell voltage of ~ 0.08 V vs. Li. Note that the greater the voltage relaxation for a fixed cell electrode area and current, the larger the impedance of the material, and the slower the electrode kinetics. When the applied current is off, the cell voltage of the control electrode relaxes to a slightly greater value than the Cu-metallized graphite electrode. The resultant ASI (at 30 seconds) of the Cu-metallized electrode is about 30 $\Omega \cdot cm^2$ or about 60% lower.

The cycling improvement for the Cu-metallized graphite over the pristine graphite was also observed by K. Guo et al. [15] in their study of electroless Cu deposited on graphite cycled in a lithium cell with a 20% PC blend electrolyte. Also, they recorded a rate capability improvement in their Cu graphite material as well. At a current density of 1.4 mA/cm^2, the cell achieved about 60% (~ 200 mAh/g) of the charge capacity measured at 0.14 mA/cm^2, compared to about 30% (~ 100 mAh/g) for the non-treated pristine natural graphite cell [15].

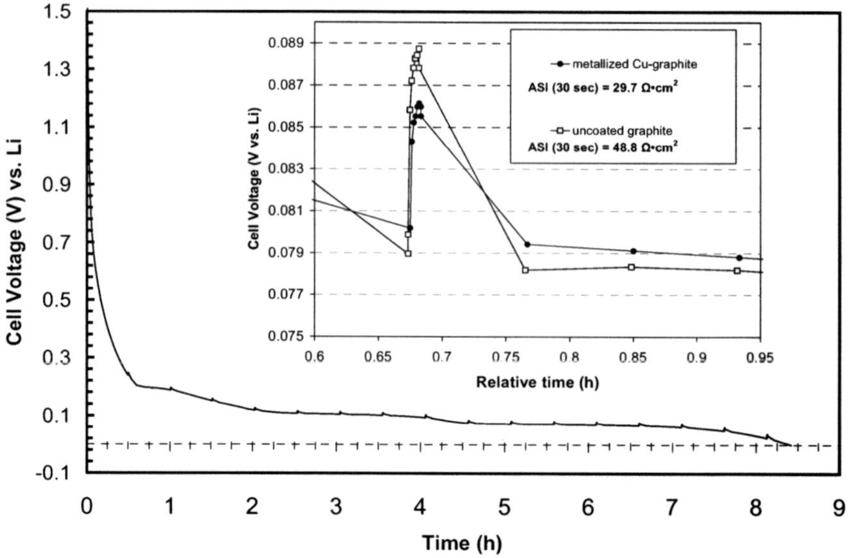

Figure 9. Typical discharge (lithiation) voltage profile of the Li/11.7%Cu-graphite cell at 50 °C in 1:1 EC:DEC (1 M $LiPF_6$; LP-40). Inset is an expanded region showing the voltage relaxation change during current interruption at about 0.08 V of the Li/11.7%Cu-graphite and Li/graphite control cell.

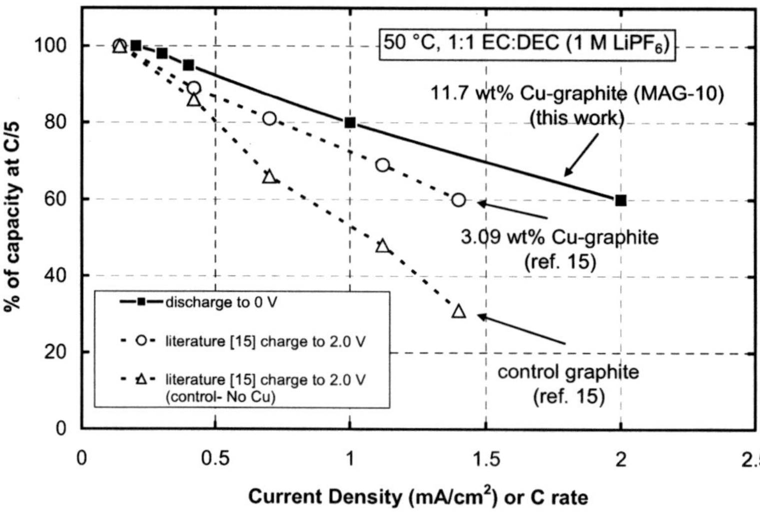

Figure 10. Rate performance plot of % capacity versus current density for Li/11.7%Cu-graphite cell at 50 °C in 1:1 EC:DEC (1 M $LiPF_6$; LP-40 and literature values ofLi/ 3.09% Cu (electroless) on graphite cell and Li/graphite cell (no Cu) at room-temperature in 20% PC blended electrolyte [15].

For comparative purposes, their data was interpolated and also plotted in Fig. 10.

The room-temperature cycling data for lithium cells containing carbon-coated MAG-10 graphite and the Superior Graphite (SGC) natural graphite with rounded edges are shown in Figure 11. For comparison the standard uncoated MAG-10 graphite cell is also plotted in the Fig. 11. The electrochemical performance of the carbon-coated graphite was superior to the uncoated but lower than that of the SGC sample. This result is consistent with other studies that have documented that carbon coating improves the performance of natural or synthetic graphite in PC containing electrolytes [13]. Interestingly, despite no carbon coating, the SGC sample performed well due to its improved morphology. The sample was mechanically milled to a spherical morphology with rounded edges in order to reduce the number of exposed edge planes. The irreversible capacity on the first cycle is only 14%, then 42%, and finally 75% for the SGC sample, carbon-coated MAG-10, and uncoated MAG-10 respectively. Thus, treatment of the pristine edge planes by milling or carbon-coating improves the first cycle performance and suggests that less solvent co-intercalation occurred during the initial lithium insertion into the graphene layers.

To cycle natural graphite in a PC electrolyte, another approach is to dissolve small quantities of film-forming additives into the electrolyte prior to cycling such as acrylic acid nitrile (AAN) [33], ethylene sulfite (ES) [34], and vinylene or vinylethylene carbonate (VC, VEC) [35-36]. These additives have been shown to reductively breakdown on the first cycle at the graphite surface thereby pre-passivating it prior to electrolyte-induced SEI film formation. This in turn can control the extent of solvent co-intercalation, gas formation, and graphite exfoliation. The carbon-coating performed in this study is another method of controlling the interface of the graphite electrode material in a beneficial way. No electrolytes additives were employed in this approach.

Interestingly, the carbon-coating which is quite high in this work at 22 wt% also provided capacity to the cell since the capacity on the 20th cycle was 320 mAh/g compared to a theoretical of 290 mAh/g (graphite; 290 mAh/g = 372 mAh/g – 82 mAh/g or 22%) if it were, instead "inactive" dead-weight. Future studies will concentrate on optimizing the amount of carbon coating while still maintaining peak electrochemical performance.

The combination of rounded edge graphite *and* metallization results in a much improved electrode material as presented in Figure 12. The electrochemical cycling data in capacity versus cycle number are presented for the Superior Graphite Co. natural graphite rounded–edge sample in the 30% PC solution at room-temperature and at 50°C under constant current conditions between 1.5 V (de-lithiation-charge) and 0 V (lithiation-discharge).

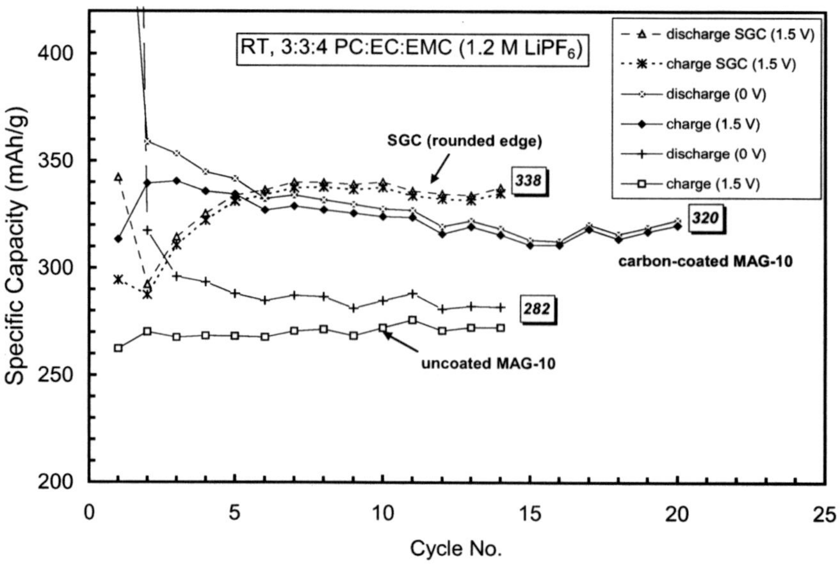

Figure 11. Specific capacity versus cycle number plots of Superior Graphite Co. natural graphite (rounded edge type), uncoated (pristine) MAG-10 graphite, and carbon-coated MAG-10 graphite in respective Li cells cycled at room temperature in 30% PC blended electrolyte.

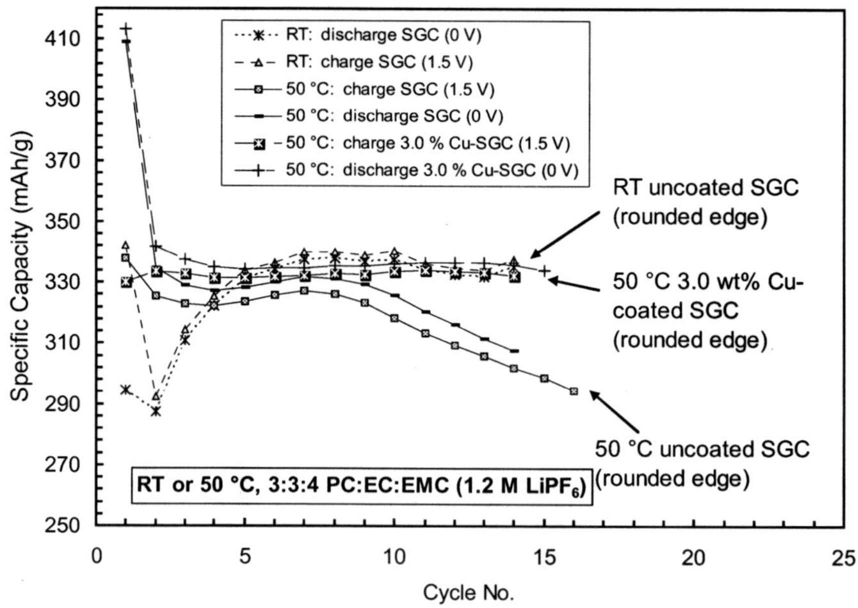

Figure 12. Specific capacity versus cycle number plots of non-metallized Superior Graphite Co. natural graphite (rounded edge type) at room temperature and 50 °C, and a 3.0 wt.% Cu-metallized Superior Graphite Co. natural graphite (rounded edge type) in Li cells cycled at 50 °C in 30% PC blended electrolyte.

Since the rounded edge morphology is designed to lessen the extent of exposed edge planes that are reactive and susceptible to exfoliation from co-intercalation of PC on the first lithiation cycle, an improved performance over the MAG-10 standard (flaky edge) was noted in Figure 11. However, it was further observed that the RT data was somewhat better than the 50°C cycling data that showed capacity degradation after 8 or 9 cycles (Figure 12). After Cu-metallization (3.0 wt%) using the fluidized bed (Wurster coating), a clear improvement in cycling was seen at 50°C where the results were similar to RT cycling and the performance was stable in 30% PC blended electrolyte. We attribute the improvement to decreased PC co-intercalation in the graphite electrode. The Cu may be acting as a de-solvation catalyst for stripping the PC coordination sphere from Li cations prior to electrochemical insertion of lithium and therefore operates more effectively at a higher temperature.

4. CONCLUSIONS

Some new results on metallization of graphite materials using a fluidized bed method (Wurster coating) of coating were presented. Cu and Ni-metallization occurs homogenously over the surface of the graphite from the formate salts. The coating method disperses the metal formate salt over the graphite surfaces in an even manner. After a 400°C low-temperature heat-treatment in flowing dilute hydrogen (4 vol.%) in helium gas, the Ni and Cu metals are directly formed as spherical 10-100 nm nanodeposits. Extensive electrochemical cycling tests were performed in lithium cells on the 11.7 wt% Cu-graphite samples in a 30% PC blend electrolyte in order to check the resistance to solvent co-intercalation and the improvement in rate capability.

At 50°C, a marked improvement was seen over the baseline cell, both in terms of more stable cycling, a higher rate capability, and less first cycle irreversible capacity loss. Reasons for the improvement in performance appear to be related to a lowering of the electrode material impedance and a smaller first cycle irreversible capacity by suppression of PC solvent co-intercalation due to a de-solvation catalyzed by Cu metal.

Also presented were data on carbon-coating of graphite powder using a propylene gas thermal decomposition processes. High weight percent amorphous carbon-coatings are possible with this method, and the process appears uniquely suited to materials that are reductively stable to 700°C. The coated materials work better in the 30% PC electrolyte solutions, thus showing better resistance to solvent co-intercalation problems versus uncoated types.

ACKNOWLEDGEMENTS

Support from the Office of Basic Energy Sciences, Division of Chemical Sciences of the U.S. Department of Energy (DOE) for the HRTEM studies and from the Office of FreedomCAR and Vehicle Technologies (also of DOE) for the surface treatment of the graphite electrodes and the electrochemical studies under Contract No. W31-109-Eng-38, is gratefully acknowledged. We would like to especially thank Dr. Igor V. Barsukov, and Dr. Joseph E. Doninger from Superior Graphite Co. for supplying the rounded-edge natural graphite powders.

REFERENCES

1. Kepler K. D., Vaughey J. T., Thackeray M. M., $Li_xCu_6Sn_5$ ($0 < x < 13$): An intermetallic insertion electrode for rechargeable lithium batteries, Electrochem. and Solid State Lett., (1999) **2** (7), 307-309.
2. Vaughey J. T., O'Hara J., Thackeray M. M., Intermetallic insertion electrodes with a zinc blende-type structure for Li Batteries: A study of Li_xInSb ($0 <= x <= 3$), Electrochem. and Solid State Lett., (2000) **3** (1), 13-16.
3. Fransson L. M. L, Vaughey J. T., Benedek R., Edström K., Thomas J. O., Thackeray M. M., Phase transitions in lithiated Cu_2Sb anodes for lithium batteries: an in situ X-ray diffraction study, Electrochem. Commun., (2001) **3** (7), 317-323.
4. Vaughey J.T, Johnson C. S, Kropf A. J, Benedek R, Thackeray M. M, Tostmann H., Sarakonsri T., Hackney S., Fransson L., Edström K., Thomas J.O., Structural and mechanistic features of intermetallic materials for lithium batteries, J. Power Sources, (2001) 97-8, 194-197.
5. Wu, Y. P., Rahm, E., and Holze, R., Carbon anode materials for lithium-ion batteries, J. Power Sources, (2003) **114**, 228-236.
6. Lu, W., Donepudi, V. S., Prakash, J., Liu, J., and Amine, K., Electrochemical and thermal behavior of copper coated type MAG-20 natural graphite, Electrochimica Acta (2002) **47**, 1601-1606.
7. Winter M., Wrodnigg G. H., Besenhard J. O., Biberacher W., Novák P., J. Electrochem. Soc., **147**, 2427 (2000).
8. Santner H. J., Wagner M. R., Fauler G., Raimann P., Veit C., Möller K. C., Besenhard J. O., Winter M., Taipei Power Forum and Exhibition (TPF2003), December 1-3, 2003, Taipei (Taiwan), Proceedings Volume.
9. Chung G. C., Jun S. H., Lee K. Y., Kim M. H., Effect of surface structure on the irreversible capacity of various graphitic carbon electrodes, J. Electrochem. Soc., (1999) **146**, (5), 1664-1671.
10. Abe, T., Kawabata, N., Mizutani, Y., Inaba, M., and Ogumi, Z., Correlation between cointercalation of solvents and electrochemical intercalation of lithium into graphite in propylene carbonate solution, J. Electrochem. Soc. (2003) **150** (3), A257-A261.
11. Yoshio, M., Wang, H., Fukuda, Abe, T., and Ogumi, Z., Soft carbon-coated hard carbon beads as a lithium-ion battery anode material, Chemistry Letters (2003) Vol. 32, No. 12, 1130-1131.
12. Yoon, S., Kim, H., and Oh S. M., Surface modification of graphite by coke coating for reduction of initial irreversible capacity in lithium secondary batteries, J. Power Sources (2001), **94**, 68-73.

13. Yoshio, M., Wang, H., Fukuka, K., Hara, Y., and Adachi, Y., Effect of carbon coating on electrochemical performance of treated natural graphite as lithium-ion battery anode material, J. of Electrochem. Soc. (2000) **147** (4) 1245-1250.
14. Lee, H-Y., Baek, J-K., Jang S-W., Lee, S-M., Hong, S-T., Lee, K-Y., and Kim, M-H., Characteristics of carbon-coated graphite prepared from mixture of graphite and polyvinylchloride as anode materials for lithium ion batteries, J. Power Sources (2001), **101**, 206-212.
15. Guo, K., Pan, Q., Wang, L. and Fang, S., Nano-scale copper coated graphite as anode material for lithium-ion batteries, J. Applied Electrochemistry (2002) **32**: 679-685.
16. Yu, P., Ritter, J. A., White, R. E., and Popov, B. N., Ni-composite microencapsulated graphite as the negative electrode in lithium-ion batteries, J. of Electrochem. Soc. (2000) **147** (4), 1280-1285.
17. King, S-S., Kadoma, Y., Ikuta, H., Uchimoto, Y., and Wakihara, M., Electrochemical performance of natural graphite by surface modification using aluminum, Electrochemical and Solid-State Letters, (2001) **4** (8), A109-A112.
18. Veeraraghavan, B., Durairajan, A., Haran, B., Popov, B., and Guidotti, R., Study of Sn-coated graphite as anode material for secondary lithium-ion batteries, J. Electrochem. Society, (2002) **149**, (6), A675-A681.
19. Holze, R. and Wu, Y. P., Novel composite anode materials for lithium ion batteries with low sensitivity towards humidity, J. Solid State Electrochem. (2003) **8**: 66-72.
20. Nishimura, K., Honbo, H., Takeuchi, S., Horiba, T., Oda, M., Koseki, M., Muranaka, Y., Kozono, and Miyadera, H., Design and performance of 10 Wh rechargeable lithium batteries, J. Power Sources (1997) **68**, 436-439.
21. Wu, Y. P. and Holze, R., Anode materials for lithium ion batteries obtained by mild and uniformly controlled oxidation of natural graphite, J. Solid State Electrochem. (2003) **8**: 73-78.
22. Shao-Horn Y., Hackney S. A., Cornilsen B. C., Structural characterization of heat-treated electrolytic manganese dioxide and topotactic transformation of discharge products in the Li-MnO_2 cells, J. Electrochem. Society, (1997) **144**, 3147-3153.
23. Sandí, G., Joachin, H., Lu, W., Prakash, J., and Tassara, G., Comparison of the electrochemical performance of carbon produced from sepiolite with difference surface characteristics, J. of New Materials for Electrochemical Systems, (2003) **6**, 75-80.
24. Wang, H., Fukuda, K., Yoshio, M., Abe, T., and Ogumi, Z., Measurement of carbon amount in carbon-coated graphite by thermal analysis, Chemistry Letters (2002) 238-239.
25. Caturla F., Molina F., Molina-Sabio M., Rodriguez-Reinoso F., Esteban A., Electroless plating of graphite with copper and nickel, J. Electrochem. Soc., (1995) **142** (12), 4084-4090.
26. Pacchioni, G., Rosch N., Carbonylated Nickel Clusters – From Molecules to Metals, Acc. of Chem. Res., (1995) **28** (9): 390-397.
27. Ravet N., Chouinard Y., Magnan J. F., Besner S., Gauthier M., Armand M., Electroactivity of natural and synthetic triphylite, J Power Sources, (2001) **97-8**, 503-507.
28. Johnson C. S. and Sandí, G, unpublished results, (2001).
29. Nelson P., Bloom I., Amine K., Henriksen G., J. of Power Sources, (2002) **110**, 437-444.
30. Suzuki J., Yoshida M., Nakahara C., Sekine K., Kikuchi M., Takamura T., Li mass transfer through a metallic copper film on a carbon fiber during the electrochemical insertion/extraction reaction, Electrochem. and Solid State Lett., (2001), **4** (1), A1-A4.
31. Klemm W., Volavsek, B. Z. Anorg. Chem. (1958) **296**, 184-187, or Bull. Alloy Phase Diagrams, (1986) **7** (2), or Binary Alloy Phase Diagrams (1990), Vol. 2, 2nd Ed., ASM International, Materials Park, Ohio, Editor-in-chief, Massalski T. B., pg. 1430.
32. Binary Alloy Phase Diagrams (1990), Vol. 3, 2nd Ed., ASM International, Materials Park, Ohio, Editor-in-chief, Massalski T. B., pg. 2450.

33. Santner H. J., Möller K. C., Ivanco J., Ramsey M. G., Netzer F. P., Yamaguchi S., Besenhard J. O., Winter M., Acrylic acid nitrile, a film-forming electrolyte component for lithium-ion batteries, which belongs to the family of additives containing vinyl groups, J. Power Sources, (2003) **119**, 368-372.
34. Wrodnigg G. H., Besenhard J. O., Winter M., Ethylene sulfite as electrolyte additive for lithium-ion cells with graphitic anodes, J. Electrochem. Soc. (1999), **146** (2), 470-472.
35. Aurbach D., Gamolsky K. Markovsky B., Gofer Y., Schmidt M., Heider U., On the use of vinylene carbonate (VC) electrolyte solutions for Li-ion as an additive to batteries, Electrochim. Acta, **47** (9), 1423-1439.
36. Vollmer J. M., Curtiss L. A., Vissers D. R., Amine K., Reduction mechanisms of ethylene, propylene, and vinylethylene carbonates - A quantum chemical study, J Electrochem. Soc., (2004) **151** (1), A178-A183.

CHAPTER 5:

NEW NANO- THROUGH MACRO-CARBONS FOR ENERGY SYSTEMS: SYNTHESIS, MODELING, CHARACTERIZATION

Chapter 5: Subject Overview

A collection of six papers in this chapter addresses subjects of synthesis, characterization and modeling of new promising carbon materials, which have not yet found application in the market of power sources, but may do so in the foreseeable future.

Thus, in the first paper on stabilization mechanisms discovered for reaction of synthesis of graphite nitrate, M. Savoskin *et al.* uncovers a number of valuable, unknown phenomena related to production of expandable graphite (GIC), an intermediate product in manufacturing of thermally exfoliated graphite. The later carbon type is currently being used in many battery systems as one of the most efficient conductive additives for electrode matrixes. Authors study intercalation of graphite via reaction of flake with nitric acid, followed by additions of a modifier. An attempt is made in the paper to group various (mostly organic) modifiers into classes. Extensive experimental data is presented in order to support the suggested mechanism of formation of stable GICs.

The paper by Kovalenko *et al.* introduces a method of cyclic voltammetry as a way to estimate stability of various carbon materials towards electrochemical oxidation. This reaction is of particular significance in rechargeable batteries with aqueous electrolytes. Authors discuss results of the electrochemical performance of expanded graphite, whose surface/structure has been modified by 5wt% B_2O_3. Reportedly, this graphite outperforms conventional graphite in terms of featuring higher resistance towards electrochemical corrosion.

Dimovski *et al.* addresses a very interesting concept of synthesis of various forms of carbon materials through reaction of chlorine treatment of iron carbides, typically found in the kish graphite ore. Currently, kish is a worthless compound, found in tailings of metallurgical plants. Recovering graphite from it could be a very beneficial idea. The authors were able to synthesize amorphous carbon structures at temperatures of chlorine below 1000°C, while at higher temperatures some well-graphitized synthetic graphite has been recovered. If commercialized, the technology may have revolutionary implications, also for the needs of energy market.

The contribution by Rouzaud *et al.* teaches to apply a modified version of high resolution Transmission Electron Microscopy (TEM) as an efficient technique of quantitative investigation of the mechanism of irreversible capacity loss in various carbon candidates for application in lithium-ion batteries. The authors introduce the "Corridor model", which is interesting and is likely to stimulate active discussion within the lithium-ion battery community. Besides carbon fibers coated with polycarbon (a candidate anode material for lithium-ion technology), authors study carbon aerogels, a known material for supercapacitor application. Besides the capability to form an efficient double electric layer in these aerogels, authors report observing some hydrogen storage capability with this type of carbon materials. Reportedly, about 1% of hydrogen can be stored and released by the micropores of an aerogel synthesized at 2600°C.

The fifth paper in this chapter is by S. Kochetova and N. Tumanova of the National Academy of Sciences of Ukraine. It addresses subjects of electrolysis of molten carbamides. Authors report on interesting results of their study of the mechanisms of reactions occurring in the carbamide and carbamide-chloride melts using techniques of cyclic voltammetry in combination with gas chromatography and IR spectroscopy.

In the last paper of this chapter, the authors M.Savoskin, M.Mochalin *et al.* of L.M. Livinenko Institute of Physical Organic and Coal Chemistry in Donetsk, Ukraine introduce simple methods of "one –step" synthesis of carbon-carbon composites, as well as relatively new to the world, carbon structures called "nanoscrolls". The authors use graphite intercalation compounds as precursors. Editors like to specifically draw the reader's attention to the open-ended carbon nanostructures shown by Figure 4 in the subject paper. While being unable to qualify the yield of these unusual nano-compounds, authors suggest the open-ended materials could be used for storage of various foreign molecules, one of them being hydrogen.

Editors hope that readers will find this chapter interesting and educating.

STABILIZATION OF GRAPHITE NITRATE VIA CO-INTERCALATION OF ORGANIC COMPOUNDS

M.V. Savoskin*, A.P. Yaroshenko, R.D. Mysyk, G.E. Whyman

L.M.Litvinenko Institute of Physical Organic and Coal Chemistry, National Academy of Sciences of Ukraine, 70 R.Luxemburg St., 83114 Donetsk, Ukraine

Abstract

New thermally stable graphite intercalation compounds were obtained via additional intercalation of a series of organic compounds into graphite nitrate. An explanation for stabilization has been proposed that involves hydrogen bonding between the protonated modifying organic compound and neutral surrounding molecules.

Keywords

Graphite nitrate stabilization; co-intercalation; exfoliation; proton affinity.

1. INTRODUCTION

The structure of acceptor-type graphite intercalation compounds (GICs) is very specific. Their quasi-two-dimensional layers of intercalant are quite labile, being not connected to a rigid graphite matrix by covalent bonds. These negatively charged layers alternate with positively charged graphene layers of the carbon matrix. The mentioned features of the GICs structure lead to their instability. For instance, even at room temperature, the α crystal phase of graphite nitrate of a stoichiometry $C_{24n}^{+}\cdot NO_3^{-}\ 3HNO_3$ with an interlayer distance of 7.80 Å turns into β phase of the same n-th stage that is characterized by a formula $C_{24n}^{+}\cdot NO_3^{-}\ 2HNO_3$ and interlayer distance of 6.55 Å. Moisture can lead to the substitution of nitric acid by water [1, 2].

* Corresponding author. E-mail: m-savoskin@yandex.ru

I.V. Barsukov et al. (eds.), New Carbon Based Materials for Electrochemical Energy Storage Systems, 381–387.

As we have recently shown [3], the stability of GIC with Brønsted acids is governed by ionization potential of the intercalated anion. Solvation of the anion substantially increases ionization potential and hence stabilizes GIC. Such a possibility was predicted by Inagaki [4] who proposed to use organic molecules as additional cointercalants for the purpose of stabilization. Afterwards, we successfully used glacial acetic acid and water to synthesize stable products [5].

The aim of the present work is to elucidate the mechanism for the stabilization of graphite nitrate via cointercalation of different kinds of organic compounds.

2. EXPERIMENTAL

Modified graphite nitrate was synthesized in a thermostatic reactor at 25°C. Nitric acid with a density of 1.502g/cm^3 was added to natural flake graphite GSM-1 (Zavalie Graphite Works, Kirovograd region, Ukraine) sample of 10g. The mixture was stirred for 10 minutes. Then an organic compound (modifier) was added and the mixture was stirred again for 10 minutes. The consumptions of nitric acid and the modifier were 0.4 and 4 cm^3 per 1g of graphite, respectively.

Further synthesis followed two ways. According to variant I, a solid phase was separated by filtration and divided into two parts. The first one was dried at 105°C for two hours (product **A**), while the second one was left at 20°C until a sample mass became constant (product **B**).

Following variant II, after treating by the modifier, the solid phase was separated by filtration and additionally washed off with 0.2 liters of water under vacuum of the water-jet pump. Further, as in variant I, one part of the product was dried at 105°C for two hours (product **C**), while the other was dried at 20°C until a sample mass became constant (product **D**). The products were investigated by the powder X-ray phase analysis (PXRD) using nickel-filtered CoK$_\alpha$ radiation.

The new compounds obtained are black-gray powders resembling initial graphite. Like other GICs, they possess a unique property of increasing their volume by a factor of tens or hundreds under heating. Usually, the so-called expansion coefficient serves as a measure of the mentioned property, $K_V=V/m$, where V is the volume in cm^3, and m is the initial mass of the sample before heating. Note that K_V is lower than the specific volume of exfoliated graphite due to the mass loss resulting from the escape of GICs-decomposition products. Values of K_V were determined in accordance with the following method. A stainless cuvette of 150cm^3 was placed into a muffle furnace preheated up to 900^{0}C. A product sample with the mass of 0.2-0.3g was inserted into the heated cuvette and kept in the furnace for 60s. A deviation between parallel measurements of K_V was less than 5%. To estimate the thermal stability of the compounds obtained, the stability index was introduced as $K_{st} = K_V^{105}/K_V^{20}$, where K_V^{105} and K_V^{20} are the expansion coefficients for the samples with the same modifier dried at 105 ^{0}C and 20 ^{0}C, respectively. Obviously, K_V^{105} is less than K_V^{20}, since a product dried at a higher temperature has to contain a smaller amount of the intercalant. Thus, the stability index K_{st} has to be less than unity, its value being closer to 1 for more stable compounds.

Listed in the Table are the expansion coefficients and stability indices for graphite nitrates cointercalated with different organic compounds, as well as with nitric acid and water. The modified graphite nitrates produced may be divided into three groups. Compounds of the first group (lines 1-8 of the Table) become stable just after modifier intercalation, and, in fact, their stability does not increase substantially after subsequent water treatment. Moreover, the stability of some compounds (lines 5, 8) decreases after the water treatment. Compounds in lines 9-21 belong to the 2-nd group. They remain unstable after modifier cointercalation, but after additional water treatment their stability rises sharply. At last, compounds of the 3-rd group in lines 22-23 remain unstable under any condition of the synthesis, as well as initial graphite nitrate.

It is interesting that ethers give GICs of the 1-st group, while esters behave in different ways. Ethyl acetate and ethyl formate lead to GICs of the 1-st group, while esters 14-17 produce those of the 2-nd one, and propylene carbonate (not included in the Table) gives the GIC of the 3-rd group. The compounds modified by carboxylic acids belong to the 2-nd group. At the same time, strong acids stabilize graphite nitrate to a smaller extent as compared with weaker acids, such as acetic and propionic acids. Various procedures of the treatment with water and nitric acid do not stabilize graphite nitrate (see lines 22 and 23).

There are some values of K_{st} greater than unity in the Table. These contradict the meaning of K_{st} as a stability index. This fact may be explained by a high reactivity of the modifier (acetone, tetrahydrofuran and dimethylacetamid), which probably causes some reactions in the interlayer space of GICs. The high reactivity is also characteristic of some other modifiers with $K_{st}<1$. This is particularly true for formic acid, ethers, esters and nitriles. In this case, some products of oxidation or hydrolysis are likely to be in the interlayer space. Structural transformations are illustrated by the PXRD analysis of the GIC modified by acetic acid. The analysis of the X-ray patterns of initial graphite nitrate in Fig. 1 A shows the presence of two GICs with the basal repeat lengths I_c equal to 9.88 and 17.73 Å. According to the well-known formula $I_c = d_i + (n\text{-}1)\cdot d_e$, these values correspond to the mixture of the second stage of the β form and fourth stage of the α form, on account of the above mentioned gallery heights, d_i = 6.55Å and 7.8Å.

Table. Expansion coefficients and stability indices of modified graphite nitrates.

No	Modifier	No water			Treated with water		
		K_V, cm^3/g		K_{st}	K_V, cm^3/g		K_{st}
		A (dried at 105 °C)	**B** (dried at 20 °C)		**C** (dried at 105 °C)	**D** (dried at 20 °C)	
1	Acetone	220	188	**1.17**	183	161	**1.13**
2	Ethyl acetate	172	229	**0.75**	216	247	**0.87**
3	Ethyl formate	279	283	**0.99**	291	303	**0.96**
4	Dioxane	127	175	**0.72**	144	172	**0.84**
5	Tetrahydrofurane	127	90	**1.41**	107	114	**0.93**
6	Ethylene glycol dimethyl ether	146	161	**0.90**	139	147	**0.95**
7	Diethylene glycol dimethyl ether	162	165	**0.98**	151	157	**0.96**
8	Dimethylacetamid	99	85	**1.17**	74	101	**0.73**
9	Formic acid	144	266	**0.54**	135	151	**0.89**
10	Acetic acid	45	308	**0.15**	317	354	**0.90**
11	Propionic acid	57	322	**0.18**	283	324	**0.87**
12	Trifluoroacetic acid	28	242	**0.12**	69	289	**0.24**
13	Dichloroacetic acid	22	256	**0.09**	85	181	**0.47**
14	Methyl acetate	50	242	**0.21**	167	252	**0.66**
15	Penthyl acetate	102	279	**0.37**	142	269	**0.53**
16	Dimethylcarbonate	27	266	**0.10**	107	248	**0.43**
17	Diethylcarbonate	48	336	**0.14**	173	308	**0.56**
18	Acetonitrile	31	229	**0.13**	81	181	**0.45**
19	Propionitrile	35	196	**0.18**	61	216	**0.28**
20	Nitromethane	35	242	**0.15**	78	198	**0.40**
21	Trichlorethylene	58	288	**0.20**	62	206	**0.30**
22	Water	47	169	**0.28**	33	178	**0.19**
23	Nitric acid	30	298	**0.10**	20	125	**0.16**

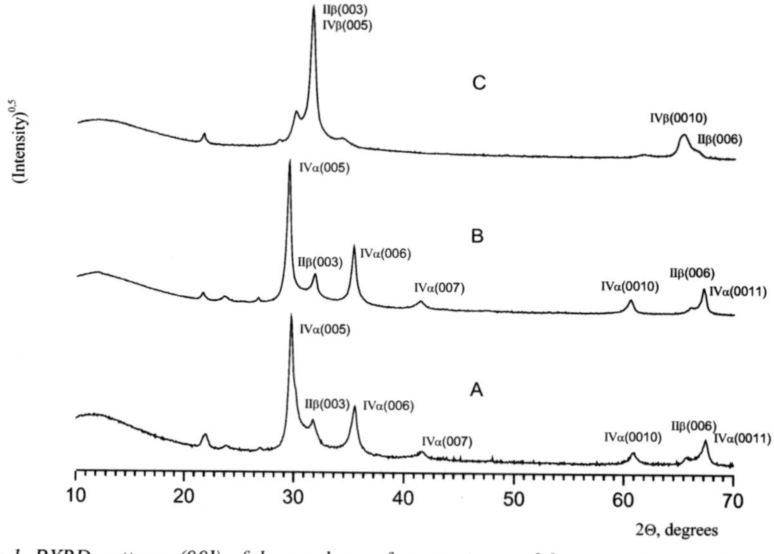

Figure 1. PXRD patterns (00l) of the products of successive modifications: A - initial graphite nitrate; B - graphite nitrate modified with glacial acetic acid; C - graphite nitrate successively modified with glacial acetic acid and water.

Heights of the empty layers d_e were found to be equal to 3.33 and 3.31 Å for the β and α forms, respectively, which are lower compared to pure graphite. Such a possibility of the contraction of an empty layer height was discussed by Inagaki and Fuzellier [1, 4].

The PXRD pattern does not change after the treatment of graphite nitrate with acetic acid (Fig. 1 B), indicating the conservation of I_c and the stage number of the GIC when nitric acid is partially substituted by acetic acid. Note that both initial graphite nitrate and that modified with acetic acid are thermally unstable (see the Table above.)

Contrary to this, the subsequent treatment with water sharply changes the PXRD pattern of graphite nitrate modified with acetic acid (Fig. 1 C). Most of peaks disappear, indicating the fundamental reconstruction of the GIC. At first sight, this pattern resembles that of pure graphite. However, this compound produced by the successive treatment of graphite nitrate with acetic acid and water contains about 20-26 wt% of the intercalant and has a greater ability of expansion and higher thermal stability compared to graphite nitrate (see line 10 in the Table.). Indeed, the detailed analysis reveals the presence of the second and fourth stages of the β phase only. The found patterns sequence from A to C is in good agreement with the known regularity of the graphite nitrate staging established in Ref. [1].

For the modifiers from the first group, the transformation is more complicated. As is shown by the corresponding X-ray patterns, the significant change occurs in the staging along with the probable change in the height of filled layers of GICs. Additional X-ray investigations are required to identify the compounds obtained.

3. DISCUSSION AND CONCLUSIONS

In the quasi-two-dimensional structure of acceptor type GICs shown schematically in Fig. 2 each anion is surrounded by six acid molecules **S** [7]. However, according to the stoichiometry each anion is solvated by only two ligands. Additional stabilization might come from formation of a continuous two-dimensional net of hydrogen bonds among the solvation ligands. However, this is impossible in the case of pure graphite nitrate as the molecules of nitric acid, which solvates the anion, cannot give rise to the hydrogen bonding among them. This is because the only proton of nitric acid has already been bound with the nitrate anion. The insertion of the second intercalant capable of the additional hydrogen bonding could realize such a possibility.

On modifying graphite nitrate, as follows from the material balance of the reaction, a major part of nitric acid is replaced by the organic intercalant. The latter, if it is protonated by the rest of nitric acid in the interlayer space, is capable of the strong hydrogen bonding with two molecules. A similar possibility for graphite bisulfate modified with acetic acid is discussed in Ref. [8]. Noteworthy that the energy of the hydrogen bond incorporating charged species, protonated modifiers in the present case, is several times greater than that for neutral molecules. It can exceed 100 kJ/mol [9]. Unprotonated organic cointercalant and water can play a role of dielectric spacers among the charged species.

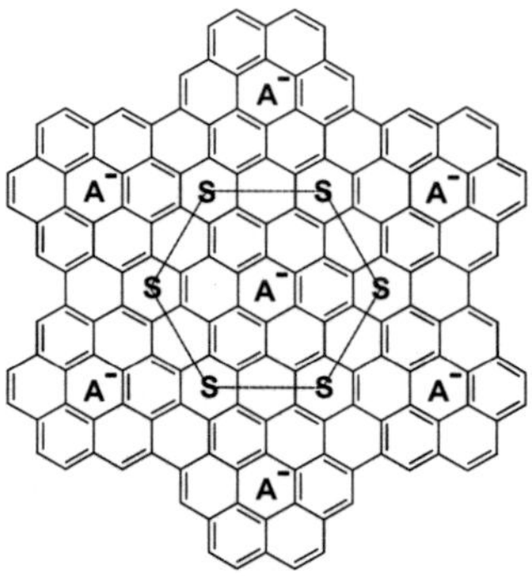

Figure 2. Fragment of idealized structure of the GIC [7]. The intercalant layer, which comprises the anions A^- *and ligands* ***S****, is placed between two positively charged graphite layers.*

To verify the mechanism presented, the quantum-chemical calculations of proton affinity, A_H, were carried out for modifiers, since the corresponding experimental data are quite rare. The calculations were performed for isolated molecules, since the properties of species in the interlayer space are probably closer to the gas phase rather than the liquid. The values of A_H were calculated as a difference in the total energy between the initial and protonated forms of the modifier. Energies were calculated using the TZV(2df, 2p) basis and MP2 electron correlation correction. Preliminarily, geometries were fully optimized in the framework of the MP2/6-31G(d, p) calculation. The GAMESS suite of *ab initio* programs was employed [10]. Comparison between the calculated at 0 K proton affinities for water (7.46 eV) and dioxane (8.50 eV) and the experimental data 7.50 eV and 8.42 eV at 298 K, respectively (see [11]), demonstrates a sufficient accuracy of the calculation.

Despite a wide range of the modifiers used, there is a clear symbasis between the calculated proton affinities and the experimental stability indices K_{st} (Figure 3). The decrease in concentration of nitric acid in the interlayer space will give rise to a poorer stability of GICs, if the proposed stabilization mechanism is valid. To verify this assumption, the treatment of graphite nitrate with 25-fold excess of acetic acid, compared to that used in the above procedure, was carried out in a glass column. Most of the molecules of nitric acid are probably removed from the interlayer space under such a treatment. It turns out that this procedure reduces the expansion coefficients, from 317 to 147 cm^3/g for product **C** and from 354 to 210 cm^3/g for product **D**. The stability indices K_{st} also decreases from 0.90 to 0.70. This substantial decrease in K_V and K_{st} additionally supports the proposed stabilization mechanism.

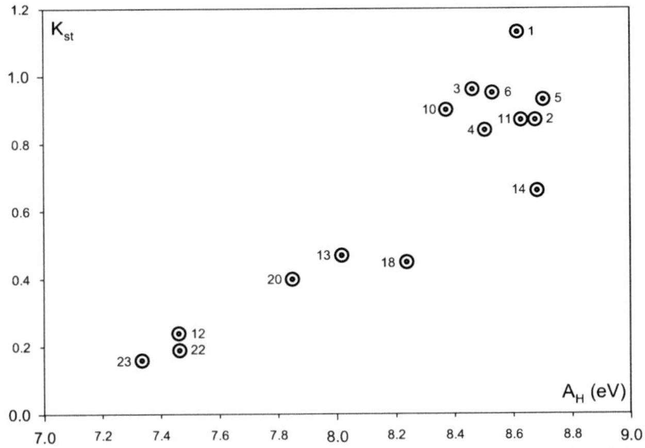

Figure 3.Comparison between the calculated proton affinities and stability indices. Numerals correspond to the line numbers in the Table.

REFERENCES

1. Fuzellier H, Melin J, Herold A. Une nouvelle variété de nitrate de graphite. Mater. Sci. Eng. 1977; 31:91-94.
2. Forsman W.C., Mertwoy H.E., Wessbecher D.E. Nonreductive spontaneous deintercalation of graphite nitrate. Carbon 1988; 26:693-699.
3. Savoskin M.V., Yaroshenko A.P., Whyman G.E., Mestechkin M.M., Mysyk R.D., Mochalin V.N. Theoretical study of stability of graphite intercalation componds with Brønsted acids. Carbon 2003; 41:2757-60.
4. Inagaki M. Applications of graphite intercalation compounds. J. Mater. Res. 1989; 4: 1560-68.
5. Savoskin M.V., Yaroshenko A.P., Mysyk R.D., Shologon V.I., Khripunov S.V. Synthesis of expandable residual graphite nitrate stabilized with glacial acetic acid. Proceedings of the International Conference on Carbon; 2003; Oviedo, Spain. P. 188-189.
6. Inagaki M. Graphite-nitrate residue compounds with a smaller interlayer spacing than graphite. Carbon 1967; 5:317-8.
7. Jiang J., Beck F., Krohn H. Electrochemical reversibility of graphite oxide. J. Indian Chem. Soc. 1989; 66:603-9.
8. Kang F., Zhang T.-Y., Leng Y. Electrochemical behavior of graphite in sulfuric and acetic acid. Carbon 1997; 35:1167-73.
9. Joesten M.D., Schaad L.I., Hydrogen bonding. New York: Marcel Dekker, 1974.
10. Schmidt M.W., Balderidge K.K., Boatz J.A., Elbert S.T., Gordon M.S., Jensen J.H., Koseki S., Matsunaga N., Nguyen K.A., Su S., Montgomery J.A. General atomic molecular electronic structure system. J. Comput. Chem. 1993; 14:1347-63.
11. Gasteiger J., Hutchings M.G. Proton affinities of different gases. J. Amer. Chem. Soc. 1984; 106:6489-95.

ELECTROCHEMICAL STABILITY OF NATURAL, THERMALLY EXFOLEATED AND MODIFIED FORMS OF GRAPHITE TOWARDS ELECTROCHEMICAL OXIDATION

Igor O. Kovalenko[1*], Mayya L. Barsukov[2], Vyacheslav Z. Barsukov[3] and Y. V. Myshko

[1]*ChimLaborReactiv, Ltd., Remontnaya Street, 8, Kiev, 02099, Ukraine*
[2]*Institute for Problems of Materials Science, National Academy of Science of Ukraine*
[3]*Kiev National University of Technologies & Design, Department of Chemistry, 2, Nemirovich-Danchenko street, Kiev, 252011, Ukraine*

Abstract

Acceptor-type graphite intercalation compounds (A-GIC) can usually be obtained through electrochemical oxidation of graphite in aqueous acidic or neutral solutions. Together with formation of A-GIC compounds, oxygen evolution usually takes place at the anode. This may lead to irreversible changes in the carbonaceous material, and, consequently, may result in a loss of reversibility of the system, caused by the partial oxidation of graphite. Thus, stability of graphite during cycling is the most important factor of electrode durability.

A special electrochemical method of prolonged sequential cyclic voltammetry is proposed as means of evaluation of stability of graphite towards oxidation. Key corrosion mechanisms for different types of graphite are cosidered. Special types of thermally expanded graphite (TEG) modified by P_2O_5, B_2O_3 and H_3BO_3 are compared. A maximal cycle life (of up to 550 cycles) is obtained with TEG modified by 5% of B_2O_3.

1. INTRODUCTION

Graphite finds wide range of applications in the electrodes for certain types of rechargeable batteries and supercapacitors, in electro-sorption/desorption electrodes, as anodes in a number of processes of

* Corresponding author. E-mail: igorkio@kiev.ldc.net, kio2001@ukr.net

I.V. Barsukov et al. (eds.), New Carbon Based Materials for Electrochemical Energy Storage Systems, 389–398.

electrochemical synthesis, etc. Nevertheless, corrosion processes during anodic polarization of such electrodes are among the main limitations for their practical use.

For example, graphite intercalation compounds (GICs) as active material for metal-free rechargeable batteries (MFRB) [1], are represented by graphite anodes, that are being reversibly intercalated-deintercalated in aqueous solutions of strong acids by the following reaction:

$$C_x + A^- + yHA \Leftrightarrow [C_xA \cdot yHA] + e^-, \tag{1}$$

where C_x - the carbon macromolecule, A^- - the anion, y - the stechiometric factor for solvated molecules of the acid, HA. Nevertheless, in parallel to formation of GICs, there usually take place side corrosion reactions [2] connected with oxygen evolution process on anode:

- *the anodic evolution of oxygen*

$$2H_2O \Rightarrow 4H^+ + 4e^- + O_2; \tag{2}$$

- *the oxidation of graphite to CO_2*

$$C_X + 2H_2O \Rightarrow CO_2 + C_{X-1} + 4H^+ + 4e^-; \tag{3}$$

- *the formation of surface groups*

$$C_X + 2H_2O \Rightarrow C_{X-1}COOH + 3H^+ + 3e^-; \tag{4}$$

- *the formation of graphite oxide*

$$C_XA + H_2O \Rightarrow C_XOH + H^+ + A^-. \tag{5}$$

This may lead to the irreversible changes in the material, caused by partial oxidation of graphite, loss of reversibility of the system along with the efficiency of reaction for intercalation-deintercalation (1).

2. EXPERIMENTAL

2.1. Electrode Preparation

We have manufactured the graphite electrodes of "plug" type. An active mass (88wt% graphite, 10wt% carbon black and 2wt% PTFE) was pressed into a polyethylene plug of 11 mm inside diameter. We have studied the characteristics of electrodes with an equal weight of 0.280g. Before pressing, the Pt current collector was inserted into an opening, pierced in the plug wall.

2.2. Graphite Materials for Investigation

We have performed comparative investigation of graphite electrodes, obtained from various types of starting natural and artificial types of graphite. The following materials were compared:

#1. Grades of natural graphite from Zavalie (NGZ) graphite deposit, Kirovograd Region, Ukraine;

#2. Natural graphite flakes (NGF) "Kropfmuhl Normalflocke" (BASF®, Munich, Germany);

#3. Thermally exfoliated graphite (TEG) and their composites from Superior Graphite (Chicago) and TEGs prepared from Ukrainian graphite.

2.3. Synthesis of Thermally Exfoleated Graphite

Starting natural graphite from Zavalie deposit in, Ukraine was chemically intercalated in sulfuric acid. Potassium persulphate was used as an oxidizer. Thermal expansion of graphite was performed at 900°C [3]. Carbon content in the experimental samples was of about 99.0%.

A United States - manufactured TEG (ABG-74, a commercial grade of Superior Graphite Co., Chicago, Illinois, USA) was manufactured in principle in a similar manner, yet with a process that used few more important processing steps, as follows.

First of all, the precursor natural flake graphite having a starting purity of approximately 95%C has been heat treated is a continuous thermal purification process (this unique process of Superior Graphite Co. has ability to achieve and sustain a relatively large zone of uniform temperature in excess of 2500°C). The purpose of heat treatment of the precursor (95% carbon natural flake graphite) was reduction of the mineral ash in graphite structure in order to achieve lowest possible resistivity (0.007 Ohm·cm) and very high pure in the final product (99.95%C), [4].

2.4. Structural and Electrochemical Investigation of Graphite

For purposes of investigation of the micro- and macro-structural characteristics of graphite, we have used methods of X-ray diffraction, SEM and we also studied the pattern of particle distribution by size with the SK laser micron sizer (Horiba, Japan).

2.4.1. Consecutive Galvanostatic Cycling with Control of the Electrode Weight Loss

We have used the method of galvanostatic cycling in the 0.5-1.75V or 0.5-2.00V potential ranges vs the Ag/ AgCl reference electrode. We have monitored the loss of electrode active mass during cycling for purposes of studying the possible mechanisms of electrochemical corrosion on graphite electrodes.

2.4.2. Consecutive Voltammetric Cycling

Electrodes were manufactured of NGZ, NGF and TEG and were tested under comparable conditions. The relative resistivity of graphite electrode to corrosion was compared on stability of CVs during the N (stands for "number") of consecutive voltammetric cycles in a given potential region. The relative resistivity to corrosion was estimated by the number of cycles for very stable cycling (N_{stable}) and for saving of electrochemical activity (N_{max}).

3. RESULTS AND DISCUSSION

3.1. Structural Investigations

3.1.1. Structural Investigations in the Beginning and Finish of Cycling for NGZ and NGF

It is possible to make conclusion about the absence of noticeable structural changes based on comparison of structural parameters of graphite in the beginning and at the end of cycling, using the XRD method.

The parameters of an elementary crystal cell for initial and finite state are practically identical. For example, interplanar distance C= $(6.718 \pm 0.007)\cdot 10^{-8}$ cm for NGZ (GAK-1 grade) and NGF dos not change during cycling.

3.1.2. Comparison the Particle Size Distribution for NGF and TEG

We have studied the key physical properties of samples of TEG, NGF and NGZ using an SK laser micron sizer and BET surface area analyzer. Unless otherwise stated, the particle size distribution and surface area measurements for this work have been carried out at the Institute for Problems of Materials Science of the National Academy of Sciences of

Ukraine. It was observed that the particle size distribution of products does not vary significantly during cycling.

The average particle size values for NGZ (GAK-1 type) and NGF was determined to be, in particular, 35-40 μm. The specific surface area of such graphites in initial and final states is also practically identical and is of approximately 6.8-7.0 m^2/g for natural graphite and approximately 20-25 m^2/g for the internally produced exfoliated graphite.

The average size of particles for TEG from Superior Graphite Co. is sufficiently less ($D_{50} \leq 10$ μm.)* [4]. The specific surface area of the ABG-74 grade of exfoliated graphite is 24 m^2/g.

Due to having finer particle size and higher surface area, one may suppose that TEG is less oxidation-resistant than NGF and NGZ. However, we have shown TEG to be much more oxidation-resistant than NGF.

The cycling data is summarized in Table 1. It becomes clear from the data in Table 1 that the initial specific capacitive parameters of some types of NGZ are higher than NGF by a factor of 1.5-2.0. Nevertheless, the values of N^G_{max} for NGF are considerably higher than those for NGZ. Thus, NGZ are less resistant to electrochemical oxidation than NGF.

3.2. Electrochemical Investigations

3.2.1. Comparison of Cyclic Voltammograms

On the first stage of investigation, cyclic voltammograms (CVs) were compared for the graphite electrodes during cycling in maximal possible potential range ΔEmax = 0-2.1V (vs. the Ag/ AgCl reference electrode) at a sweep potentials rate of 10 mV/s.

As our cyclic voltammety investigations show, the minimal resistivity to corrosion (Nstable=20-30 cycles; Nmax=50-90 cycles) is demonstrated by the natural Zavalie graphites (group #1).

It can be clearly seen from the curves presented for TEG (Figure 1) that there is not any evidence for the electrode surface oxidation, at least during the first N_{stable} = 190 cycles, whereas the oxygen evolution rate is quite high under these conditions. After 190 cycles a gradual decrease of peak currents can be noticed (see curves corresponding to 200 and 250 cycles, respectively). The service life of the electrode composed of TEG under these conditions is N_{max}=250 cycles.

---

* 50% particles less than 10μm.; 90% particles less than 24 μm.

Table 1. Summary of main characteristics of NGF and different NGZ materials (per results of galvanostatic cycling)

#	*Type of Graphite*	N^{G}_{max} Cycles	α_{max}, %	q_{max}, A·h/kg	μ_{max} %
1	Natural graphite flakes (NGF) Kropfmuhl Normalflocke, Germany	>100	67.1	5.77	20.5
Natural graphites from Zavalie (NGZ) graphite deposit, Kirovograd Region, Ukraine:					
2	Graphite for Zn-MnO_2 cells (Standard #7478-75):				
2.1	GE-1	39	90	10.5	37.3
2.2	GE-2	35	90	9.3	33.1
3	Graphite for motor carbon brushes, EUZ-M type (Standard #10274-79)	26	31	12.8	40
4	Graphite for electric motors, EUZ-E type (Standard # 21-25-156-75)	16	63.5	10.8	38.4
5	Graphite for alkaline accumulators (Standard #10273-73):				
5.1	GAK-1	10	35.2	11.1	39.5
5.2	GAK-2	7	27	8.29	25.9
6	GMS-1, low ash graphite for special applications (Standard # 18191-78)	5	29	8.52	30.3
7	Special graphite, C-1 type (Standard # 113-08-48-63-90)	4	18	6.1	19.1

In this table:

N^{G}_{max} *- is the maximum number of cycles before the discharge electrode capacity Qd will reduce to the minimal level (*$Qd^{min}=4Ah/kg$ *in our case);*
$\alpha_{max} = (Qd/Qch)\cdot 100\%$ *- is the maximum value of current efficiency; Qch- charge electrode capacity);*
q_{max} *- is the maximum value of specific discharge capacity (A·h/kg) during the cycling;*
$\mu_{max} = (Qd/Q^{Th}d)\cdot 100\%$ *- is the utilization factor (%);* $Q^{Th}d$ *- theoretical discharge capacity.*

When cycling a NGF electrode under the same conditions (Figure 2), the deintercalation peak current values are much higher. At the same time, under comparable conditions, a noticeable surface oxidation of NGF starts to be observed significantly earlier, after Nstable = 50-60 cycles. The service life of the NGF electrode is Nmax = 100 cycles.

On the second stage of investigation (Fig. 3) cyclic voltammograms were compared for the NGF (a), TEG (b) and TEG modified by P_2O_5 (c), B_2O_3 (d) and H_3BO_3 (e) during cycling in more realistic and narrower range of potentials $\Delta En = 0.8\text{-}2.1V$ at the same sweep potentials rate of 10 mV/s.

In such ΔEn, the values of Nstable and Nmax becomes higher (for instance, Nmax = 180 and 350 cycles for NGF and TEG, accordingly).

Thus, considering the results of the above comparison, it is easy to conclude that the TEG electrode is more resistant to the electrochemical corrosion (oxidation). A maximal cycle life in such conditions (up to 550 cycles) is obtained with TEG modified by 5% of B_2O_3.

3.3 Investigation of Mechanisms of Electrochemical Corrosion of Graphite Electrodes

3.3.1. Comparison of Electrode Capacity Loss with the Loss of Graphite Electrodes Mass

There exists a point of view that the main mechanism of electrochemical corrosion is assoceated with a full step by step oxidation of graphite to CO_2 during the cycling, according with reaction (3), [2].

For purposes of studying of the possible mechanisms of electrochemical corrosion we have compared first of all the capacity loss for graphite electrodes with the loss of their mass during prolonged galvanostatic cycling (Table 2).

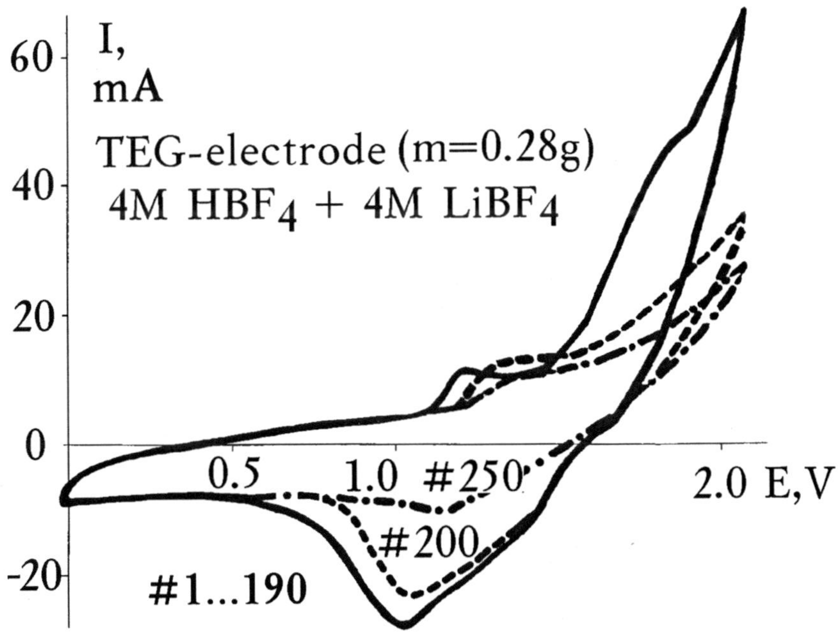

Figure 1. CVA for the TEG - electrode (m=0.28g); Emax=2.1V; v=10mV/s.

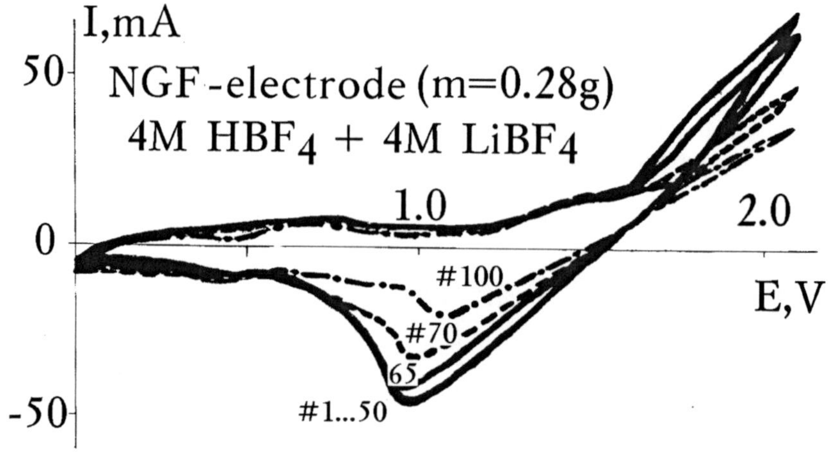

Figure 2. CVA for the NGF - electrode (m=0.28g). Emax=2.1V; v=10mV/s.

Table 2. The Capacity and the Electrode Weight Loss for NGZ (GAK-1 type) and NGF ("Kropfmuhl Normalflocke") During the Consecutive Galvanostatic Cycling up to $Umax_1$=1.75V and $Umax_2$=2.00V.

Graphite Type	U_{max}, V	Electrode capacity loss			Weight loss of electrode active mass		
		Q_0, mA·h	Q_{min}, mA·h	δ, %	M_0, g	M_{min} g	δ, %
NGZ	1.75	5.00	1.66	66.8	2.00	1.00	50.0
(GAK-1)	2.00	9.00	0.48	94.7	2.00	0.87	56.5
NGF	1.75	5.70	3.33	67.5	1.14	1.11	2.51
	2.00	9.11	3.33	68.0	1.13	1.11	2.49

The comparison of electrode capacity loss with loss of their mass during the cycling unambiguously shows, that degradation of electrical characteristics cannot be explained only by CO_2 derivation with appropriate decrease of an active material. The main mechanism depends strongly on the type of graphite.

For example, for NGZ, the loss of capacity is higher but comparable with the loss of graphite active mass (66.8 and 50.0%; 94.7 and 56.5%). Thus, corrosion mechanism (3) is a predominant one for such an unstable to oxidation graphite class.

For NGF, the loss of capacity is much higher (approximately 68%) than the loss of graphite active mass (approximately 2.5%). Thus, for this type of graphite the corrosion mechanisms (4) and (5) with formation of surface groups may be considered predominant.

Similar mechanisms (please see equations (4) and (5)) are typical also for TEG and modified TEG. As results of additional acid-base titration shows*, even in the initial state of cycling, the highly developed internal surface of TEG is covered by thin layer of -COOH and -OH groups, which are the result of heat-treatment during preparation of expanded graphite. This makes TEG more stable towards the electrochemical oxidation during cycling. It is well known also that chemical activity of any carbon material is less if this carbon material was heat treated at higher temperature.

The stabilizing influence of oxides, which cover the surface of graphite (result of heat treatment during purification and exfoliation) is usually associated with enhancement of its acceptor properties. This influence in the case of TEG was seen to produce highly stable to oxidation carbonaceous material.

Furthermore, we believe that the stabilizing influence of boron in the structure of graphite is connected with enhancement of its acceptor properties, which manifest themselves when Boron atoms substitute carbon atoms in the crystalline structure (hexagon ring) of carbon. Such effects are mentioned in the literature for some types of carbon materials [3] and the influence of boron on TEG can be the similar.

4. CONCLUSIONS

The minimal resistance to oxidation (few dozens of cycles) was demonstrated by the natural graphite from Zavalie deposit, Ukraine. Natural graphite flakes «Kropfmuhl Normalflocke» proved to possess significantly higher resistance to oxidation (up to 100 cycles). Furthermore, TEGs demonstrated even greater resistance to oxidation (250 cycles).

The main corrosion mechanism for the natural graphite from Zavalie deposit, as proposed by the authors, is associated predominantly with a full step by step oxidation of graphite to CO_2 upon cycling, in accordance with reaction (3). The main mechanism of corrosion of graphite electrodes based on NGF and TEG is believed to be the derivative accumulation of surface groups (reactions (4), (5)), which gradually block "entrances" into the crystalline lattice and thus hinder intercalation of anions in interlayer spaces. Our investigations show that the most probable areas, where the redundant acid centers are located, are the initial sites and "entrances" in the interlayer spacing.

Application of new types of graphite, found to be more oxidation-proof (in particular, TEG and TEG modified by boron), can largely increase the electrochemical stability of materials used in aqueous electrolyte media. Their high resistance to oxidation and enhanced long-term cycling stability create realistic prerequisites for wide range of applications for such graphite

* The acid-base titration was performed by a Ph.D. V. Lucenko and an engineer V. Drozdik of Kiev National University of Technologies and Design, Kiev, Ukraine.

materials in rechargeable batteries, electrosorption, electrochemical synthesis, etc. applications.

REFERENCES

1. I.V. Barsukov, T.I. Motronyuk, V.Z. Barsukov, V.I. Drozdik. Metal-Free 1.5V Rechargeable Batteries: Steps of Optimization and Prospects for the Practical Application// *In book: Batteries for Portable Application and Electric Vehicles,* C.F. Holmes and A.R. Landgrebe eds., The Electrochemical Society, Inc., Pennington, NJ, 592-596 (1997).
2. H. Krohn, F. Beck, and H. Junge, Reversible Electrochemical Graphite Salt Formation from Aqueous Salt Electrilytes, *Ber. Bunsenges. Phys. Chem.*, **86**, 704 - 710 (1982).
3. I.G. Chernish, I.I. Karpov, G.P. Prikhodko, and V.M. Shai, *Phisical Chemical Properties of Graphite and its Compounds*, Naukova Dumka, Kiev (1990) */in Russian/.*
4. Product Information Bulletin – Expanded Graphite ABG-74. Superior Graphite Co. (1999).

LOW TEMPERATURE SYNTHESIS OF GRAPHITE FROM IRON CARBIDE

Svetlana Dimovski, Alexei Nikitin, Haihui Ye and Yury Gogotsi*

Drexel University, Department of Materials Science and Engineering
A. J. Drexel Nanotechnology Institute, 3141 Chestnut Street, *Philadelphia, PA 19104, USA*

Abstract

Synthesis of graphite by extraction of iron from iron carbide by chlorine is discussed in this work. This process is attractive because it can produce well-ordered graphite at temperatures as low as 600°C, providing an opportunity for low-temperature solid-state synthesis of graphitic materials for batteries. The composition of the reaction products under equilibrium conditions was determined by thermodynamic simulation, and the initial process parameters such as temperature and chlorine/carbide molar ratio were selected accordingly. Crystalline parameters of the graphitic reaction products, such as the interlayer spacing and crystalline sizes, were calculated from X-ray diffraction measurements. The degree of orientation of the graphitic layers was determined by Raman spectroscopy. Three temperature regimes have been identified. At temperatures below 500°C, amorphous or disordered carbon is formed as shown by Raman spectroscopy and TEM studies. Well-ordered graphite microcrystals are formed by solid-state growth between 600°C and 1100°C. Above the eutectic temperature 1130°C in the Fe/Fe_3C system, the growth of large graphite crystals occurs from the liquid phase, similar to the formation of kish graphite by precipitation of carbon at high temperatures from supersaturated molten iron. Iron chlorides, the main impurities in the material synthesized by the solid-state growth, can be removed by using excess chlorine gas or by a separate wet chemical purification step.

1. INTRODUCTION

It has been previously shown that selective etching of carbides is an attractive technique for the synthesis of various carbon structures. Carbon produced by extraction of metals from carbides is called carbide-derived

---

* Corresponding author. E-mail: gogotsi@drexel.edu; Web: http://nano.materials.drexel.edu

I.V. Barsukov et al. (eds.), New Carbon Based Materials for Electrochemical Energy Storage Systems, 399–410.

carbon (CDC).[1] Leaching in supercritical water, high-temperature treatment in halogens and vacuum decomposition can be used to remove metals from their carbides, producing carbon coatings, powders or bulk shapes. For the chlorination reaction to proceed, it is necessary that molecules of $Cl_2(g)$ are transported to the carbide/carbon interface and molecules of metal chloride are transported away. Linear reaction kinetics have been demonstrated for the chlorination of SiC[2], implying that the carbon film formed during the process on the carbide surface is nanoporous[3], and as such allows easy permeation of Cl_2 molecules. This phenomenon allows transformation of a carbide particle or a bulk component to any depth until it is fully converted to carbon. However, the microstructure of the CDC formed by selective etching in halogens can vary significantly with experimental conditions. Thus amorphous and nanocrystalline graphitic carbon,[4] nanotube-like structures[5] and nanocrystalline diamond[2] were reported. Ordered graphite was only observed in small amounts as a thin film on the surface of SiC crystals, because a solid state synthesis of graphite[6,7] usually requires very high temperatures (exceeding 2500°C). This paper describes the formation and evolution of graphite in CDC produced by extraction of iron from iron carbide using chlorine gas. It indicates the necessity of transition metals in the system for low temperature graphite synthesis. An increased degree of graphitization in CDC produced by chlorination of TiC in the presence of small amounts of Fe, Ni, and Co has been recently demonstrated.[8] However, the interaction of iron carbide with chlorine or other halogens has not been reported in the existing literature.

2. MATERIALS AND EXPERIMENTAL

2.1. Materials

Iron carbide powder for this study was provided by Hazen Research, Inc. Chlorination experiments were done on the as-received powder. Scanning electron microscopy, energy-dispersive X-ray spectroscopy (EDS), wide angle powder X-ray diffraction (XRD), and Raman spectroscopy analyses were conducted to characterize the original material. The average particle size was measured to be ~0.5 μm. These particles formed grain-like iron carbide clusters (av. size ~350 μm). The XRD pattern corresponds to that of synthetic Fe_3C (ICDD form 35-0772). No other compounds were detected by the XRD of raw powder. However, EDS analysis showed traces of oxygen, probably originating in the very thin surface layer of intrinsic iron oxide, and/or CO_2 and water absorbed from the air, and aluminum, most likely from the sample holder.

Chlorine (BOC Gases, 99.5%) was dried with concentrated sulfuric acid and anhydrous calcium sulfate. Purging was done with 99.998% pure argon (BOC Gases).

2.2. Thermodynamic Simulation

The ChemSage v4.14 advanced SOLGASMIX-based Gibbs energy minimization program from GTT, Germany, was used to estimate equilibrium amounts of reaction species as functions of the incoming amount of chlorine gas (per one mole of Fe_3C) and chlorination temperature for different amounts of chlorine. The calculations were done for a closed system with a constant total pressure. The solid phase was treated as a mechanical mixture of compounds with unit activities, and the gas phase was considered a mixture of ideal gases.

2.3. Synthesis

The apparatus used to treat Fe_3C with chlorine at atmospheric pressure has been described elsewhere.[1] Fe_3C was loaded in a quartz boat and placed in a horizontal quartz tube reactor, which was purged with argon gas until the desired reaction temperature was reached. The sample was then exposed to flowing Cl_2 gas for 3 h, and purged again with argon during the cooling period, as described in literature[1,9]. Constant reaction temperature and gas flow rate were maintained throughout every run. The interaction of iron carbide with chlorine at temperatures ranging from 400°C to 1200°C was investigated. Carbon collected from the quartz boat was purified, when necessary, by washing away residual impurities with 10% HCl followed double distilled water 3 times in sequence, and dried carefully at a temperature of about 100 - 150°C.

2.4. Characterization

The chlorinated samples were analyzed using a Philips XL-30 environmental field-emission (FE) SEM. SEM analysis was always accompanied by EDS analysis to verify that the analyzed material was carbon. TEM samples were prepared by dispersing the synthesis products in isopropyl alcohol over a copper grid with a lacey carbon film, and transmission electron microscopy (TEM) analysis was performed using a JEOL JEM-2010F (200kV). A Gatan image filter for energy filtered imaging and electron energy loss spectroscopy (EELS) was used to identify carbon areas and impurities. Raman microspectroscopy (Renishaw 1000) with the excitation wavelength of 514.5 nm (Ar ion laser) with a range of 800-2000 cm^{-1} and exposure time of 10 s was used to analyze the degree of carbon ordering. The spectra were collected from at least 5 different points of each sample with a 2-μm spot size. Low laser power was applied to avoid damaging and heating the sample. GRAMS 32 v 5.2 spectral analysis software was used for peak fitting and deconvolution. The in-plane crystallite size of graphite, L_a is shown to be inversely proportional to the ratio of D band at 1350 cm^{-1} and the G band at 1580 cm^{-1} (I_D/I_G ratio). It can be calculated as: $L_a=4.4(I_D/I_G)^{-1}$ (nm) according to Tuinstra and Koenig's work[10] for 2.5 nm < L_a < 3000 nm.

X-ray diffraction was done using a Siemens D-500 diffractometer utilizing CuKα radiation (1.406 Å). The data were collected as step scans, with a step size of 0.05° 2θ and a count time of 2 s/step between 10° and 80° 2θ.

3. RESULTS AND DISCUSSION

3.1. Reaction Thermodynamics

Since iron chlorides are more thermodynamically stable than CCl_4, at high temperatures chlorine reacts selectively with the metal at the carbide surface by the reactions:

$$2Fe_3C + 9Cl_2 = 6FeCl_3 + 2C \text{, or} \quad (1)$$

$$Fe_3C + 3Cl_2 = 3FeCl_2 + C \quad (2)$$

The formation of chloride dimers such as Fe_2Cl_6 and Fe_2Cl_4 in the gas phase is also possible under certain conditions (Figure 1). Thermodynamic modeling can be utilized to estimate the effect of various process parameters on CDC formation and predict optimal synthesis conditions.[11] Thermodynamic analysis (Figure 1) shows that a small amount of chlorine results in the formation of solid $FeCl_2$ in the sample, and relatively high temperatures (above 1000°C) are needed to remove iron salts from CDC. Excess chlorine shifts the reaction toward the formation of Fe_2Cl_6 (g), while carbon may be consumed to form CCl_4 (g) at temperatures below 400°C. At higher temperatures, CCl_4 (g) is not stable, which enables the formation of a solid carbon phase. Intermediate regimes result in the removal of iron compounds as gaseous reaction products. However, some solid $FeCl_2$ may remain in the system. Further reduction of iron from its salts to metallic iron using H_2 gas may enable the design of CDC – metal catalyst systems.

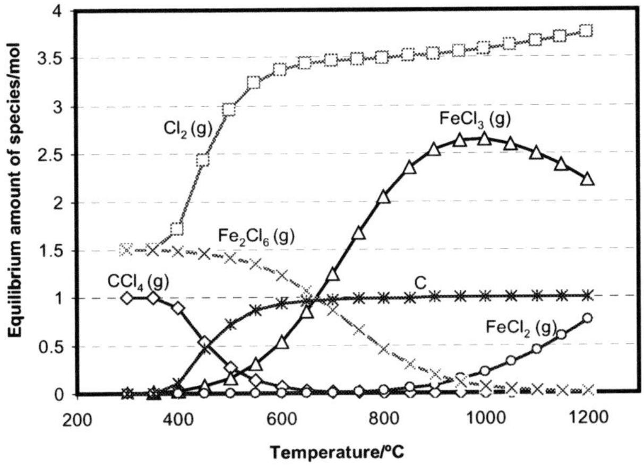

Figure 1. Thermodynamic analysis of iron carbide chlorination: a.) Equilibrium amount of species vs. chlorination temperature for n = 8 moles of Cl_2 (g).

Thermodynamic calculations of the SOLGASMIX type, such as shown here, provide information about closed systems in equilibrium. The experimental system used in this work was an open system, and gaseous reaction products were continuously removed with the gas flow, shifting the reactions (1, 2) equilibrium to the right and favoring the formation of carbon. Caution must be taken in the interpretation of the thermodynamic calculation results, in which further refinement is clearly possible; however, these results do provide some general guidelines and are generally in good agreement with experimental results.

3.2. Raman Spectroscopy Analysis

Typical Raman spectra of Fe_3C-derived carbon powders synthesized at various temperatures are shown in Fig. 2. Raman analysis shows that Fe_3C was successfully converted into carbon with various degrees of ordering, depending on the reaction temperatures. Three main temperature regimes can be identified here. At temperatures below 500°C, chlorination yields amorphous or disordered carbon. Broad disorder-induced D (1353 cm^{-1}) and graphite G (1596 cm^{-1}) Raman bands can be found in the spectrum from the sample obtained by chlorination at 400°C (Fig. 2a). A relatively high I_D/I_G ratio of ≈0.90, corresponding to in-plane crystalline size of graphite L_a<4 nm,[10] indicates that amorphous carbon is produced under these conditions.

Although the D and G bands are the most prominent features in Raman spectra of CDC, the actual Raman scattering data cannot be fitted with 2 peaks in the case of highly amorphous structures. Such difficulties were observed for Fe_3C-derived CDCs produced below 500°C. Instead, for an adequate representation of these spectra, three additional broad peaks of relatively low intensities located in the vicinity of 1225 cm^{-1} (unnamed) 1620 cm^{-1} (D') and 1500 cm^{-1} (D") were included. Peak fitting was conducted following the procedure described in reference.[9]

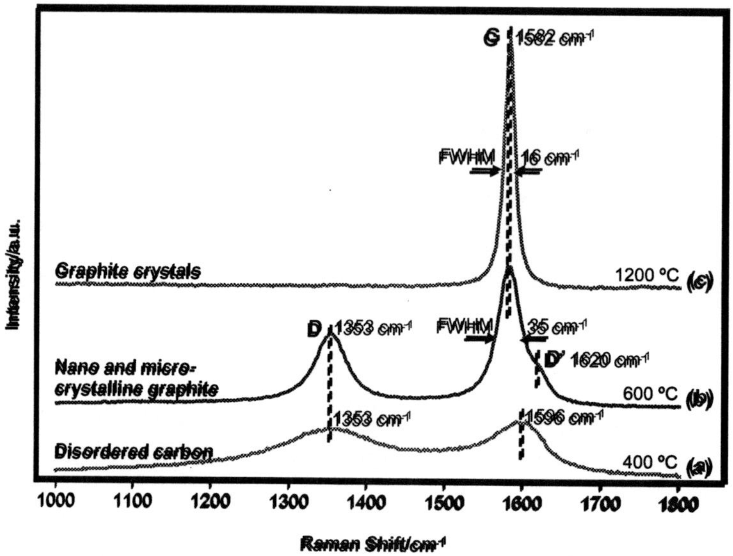

Figure 2. Raman spectra of carbon produced by chlorination of Fe_3C for 3 hours at different temperatures.

At reaction temperatures of 600°C - 1100°C, the D and the G bands become well-defined and clearly separated (D at ~1353 cm^{-1}, G at 1582 cm^{-1}). In addition, a weak D' band appears as a shoulder around 1620 cm^{-1} (Figure 2b). The shift of the G peak position towards that of ordered graphite (1582 cm^{-1}) and decrease of its full width at half maximum (FWHM) value to ~35 cm^{-1} indicates the onset of graphitization. Increasing the reaction temperature increases the intensity of the G band relative to the intensity of the disorder-induced D band, decreasing the I_D/I_G ratio from 0.44 (in average) at 600°C to about 0.13 at 1000°C, which corresponds to $L_a \approx 30$-55 nm. For comparison, the I_D/I_G ratio of CDCs produced from SiC at the same reaction temperature is close to one.[9] Thus, the presence of iron in the system facilitates graphite nucleation and growth at much lower temperatures, which is in agreement with previous publications showing the catalytic effect of iron on graphitization.[8]

Above the eutectic temperature in the iron-Fe_3C system (1130°C)[12], growth of large graphite plates and flakes occurs from the liquid phase. Carbon precipitates in the form of highly ordered graphite crystals from molten iron supersaturated with carbon. The Raman spectrum for chlorination at 1200°C is shown in Fig. 2c. A very strong and narrow

(FWHM = 16 cm^{-1}) G band at 1582 cm^{-1} suggests that graphite crystals are formed under these conditions.

3.3. Powder X-ray Diffraction

Typical X-ray diffraction patterns of three different carbon powder samples are shown in Fig. 3. Two 00*l* and two *hk*0 diffraction peaks can be distinguished in the patterns of samples produced at 800°C and 1000°C. The 002 ($2\theta \approx 26.9°$) and 004 ($2\theta \approx 54.9°$) peaks correspond to the parallel graphene layers. The 100 ($2\theta \approx 43°$) and 110 ($2\theta \approx 77.8°$) diffraction peaks are characteristics of the 2D in-plane symmetry along the graphene layers. Based on its XRD pattern, the powder synthesized at 500°C is not graphitized, which is in agreement with Raman analysis. This low temperature sample also contains traces of iron chlorides.

Further analysis of these XRD profiles shows that the width of the 002 peak decreases as the reaction temperature increases. Its position is approximately the same for both 800°C and 1000°C, which indicates that the graphitic interplanar spacing (0.334 nm) is developed in these powders at temperatures as low as 800°C. This is also in agreement with the TEM and Raman analysis. As the reaction temperature increases, the 101 band also splits into 100 and 101 peaks.

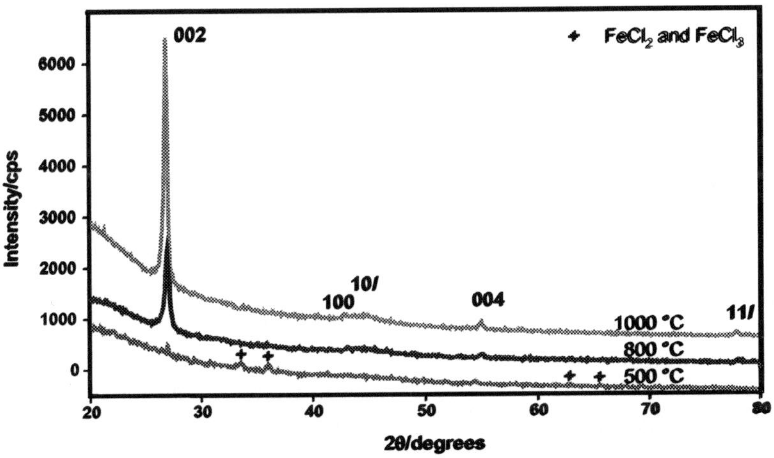

Figure 3. XRD patterns of some carbon samples synthesized at different temperatures.

XRD data for selected samples are shown in Table 1. The interplanar spacings, d_{002} and d_{004}, were evaluated from the positions of the 002 and 004 peaks respectively by applying Bragg's equation. The crystallite size L_c along the c-axis was calculated from the 002 peak using the Sherrer formula with a value of constant K=0.9.[13] The crystallite size L_a along the basal planes was calculated from the 100 peak, using the Warren-Bodenstein formula with a value of K=1.77.[14] L_a values of 45-50 nm were obtained from the XRD pattern analysis. They agree with the values of 30-55 nm estimated from the Raman spectroscopy using Tuinstra and Koenig's equation.

Table 1. XRD data for the carbon samples.

Reaction temperature	d_{002} (d_{004}) [nm]	$FWHM_{002}$ [°]	L_c [nm]	L_a [nm]
500 °C	-	-	-	28.1
800 °C	0.331 (0.334)	0.3657	22.4	45.5
1000 °C	0.333 (0.334)	0.2991	27.3	47.7

3.4. Scanning and Transmission Electron Microscopy

The effect of the reaction temperature on CDC microstructure is illustrated in Figs. 4 and 5. An SEM micrograph of 'as-received' Fe_3C is shown in Figure 4a. At lower reaction temperatures (<500°C), overall volume of the original Fe_3C grains is converted to a fluffy, disordered carbon by formation and evaporation of volatile iron chlorides. Chlorination at mild regimes yields a grainy (for 600°C, Fig. 4b) or striated (for 800°C, Fig. 4c) texture suggesting formation of graphite lamellae. Conversion of Fe_3C to carbon is possible without affecting the original Fe_3C grain morphology up to a certain temperature, between 600°C and 800°C. Samples obtained by chlorination at 1000°C show features of expanded graphite (Fig. 4d).

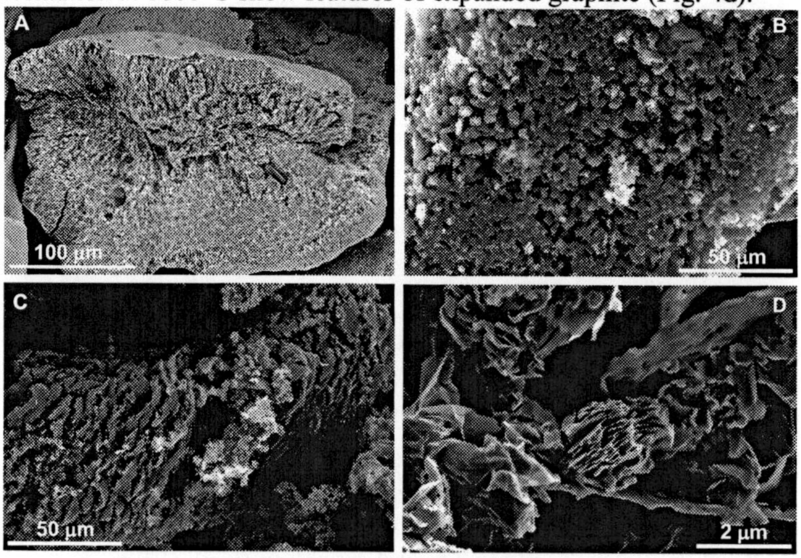

Figure 4. SEM micrographs: a.) As-received Fe_3C powder, Sample chlorinated at: b.) 600°C, c.) 800°C, and d.) 1000°C.

A very different carbon microstructure was achieved after chlorination at 1200°C, as shown in Fig. 5. Several millimeter-size graphitic spheres (one shown in Fig. 5a) were observed. Opening the spheres revealed that they are hollow inside, and their walls are built of very thin graphite flakes (Fig. 5b). The sphere's inner surface was found to be overgrown with graphite crystals. An example is shown in Fig. 5c. Figures 5d-e show the morphology of very thin graphite layers, which are formed during the synthesis process due to the removal of iron salts that intrinsically intercalate graphite. However, it is important to note that neither XRD nor TEM and Raman analysis found intercalation[15] compounds present in CDCs after their synthesis. Above the eutectic temperature in the iron-Fe_3C system, the liquid phase is formed producing melt droplets. Upon interaction with chlorine, iron is easily removed from the iron-carbon melt that becomes supersaturated with carbon. Planar graphite nucleates and grows from such melt. The surface of a sample synthesized at 1200°C is shown in Fig. 5c.

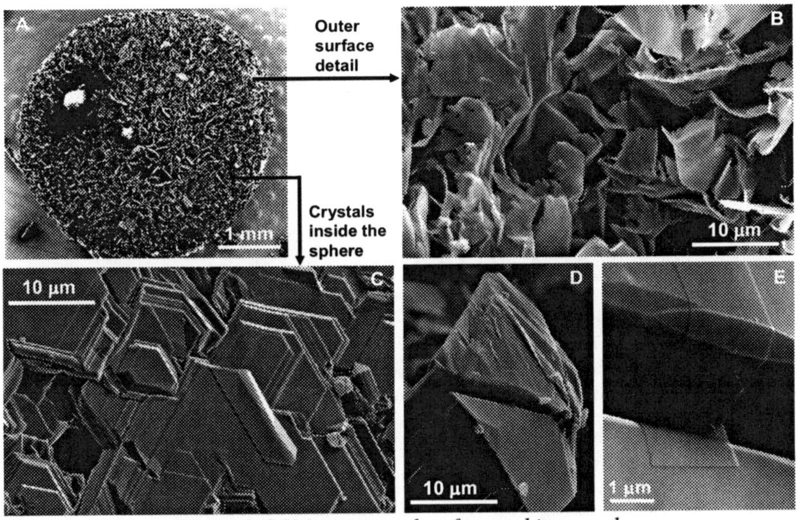

Figure 5. SEM micrographs of a graphite sample produced by chlorination of Fe_3C at the temperature of 1200°C.

TEM images of CDC powders produced at different temperatures are shown in Figs. 6-8. Similar to the SEM image in Fig. 4b, Fig. 6a shows the agglomerated microstructure of the CDC powders produced at low temperatures (400 - 600°C). The HRTEM image (Fig. 6b) shows that the particles produced at 400°C are mainly built of amorphous carbon, while minor local ordering and formation of turbostratic graphite can be found at the edge of the particles even at this very low temperature. For comparison, in CDC produced from other carbides, such as Ti_3SiC_2, graphitic carbon was not found below 700°C.[3] The measured crystallite size along graphite basal planes for 400°C is on average less then 5 nm, which is in good agreement with the L_a value estimated from the Raman spectra.

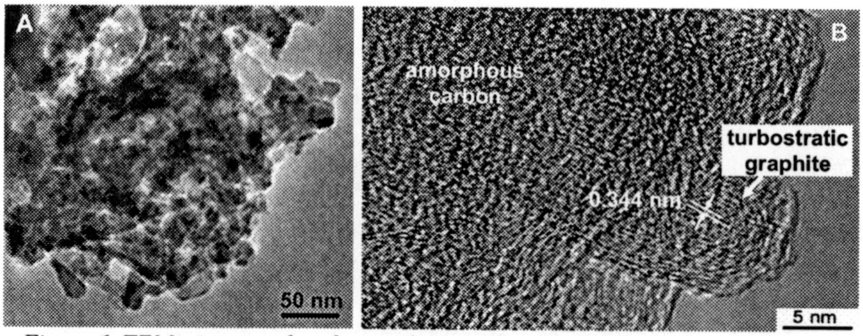

Figure 6. TEM micrographs of amorphous carbon and turbostratic graphite in a sample produced by chlorination of Fe_3C at 400°C.

Pronounced graphitization has been observed during high temperature CDC synthesis. CDC powders produced at 800°C consist of thin sheets of graphite (Fig. 7). Amorphous carbon is still present in the samples, but in significantly smaller quantities (less then 10 % vol.). For synthesis at 1200°C, TEM analysis (Fig. 8) shows that perfect graphite crystals with an interplanar distance of 0.334 nm are produced, indicating complete graphitization.

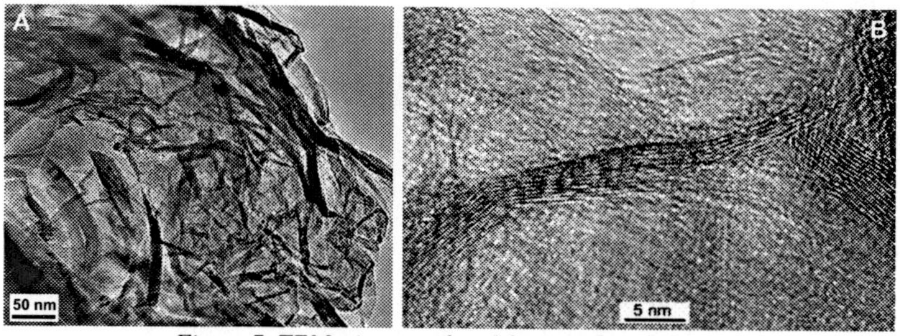

Figure 7. TEM micrographs of thin graphite flakes in a sample chlorinated at 800°C.

Graphite materials produced at 600-1100°C may find applications in lithium batteries and supercapacitors. Currently, similar flakes are produced in a complex process including graphitization at above 2500°C,[16] followed by intercalation and exfoliation of graphite[15]. Here we demonstrate that synthesis of graphite from iron carbide can be done in one step at moderate temperatures.

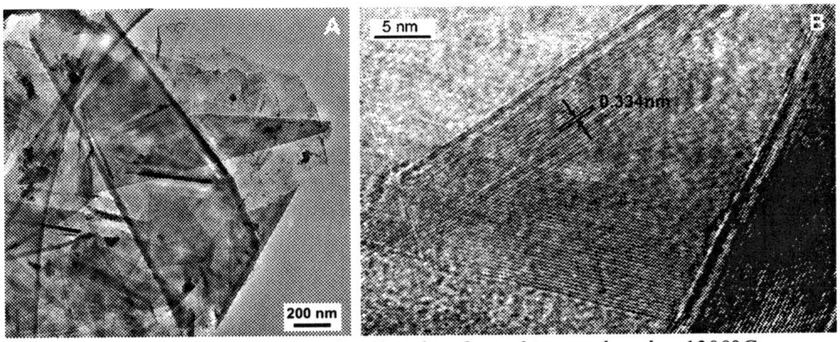

Figure 8. TEM micrographs of well-ordered graphite produced at 1200°C.

4. CONCLUSIONS

Chlorination of Fe_3C at temperatures of 400°C and above results in the formation of carbon and solid or gaseous iron chlorides. Three temperature regimes have been defined. Amorphous carbon is formed at temperatures of 400 – 500°C. Flakes and ribbons of nanocrystalline graphite form at 600 – 1100°C. Graphite ordering increases with temperature. Formation of microcrystalline graphite and millimeter-size graphite crystals occurs above the eutectic point in the Fe/Fe_3C system.

ACKNOWLEDGEMENTS

We thank Julius Chang and Jerome P. Downey for providing the iron carbide sample. This work was supported by DARPA via ONR contract. The purchase of the Raman spectrometer and environmental SEM was supported by NSF grants DMR-0116645 and BES-0216343.

REFERENCES

1 A. Nikitin and Y. Gogotsi, 'Nanostructured Carbide-Derived Carbon (CDC)', ed. H. S. Nalwa, American Scientific Publishers, 2003.
2 Y. Gogotsi, S. Welz, D. A. Ersoy, and M. J. McNallan, *Nature*, 2001, **411**, 283.
3 Y. Gogotsi, A. Nikitin, H. Ye, W. Zhou, J. E. Fischer, B. Yi, H. C. Foley, and M. W. Barsoum, *Nature Materials, in press*, 2003.
4 W. A. Mohun, Proc. 4th Biennial Conf. Carbon, Oxford, 1959, p. 443.
5 J. Leis, A. Perkson, M. Arulepp, M. Kaarik, and G. Svensson, *Carbon*, 2001, **39**, 2043.
6 H. O. Pierson, 'Handbook of Carbon, Graphite, Diamond and Fullerenes: Properties, Processing and Applications', ed. R. F. Bunshah, G. E. McGuire, and S. M. Rossnagel, Noyes Publications, 1993.
7 M. Inagaki, 'New Carbons: Control of Structure and Functions', Elsevier, 2000.
8 J. Leis, A. Perkson, M. Arulepp, P. Nigu, and G. Svensson, *Carbon*, 2002, **40**, 1559.

9 D. A. Ersoy, M. J. McNallan, and Y. Gogotsi, *Mat. Res. Innovat.*, 2001, **5**, 55.
10 F. Tuinstra and J. Koenig, *J. Chem. Phys.*, 1970, **53**, 1126.
11 N. S. Jacobson, Y. G. Gogotsi, and M. Yoshimura, *J. Mater. Chem.*, 1995, **5**, 595.
12 J. M. Tanzosh, in 'Piping materials', ed. M. L. Nayyar, 2000.
13 J. Biscoe and B. Warren, *J. Appl. Phys.*, 1942, **13**, 346.
14 B. Warren and P. Bodenstein, *Acta Crystallogr.*, 1966, **5**, 602.
15 A. Yoshida, Y. Hishiyama, and M. Inagaki, *Carbon*, 1991, **29**, 1227.
16 D. Gonzalez, M. A. Montes-Moran, and A. B. Garcia, *Energy & Fuels*, 2003, in press.

HIGH RESOLUTION TRANSMISSION ELECTRON MICROSCOPY IMAGE ANALYSIS OF DISORDERED CARBONS USED FOR ELECTROCHEMICAL STORAGE OF ENERGY

Jean-Noël Rouzaud*, Christian Clinard, Frédéric Chevallier, Alexandre Thery and François Béguin

CRMD, CNRS-Université, 1B Rue de la Férollerie, 45071 Orléans Cedex 02, France

1. INTRODUCTION

Numerous forms of more or less disordered carbons (soft carbons, hard carbons, fibers, nanotubes, mesophase microbeads, etc.) are presently tested for different energy storage applications. For instance, microporous carbons i.e. hard carbons, are frequently studied as anodes for lithium-ion batteries, whereas microporous/mesoporous solids such as carbon areogels seem to be optimal for supercapacitor applications, or hydrogen storage[1,2]. If the mechanism of lithium intercalation between the infinite and perfect graphene layers of a graphite crystal is well understood[3], it is not the case with disordered carbons. These carbons are made of short and poorly stacked polyaromatic layers with a large distribution of layer length and interlayer spacings, and, consequently various sites are available for lithium ion trapping.

Disordered carbons usually exhibit a multiscale organization (structure, microtexture, texture)[4]. Structurally, they are made of more or less distorted polyaromatic layers, nanometric in size. The spatial association or the layers, from the nanometric to the micrometric scales, gives rise to different microtextures (lamellar, porous, concentric, fibrous, etc.) forming the carbons skeleton[4]. The multiscale organization is the fingerprint of the kind of precursor and of the formation conditions (temperature, pressure, strains, time, etc.) met either in laboratory experiments or in Nature, and is directly related with numerous properties.

As far as the diffraction data (as well electron and X-ray diffraction) of disordered carbons are concerned, only averaged data can be obtained from the broad and faint bands due to the reflections on the nanometer-size

* Corresponding author. E-mail: rouzaud@cnrs-orleans.fr

I.V. Barsukov et al. (eds.), New Carbon Based Materials for Electrochemical Energy Storage Systems, 411–424.

coherent domains. Only the 002 reflection on the aromatic layers stacks is usually studied. The average interlayer spacing d_{002} and the height L_c of the coherent domains are determined from the position and peak width, respectively. Unfortunately, neither de-averaged structural data, nor microtextural information can generally be obtained. By contrast, High Resolution Transmission Electron Microscopy (HRTEM) allows the organization to be determined by imaging directly the profile of the polyaromatic layers[5]. Two particular examples of disordered carbons are shown in this paper : (i) microporous carbon fibers (Figure 1), (ii) carbon aerogels (see Figure 8). However, up to now, the information from such complex images is mainly qualitative, whereas quantitative structural and microtextural data are required to precisely describe the organization, to propose models and to foresee the properties, for example energy storage properties.

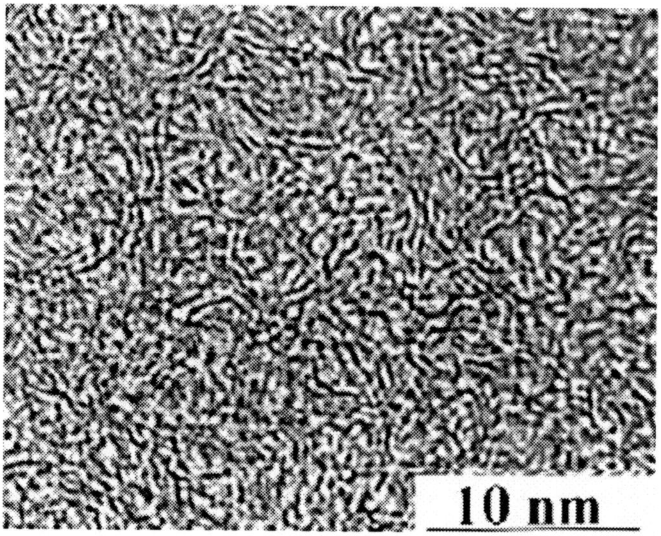

Figure 1. HRTEM image of a microporous carbon fiber obtained from the pyrolysis of viscose at 1000°C.

Various models were recently proposed to interpret the lithium storage performance of disordered carbons, especially the house of cards model from Zheng *et al*[6] and the porous carbon made of 'misoriented graphitic-like crystallites' from Han *et al*[7]. They are usually based on X-ray techniques (diffraction and diffusion). Some structural features of these models disagree with the HRTEM images (see Figure 1): only small and "single" layers in the house of card model, too ordered and large "graphitic crystallites" in the porous model. Furthermore, the models are rather qualitative, providing no data on the layer lengths and interlayer spacing for instance. Consequently, image analysis techniques are required to obtain more quantitative data, as it was recently successfully shown for chars or soots[8]. We have undertaken this challenging task by developing an in-house HRTEM image analysis procedure allowing useful structural data to be extracted[4]. The results obtained on two types of disordered carbons used for electrochemical energy storage, e.g. a carbon-carbon composite based on an isotropic carbon fiber coated with pyrocarbon and carbon aerogels, will be presented in this paper.

2. HRTEM IMAGE ANALYSIS OF DISORDERED CARBONS

The multiscale organization (structure, microtexture, texture) was studied by HRTEM, using a Philips CM 20 microscope, operating at 200 kV. Adequate under focusing was selected in order to be on the transfer function plateau. Interferences between the transmitted electron beam and hkl diffracted beams produce fringes, which directly image the profile of the hkl planes in the Bragg angle, i.e. quasi-parallel to the incident electron beam. The optimal resolution of 0.14 nm allows seeing the profile of the aromatic layers on the Bragg angle, even for weakly organized carbons (in this case, due to the reciprocal node elongation, the angle tolerance on the Bragg angle can reach +/- 10°). Since HRTEM images are 2D projections, objects thinner than 10 nm are required to avoid fringe superimposition. So thin sections are almost impossible to be obtained by ultramicrotomy or ion thinning, therefore we used classical grinding under alcohol in an agate mortar. After sonifying of the alcoholic suspension, a droplet is deposited on a lacey carbon grid and dried. Using the edges of the thinnest fragments deposited across the holes minimizes risks of fringe superimposition.

Our image analysis procedure especially developed for disordered carbons is presented in detail in the reference[4]. At first the images are skeletonized as follows. 20 x 20 nm representative areas of raw HRTEM images obtained with a 300,000 x magnification are sampled and digitalized on 1024 x 1024 pixels with a resolution of 4000 ppi; one pixel then corresponds to 0.019 nm. The first step of the image analysis process is to obtain a homogeneous image without background noise. Specific software modules added to the Visolog (Noesis) commercial software were developed in order to filter the image at the Fourier transform (FT) level. The 'clean' resulting image, always characterized by 256 gray levels reproducing the intensity of each pixel, is then threshold and binarised. Finally, the binary images are skeletonized giving one-pixel width lines for fringes. The skeletonized pixel-based image is transformed into a vectorial image, where the points coordinates determine each fringe which can be analyzed individually, or/and in relation with its neighbours. Some criteria have to be taken into account to avoid artefacts: (1) since the size of each aromatic ring is 0.246 nm, all the fringes shorter than this value have to be eliminated; (2) a fringe crossed by segments smaller than the C-C bond length, i.e. 0.142 nm, will be considered rectilinear; (3) the van der Waals interaction is supposed to be negligible between graphene layers spaced by more than 0.7 nm (twice the d_{002} value of graphite); consequently, pairs of more spaced fringes and/or forming an angle larger than 15° are considered non-stacked and independent.

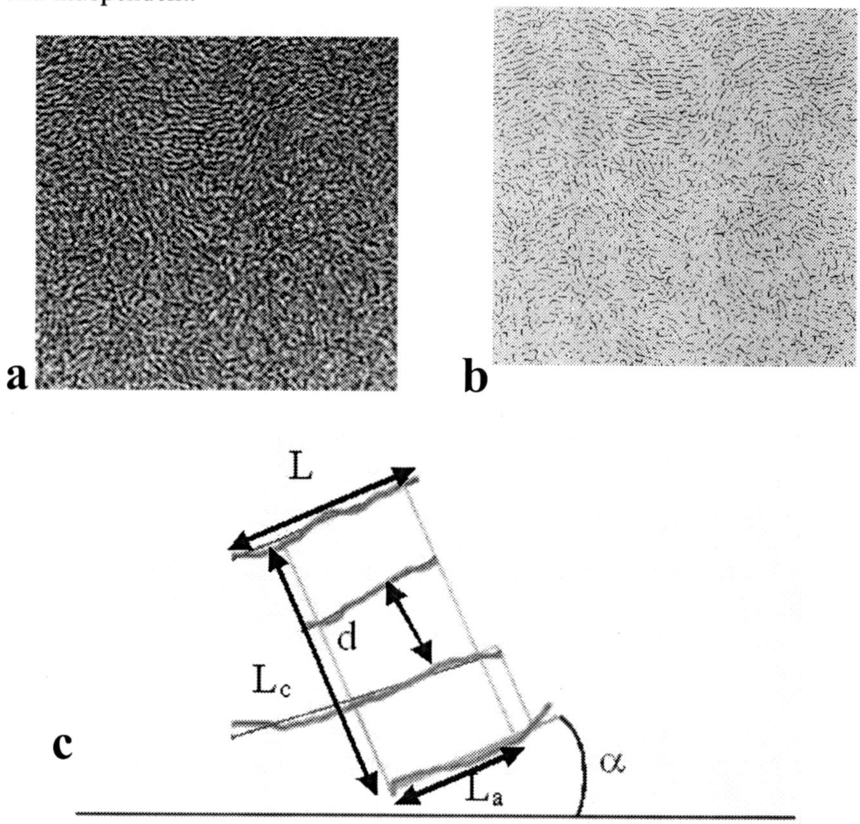

Figure 2. HRTEM image analysis. a: raw HRTEM image; b: corresponding skeletonized image; c: limits of a coherent domain, definition of the structural (L, L_a, L_c, d) and microtextural (α) parameters; N is the number of layers stacked within a domain.

Usually, at least several hundreds of fringes are measured and characterized. Taking the classical example of X-ray diffraction, coherent domains are defined by the stacks of polyaromatic layers (Figure 2). The coherent domains are distinguished from the single layers and their relative abundance is determined. They are characterized by their diameter L_a and their height L_c along c axis, the number N of stacked layers and the interlayer spacings d. Whereas X-ray diffraction only gives access to averaged data, the distribution of each structural parameter can be represented by bar diagrams and average values can be calculated.

3. RESULTS AND DISCUSSION

3.1 Application to the Study of Carbon Anodes for Lithium-Ion Batteries

Most of the commercially available lithium-ion batteries are based on a graphite negative electrode. In standard conditions, the reversible capacity of graphite is limited to 372 mAh/g, that corresponds to the formation of the saturated 1st stage intercalation compound LiC_6[3]. Higher reversible capacities were demonstrated by disordered carbons giving up to twice the reversible capacity of graphite[6]. However, these carbons are characterised by an important irreversible capacity due to the electrolyte decomposition on the carbon surface and a polarization between charge and discharge, so-called hysteresis. Recently, few papers claimed that the extent of electrolyte decomposition is efficiently depressed by a pyrolytic carbon coating[9-13].

For our experiments, a carbon fiber cloth (Figure 3a) was prepared by carbonization of viscose under neutral atmosphere for 15 minutes, successively at 400, 700 and 1000°C. The carbon cloth was coated with pyrolytic carbon, using chemical vapor decomposition of propylene (2.5 ml/mn) diluted in nitrogen (100 ml/mn) during 10 minutes at 900°C. The resulting composite carbon material exhibits a very low irreversible capacity and 1.5 times the reversible capacity of graphite[9-11].

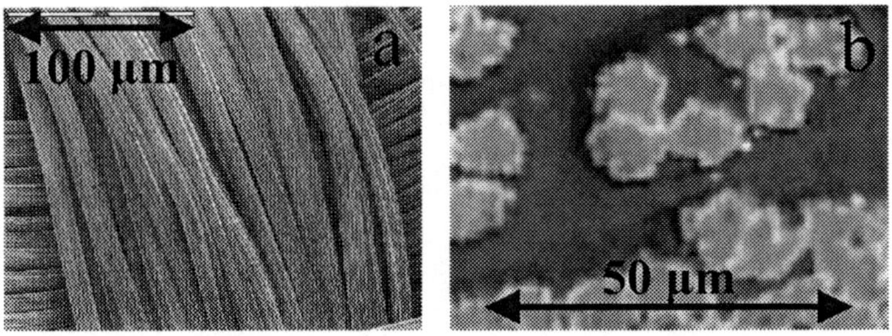

Figure 3. Images of a cross-section of carbon fibers after propylene pyrolysis. 3a : Scanning Electron Microscopy of a piece of the carbon cloth. 3b : optical microscopy (crossed polarizers with a wave retarding plate).

Figure 3 shows the images of a cross-section obtained by ultramicrotomy of the coated fibers. Figure 3b shows an optical microscopy image obtained in reflexion mode, between crossed polarizers and with a wave retarding plate; it gives textural information (micrometric scale). The fibers, which are optically isotropic at this scale, show a characteristic magenta tint, whereas the yellow and blue tints indicate an anisotropic pyrocarbon coating (due to a preferential orientation of the aromatic layers along the fiber axis).

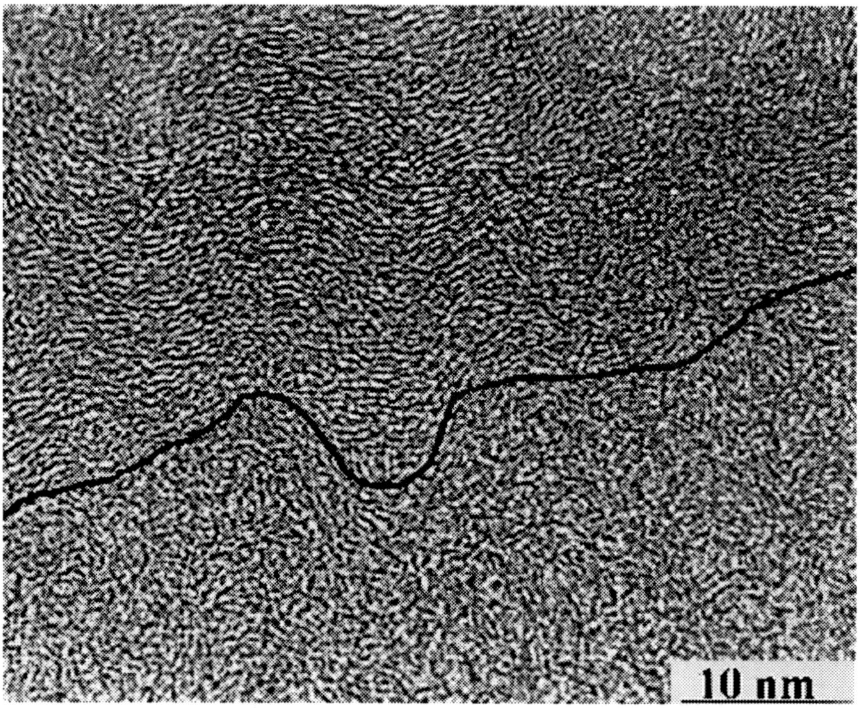

Figure 3c. HRTEM image of a cross-section of a carbon fiber after propylene pyrolysis: the black line represents the boarder between the lamellar pyrocarbon (at the top) and the microporous fiber (at the bottom).

HRTEM allows to image directly the microtexture and the structure of each component of the fiber-pyrocarbon composite (Figure 3c). The aromatic layers of the fiber core are highly misoriented and frequently non-stacked, as shown by the lower part of the image; this explains the isotropy observed by optical microscopy. By contrast, the layers of the pyrocarbon coating are more frequently stacked in Basic Structural Units (BSUs), as shown on the upper part of the image. The BSUs are rather parallel to the external surface of the fiber, being responsible for the lamellar microtexture and the anisotropic rim in optical microscopy; the coating thickness is estimated here to about 20 nm. Image analysis indicates that the amount of non-stacked layers is 40% and 30% for the fiber and pyrocarbon, respectively. Furthermore, about 90% of the BSUs in the fiber are only formed of 2 or 3-stacked layers, whereas more than 25 % of the BSUs in the coating contain more than 3 layers. In the histograms of interlayer spacings, a well visible maximum is found at about 0.4 nm for pyrocarbon, whereas the values range mainly from 0.4 to 0.6 nm for the fiber (Figure 4). Hence, the microtexture of the pyrocarbon is denser than that of the fiber.

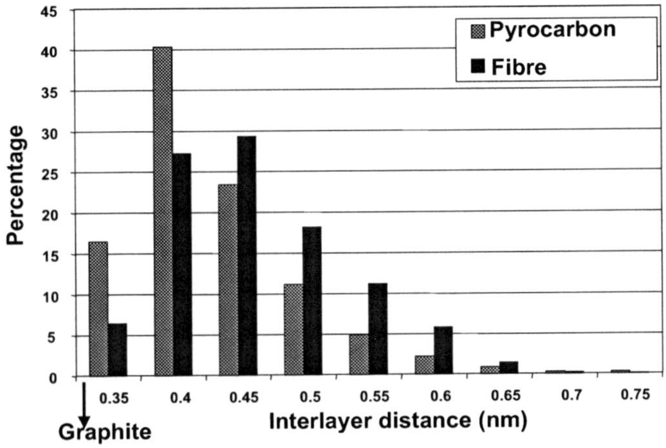

Figure 4. Histogram of the interlayer spacings d obtained by computer analysis of the HRTEM image of a cross-section of a carbon fiber coated by pyrocarbon.

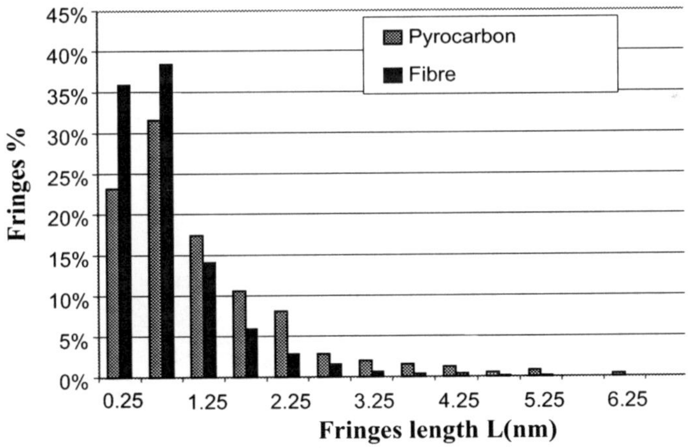

Figure 5. Histogram of the fringe length L obtained by computer analysis of the HRTEM image of a cross-section of a carbon fiber coated by pyrocarbon.

The histogram of Figure 5 shows that the layers of both kinds of carbon are short, especially in the case of the fiber where about 75 % of the layers are smaller or equal to 0.75 nm, i.e. to 3 cycles. By contrast, this proportion is only about 55 % for the pyrocarbon.

The tortuosity of each fringe (and consequently of the layer) was estimated with a structural parameter called the "order degree". The layers forming the fiber are more distorted than the layers of the pyrocarbon coating. It must be noted that, even if the distorted layers appear discontinuous by TEM (some parts can be out of the Bragg angle), they are in fact continuous, as proved by the total insolubility of these carbons. Continuous layers are formed by small pieces of flat layers connected each other by defects (pentagons, tetrahedral carbons, etc.), which limiting their extent. The higher the density of defects, the smaller the order degree and the extent of planar layers. In fact, the image analysis shows that the shortest layers are usually characterized by the highest interlayer spacings (>0.4 nm). It is important to underline that large interlayer spacings, generally formed by more or less oblique pairs of layers, are not easily detected by XRD (broad and faint corresponding 002 reflections not be easily distinguishable from the background noise). Two extreme cases are observed in these disordered carbons: (i) large stacks of high order degree with a noticeable number (> 3) of long (> 2 nm) and planar layers; and (ii) pairs of short (< 1 nm) and misoriented layers, the latter forming slit-shaped pores. Both fiber and pyrocarbon contain the two types of stacks, but large stacks are rather present only in the pyrocarbon coating, whereas the slit-shape pores are especially frequent in the fiber.

Based on these HRTEM data, a sketch of the multiscale organisation of the composite is proposed[13] in Figure 6. Even if this representation is very schematic, it illustrates the two types of carbon: short and very disordered layers mainly in the fiber, and a developed preferential orientation of longer and better stacked layers in the pyrocarbon coating.

Figure 6. Sketch of the cross-section of a carbon fiber coated by pyrocarbon, taking into account the HRTEM image analysis data.

3.2. Consequences on the Electrochemical Properties

The kind of model presented in Figure 6 allows the electrochemical characteristics of the carbon-carbon composite used as anodes of lithium-ion battery to be better understood. First, let us try to interpret the high irreversible capacity of the carbon fiber and its decrease after coating by pyrocarbon. The active sites responsible for the electrolyte decomposition during the first reduction process correspond to the edge planes and the defects developed in the graphene planes. Such defects are numerous in the as-received fiber, therefore the irreversible capacity is important. By contrast, the electrochemical characterization[11] of the composite shows that the passivation layer is still formed, but is less developed. In fact, the pyrocarbon coating, which is directly exposed to the electrolyte, contains much less defects and, due to its lamellar microtexture, only few edge planes are accessible for adsorbed species. The pyrocarbon coating forms a barrier preventing the diffusion of the large solvated lithium ions to the active sites of the fibers that results in a low value of irreversible capacity. This observation agrees well with the CO_2 adsorption measurements, which show that the micropore surface area strongly decreases from 480 to 60 m^2/g after pyrocarbon deposition.

As far as the reversible capacity is concerned, *in situ* ^{7}Li NMR experiments[12-13] reveal the presence of 2 sites for the lithium trapping in this composite: "truly" intercalated lithium with a characteristic Knight shift of 18 ppm, and "quasi-metallic" lithium clusters characterized by a shift increasing up to 110 ppm. The latter could correspond to lithium trapped in the nanometer-sized voids between misoriented BSUs. However, the HRTEM image analysis shows that the proportion of these kinds of micropores do not represent more than 10 % of the fiber image, i.e. too limited to explain the reversible capacity 50% higher than the graphite one. Consequently, we assume that the "quasi-metallic" lithium is trapped in subnanometric bidimensional clusters between the pairs of spaced graphene layers forming slit-shaped pores.

Reminding that image analysis revealed the presence of two extreme types of stacking: (i) large stacks of long and planar graphene layers, (ii) pairs of short and misoriented layers forming slit-shaped pores, let us consider the electron affinity of the aromatic layers adjacent to inserted lithium. As reference values, the electron affinity is 0 eV for benzene, 1.8 eV for anthracene (3 condensed rings - molecule length: about 0.75 nm) and reaches 4.39 eV in the case of graphite. Consequently, lithium can intercalate, with a charge transfer to carbon, in the well-ordered stacks of long layers (interlayer spacing smaller than 0.4 nm) presenting a high electron affinity. By contrast, the electron affinity of the "slit-shaped pores", with short graphene layers, is very low. Hence, lithium located in these rather large interlayer spacings (> 0.4 nm) undergoes a very weak charge transfer, being in a state close to metal which we call "quasi-metallic".

Coupling HRTEM image analysis, *in situ* ^{7}Li NMR experiments and galvanostatic cycling of lithium/composite carbon cells, we propose the "Corridor model" shown in Figure 7[12-13]. During reduction, lithium intercalates at first in the smallest intervals between the nanometer size graphitic type layers, giving a maximum shift of 18 ppm. Then, lithium diffuses into the slit-shaped pores where the ions are screened by delocalized electrons through a backdonation process, forming the quasi-metallic clusters with a shift progressively increasing up to 110 ppm. During the galvanostatic oxidation, the reverse process occurs, with a complete disappearance of the two NMR peaks. When oxidation is fully completed, there is only one peak at ca. 0 ppm characteristic of the Solid Electrolyte Interface (SEI).

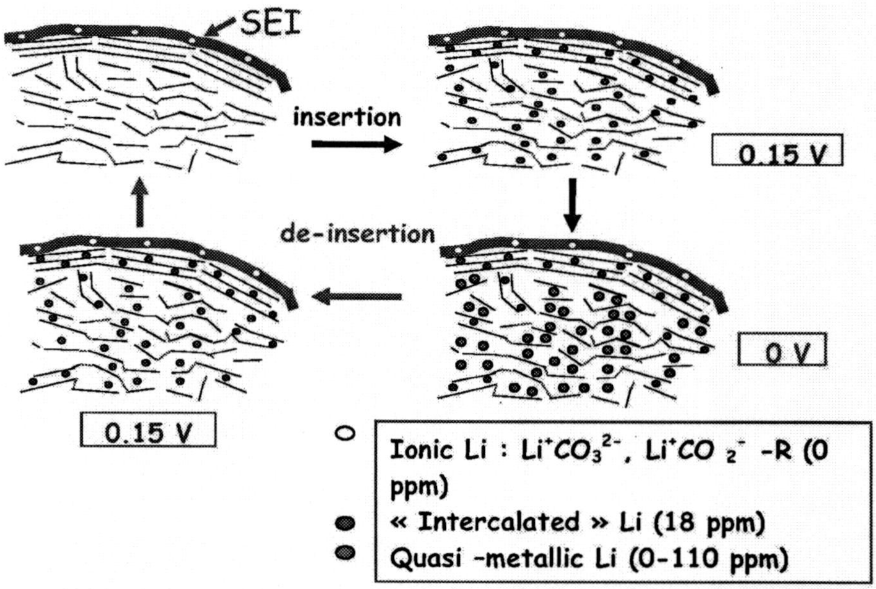

Figure 7. Schematic model based on the TEM image analysis and on in situ ^{7}Li-NMR during galvanostatic reduction/oxidation of the carbon composite. During insertion, ionic lithium penetrates at first in the smallest interlayer spacings, then it diffuses in the slit-shaped pores where quasi-metallic clusters are formed.

In conclusion, the composite formed by a carbon fiber cloth coated with pyrocarbon is a very promising material for the negative electrode of lithium batteries, since it couples a high reversible capacity due to the disordered fiber (1.5 time the graphite value) with a small irreversible capacity due to the dense pyrocarbon coating which blocks the diffusion of solvated lithium to the active sites of the fiber.

3.2. Characterization of Carbon Aerogels

New forms of carbon, such as carbon aerogels, were designed in order to get materials with a controlled pore size distribution adapted for galvanostatic energy storage in aqueous medium[14]. In a first step, organic aerogels are synsethised from phenolic resins using a soft chemistry approach[15-17]. Then, they are pyrolysed under an inert atmosphere to form carbon aerogels[18-19]. Different organic aerogels can be obtained over an enough large density range (from 0.05 to 0.3) by playing on the precursor (resorcinol, phloroglucinol) and the nature and amount of catalyst (Na_2CO_3, $Ca(OH)_2$, etc.). Moreover, using the same organic aerogel, but changing the pyrolysis temperature (between 800 and 2600°C) allows different carbon aerogels to be obtained. The multiscale organization of the carbon aerogels was directly imaged by HRTEM over 3 orders of magnitude (μm – nm). The example of the carbon aerogel obtained at 1050°C from a $Ca(OH)_2$ based organic aerogel (Resorcinol/Catalyst = 45) is shown in Figure 8.

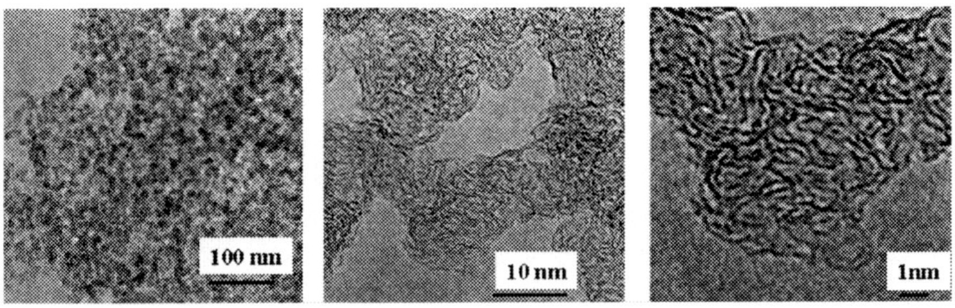

Figure 8. Multiscale organization of carbon aerogels imaged by TEM. Example of the $Ca(OH)_2$ based organic aerogel (Resorcinol/Catalyst = 45) heated at 1050°C.

At the textural scale, this carbon is constituted of the arrangement of round-shaped nodules, about 20 nm in diameter; mesopores are formed by the voids between the nodules. The nodules are made of more or less distorted layers, with numerous non-stacked layers. The HRTEM images suggest that slit-shaped pores can be formed by pairs of spaced polyaromatic structures, whereas the misorientation of these units could be responsible for a small mesoporosity. The final density of the carbon aerogel depends on the antagonistic effects of weight loss and shrinkage, both increasing with the highest temperature of treatment (HTT). The micropore and mesopore volumes were determined from the gas adsorption data. As the carbon aerogel density increases, the micropore volume determined by CO_2

adsorption at 293K (giving pore size smaller than 0.7 nm) decreases, whereas the mesopore volume (pore size larger than 20 nm) determined from the N_2 desorption isotherm at 77K increases.

After heat-treatment at 2600°C, the mesopores become clearly visible, with walls formed of few well-staked layers (2-3 for Na_2CO_3 based carbon aerogels, and 3-5 for the $Ca(OH)_2$ based ones). As for carbon blacks, the diameter of the closed pores obtained above 2000°C is strongly determined by the initial size of the nodules[5] in the low-temperature carbon aerogels. The following table reports structural data determined by HRTEM image analysis on carbon aerogels obtained at 1050°C and 2600°C from $Ca(OH)_2$- and $Na_2(CO)_3$-based organic aerogels.

Table. Structural data from computer analysis of the HRTEM images obtained on carbon aerogels formed by pyrolysis at 1050 and 2600°C of $Na_2(CO)_3$ and $Ca(OH)_2$ based aerogels heated.

Catalyst, pyrolysis temperature	L (nm)	% nsl*	N	d (nm)
$Ca(OH)_2$, 1050°C	0.7	49	2.4	0.47
$Na_2(CO)_3$, 1050°C	0.7	56	2.4	0.51
$Ca(OH)_2$, 2600°C	1.3	33	2.8	0.41
$Na_2(CO)_3$, 2600°C	0.8	42	2.5	0.45

* nsl = non stacked layers

Both carbons formed at 1050°C are highly disordered: about half of the layers are non-stacked; the layer extent is small (about 0.7 nm) and the interlayer spacing important (about 0.5 nm). As far as the catalyst effect is concerned, the Na_2CO_3 based carbon aerogels are slightly more disordered, as shown by higher amounts of non stacked layers and larger interlayer spacing. This confers a strong microporous character to this aerogel with numerous and well-defined slit-shaped pores. As expected, the heat-treatment at 2600°C leads to a noticeable structural improvement: the layer length increases, the proportion of stacked layers increases and the interlayer spacing decreases; consequently the micropore amount strongly decreases. All these features are especially verified in the case of the $Ca(OH)_2$ based carbon aerogel.

Attempts of galvanostatic cycling have been performed on carbon aerogels obtained at 1050°C, 1300°C, 1700°C, 2600°C in a potential range from 0 to –1.5 V *vs* NHE. Two effects are observed successively upon decreasing potential: the formation of the electrochemical double layer and water electrolysis at potential lower than –0.9 V *vs* NHE with hydrogen storage in the micropores[14]. The capacitive contribution decreases with the HTT. Multi-sites occupation of hydrogen is observed for the low-temperature carbons and a single peak in the voltammetry curve of the 2600°C treated sample seems to indicate a phenomenon close to hydrogen intercalation. About 1% of hydrogen can be stored then released from these porous materials. As expected, the values are slightly higher in the case of the most disordered material, i.e. the Na_2CO_3 carbon aerogels.

4. CONCLUSIONS

Our procedure of computerized HRTEM images analysis allows quantitative and de-averaged structural data on carbons to be obtained. Such data give the solid experimental basis required to build reliable models of organization. Moreover, the relationships between formation conditions - carbon material organization and properties can thus be specified. Such approach was successfully applied to two types of disordered carbons presently tested for the electrochemical storage of energy: a carbon-carbon composite material (a microporous carbon fiber-based cloth coated with a pyrocarbon) used as anode for lithium-ion battery, and carbon aerogels with a controlled pore size distribution (micropores and mesopores) adapted for galvanostatic energy storage in aqueous medium. In the first case, coupling *in situ* ^{7}Li-NMR- potentiometry-HRTEM image analysis leads to a relevant model of lithium insertion and de-insertion in these anodes ('Corridor model'). For the second case, different ways of energy storage in aqueous medium are presently tested on the designed aerogels; in the future, the storage capacity will be discussed taking to account the data on their multiscale organization quantified by HRTEM image analysis.

REFERENCES

1. Frackowiak E., Béguin F. Carbon materials for the electrochemical storage of energy in capacitors. Carbon 2001; 39: 937-50.
2. Frackowiak E., Béguin F. Electrochemical storage of energy in carbon nanotubes and nanostructured carbons. Carbon 2002; 40:1775-87.
3. Guérard D., Hérold A. Intercalation of lithium into graphite and other carbons. Carbon 1975; 13:337-45.
4. Rouzaud J.N. and Clinard C. Quantitative high-resolution transmission electron microscopy: a promising tool for carbon materials characterization. Fuel Proces. Technol. 2002; 77-78:229-35.
5. Oberlin, Agnès. "High Resolution TEM studies of carbonization and graphitization". In *Chemistry and physics of carbon*, Vol 22, Thrower P.A. (Ed.), Marcel Dekker, New-York, 1989, p. 1-143.
6. Zheng T, Dahn JR "Applications of carbon in Lithium-ion batteries". In *Carbon materials for advanced technologies*, Burchell T.D. Ed., Elsevier, Oxford, 1999, pp 341-88
7. Han Y.S., Yu J.S., Park G.S., and Lee J.Y. Effects of synthesis temperature on the electrochemical characteristics of pyrolytic carbon for anodes lithium-ion secondary batteries. J. Electrochem. Soc. 1999; 146:3999-4004.
8. Shim H.S., Hurt R.H. and Yang N.Y.C. A methodology for analysis of 002 lattice fringe images and its application to combustion-derived carbons. Carbon 2000; 38:29-45.

9. Chevallier F., Gautier S., Salvetat J.P., Clinard C., Frackowiack E., Rouzaud J.N. and Béguin F. Effects of post-treatments on the performance of hard carbons in lithium cells. Journal of Power Sources 2001; 97-98:143-5.
10. Yoon S., Kim H. and Oh S.M. J. Power Sources. 2001, 94.
11. Chevallier F., Vix C., Saadallah S., Rouzaud J.N., Frackowiak E. and Béguin F. A better understanding of the irreversible lithium insertion mechanisms in disordered carbons, J. Phys. Chem. Solids 2004; 65:211
12. Letellier M., Chevallier F., Clinard C., Frackowiak E., Rouzaud J.N., Béguin F. The first *in situ* ^{7}Li nuclear magnetic resonance study of lithium insertion in hard-carbon anode materials for Li-ion batteries, J. Chem. Phys. 2003; 118:6038-45
13. Chevallier F., Letellier M., Morcrette M., Tarascon J.M., Frackowiak E., Rouzaud J.N., Béguin F. *In situ* ^{7}Li-Nuclear Magnetic Resonance Observation of Reversible Lithium Insertion into Disordered Carbons, Electrochem. Solid State Lett. 2003; 6:A225-8.
14. Jurewicz K., Frackowiak E., Beguin F. Towards the mechanism of electrochemical hydrogen storage in nanostructured carbon materials. Appl. Phys. A, in press (2004).
15. Pekala RW. Organic aerogels from the polycondensation of resorcinol with formaldehyde J. Mat. Sci. 1989; 3221-7.
16. Pekala R.W., Alviso C.T., Kong F.M., Hulsey S.S. Aerogels derived from multifunctional organic monomers. J. Nn-Cryst Solids 1992; 145:90-8.
17. Pekala RW and F.M. Kong FM. A synthetic route to organic aerogels-mechanism, structure, and properties. Revue de physique appliquée. ColloqueC4, tome 24, avril 1989.
18. Pekala R.W., Alviso C.T. Carbon aerogels and xerogels. Mater Res Soc Symp Proc 1992; 270-3.
19. Tamon H, Ishizaka H, Mikami M, Okazaki M. Porous structure of organic and carbone aerogels synthesized by sol-gel polycondensation of resorcinol with formaldehyde. Carbon 1997; 35:791-6

ELECTROLYSIS OF CARBAMIDE-CHLORIDE MELTS AT INERT ELECTRODES

S.A. Kochetova and N.Kh. Tumanova*

Institute of General and Inorganic Chemistry, National Academy of Sciences of Ukraine, Pr. Palladina 32/34 Kiev, 03680, Ukraine.

Abstract

Electrode reactions in carbamide and carbamide-chloride melts have been studied by cyclic voltammetry in combination with chromatography and IR spectroscopy of melts and gases for determining the electrolysis products. The interest in this process is accounted for by application of these melts as supporting electrolytes for electrochemical investigations. The peculiarities of carbamide-NH_4Cl melt electrochemical behavior have been established. A new compound is formed during electrolyses at the potential of 1V, which is present in melt and in anodic gases (IR Spectroscopy). The investigation of electrochemical properties of this compound showed that it is electrochemically active (cyclic voltammetry) and is adsorbed (C-E method) on inert electrodes (Pt, glassy carbon). The formation of this compound, characterized by N-Cl bond, explains the depassivation effect of refractory metals under anodic dissolution if polarizable electrode reaches the transpassive region.

Keywords

Electrolysis; cyclic voltammetry; molten salts; carbamide; potassium chloride; ammonium chloride; compound; adsorbtion.

1. INTRODUCTION

Molten carbamide is known to be a good solvent for the salts of many metals [1] and was used as a supporting electrolyte in polarographic investigations both at a dropping mercury electrode and at a stationary

* Corresponding author. E-mail: tumanova@ionc.kar.net

I.V. Barsukov et al. (eds.), New Carbon Based Materials for Electrochemical Energy Storage Systems, 425–432.

platinum electrode [2,3]. Because of the low electrical conductivity of molten carbamide, mixed carbamide-nitrate [4] and carbamide-halide [5] melts were used more often. When investigating the electrochemical behavior of refractory metals (Nb, Ta, Ti) in a carbamide-chloride melt, it was found that these metals are passivated in chloride-containing melts, and that their electrochemical dissolution (Nb, Ta) is impossible if electrode polarization does not reach 1V. It was suggested [6,7] that during melt electrolysis at the above potential, some compound is formed, which functions as depassivator. To confirm this, a series of investigations of the electrochemical properties of a melt – solvent at inert electrodes has been carried out, and the products of electrode reactions have been investigated by the spectroscopic and chromatographic methods. This article presents the results of the investigation carried out.

2. EXPERIMENTAL PROCEDURES

In order to determine the electrochemical properties of the solvent, the electrode process in molten carbamide and in carbamide-MeCl (where Me - NH_4, K) mixtures on inert electrodes (platinum, glassy carbon) were investigated using cyclic voltammetry. The electrode reaction products were analysed by spectroscopic methods. The adsorbtion of carbamide- NH_4Cl anodic product was investigated by differential capacity method.

The electrochemical study was conducted by recording cyclic voltamperic curves with a PI-50-1 potentiostat and C-E curves were taken with Resistance Bridge P-5021. A platinum electrode with a working surface area of $5\text{-}8\times10^{-3}$ cm^2 was used as a working electrode. The auxiliary electrode was platinum plate and for obtaining the differential capacity curves – the cylinder, coaxial located to cathode. The potential was measured against a silver reference electrode (Ag/carbamide + NH_4NO_3 (16 mol.%) + 0.07 mol kg^{-1} $AgNO_3$). Experiments were carried out in the temperature range 100 to 140^0C under argon.

The state of carbamide melts taken in "quenched" melts were investigated by IR spectroscopy with UR-20 and Specord M-80. The validity of extending data for "quenched" melt to the process in melt has been proved for a number of systems in work. The investigations of the gas phase composition were carried out on a SELMIKHROM-1 chromatograph of the SELMI joint-stock company, which consisted of two units: analysis unit and control-monitoring unit. Control of the chromatograph, obtaining and processing of the incoming information were performed by means of an IBM-486 computer. Two series of experiments have been carried out using first argon and then helium as the carrier gas. The replacement of argon by helium was due to the necessity to separate and identify hydrogen and carbon monoxide.

Reagent grade (carbamide, MeCl) and spectral grade (NH_4Cl) salts and dried argon were used in the study. The salts were predried in vacuum at 80^0C (carbamide, ammonium chloride) and 200 to 300^0C (alkali chlorides).

3. RESULTS AND DISCUSSION

In an individual molten carbamide, the electrode processes are feebly marked at melt decomposition potentials (Fig. 1) because of its low electrical conductivity. Both electrode processes are accompanied by gas evolution.

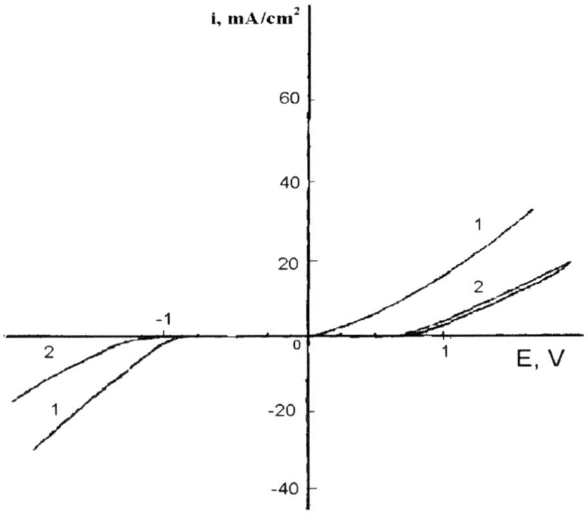

Figure 1. Cyclic voltamperogram of Pt (1) and glassy carbon (2) electrodes in molten carbamide.

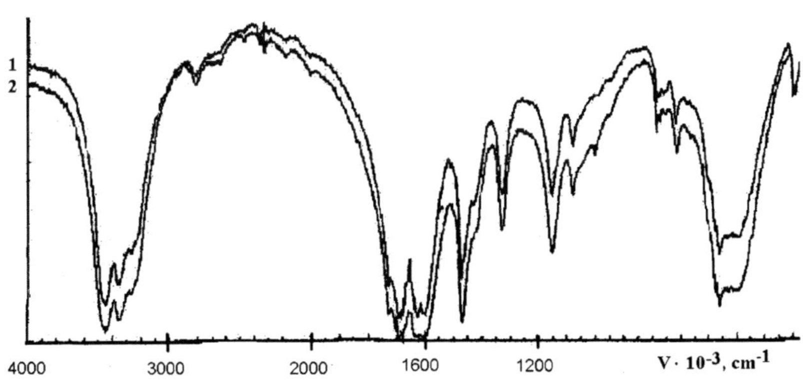

Figure 2. IR spectra of the "quenched" carbamide: anolyte (1) and catholyte (2).

A large amount of ammonia and a small amount of CO and CO_2 have been found in the cathode gases of carbamide melt by means of IR spectroscopy and gas chromatography. An intense CO_2 band is also present in IR spectra of anode gases; in addition, traces of CO and a number of unidentified bands in the range 1500-1700 cm^{-1} have been found [8]. Using chromatographic analysis of anode gases, the presence of N_2 in them in melt electrolysis has been shown. It has been found that the IR spectra of the catholyte and anolyte are identical (Figure 2).

They have a signal from the carbonyl ion of carbamide, a signal that corresponds to ammonium ion, and a signal that characterizes the C-N bond. On the basis of the results obtained by us, only a rough scheme of cathode and anode reactions can be proposed because a number of bands in the IR spectra have not been identified:

$$2CO(NH_2)_2 \rightarrow 2NH_3 + NH_2CN + CO + O^{2-}. \quad (1)$$

$$2CO(NH_2)_2 \rightarrow CO_2 + NH_2CN + N_2 + 6H^+ + 6e. \quad (2)$$

Then the overall reaction is:

$$8CO(NH_2)_2 \rightarrow 3NH_3 + 3CO + CO_2 + N_2 + 4NH_2CN + 3NH_4OH \quad (3)$$

If it is assumed that electrolysis product may be cyanamide (as the simplest substance in terms of IR spectroscopic data for the melt), then the chemical instability of NH_2CN and the possibility of the reversible reaction:

$$NH_2CN + 2NH_3 + CO_2 \rightarrow 2CO(NH_2)_2, \quad (4)$$

which probably proceeds during melt electrolysis in a cell with unseparated compartments, are worth taking into account in the analysis of the above overall-reaction scheme.

Our investigations showed that in mixed melts of eutectic composition carbamide-NH_4(K)Cl, the oxidation and reduction of melt constituents take place mainly independently of each other. The anodic process at platinum electrodes in the range of potentials below 0.9V is associated with the direct oxidation of carbamide to secondary and tertiary amide compounds, accumulation of ammonium ions in the melt, and evolution of the same gaseous products as in carbamide electrolysis [8]. The cathodic process is accompanied by the formation of ammonia, CO, and CO_2, i.e. of the same products as in pure- carbamide electrolysis. In contrast to carbamide melt, a large amount of hydrogen appears in the cathode gases of the mixed melt, and in the anode gases of the carbamide-KCl melt, the presence of chlorine has been established at potentials above 0.9V. In the carbamide-NH_4Cl melt, a new compound appears instead of chlorine, which is characterized by an intense signal at 630 cm^{-1} in the IR spectra of anode gases and, to a smaller extent, in the IR spectra of the anolyte. This signal corresponds to the N-Cl bond. At first it was suggested that the substance that is formed is NCl_3. By analogy with aqueous solutions [9], the formation of NCl_3 may take place according to the scheme:

$$NH_4Cl + 3Cl_2 = NCl_3 + 4HCl. \quad (5)$$

However, the physicochemical properties of nitrogen chloride give no grounds for the explanation of its stability under melt electrolysis conditions. The possibility of the formation of a chlorine derivative of carbamide according to the scheme:

$$2NH_4Cl + Cl_2 + CO \rightarrow CO(NHCl)_2 + 6HCl, \quad (6)$$

appears to be more probable. The evolution of HCl is supported by the IR spectroscopy of the anode gases of the carbamide-NH_4Cl melt and is observed visually after long electrolysis.

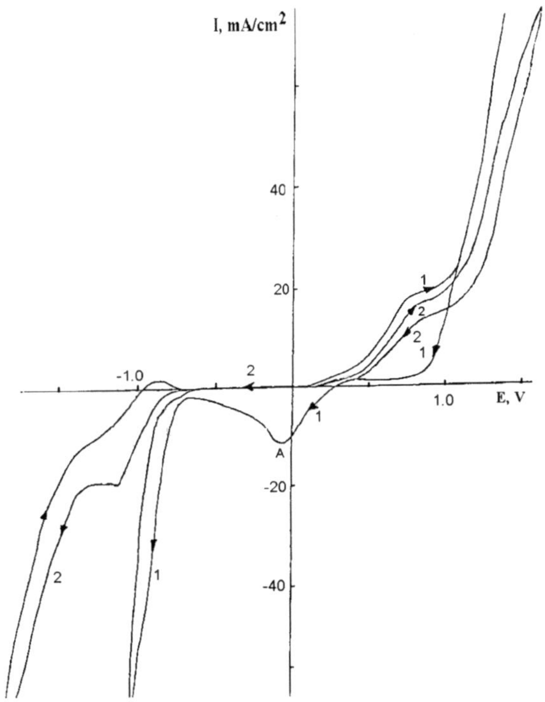

Figure 3. Cyclic voltamperogram of Pt-electrode in carbamide-NH_4Cl (curve 1) and in carbamide-KCl (curve 2) melts.

However, the compounds formed by the electrolysis of the carbamide-NH_4Cl melt exhibit a strong depassivating action, which is much stronger than that of Cl^- (this was proved when studying the electrochemical behavior of titanium in this melt) [10]. Therefore, it appears not very correct to consider that the chlorine derivatives of carbamide can have such properties. If one bears in mind that cyanamide is formed by carbamide electrolysis and taking into account that the formation of the compound is only observed if the melt contains ammonium ions, then the following reaction is possible at the anodic discharge potential:

$$NH_2CN + 4NH_4Cl + 2CO_2 = 3NH_2CHNCl + HCl + 4H_2O. \qquad (7)$$

Chlorine derivative of cyanamide possesses a high reactivity and can show depassivating properties, but it is difficult to unambiguously identify the compound that is formed because the presence of all possible compounds in the melt will be reflected by the same signal at 630 cm^{-1} [11].

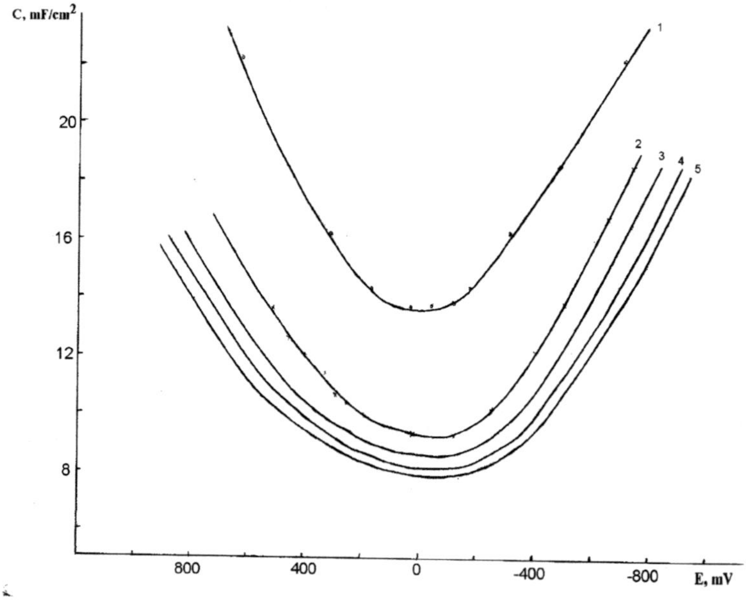

Figure 4. Differential capacity of Pt electrode in carbamide-NH_4Cl melt (curve 1) and after 0.5 (curve 2), 1 (curve 3), 1.5 (curve 4), 2 (curve 5) hours of electrolysis.

The compound formed is electroactive. The cyclic voltammogram of a Pt electrode in the carbamide-NH_4Cl melt exhibits a maximum (wave A) in the region of the zero value of potential on the reverse anodic sweep, which reflects the electrochemical reduction of the compound formed (Fig. 3). For comparison, a voltammetric curve is shown in the same figure, which was obtained in the carbamide-KCl melt. The absence of maxima on the reverse sweep supports the conclusion that maximum A does not relate to the reduction of Cl^- ions formed by oxidation but reflects the reduction of a new compound, which is formed at potential of 1V in the carbamide-NH_4Cl melt.

Adsorption of the compound formed on the electrode surface has been established by the examination of differential capacity curve [12]. The investigation carried out is of qualitative character because the concentration of the compound is unknown: it is formed during electrolysis. As is evident from the examination of C-E curves (Figure 4), the compound is adsorbed on the electrode and decreases the double-layer capacity at the zero-charge point. The degree of coverage of the electrode surface by the adsorbed compound (θ) was calculated taking as a basis the fact that the time of electrolysis at the compound formation potential (1V) is proportional to the concentration of the compound. The calculated θ value are 0,61; 0,76; 0,84 and 0,95, which correspond to 0.5, 1, 1.5 and 2 hours of electrolysis. The θ value are high, i.e. the electrode surface is almost completely screened by adsorbate molecules.

Based upon the concepts of the adsorption of the anode reaction product, the share of the anodic curve, on which the carbamide oxidation processes is reflected as a wave, can be explained. It may be assumed that the adsorption of the reaction product inhibits the direct oxidation of carbamide. To verify this conclusion, the anode was polarized to the electrolysis product formation potential, and the reverse sweep was stopped before the electrolysis product was reduced at the electrode. Then the carbamide oxidation process was completely inhibited on the subsequent forward sweep, and the curve exhibited only a current increase at the chlorine ion oxidation potential.

4. CONCLUSIONS

In an individual molten carbamide, the electrode processes are feebly marked at melt decomposition potentials because of its low electrical conductivity. Both electrode processes are accompanied by gas evolution (NH_3, CO, CO_2, N_2) and NH_2CN (approximately) is formed in melt. In eutectic carbamide-chloride melts electrode processes take place mainly independently of each other. The chlorine must evolve at the anode during the electrolysis of carbamide - alkali metal and ammonium chloride melts, which were revealed in the electrolysis of the carbamide-KCl melt. But in the case of simultaneous oxidation of carbamide and NH_4Cl, however, a new compound containing N-Cl bond has been found in anode gases instead of chlorine. It is difficult to fully identify this compound by the experimental methods employed in the present work, but it can be definitely stated that this compound is formed in the carbamide- NH_4Cl melt (IR spectroscopy of melts and gases), is electrochemically active (cyclic voltammetry), and is adsorbed on the electrode surface (C-E curves). This compound possesses a high reactivity and exhibits a strong depassivation active in the time of electrodissolution of refractory metals in carbamide- NH_4Cl melt.

REFERENCES

1. Burk W. Geschmozenez Harnstoff als Losungsmi ttel fur Alkalichloride. Chemie, 1964; 3: 354-5.
2. Vecchi E., Zuliani G. Molting Carbamide Supporting Electrolyte in Polarography. Ric. Sci., 1955; 25: 2667-69.
3. Delimarskii Y.K., Tumanova, N.Ch., Shilina G.V., Barchuk L.P. *Polarography of Ionic Melts*. Kiev: Naykova Dumka, 1978 (in Russian).
4. Tkalenko D.A. *Electrochemistry of nitrate melts*. Kiev: Naykova Dumka, 1983 (in Russian).
5. Tumanova N. Electrochemistry of Nb, Ta and Ti in carbamide based melts. Proceedings of the Green Industrial Applications of Ionic Liquids. NATO Science Series. Ed. R. Rogers, K. Seddon, S. Volkov. Kluwer Academic Publishers, 2003.
6. Tumanova N., Volkov S., Kochetova S., Triphonova T., Buryak N. Passivation and red-ox processes of refractory metals in carbamide-chloride melt. J. Mining and Metallurgy, 2003; 39: 69-81.
7. Tumanova N., Babenkov E., Boyko O., Buryak N. The peculiarities of electrochemical niobium dissolution in carbamide based low-temperature melts. Abstracts of the 195-th Meeting of Electrochem. Society, 1999 May 2-6.
8. Tumanova N., Babenkov E., Chernuhin S. and others. Electrochemical properties of carbamide-halide melts. Ukr. Chim. J., 2000; 66: 35-9 (in Russian).
9. *Comprehensive inorganic chemistry*. Ed. Bailar J.C. Oxford - N. Y.: Pergamon Press, 1973.
10. Volkov S., Tumanova N., Kochetova S., Buryak N. The polysurface Mechanism of Ta and Ti Anodic Dissolution in Low-Temperature Carbamide-NH_4Cl Melt. Z. Naturforsh, 2001;56a: 761-3.
11. Nakamoto K. *Infrared and Raman Spectra of Inorganic and Coordination Compounds*, New York: Wiley, 1986.
12. Damaskin B., Petriy O. *Adsorption of organic compounds on electrodes*. M.: Science, 1968 (in Russian).

GRAPHITE INTERCALATION AS A WAY TO CARBON-CARBON COMPOSITES AND CARBON NANOSCROLLS

M.V. Savoskin*, A.P. Yaroshenko, V.N. Mochalin*, N.I. Lazareva, T.E. Konstantinova

L.M. Litvinenko Institute of Physical Organic and Coal Chemistry, National Academy of Sciences of Ukraine, 70 R.Luxemburg St., 83114 Donetsk, Ukraine

Abstract

New simple methods for synthesis of carbon-carbon, as well as carbon-carbon-TiO_2 composites and carbon nanoscrolls were proposed making use of readily available graphite intercalation compounds.

Keywords

Carbon composites; Anatase; TiO_2; Nanoscrolls; Exfoliated graphite.

1. INTRODUCTION

Exfoliated graphite – amorphous carbon (EG-AC) composites are widely used in a broad range of applications like adsorption of gases and liquids, separation, and high-surface catalyst supports. They also seem to be attractive for energy- and gas- storage systems.

There are different methods of producing such materials available today. Greinke and Howard[1] have described a method of production of EG-AC composites starting from exfoliated graphite preformed into some article, for example a ring. Such article is impregnated by liquid polymer or polymer solution in organics, followed by heating in order to cure polymer or evaporate organic solvent. The semi-finished product is then carbonized to form an amorphous carbon phase by heating in a nitrogen atmosphere at a

* Corresponding authors. E-mails: m-savoskin@yandex.ru, mail2mochalin@yahoo.com

I.V. Barsukov et al. (eds.), New Carbon Based Materials for Electrochemical Energy Storage Systems, 433–439.

rate of 60°C/hour from 50 to 520°C with a one-hour hold time at 520°C. Finally the article is compressed yielding the end product.

Another method of producing EG-AC composites has been proposed recently by Mareche *et al.*[2] These methods consists of the following stages: 1) obtaining EG by heating the exfoliatable graphite intercalation compound (GIC); 2) multiple impregnation of EG particles by an amorphous carbon precursor - polyfurfuryl alcohol; 3) drying at a relatively low temperature (of about 100°C); 4) carbonization of material obtained at the heating rate of 5°C/min with two-hours hold time at final temperature of 520°C; 5) activation of EG-AC composite obtained by water steam at 800°C.

As could be seen from the above, the modern methods of synthesis of EG-AC composites are multistage long-term procedures requiring as a rule the use of some organic solvents or inert-gas atmosphere which makes them difficult to be scaled up by the industry.

From this point of view a one-stage process of simultaneous GIC exfoliation, impregnation of EG by amorphous carbon precursor and its subsequent carbonization would be very attractive. But up to now this simple one-stage process could not be realized because of high exfoliation temperature of the ordinary GICs needed (900 – 1200°C), which leads to almost complete burnout of AC precursor at the atmospheric air conditions.

In our laboratory we have elaborated the method of producing residual graphite hydrosulfate, which could be exfoliated efficiently at temperatures of about 500°C[3]. Having such GIC we tried to realize one-stage procedure of producing EG-AC composites.

2. EXPERIMENTAL

We used residual graphite hydrosulfate produced from GSM-1 graphite (which is a purified natural flake graphite from Zavalye Graphite deposit, Kirovograd region, Ukraine) and sucrose from Merck as an AC precursor. Sucrose was ground into fine powder and mixed with residual graphite hydrosulfate in a glass reactor for 10 min. As a function of experiment, the content of components in the mixture varied, however, the total weight of mixture was kept constant (25g). After being mixed, a portion of about 3g of GIC – sucrose mixture was heated in a Fisher Scientific Isotemp® Model 650 Programmable Muffle Furnace at 550°C during 4 min. The conditions of heat treatment, particularly temperature and heating time, strongly influence the properties of the product. In special experiments we have found the optimal parameters of heat treatment mentioned above. Heating the mixture of GIC with amorphous carbon precursor results in exfoliation of residual graphite hydrosulfate accompanied by caramelization of sucrose and soaking of the EG honeycomb structure by melted caramel with simultaneous sucrose carbonization, yielding EG-AC composite. The particles of final product were very similar to those of individual exfoliated graphite in their worm-like shape.

SEM studies of composites were carried out using a JEOL JSM-T300 scanning electron microscope. Energy dispersive X-ray (EDX) analysis was performed by energy dispersive device LINK 860-500 attached to the SEM unit. JEOL-JEM-200 transmission electron microscope was used for TEM investigations.

3. DISCUSSION

Typical SEM images of EG-AC composites obtained by the one-stage method proposed are presented in Figure 1.

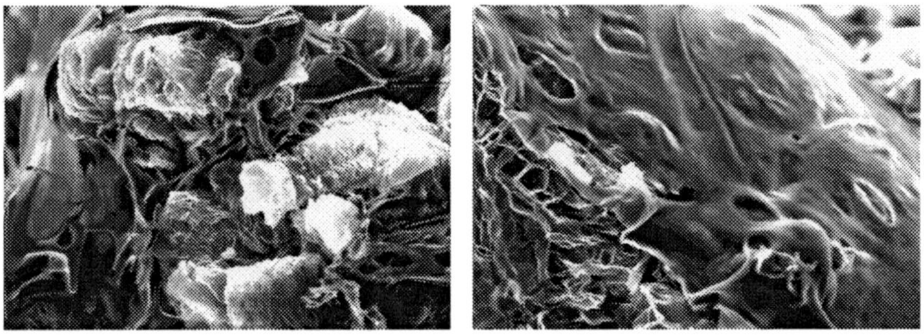

$529\cdot10^{-6}$ m $\qquad$ $187\cdot10^{-6}$ m

Figure 1. Typical SEM photographs of EG-AC composites.

These images clearly show the macroporous structure of composite with its large pore diameter inherited from the EG matrix and the amorphous carbon phase covering the surface of exfoliated graphite worm-like particles. During the heating process, the AC precursor has partially fused and wet its graphite support leading to an intimate contact between the two solids. This structure should lead to good mechanical strength and very efficient thermal contact between the amorphous carbon and its graphite support. Note that these images are similar to those presented by Menard, Py et al.[4,5] for composites produced from exfoliated graphite and in situ prepared activated carbons.

Figure 2 represents specific surface area of composites obtained as measured by low-temperature adsorption-desorption of argon in comparison to that calculated by the additive model. Over the entire range of sucrose content, excluding the end points corresponding to individual components the measured specific area of the composites obtained exceeds the values calculated by the additive model. The higher the amorphous carbon content in the composite, the more pronouncing difference has been observed. Such behavior could be attributed to the acidic catalysis of condensation reactions during the heating and activation of AC by the acidic gaseous species (H_2SO_4, SO_2, SO_3 etc.) and water evaporated during the GIC exfoliation process.

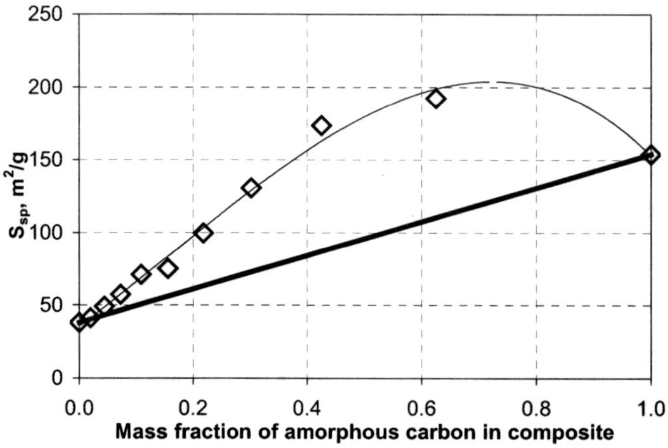

Figure 2. Specific surface area of EG-AC composites produced. Points – experiment, straight line – calculated by the additive model.

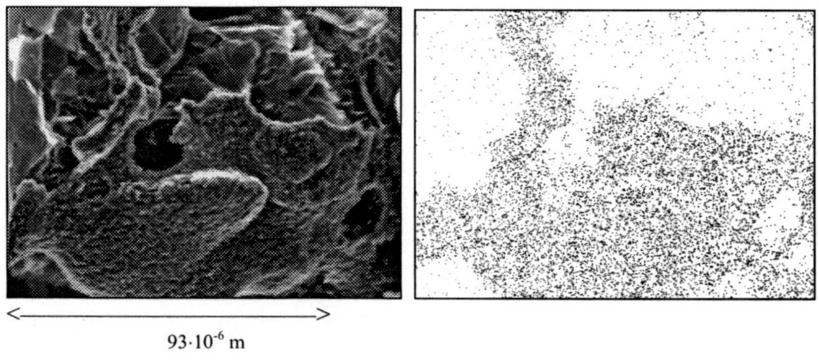

Figure 3. SEM image of an EG-AC-TiO2 composite (left) and corresponding EDX mapping of titanium (right).

One of the main advantages of the method of EG-AC composite synthesis proposed here is a possibility of using this technique for obtaining the tri-component composites, such as EG-AC-TiO_2. Composites between EG and photoactive anatase-type TiO_2 are effective buoyant photocatalysts for degradation of water contaminants[6,7]. Their synthesis is typically a multistage process, which has essentially the same disadvantages described above for the known methods of EG-AC composites production.

Therefore we tried to obtain EG-AC-TiO_2 composites using the procedure similar to that elaborated for production of EG-AC composites. The only difference here from the scheme described above is the preparation of initial mixture for now containing three components: residual graphite hydrosulfate, sucrose and anatase-type TiO_2 (Aldrich).

Typical SEM images of EG-AC-TiO_2 composites obtained are presented in Figure 3. The rough surface sites on the left image correspond to the TiO_2 localization sites as could be seen from comparison of SEM and EDX images obtained from the same part of composite sample. It is clear that in this case amorphous carbon acts not only as adsorbent but also as a binder allowing skipping the use of additional binder needed in conventional methods. We believe that EG-AC-TiO_2 composites obtained should be very attractive and low cost buoyant photocatalysts so their photocatalytic activity will be investigated in the near future.

Another application of GICs in modern industry is the rapid growing field of nanomaterials. In the recent papers of Kaner et al.[8] and Shioyama and Akita[9] graphite intercalation and exfoliation were used as the key processes in the so-called "chemical" route to the new kind of carbon nanomaterials – carbon nanoscrolls. The method they propose differs in advance from the known "physical" methods in comparatively low energy consumption and low-cost simple equipment, therefore it is more convenient to realize this technology in large-scale applications. In the study of Kaner *et al.* [8] and Shioyama and Akita [9] the first stage graphite intercalation compound with potassium (KC_8) was used as the initial compound. But this compound is expensive and difficult to obtain and operate in the large-scale

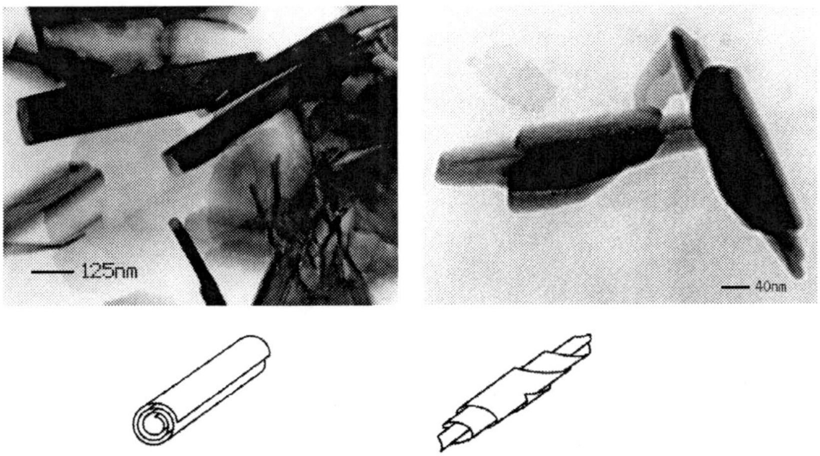

scroll-shaped skew-shaped

Figure 4. TEM images of nanoscrolls prepared from graphite nitrate.

applications. We succeeded in producing of nanoscrolls using the graphite nitrate which is much more attractive from the industrial point of view. Our procedure of nanoscrolls production could be easily transferred into commercial large-scale application. Typical TEM images of the nanoscrolls produced are shown in Figure 4.

On these images, one could see scrolled structures formed from one or multiple graphitic sheets as well as non scrolled transparent graphitic sheets in the background. The diameter of nanoscrolls produced ranges from 40 to 130 nm and length comes up to 750 nm. Based on an interlayer spacing of graphite (3.35 Å) it could be concluded that nanoscrolls typically consist of 10 to 40 graphene layers. These values are in good agreement with data reported in [8-9]. According to the two possible types of scrolling process we have found two types of nanoscrolls shown on this picture: scroll- and skew-shaped. Both types have opened tips so their internal surface could be accessible for small molecules.

Visual estimation of nanoscrolls yield is difficult because they concentrate near the edges of copper grid support used in TEM studies. Nevertheless TEM images provide clear evidence of nanoscrolls presence in large amount.

4. SUMMARY

Graphite intercalation compounds are prospective precursors of the new carbon materials, which are in high demand by the market today.

Combined with appropriate amorphous carbon precursors graphite intercalation compounds could be used in one-stage process of production of carbon-carbon composites, which could possess attractive properties for such applications as supercapacitors elements, sorbents as well as catalyst supports and materials for energy- and gas-storage systems.

The process of producing the new kind of carbon nanomaterials namely carbon nanoscrolls from graphite nitrate is proposed which differs in advance from other techniques in its simplicity, availability of raw material, low level of energy consumption.

REFERENCES

1. US Pat. 5228701 Flexible graphite particles with an amorphous carbon phase at the surface. Greinke R.A., Howard R.A.; UCAR Carbon Technology Corporation. 20.07.1993.
2. Mareche J.F., Begin D., Furdin G., Puricelli S., Pajak J., Albiniak A., Jasienko-Halat M., Siemieniewska T. Monolithic activated carbons from resin impregnated expanded graphite. Carbon 2001; 39:771-3.
3. Yaroshenko A.P., Savoskin M.V., Magasinski A.N., Savsunenko O.B. Synthesis and thermal expanding of residual graphite bisulfates produced in $Na_2Cr_2O_7$-H_2SO_4 system.

Proceedings of 12th International Symposium on Intercalation Compounds; 2003 June 1-5; Poznan, Poland. Poznan University of Technology, 2003:134.
4. Menard D., Py X., Mazet N. Development of thermally conductive packing for gas separation. Carbon 2003; 41:1715–27.
5. Py X., Daguerre E., Menard D. Composites of expanded natural graphite and in situ prepared activated carbons. Carbon 2002; 40:1255–65.
6. Modestov A., Glezer V., Maryasin I., Lev O. Photocatalytic degradation of chlorinated phenoxyacetic acids by a new buoyant titania - exfoliated graphite composite photocatalyst. J Phys Chem B 1997; 101:4623-9.
7. Tsumura T., Kojitani N., Umemura H., Toyoda M., Inagaki M. Composites between photoactive anatase-type TiO_2 and adsorptive carbon. Appl Surface Sci 2002; 196:429-36.
8. Viculis L.M., Mack J.J., Kaner R.B. A chemical route to carbon nanoscrolls. Science 2003; 299:1361.
9. Shioyama H., Akita T. A new route to carbon nanotubes. Carbon 2003; 41:179-81.

CHAPTER 6:

CARBONS IN THE CATHODES OF LITHIUM-ION BATTERIES; ALTERNATIVE FORMS OF MnO_2, CATHODE / CARBON MODELING

Chapter 6: Subject Overview

The seven papers in Chapter 6 are focused on cathode materials for lithium and lithium-ion batteries. Carbon is used as a conductive additive in composite electrodes for batteries. The type of carbon and the amount can have a large effect on the electrochemical performance of the electrode.

The first paper in this chapter is a contribution from R. Kostecki and F. McLarnon of Lawrence Berkeley National Laboratory and consists of a diagnostic study of the power fade issues related to high-power Li-ion batteries that contain a $Li(Ni_{0.8}Co_{0.15}Al_{0.05})O_2$ cathode. Li-ion batteries are prone to poor life and experience problems with impedance rise during elevated temperature aging. SEM, Raman and current-sensing AFM techniques were used to address the mechanism of power fade at the cathode with aged cells. From the results, the authors describe a process of carbon retreat/redistribution in the composite electrode upon accelerated aging. The carbon retreat/redistribution is said to have a major impact in the contribution to the cathode impedance rise and loss of power. They finish with a discussion on the loss of conductivity across the composite cathode as a result of carbon retreat/redistribution.

In the paper from V. Matveyev of the Ukrainian State University of Chemical Engineering, an examination of the role of conductive carbon additives in a composite porous electrode is conducted. A model for calculation of the local electrochemical characteristics of an electrode is presented. A comparison on the polarization of the electrode as a function of the redox state of the electroactive species is emphasized in the model. The electrochemical reaction of chloranil (tetrachlorobenzoquinone) was measured and results compare favorably to calculations derived from the model.

S. A. Kirillov *et al.* describe the design, synthesis and application of amorphous manganese dioxide in lithium batteries in the third paper of this chapter. A synthetic route of oxidation of Mn(II) aqueous solution with hydrogen peroxide was used to prepare the materials. The prepared MnO_2 material remains amorphous up until 600°C. The amorphous MnO_2 delivers higher capacity of 247 mAh/g versus 189 mAh/g for Russian commercial crystalline EMD when discharged to 2.0 V. Further testing has shown that the amorphous phase demonstrates reversible cycling with lithium. These results bode well for the development of a new class of amorphous manganese oxides for charge storage in lithium or perhaps lithium-ion batteries in the future.

The performance in Zn-MnO_2 alkaline cells of electrolytic manganese dioxide produced in the presence of fluoride ions is compared with commercial EMD materials by I. Makyeyeva *et al.* The material produced in the presence of fluoride ions was shown to have superior utilization to the commercial EMD. The authors state that the improved performance of the manganese oxide synthesized in the presence of fluoride ions is due to an increased or larger number of defects in the crystalline structure. The defects are associated with a higher non-stoichiometry in the material due to the presence of more Mn(III). In addition, the authors explain that a higher percentage of hydroxide (OH^-) moieties in their material are associated with an improved ionic conductivity.

In the fifth paper of this chapter on cathodes, an investigation of thin-film oxide-hydroxide electrodes containing Cr, Ni, and Co compounds was authored by N. Vlasenko *et al.* The thin-films were produced by electrochemical deposition from transition metal aqueous fluorine-containing electrolytes onto steel substrates. These thin-films were tested in Li coin cells. Electrochemical activity appears to scale with the amount of fluoride used in the deposition; the larger concentration of fluoride in the bath, the greater the capacity. One Ni oxide-hydroxide film electrode showed greater than 175 mAh/g reversible capacity on the 50th cycle with excellent coulombic efficiency.

In a second paper from Kirillov *et al.*, synthetic aspects of liquid phase precipitation reactions are discussed in relation to the factors responsible for metal-oxide formation. In particular the Pechini reaction, a sol-gel process, was examined. A good reference section is provided to introduce the reader to previous literature. The authors stress that more work needs to be done to establish what factors are important for producing the highest quality metal-oxide powders from such reactions. The conclusion specifically lists the criteria needed in order to accomplish this task.

In the final paper of this chapter, authors S.-H. Kang and K. Amine from Argonne National Laboratory discuss high-power lithium ion batteries and present a new lithium nickelate material: $Li(Ni_{0.4}Co_{0.2}Mn_{0.4})O_2$ with a well-formed layered structure. The electrochemical performance in lithium cells is compared for materials that are uncoated versus coated with aluminum oxide. The capacity fade is less for coated material versus uncoated both at room-temperature up to 100 cycles and 50°C up to 70 cycles. The improvement was found to be linked to an interfacial impedance drop for electrodes of the coated material. The authors state that the surface coating suppresses the degradation of the cathode/electrolyte interface and that the mechanism may be involved with the scavenging of HF in the electrolyte by the coating. Thus the coating would protect the underlying layered oxide material to acid attack and possible dissolution/corrosion effects during cycling in the $LiPF_6$ organic electrolyte.

The editors hope that you enjoy reading the last chapter in this NATO series volume, and sincerely thank you for your interest in all of the papers from the NATO-Carbon Advanced Research Workshop and Conference (NATO-CARWC).

DIAGNOSTIC EVALUATION OF POWER FADE PHENOMENA AND CALENDAR LIFE REDUCTION IN HIGH-POWER LITHIUM-ION BATTERIES

Robert Kostecki* and Frank McLarnon

Environmental Energy Technologies Division
Lawrence Berkeley National Laboratory; University of California, Berkeley, CA 94720, USA

Abstract

High-power Li-ion cells with graphite anodes and $LiNi_{0.8}Co_{0.15}Al_{0.05}O_2$ cathodes that were cycled and stored at elevated temperatures showed a significant impedance rise and capacity fade, which were associated primarily with the $LiNi_{0.8}Co_{0.15}Al_{0.05}O_2$ cathode. A combination of electrochemical, physical, and chemical diagnostic techniques, including Raman, SEM, and current-sensing AFM, was used to characterize the cathodes from these cells in order to produce a clear picture of the mechanism for cell degradation. Systematic Raman mapping of 50 x 80μm areas at 0.9μm spatial resolution produced semi-quantitative composition maps of cathode surfaces. Raman microscopy surface composition maps and SEM images of cathodes from tested cells revealed that cell cycling or storage at elevated temperatures led to significant changes in the $LiNi_{0.8}Co_{0.15}Al_{0.05}O_2$/elemental-carbon surface concentration ratio. The loss of conductive carbon correlated with the power and capacity fade of the tested cathodes and the loss of surface electronic conductivity.

Keywords

Li-ion battery; cathode; carbon retreat; conductivity; Raman; AFM.

1. INTRODUCTION

Lithium-ion batteries are being seriously considered for application in all-electric vehicles (EV) and hybrid electric vehicles (HEV's) because of their high power and energy densities [1, 2]. The U.S. Department of

* Corresponding author. E-mail: R_Kostecki@lbl.gov

I.V. Barsukov et al. (eds.), New Carbon Based Materials for Electrochemical Energy Storage Systems, 445–452.

Energy's Advanced Technology Development (ATD) Program supports the development of high-power $LiNi_{0.8}Co_{0.15}Al_{0.05}O_2$/ EC-EMC, 1.2 M $LiPF_6$/graphite batteries for hybrid electric vehicle applications [3]. A major goal of the ATD Program is to develop a detailed understanding of the power and capacity fade mechanisms of a single cell chemistry designed for the HEV application and to develop predictive models for the accelerated life test behavior based on these mechanisms. For this purpose, this cell chemistry has been subjected to elevated-temperature calendar life and lifecycle testing with a pulse-type profile over a very limited range of state-of-charge (SOC) [4].

Diagnostic evaluations of Li-ion cells that were aged and/or cycled under various conditions were carried out to determine the mechanisms responsible for the cell power loss that accompanies life tests at elevated temperatures. Many cells have been examined to determine the mechanisms for capacity and power fade, using a wide range of diagnostic techniques that were developed and refined during studies of the ATD cells [5]. Impedance measurements of the cell components indicated that the $LiNi_{0.8}Co_{0.15}Al_{0.05}O_2$ cathode is primarily responsible for the observed cell power loss at elevated temperatures, similarly to the $LiNi_{0.8}Co_{0.2}O_2$ cathode, which was studied in our previous work [6]. Possible causes of the increase in cathode impedance include formation of an electronic and/or ionic barrier at the cathode surface. However, the observation that, in these cells, power fade is always accompanied by a loss of cell discharge capacity suggests that the mechanism of degradation is more complicated and may consist of multiple processes, which may be a somewhat general phenomenon that affects several types of cathodes. X-ray diffraction spectroscopy failed to detect noticeable changes in the bulk structure of the tested cathodes, and no evidence of material structural degradation or formation of new phases was found. In this study we focused our attention on cathode surface processes, which we believe have a dominant effect on impedance behavior. A systematic diagnostic evaluation of $LiNi_{0.8}Co_{0.15}Al_{0.05}O_2$ cathodes removed from high-power Li-ion cells, which were stored and/or cycled at elevated temperatures, was carried out. The application of local probe techniques such as current-sensing AFM and Raman microscopy to characterize physico-chemical properties of the electrode surface at nanometer resolution can provide unique insight into the complex mechanism of the detrimental processes that are likely responsible for the composite cathode capacity fade and impedance increase.

2. EXPERIMENTAL

High-power Li-ion cells with a $LiNi_{0.8}Co_{0.15}Al_{0.05}O_2$ cathode, a synthetic graphite anode, 1.2 M $LiPF_6$ + ethylene carbonate + ethyl-methyl carbonate (EC/EMC) electrolyte, and a Celgard® 2300 separator, were manufactured, aged, cycled, and/or abused and then characterized under the ATD Program [3, 4]. We compared a fresh cathode with cathodes taken from cells that were aged or cycled at elevated temperatures for up to 68 weeks, losing up to 52% of their initial power and 24% of their initial capacity.

The samples were collected from the cathodes ~ 2.5 cm away from the current collector tab, washed in pure dimethyl carbonate (DMC), and soaked in DMC for 30 minutes after removal from Li-ion cells inside an argon-filled glove box. This procedure removed electrolyte salt from the electrode to prevent its reaction with air and moisture. An integrated Raman microscope system "Labram" made by ISA Groupe Horiba was used to analyze and map the cathode surface structure and composition. The excitation source was an internal He-Ne (632 nm) 10 mW laser. The power of the laser beam was adjusted to 0.1 mW with neutral filters of various optical densities. The size of the laser beam at the sample was ~1.2 μm.

We used current-sensing atomic force microscopy (CSAFM) to measure and image-map the electronic conductivity of individual grains of the $LiNi_{0.8}Co_{0.15}Al_{0.05}O_2$ powder (Fuji Chemical) that was used to fabricate ATD Program composite cathodes. The powder was pressed into a gold foil to produce randomly scattered particles of $LiNi_{0.8}Co_{0.15}Al_{0.05}O_2$ in good electronic contact with the Au substrate. The microscope consisted of a Molecular Imaging (MI) scanning probe microscope coupled with a Park Scientific Instruments (PSI) AutoProbe Electronic Module. The Si atomic force microscope (AFM) tips were coated with a thin conductive layer of W_2C. All CSAFM experiments were performed in constant-force mode with controlled oxide-tip voltage difference. A single scan of the tip over the sample surface simultaneously produced two images: a topographic image and a conductance image; the latter represents $Au/LiNi_{0.8}Co_{0.15}Al_{0.05}O_2$ - tip current variations during scanning at a given sample-tip voltage difference. CSAFM imaging was conducted in a small glove box specially designed for scanning-probe microscopy tests under a controlled N_2 atmosphere.

Scanning electron microscope (SEM) images were recorded using a JEOL field emission microscope, model JSM 634OF. Sample preparation was carried out in the inert-atmosphere glove box. The samples were then sealed in a small bottle for transportation to the AC sputtering chamber where they were coated with gold-palladium prior of being transferred to the SEM vacuum chamber.

3. RESULTS AND DISCUSSION

Cathode surface-average Raman spectra were produced with the Raman microscope. Raman spectra were collected from multiple 52x75μm sections of the cathode surfaces at 0.7μm resolution and averaged into one representative spectrum for each cathode. MicroRaman surface-average spectra of the cathode from the virgin cell, and cathodes removed from cells, which lost 10, 24, 34 and 52 % of power are shown in Figure 1. The Raman spectra of all cathodes are dominated by two groups of bands: a broad maximum centered at ~510 cm^{-1}, characteristic for $LiNi_{0.8}Co_{0.15}Al_{0.05}O_2$ oxide, and two peaks at ~1350 and ~1600 cm^{-1}, which correspond to the D and G bands of elemental carbon, respectively. Raman spectroscopy is a particularly useful tool for characterizing the near-surface structure of carbons because of its relatively large Raman scattering cross-section. Recorded average spectra of the composite cathodes display strong carbon bands, which are predominant at almost all locations and originate from graphite and carbon black conductive additives.

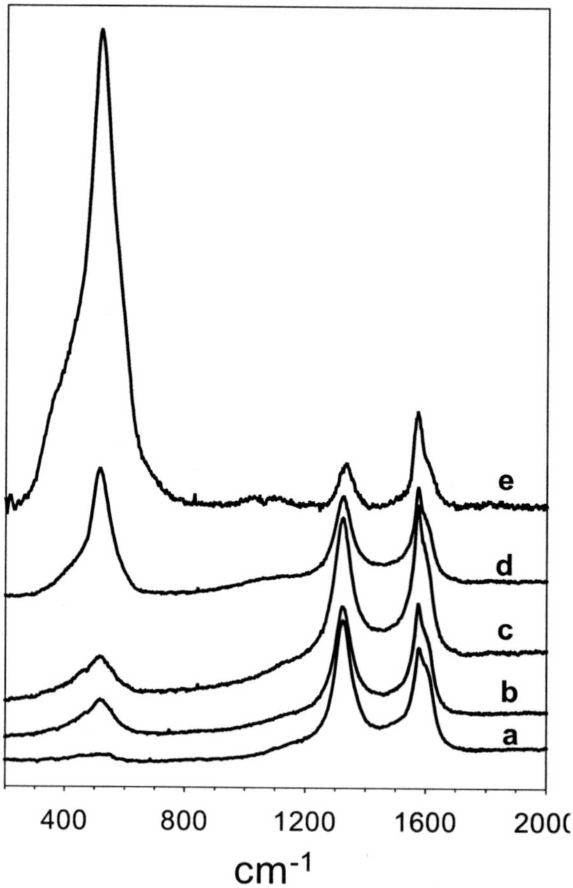

Figure 1. Average Raman microscope spectra of the fresh composite $LiNi_{0.8}Co_{0.15}Al_{0.05}O_2$ cathode (a), and the cathode from cells that exhibited 10 (b), 24 (c), 34 (d), and 52% (e) power loss.

The intensity ratio between the three major bands at ~510 cm^{-1} ($LiNi_{0.8}Co_{0.15}Al_{0.05}O_2$), and 1350, 1580 cm^{-1} (carbon) represent a semi-quanti-tative comparison because of the different Raman scattering cross-sections of $LiNi_{0.8}Co_{0.15}Al_{0.05}O_2$ and carbon. However, it is clear from Figure 1 that the active material/elemental-carbon surface concentration ratio increases with the increasing extent of cell and cathode degradation and is substantially higher for the cells that lost 10, 24, 34% of power compared with the fresh cell. The most drastic change in the oxide/carbon surface concentration ratio was observed in the cathode from the cell that lost 52% of its original power. Interestingly, the cathode surface state of charge varied between and within large individual agglomerates of active material and, at some locations, the spectra correspond to fully charged material, even though the cell was fully discharged at the end of testing and before disassembly.

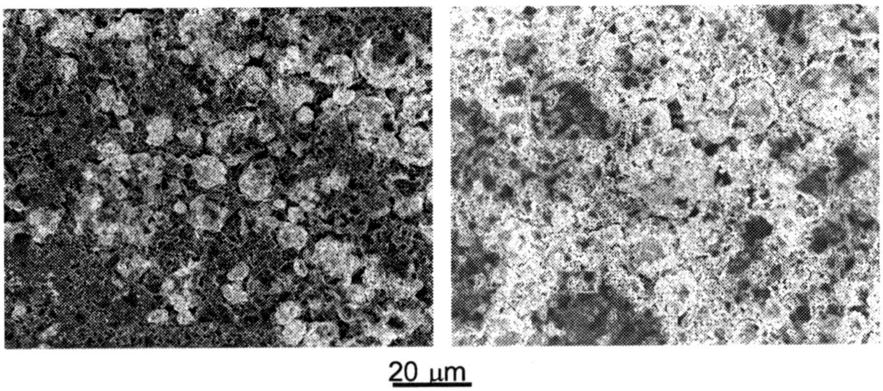

Figure 2. SEM images of the fresh composite $LiNi_{0.8}Co_{0.15}Al_{0.05}O_2$ cathode (a), and the cathode removed from a cell that lost 52% of its original power.

Figure 2 shows SEM images of a composite $LiNi_{0.8}Co_{0.15}Al_{0.05}O_2$ cathode extracted from a fresh cell and a cell that was cycled at 45°C for 68 weeks and lost 52% of its original power. Note that the surface of the fresh cathode consists of 5-20 µm agglomerates of $LiNi_{0.8}Co_{0.15}Al_{0.05}O_2$ (bright particles) and a dark fine particle coating, most likely carbon black, which is non-uniformly distributed across the surface. Interestingly, the surface morphology of the cathode from the tested cell reveals a noticeable change. There remain a few areas that show no or little change compared to the original morphology, whereas in most regions the black carbon coating has almost entirely disappeared from the surface leaving the active material fully exposed. This observation is in full agreement with our Raman spectra, which showed a dramatic decrease of the carbon signal in the tested cathodes.

In order to better understand the potential role of carbon retreat/redistribution on the cathode impedance rise we measured the surface electronic conductivity of these cathodes with CSAFM. Figure 3 shows CSAFM images of surface conductance (right-hand panel) and topography (left-hand panel) of a representative 5 x 5 μm region of the composite $LiNi_{0.8}Co_{0.15}Al_{0.05}O_2$ cathode surface at 1.0 V tip-sample voltage difference for (A) a fresh cell; and (B) a cathode from the cell that lost 34% of its power. The surface morphology images of both cathodes (left-hand panel) show large polycrystalline agglomerates and reveal no significant changes in the cathode surface topography in tested cells. This result is contrary to our earlier study of $LiNi_{0.8}Co_{0.2}O_2$ cathodes, which showed that considerable amounts of nanocrystalline deposits accumulated in the cathode inter-granular spaces and across the crystal planes [6,7].

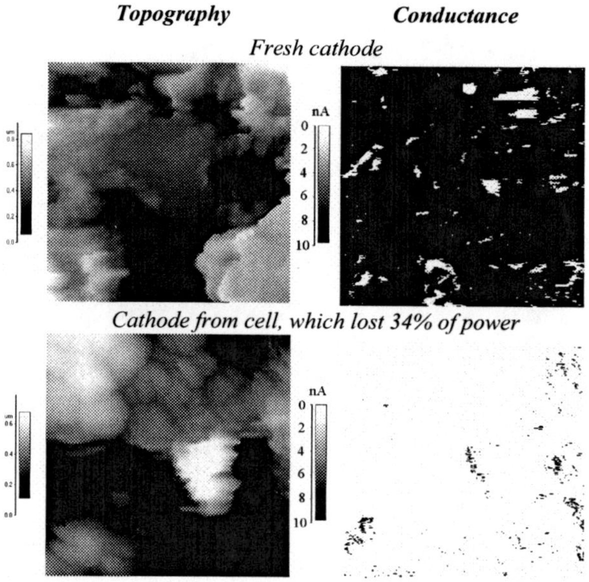

Figure 3. CSAFM images of surface conductance (right-hand panel, black areas are conductive) and topography (left-hand panel) of a 5 x 5 μm region of the cathode surface at 1.0 V tip-sample voltage difference.

In the conductance image, a dark color represents high electronic conductance, whereas a white color corresponds to areas of low or zero electronic conductance. Taking into account that the tip is in physical contact with the oxide, the magnitude of the current is determined by the local electronic properties of the electrode and the tip, and the tip-sample voltage difference. The surface conductance image of the cathode from the fresh cell exhibits areas of mostly excellent electronic conductance and only few insulating regions. Highly conductive graphite and acetylene black certainly contribute to the observed high tip current. The $LiNi_{0.8}Co_{0.15}Al_{0.05}O_2$ may also contribute to the tip current, but its contribution is substantially lower than that of carbon. The insulating areas on the cathode surface are most likely associated with the presence of PVDF binder, a solid electrolyte interphase (SEI), and/or unusual surface morphology (*e.g.*, deep crevices).

The conductance image of the tested cathode reveals that almost the entire electrode surface became insulating except for a few locations, mainly in the deep crevices and intergranular spaces, which remained conductive.

Previous results for $LiNi_{0.8}Co_{0.2}O_2$ cathodes [6,7] also showed a complete lack of conductivity of the cathode surface, consistent with loss of electronic contact between particles at the electrode surface and the remainder of the electrode. Another plausible explanation for the observed phenomena could be formation of a very thin SEI layer. A relatively thick layer of Li_2CO_3 and P-, O- and F- containing compounds forms upon storage of $LiNi_{1-y}Co_yO_2$ in $LiPF_6$-containing electrolyte at 80-90°C, as found in [8,9]. However, our Raman and infrared measurements revealed no observable SEI layer on the cathode surface. On the other hand, the presence of charged material at the cathode surface may also indicate that some particles of $LiNi_{0.8}Co_{0.15}Al_{0.05}O_2$ became electrically disconnected from the remaining part of the cathode due to mechanical stress and/or carbon additive retreat.

A reduction in conductance of portions of the cathode can easily explain the observed increase in cell impedance as well as loss of cathode capacity *via* isolation of oxide active material. Particle isolation is also in concert with our observations of carbon retreat or rearrangement in tested cathodes. These results represent the most clear and obvious difference (compared to other diagnostic studies) in cathode characteristics after prolonged cell tests at elevated temperatures.

ACKNOWLEDGEMENTS

This work was supported by the Assistant Secretary for Energy Efficiency and Renewable Energy, Office of FreedomCAR and Vehicle Technologies of the U.S. Department of Energy under Contract No. DE-AC03-76SF00098. The authors gratefully acknowledge the tested cells, help, and advice provided by the ATD Program participants.

REFERENCES

1. G. Nagasubramanian, R.G. Jungst, D.H. Doughty, *J. Power Sources*, **83**, (1999) 193.
2. Q. Wu, W. Lu, J. Prakash, *J. Power Sources*, **88** (2000) 237.

3. "FY 2000 Progress Report for the Advanced Technology Development Program," U.S. Department of Energy, Office of Advanced Automotive Technologies, Washington, D.C. (December 2000).
4. "FY 2001 Progress Report for the Advanced Technology Development Program," U.S. DOE OAAT, February 2001.
5. "Handbook of Diagnostic Techniques", Lawrence Berkeley National Laboratory Report no. LBID-2464, April 2003.
6. A. X. Zhang, P. N. Ross, Jr., R. Kostecki, F. Kong, S. Sloop, J. B. Kerr, K. Striebel, E. Cairns, and F. McLarnon, *J. Electrochem. Soc.* **148**, A463 (2001).
7. R. Kostecki and F. McLarnon, *Electrochem. Solid State Lett.*, **5**, A164 (2002).
8. A. M. Andersson, D. P. Abraham, R. Haasch, S. MacLaren, J. Liu, and K. Amine, *J. Electrochem. Soc.*, **149**, A1358 (2002).
9. D. Ostrowskii, F. Ronci, B. Scrosati, and P. Jacobsson, J. Power Sources, **94**, 183 (2001).

MODELING OF ELECTROCHEMICAL PROCESSES IN THE ELECTRODES BASED ON SOLID ACTIVE REAGENTS AND CONDUCTIVE CARBON ADDITIVES

Vadym V. Matveyev*

Ukrainian State University of Chemical Engineering
8, Gagarin avenue., Dniepropetrovsk, 49005, Ukraine

Abstract

The processes that occur at single crystals of an active substance embedded in an electroconducting porous matrix are considered. Conversion of solid reagents is assumed to proceed via a liquid-phase scheme as follows: dissolution–electrochemical reaction–crystallization. A set of equations is proposed. The set having been solved, an analytical expression is derived for a local polarization characteristic as a function of polarization, oxidation state of reagents, and structural and physicochemical parameters of the system. In order to check the model, galvanostatic characteristics of the electrodes made of the mixture of chloranile ($C_6Cl_4O_2$) and carbon black have been investigated. The theoretical and experimental characteristics show very good agreements.

Keywords

Theoretical model; porous electrode; solid reagent; dissolution; electrochemical reaction; crystallization; polarization characteristic; chloranile; carbon black.

1. INTRODUCTION

Most types of electrodes for electrochemical power sources have a porous structure [1]. Such electrodes are typically composed of electrochemically active material and additional components. The basic

* E-mail: vavlma@mail.ru

I.V. Barsukov et al. (eds.), New Carbon Based Materials for Electrochemical Energy Storage Systems, 453–471.

additional component ensures the essential electronic conductivity of the electrode. It builds a permanent continuous conducting network in the bulk of electrode matrix. Other additions are impregnated into active mass in small amounts. They are auxiliary agents, for example, a binder. Carbon materials such as graphite or carbon black are frequently used as the electrically conductive additives in battery electrodes.

In recent years, the electrodes of similar structure have also found application in the electroanalytical chemistry, for purposes of researching the electrochemical behavior of solid substances, etc. These electrodes are known as the modified carbon paste electrodes or carbon paste electroactive electrodes [2].

Given a widespread and growing application of such electrodes in electrochemical industry, a theory is required to describe the behavior of the electrochemical cells based on them. Such a theory would have to take into account, and to be able to distinguish between the individual contributions of processes proceeding at micro- to macro-levels in the electrochemical cell, as well as to furnish a multifunctional description of the whole system.

Several levels may be defined in the organization of such systems. We shall distinguish between the following levels [3]: an electrochemical cell as a whole; an electrode; a local section of the electrode; and an "elementary cell". The "elementary cell" is being defined as a separate crystal of active substance enclosed by a part of the conducting matrix. The “local section” is defined as being part of the electrode, which is contained in the active mass in the sufficient amount and with a set of properties, which reflect the average properties of the entire active mass; also, we shall assume that within this local section it may be possible to neglect spatial macrodistribution of the processes and the reagents concentration. Other levels of the system organization do not need to be explained due to their obviousness.

The theory on the level of the electrode and on the electrochemical cell is sufficiently advanced [4-7]. In this connection, it is necessary to mention the works of J.Newman and R.White's group [8-12]. In the majority of publications, the macroscopical approach is used. The authors take into account the transport process and material balance within the system in a proper way. The analysis of the flows in the porous matrix or in the cell takes generally into consideration the diffusion, migration and convection processes. While computing transport processes in the concentrated electrolytes the Stefan-Maxwell equations are used. To calculate electron transfer in a solid phase the Ohm's law in its differential form is used. The electrochemical transformations within the electrodes are described by the Batler-Volmer equation. The internal surface of the electrode, where electrochemical process runs, is frequently presented as a certain function of the porosity or as a certain state of the reagents transformation. To describe this function, various modeling or empirical equations are offered, and they are for the most part solved. However, such an approach does not permit to adequately take into account the processes proceeding at a microlevel, while describing the macrosystem.

There appears to be a more adequate approach when a local polarization characteristic is obtained as a result of analysis of the processes in the elementary cell and the local section of the electrode. This characteristic depends on the state transformation of the solid reagents and the concentrations of the electrolyte components. It further may be introduced into the equations describing the macrokinetic processes in an electrode, and may be used to model the behaviour of the system as a whole.

The model for calculation of the local electrochemical characteristics of an electrode is presented below.

2. MODEL

Herein, we consider the case when a porous conducting matrix with inclusion of active solid reagents represents the electrode. It is supposed, that both the reagent and the product are nonconductive. The conversion of the solid reagents is assumed to proceed via a liquid-phase mechanism in the following way: dissolution – electrochemical reaction – crystallization. Figure 1 shows the structure of the electrode and its model. The model has been developed on the bases of several assumptions.

2.1. Assumptions for the Model

Assumption #1. The elementary cell is spherically symmetric. Assumption #2. The reagent and product crystals are of low solubility. Their solubility is the following

$$c_1^{\,0}, c_2^{\,0} < 10^{-3}\mathrm{mol}\cdot \mathrm{l}^{-1} \tag{1}$$

The following designations shall be used in the proposed model (please see a schematic representation of the model as introduced by Figure 1): R_c is the radius of the elementary cell, which includes a conducting porous layer and a pore filled with crystals. R_0 is the radius of the pore containing the reagent and the product. $r_i(t)$ is the crystal radius. It is assumed that the radii of the crystals may vary with the time. The index *i* indicates which of the reagents form, either oxidized (*i* =1) or reduced one (*i* =2), is taken into account.

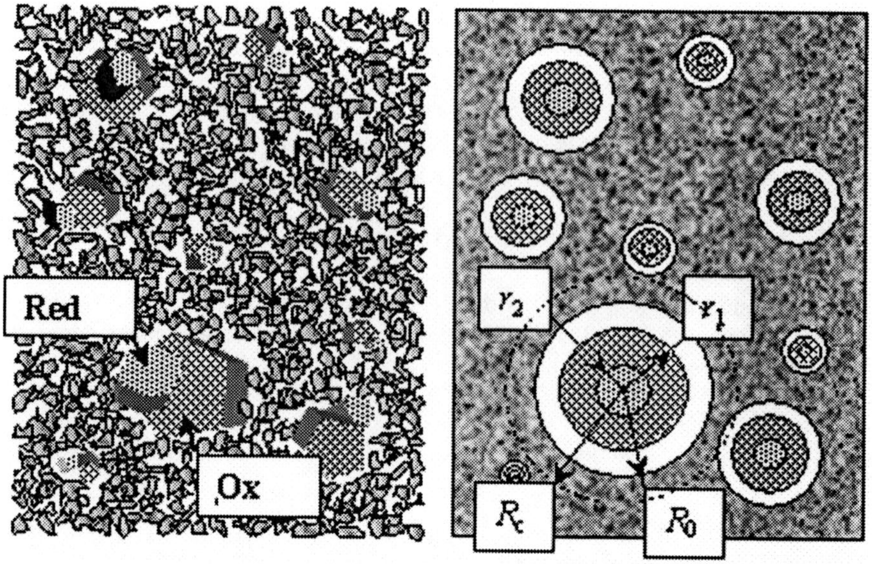

Figure 1. Typical structure of the porous electrode with solid reagents and schematic representation of the proposed physical model.

Assumption # 3. The reagent's transfer from one cell into another electrode is not taking place. During the electrochemical process, the solid reagent is being dissolved and diffuses into the conducting porous matrix. The solution gets reduced (oxidizes) electrochemically and the reaction product crystallizes within the same pore. It is supposed that the flow of the dissolved solid reagents through the border of the elementary cell does not take place.

$$\left.\frac{d\tilde{c}_i}{dr}\right|_{r=R_c} = 0, \quad i = 1, 2 \tag{2}$$

$c_i = c_i(r)$ is the concentration of the dissolved solid reagents in the pore around the crystals. The tilde (~) specifies that the parameter relates to a porous matrix.

Assumption #4. The electrochemical reaction is taking place within a porous electrode layer and it can be described by the Batler-Volmer equation (3).

$$i_V = S_V i_0 \left[\left(\frac{\tilde{c}_2}{c_{20}}\right) \prod \left(\frac{\tilde{c}_j}{c_{j0}}\right)^{k_j} e^{\alpha \frac{nF}{RT}\eta} - \left(\frac{\tilde{c}_1}{c_{10}}\right) \prod \left(\frac{\tilde{c}_j}{c_{j0}}\right)^{l_j} e^{-\beta \frac{nF}{RT}\eta} \right] \tag{3}$$

Here i_V is a local polarization characteristic, referred to as a unit of electrode volume. S_V, i_0 are the specific surface and the exchange current density, respectively. $\frac{c_i}{c_{i0}}$ is the concentration of solid-phase reagents (*i*) in the solution and this concentration is normalized to its equilibrium value; c_{i0} is usualy equal to the standard solubility c_{i0}^0. $\prod\left(\frac{c_j}{c_{j0}}\right)^{k_j(l_j)}$ is the product of the concentrations of the remaining solution components divided by their initial values. $n, F, R, T;\ \alpha, \beta$ are a number of electrons transferred in elementary electrode reaction; Faraday's constant, universal gas constant; temperature (°K) and transfer coefficients, respectively.

Assumption #5. Transfer processes as within the cell have been regarded as quasistationary. The typical time of the processes in the electrode (time of a charging or discharging being $t_0 \sim 10^4$ s) is longer than the time of the transitional diffusion process in the elementary cell $t_c \sim R_c^2/D \sim 10^{-1}$ s (radius of the cell is $R_c \sim 10^{-5}$ m, diffusion coefficient of dissolved reagents is $D \sim 10^{-9}$ m^2/s). Therefore, the quasistationary concentration distribution is quickly stabilized in the cell. It is possible to neglect the time derivatives in the transport equations.

Assumption #6. On the surfaces of the crystal, concentration of the corresponding substance in the solution corresponds to a saturated solution.

$$c_i\Big|_{r = r_i(t)} = c_i^0, \quad i = 1, 2 \tag{4}$$

Assumption #7. The reagent and the product crystals are considered to be independent.

Due to tridimentionality of the elementary cell, the reagent concentration in the solution decreases quickly (as $c_i \sim 1/r$) near the crystal surface during crystal dissolution. Therefore, the obstacles, namely crystals of a new phase, deform insignificantly the distribution of the concentration around of the crystal discussed. Figure 2 shows this peculiarity, which has been calculated in [13].

2.2. Basic Equations

Main processes in the elementary cell are schematically shown by Figure 3.

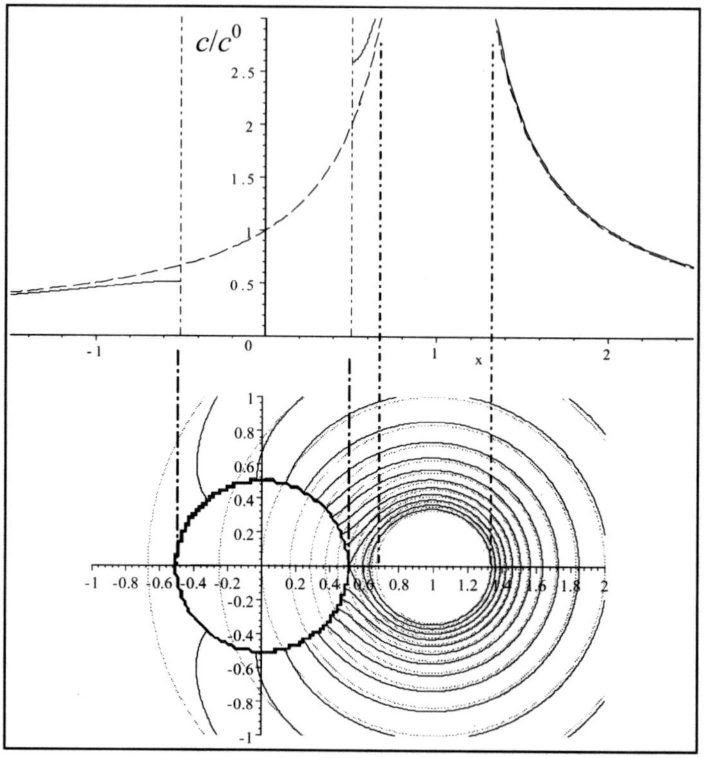

Figure 2. Allocation of the concentration in the case when the reagent diffuses into the free volume and when the other phase crystals present.

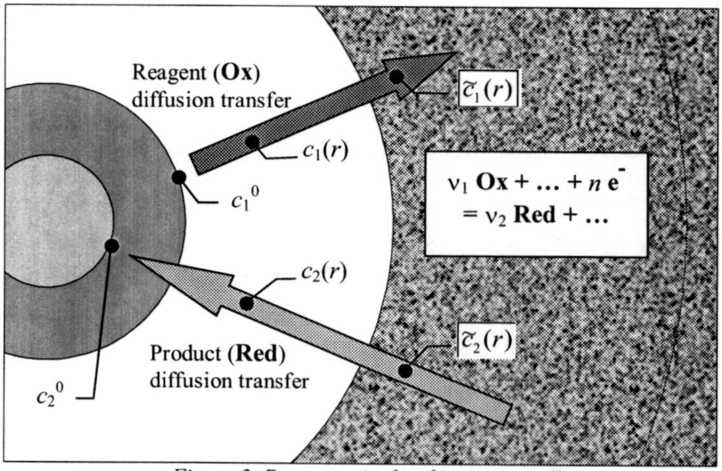

Figure 3. Processes in the elementary cell.

Taking into account the spherical symmetry and the quasistationarity, we have the following transition equations for the dissolved reagent and product in a pore ($r<R_0$)

$$\frac{d^2c_1}{dr^2}+\frac{2}{r}\frac{dc_1}{dr}=0 \qquad \frac{d^2c_2}{dr^2}+\frac{2}{r}\frac{dc_2}{dr}=0 \tag{5}$$

The equations in the porous matrix are the following:

$$\frac{d^2\tilde{c}_2}{dr^2}+\frac{2}{r}\frac{d\tilde{c}_2}{dr}=-\left(\frac{\nu_2}{nF\tilde{D}_2}\right)i_V \qquad \frac{d^2\tilde{c}_1}{dr^2}+\frac{2}{r}\frac{d\tilde{c}_1}{dr}=-\left(\frac{\nu_1}{nF\tilde{D}_1}\right)i_V \tag{6}$$

Here i_V is the local polarization characteristic (3). ν_i is the stoichiometric coefficient of substance i ($\nu_1>0, \nu_2<0$). With these coefficients, the reagent and product are included into the equation of the electrode reaction. D_i, $\tilde{D}_i$ is the diffusion coefficient of the dissolved solid reagent within the solution in the pore and within the porous matrix, respectively.

The boundary conditions on the pore surface are the following

$$c_1=\tilde{c}_1\big|_{r=R_C} \qquad c_2=\tilde{c}_2\big|_{r=R_C}$$

$$D_1\frac{dc_1}{dr}=\tilde{D}_1\frac{d\tilde{c}_1}{dr}\bigg|_{r=R_C} \qquad D_2\frac{dc_2}{dr}=\tilde{D}_2\frac{d\tilde{c}_2}{dr}\bigg|_{r=R_C} \tag{7}$$

On the boundaries of the elementary cell and the crystals, the boundary conditions are described by equations (2) and (4), respectively.

2.3. Polarization Characteristic of the Elementary Cell

Solving the system of the equations tacking into account the boundary conditions according to [3] we receive the following expression for the dimensionless polarization characteristic of the elementary cell.

$$\bar{i}=\frac{C_{20}C_k e^{\alpha\bar{\eta}}-C_{10}C_l e^{-\beta\bar{\eta}}}{\beta_{20}\left[1-\frac{1}{\xi_2}-\frac{N(\lambda)}{g_2}\right]C_k e^{\alpha\bar{\eta}}-\beta_{10}\left[1-\frac{1}{\xi_1}-\frac{N(\lambda)}{g_1}\right]C_l e^{-\beta\bar{\eta}}}. \tag{8}$$

Here $C_{i0} = \frac{c_i^0}{c_{i0}^0}$ is the solubility c_{i0} of the solid-phase reagents i normalized to its standard value. $C_{k\,(l)}$ equals $\prod \left(\frac{c_j}{c_{j0}} \right)^{k_j\,(l_j)}$.

$\bar{i} = \frac{i_c}{i_{c0}}$, $\bar{\eta} = \frac{nF}{RT}\eta$, $\xi_i = \frac{r_i(t)}{R_0}$. are the dimensionless values of the current, the polarization, the crystal radii, respectively. $i_{c0} = 4\pi nFR_0 \frac{c_0 D_0}{\nu_0}$ is the normal (characteristic) value of the current flowing from the elementary cell. $\beta_{i0} = \frac{\nu_i}{c_{i0}^0 D_i} \frac{c_0 D_0}{\nu_0}$, $g_i = \frac{D_i}{\tilde{D}_i}$ are the dimensionless parameters; c_0, D_0, ν_0 are the normalizing values of solubility, the diffusion coefficient and the stoichiometric coefficient ($\nu_0 = 1$), respectively. They are usually accepted as equal to the appropriate parameters of the solid phase in an oxidized (i=1) or reduced state (i=2).

2.4. Structural-Kinetic Factor

Function

$$N(\lambda, \xi_c) = \left[1 + \lambda \frac{\frac{\xi_c \lambda - 1}{\xi_c \lambda + 1} e^{2\lambda(\xi_c - 1)} - 1}{\frac{\xi_c \lambda - 1}{\xi_c \lambda + 1} e^{2\lambda(\xi_c - 1)} + 1} \right]^{-1} \tag{9}$$

is the factor taking into account the restrictions superimposed on the polarization characteristics (8) by the kinetics of the electrochemical transformations within the porous matrix
Here

$$\lambda = \sqrt{i_0 S_V \frac{R_0^{\;2}}{nFc_0 D_0} \left(\beta_1 C_l e^{-\beta\bar{\eta}} - \beta_2 C_k e^{\alpha\bar{\eta}} \right)} \tag{10}$$

is the kinetic factor. $\beta_1 > 0, \beta_2 < 0$ since $\nu_1 > 0, \nu_2 < 0$.

$\xi_c = \dfrac{R_c}{R_0}$ is the dimensionless radius ($\xi_c > 1$) of the elementary cell

2.5. Oxidation State

The state of oxidation $q(t)$ is the relation of the oxidized crystal mass at a given moment to the crystal mass, provided that the active substance is completely oxidized:

$$q(t) = \frac{m_1(t)}{m_1^0} = 1 - \frac{m_2(t)}{m_2^0}. \qquad (11)$$

In equation (11), $m_i(t), r_i(t)$ are the mass and the radius of solid reagent at a given moment *t*; m_i^0, r_i^0 are the crystal mass and radius provided that the active substance is completely oxidized (*i*=1) or reduced (*i*=2). Taking into account that $m_i \sim (r_i)^3$, the dimensionless radii of the crystals can be expressed via the oxidation state:

$$\xi_1(t) = \frac{r_1(t)}{R_0} = \frac{r_1^0}{R_0}\left[\frac{m_1(t)}{m_1^0}\right]^{\frac{1}{3}} = [\delta_1 q_1(t)]^{\frac{1}{3}}, \quad \xi_2(t) = [\delta_2(1 - q(t))]^{\frac{1}{3}} \qquad (12)$$

Here $\delta_i = \left(\dfrac{r_i^0}{R_0}\right)^3$ characterizes the relation between of the completely oxidized (reduced) crystal volume and the pore volume in which the crystal is placed.

2.6. Evolution Equation

In order to find the evolution equation, we shall consider the relation between the current and the oxidation state of a separate crystal.

The amount of electricity used for oxidation (reduction) of the crystal can be expressed through the change of the oxidation state in the following way:

$$dQ = \frac{nF}{M_i \nu_i} dm_i = \frac{nF m_i^0}{M_i |\nu_i|} dq \qquad (13)$$

Here M_i is the molecular weight of solid reagent *i*.

The current fall-off at this crystal is equal to the charge used for the oxidation of a single crystal within a unit of time (the anodic current is considered as a positive one)

$$\frac{dQ}{dt} = i_c \tag{14}$$

Hence, from the above, one can receive the equation of evolution, as follows:

$$\frac{\nu_0}{4\pi c_0 D_0 R_0} \frac{m_i^0}{M_i|\nu_i|} \frac{dq}{dt} = \bar{\imath}\left(q, \{c_j\}, \eta\right), \quad \eta = E(t) - E_0. \tag{15}$$

where $\bar{\imath}\left(q, \{c_j\}, \eta\right)$ denotes the expression of the dimensionless polarization characteristic of the elementary cell (8), where the radii of the crystals are expressed through the oxidation state.

2.7. Polarization Characteristic Relating to a Unit of the Solid Reagent Mass

2.7.1. Monodispersed Case

Let us assume, that the electrode includes the reagent crystals having identical size, which are placed at regular intervals in the conducting matrix. For such case,

$$N = \frac{3m_i}{4\pi\left(r_i^0\right)^3 \rho_i} \tag{16}$$

would be the number of crystals in a unit of the solid reagent mass. Here m_i, r_i^0, ρ_i are the mass, the averaged crystal radius and the solid phase density of the component i. Then

$$i_m = \frac{Ni_c}{m_i} = \frac{3nFc_0D_0\sqrt[3]{\delta_i}}{\rho_i\left(r_i^0\right)^2}\bar{\imath}\left(q, \{c_j\}, \eta\right) \tag{17}$$

is the polarization characteristic relating to a unit of the solid reagent mass in a monodispersed case.

2.7.2 Polydispersed Case

While performing dispersion analysis, the crystal aggregate is divided into several fractions; the crystals of each fraction have the sizes within a given range, and the quantity of the reagent crystals from each fraction N_n can be determined. The current relating to a unit of the active component mass may be presented as

$$i_m = \frac{1}{m_i}\sum N_n i_{cn}, \quad i_{cn} = \frac{4\pi c_0 D_0 R_{0n}}{\nu_0}\bar{i}\left(q_n, \{c_j\}, \eta\right) \tag{18}$$

Here it is supposed, that the crystals of each fraction contribute to the total current independently, i.e. the mass interchange between separate cells is disregarded. The latter is possible in the case when the packing density of the reagent crystals in a porous matrix is low and the distance between the crystals is much greater than the average radius of the crystals.

Works [14, 15] consider the case when parameter δ_i does not depend on the crystal radius, e.g. when $\delta_i \to 1$. The first case may take place when the initial solid reagent crystallizes from a solution within pores of the matrix. The second one is possible when the electrode is prepared from a mixture of the active crystal substance and the conducting additive particles having considerably smaller size than that of the crystals. Carbon black is usually used as such additive. The common expression for the current, relating to a unit of active component mass in this case is the following

$$i_m = \frac{nFc_0 D_0}{\nu_0 \rho_i \sqrt[3]{\delta_i}} \frac{\int_0^\infty f\left(r_i^0\right)\bar{i}\left[q, \{c_j\}, \eta(t)\right] r_i^0 dr_i^0}{\int_0^\infty \left(r_i^0\right)^3 f\left(r_i^0\right) dr_i^0} \tag{19}$$

Here, $f\left(r_i^0\right)$ is the size distribution function of the initial solid reagents

2.8. Peculiarities of the Polarization Characteristic

2.8.1. The Main Factor

In case the structural-kinetic factor $N(\lambda, \xi_c) \to 0$ the polarization characteristic does not depend on the kinetics of the electrochemical reaction. In this case

$$\bar{i} = \frac{C_{20}C_k e^{\alpha\bar{\eta}} - C_{10}C_l e^{-\beta\bar{\eta}}}{\beta_{20}\left[1-\left(\delta_2\left(1-q(t)\right)\right)^{-\frac{1}{3}}\right]C_k e^{\alpha\bar{\eta}} - \beta_{10}\left[1-\left(\delta_1 q(t)\right)^{-\frac{1}{3}}\right]C_l e^{-\beta\bar{\eta}}}. \tag{20}$$

The electrode current density is depends mainly on the solid reagents solubility and on the initial size of crystals

$$i_m \sim \frac{c_0}{\left(r^0\right)^2} \tag{21}$$

During the process of charge-discharge of the electrode, polarization varies with time due to modification in the crystal size.

2.8.2. On the Peculiarities of the Structural-Kinetic Factor

Factor $N(\lambda, \xi_c)$ decreases with increase of the amount of conductive additive (ξ_c), and the values of parameter $\lambda(\eta)$ increase, as shown by Figure 4.

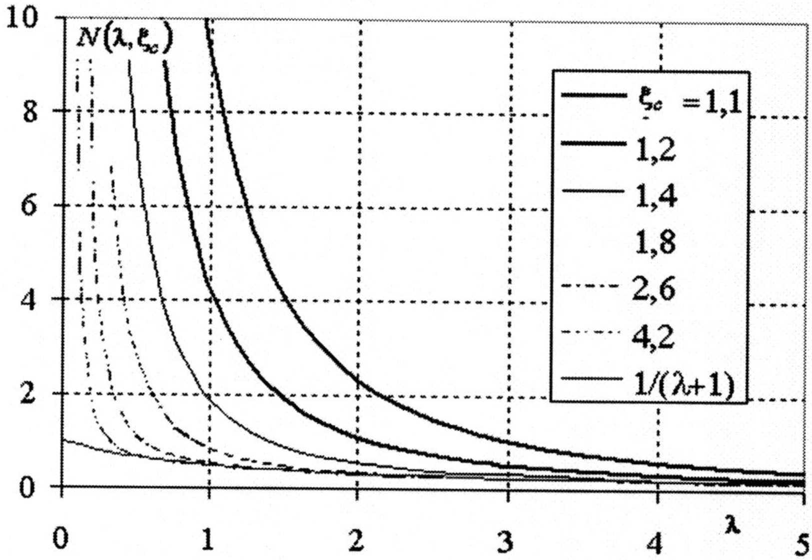

Figure 4. Dependence of the structural-kinetic factor on parameter λ at various values of ξ_c.

Function $\lambda(\eta)$ (see equation (10)) achieves its minimum

$$\lambda_{\min} \approx R_0 \sqrt{\frac{i_0 S_V}{nFc_0 D_0}} \tag{22}$$

when

$$\overline{\eta}_{\min} = \frac{1}{\alpha+\beta}\ln\left(\frac{\beta}{\alpha}\frac{|\beta_1|C_l}{|\beta_2|C_k}\right) \sim 0 \qquad (23)$$

and increases quickly with a shift of polarization both to a positive and to a negative side.

2.8.3. Estimations for a Case When the Kinetics Has no Effect on the Polarization Characteristics

The kinetics of processes has no effect on the electrochemical characteristic when factor $N(\lambda, \xi_c) < 1$. As Figure 4 shows, this is possible when the conditions $\lambda > 1$ and $\xi_c > 1.2 - 1.4$ are satisfied.

Let us estimate the specific surface of the conducting additive to get $\lambda > 1$. From (22) we have

$$\lambda \approx R_0 \sqrt{\frac{i_0 S_V}{nFc_0 D_0}} \approx r_i^{\,0} \sqrt{\frac{k_0 S_V}{D_i}} \qquad (23)$$

where k_0 is the electrochemical reaction rate constant

Usually $k_0 \sim 10^{-3} \div 10^{-9}$ m/s [27], the diffusion coefficient in solutions is in the range of $D_i \sim 10^{-9}\,\mathrm{m^2/s}$, while the average radius of the active reagent crystals is on the order of $r_i^{\,0} \sim 10^{-5}$ m.

Since the specific surface area of modern graphite materials is $S_V \sim 10^4 \div 10^{10}\,\mathrm{m^{-1}}$ ($S_m \sim 10^{-2} \div 10^4\,\mathrm{m^2 g^{-1}}$), it is almost always possible to make an electrode characterized by $\lambda > 1$ from a mixture of a carbon material and a solid reagent.

In this case, the kinetics has no effect on electrochemical performance characteristics.

3. GALVANOSTATIC CYCLING MODE

3.1. Theory

In the case when $N(\lambda, \xi_c) \to 0$, it is easy to receive expression for dependence of electrode polarization on the current, and the oxidation state of solid reagents, using polarization characteristic (20):

$$\overline{\eta} = \ln\frac{C_l}{C_k} + \ln\frac{C_1 - \bar{i}\beta_{10}\left[(\delta_1 q)^{-\frac{1}{3}} - 1\right]}{C_2 - \bar{i}\beta_{20}\left[(\delta_2(1-q))^{-\frac{1}{3}} - 1\right]} \tag{24}$$

The theoretical charge (discharge) curves calculated with the help of equation (24) at the following values of $\beta_{10} = \beta_{20} = 1$ and $\delta_1 = \delta_2 = 0.7$ are given by Figure 5. A high value of the initial polarization is due to small sizes of new phase crystals. The flow of the dissolved substance from a crystal to a matrix also decreases because the size of reagent crystals gets decreased during the dissolution process. In certain moment, the flow cannot any more provide the specified current and polarization sharply increases. The dependence of the limiting values and the oxidation state on the current is described (for anode process) by the following equation

$$q_{\max}(anode) = 1 - \frac{1}{\delta_2}\left(1 + \frac{C_2}{\bar{i}\beta_{20}}\right)^{-3} \tag{25}$$

and is shown by Figure 6.

3.2. Experimental

In order to check validity of the proposed theoretical model, galvanostatic characteristics of thin electrodes made out of the mixture of an organic active compound, chloranile ($C_6Cl_4O_2$) and acetylene carbon black with a weight ratio 1 : 0.75 have been assembled and their electrochemical performance has been investigated. A three-electrode cell made of ebonite was used to perform the electrochemical experiments. The working electrode was prepared from active mass ($m_{am} = 9.3 \cdot 10^{-5}$ g) by the way of compacting (P=300 MPa) it on a paraffin impregnated graphite disk (S = 1 cm^2). A counter electrode was Pt wire. The saturated Ag/AgCl electrode was used as reference electrode. A 0.5 M solution of H_2SO_4 was used as electrolyte. The study was conducted at 18°C. The current-generating reaction for this process is:

$$C_6Cl_4O_2 + 2H^+ + 2e^- \leftrightarrow C_6Cl_4(OH)_2 \tag{26}$$

At the beginning of cycling, the electrode was discharged and charged with a current of 0.25 mA. Both charge and discharge capacities were equal to Q_1^0 = 4.43 C. Then the electrode was cycled at a constant oxidation current of 0.25 mA (charge) and at different values of reduction (discharge) current (from 0.25 mA to 10 mA). Typical discharge curves as the function of oxidation state, which ware calculated in the following way

$$q = 1 - \frac{It}{Q_1^{\ 0}} \tag{27}$$

are shown in Figure 7.

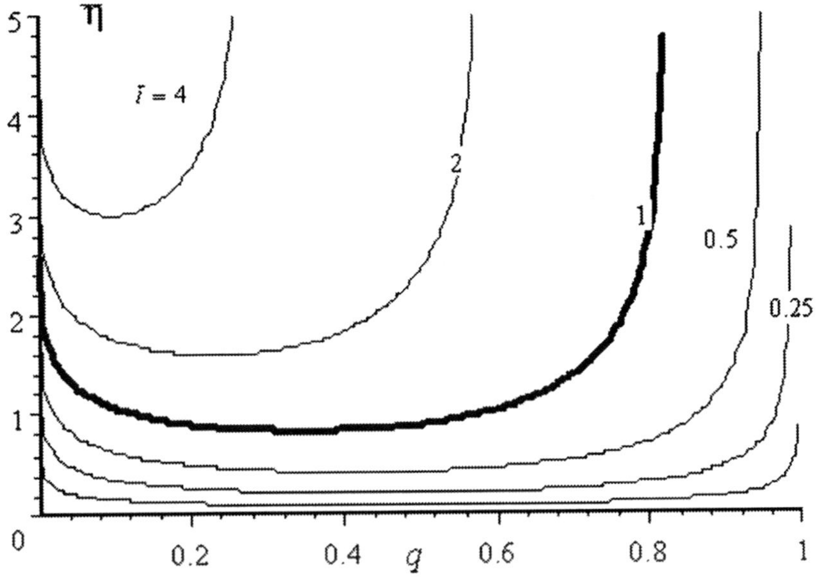

Figure 5. Dimensionless polarization as a function of the oxidation state.

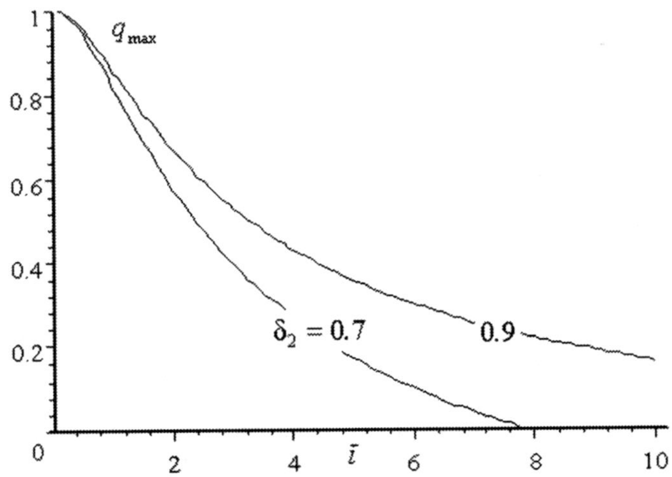

Figure 6. The limiting value of the oxidation state as a function of the dimensionless current in a galvanostatic mode.

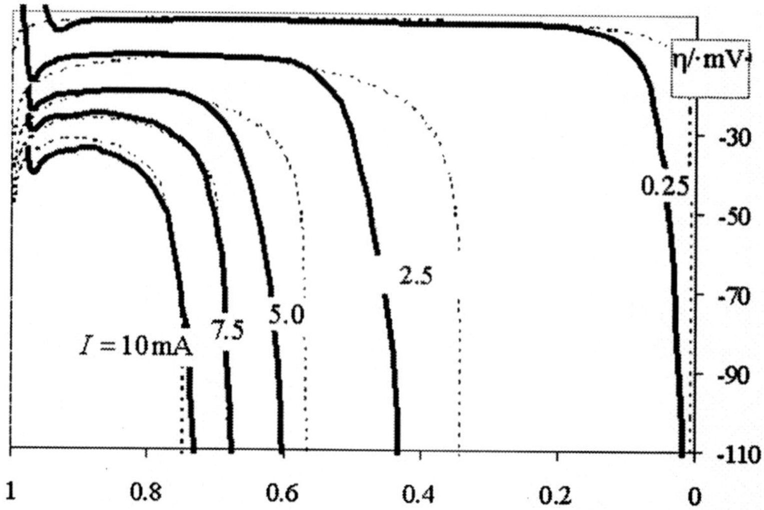

Figure 7. Comparison between the measured (solid lines) and computed (dotted lines) discharge curves as the function of the oxidation state at different currents.

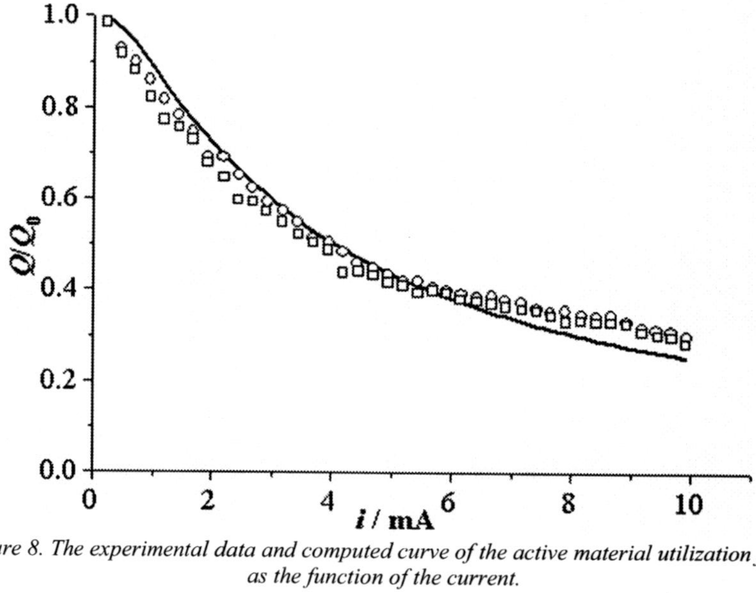

Figure 8. The experimental data and computed curve of the active material utilization factor as the function of the current.

Figure 8 provides a comparison of theoretically computed vs experimental dependences of the active material utilization factor for the investigated electrode. Analytical equations (24) and (25) were used to calculate polarization as a function of the oxidation state, and to calculate the limiting value of the oxidation state as the function of the discharge current (see Figures 7 and 8).

The dimensionless current $\bar{i}$ was calculated by using equation (17) in the following way

$$\bar{i} = \frac{I}{I_0}, \; I_0 = \frac{3nFc_0 D_0 \sqrt[3]{\delta_i}}{\rho_i \left(r_i^0\right)^2} m_i^0 \tag{28}$$

We have taken c_1^0, D_1 parameters for chloranile as the normalizing values ($c_0 = c_1^0$, $D_0 = D_1$). As the initial value, we used the following parameters of chloranile (Ox; i=1; m_1^0, ρ_1, r_1^0, δ_1). The initial chloranile mass m_i^0 was calculated as $m_i^0 = \frac{\nu_1 M_i}{nF} Q_1^0 = 5.5 \cdot 10^{-3}$ g, where M_i is the molecular weight of chloranile, ν_1=1, n=2. At small amount of the active mass the appreciable deviations in the values of the ratio of an active reagent and carbon black are probable.

We used the following values of other parameters: the density of chloranile ρ_i=1.79 g·cm^{-3} [16], the solubility of $C_6Cl_4O_2$ c_1^0=1.58 10^{-4} mol /l [17], the solubility of $C_6Cl_4(OH)_2$ c_2^0= 2.58 10^{-4} mol/l [17] and the diffusion coefficient of chloranile D_1=.7 10^{-9} m^2 /s [18]. We considered that $D_2 = D_1$. As the size of carbon black particles is significantly less than that of chloranile crystals, we believed that the parameters δ_1 and δ_2 are close to 1. In calculations the values of $\delta_1 = \delta_2$= 0.98 were used. The average size of the crystals ($r_1^0 \approx$ 6.2μm) have been defined with the help of a photomicrography of the active mass. As concentration of electrolyte did not change during the reaction, then C_k=1, C_l=1. The change of solid reagents solubility was not taken into account either. Hence C_1=1, C_2=1.

The parameters, which are used in equations (24), (25) and (28), had the following values: I_0= 1.2mA, β_{10}=1, $\beta_{20} \approx c_{10}^0 / c_{20}^0$ = 0.61. As one can see from Figures 7 and 8, the theoretical and experimental curves show very good agreements.

4. CONCLUSION

1. The proposed model generally describes the electrochemical process, when the solid reagent and the product have precise phase borders and the electrochemical reaction taking place on the surface, which does not vary essentially with the time.

2. The analytical expression is derived for a local polarization characteristic as a function of polarization, the oxidation state of reagents, and the structural and physicochemical parameters of the system.

3. A simple expression has been derived for the polarization as a function of the current and the oxidation state in the galvanostatic mode.

4. The expression for the active material utilization factor shows that in the process considered, it is impossible to achieve full utilization of the active reagents in the galvanostatic mode.

5. By reducing the crystal size, it is possible to achieve such a state when even poorly soluble substances become electrochemically active.

ACKNOWLEDGMENTS

This work has been supported by the State Foundation for Basic Research of the Ministry for Education and Science of Ukraine (grant No. 03.07/00168). The author expresses his gratitude to the U.S. Civilian Research and Development Foundation for financial support of his participation in the NATO-CARWC Wokrshop and Conference; to the Argonne National Laboratory, Superior Graphite Co., and personally to Dr. C. Johnson and Dr. I. Barsukov. for the organizational support of the visit.

REFERENCES

1. Bogotzky V.S., Skundin A.M. *Chemical Power Sources*. London-New York: Academic Press, 1980.
2. Grygar T., Markenand F., Schroder U. and Scholz F. *Collect. Czech. Chem. Commun*, 2002; 67: 163-205.
3. Matveev V.V. Porous Electrode with Weakly Soluble Nonelectroconducting Reagents: Polarization Characteristic of an Elementary Cell. *Russ. J. Electrochem*, 1997; 33: 839-46.
4. Ksenzhek 0.S. Macrokinetics of Processes on Porous Electrodes. *Electrochem. Acta*. 1964; 9: 629-37.
5. Micka K., Rousar I. Theory of Porous Electrodes. XII. The Negative Plate of the Lead-Acid Battery. *Electrochim. Acta*. 1974; 19: 499-502.
6. Micka K., Rousar I. Theory of Porous Electrodes. XVI. The Nickel Hydroxide Electrode. *Electrochim. Acta*. 1980; 25: 1085-90.
7. Vest X., Jacobsen T., Atiung S. Modeling of Porous Insertion Electrodes with Liquid Electrolite. *J. Electrochem. Soc*. 1982; 129: 1480-85.
8. Dunning J.S., Bennion D.N., Newman J. Analysis of Porous Electrodes with Sparingly Soluble Reactant. *J. Electrochem. Soc*. 1971; 118: 1251-56.
9. Newman J., Teedemann W. Porous-Electrode Theory with Battary Application. *A.I. Ch. Journal*. 1975; 21: 25-41.
9. Srinivasan F., Weidner J. W., and Newman J. Hysteresis during Cycling of the Nickel Hydroxide Electrode. *J. Electrochem Soc*. 2001; 148: A969-80.
10. Zheng G., Popov B.N. and White R. E. Application of Porous Electrode Theory on Metal-Hydride Electrodes in Alkaline Solution. *J. Electrochem Soc*. 1996; 143: 435-41.

11. Subramanian V. R. and White R. E. New Separation of Variables Method for Composite Electrodes with Galvanostatic Boundary Conditions. *J. Power Sources*. 2001; 96: 385-95.
12. Wu B. and White R. E., Modeling of a Nickel-Hydrogen Cell. Phase Reactions in Nickel Active Material. *J. Electrochem Soc*. 2001; 148: A595-609.
13. Matveev V.V. The Influence of the Other Phase Crystals Present on the Crystal Dissolution. (Rus.). *Problems of Chemistry and Chemical Technology /Voprosy Khimii i Khimicheskoy Tekhnologii/* (Ukrainian Journal); in press.
14. Matveev V.V. Calculation of the Electrochemical Characteristics of Porous Electrodes with a Polydisperse Crystal Structure (Rus.), *Problems of Chemistry and Chemical Technology /Voprosy Khimii i Khimicheskoy Tekhnologii/* (Ukrainian Journal). 2001; 3: 96-111.
15. Matveev V.V. Calculation of the Electrochemical Characteristics Porous Electrodes with Polydisperse Crystal Structure. 53rd Meeting ISE, Abstracts, Dusseldorf, Germany 15-20 September. 2002.
16. Chu S.S.C., Jeffrey G.A., Sakurai T. The Crystall Strycture of Tetracloro- p - Benzoquinono (Chloranil). *Acta crystallogr*. 1962; 15: 661-71.
17. Dinkevich F.E., Juravel T.A., Ksenzhek 0.S. and Lobach G.A. The Solubility of Some Quinon Compounds in the Sulphuric Acid Solvents. (Rus.), *Problems of Chemistry and Chemical Technology /Voprosy Khimii i Khimicheskoy Tekhnologii/* (Ukrainian Journal). 1979; 55: 10-12.
18. Lobach G.A. The Disk Microelectrode: Theory and Application. *Ph.D. Thesis*. Dniepropetrovsk: DChTI, 1982.

ON THE OPTIMAL DESIGN OF AMORPHOUS MANGANESE OXIDE FOR APPLICATIONS IN POWER SOURCES

S.A. Kirillov[1*], T.V. Lesnichaya[1], N.M. Visloguzova[1], S.A. Khainakov[1], O.I. Pendelyuk[1], D.I. Dzanashvili[2], T.A. Marsagishvili[2], V.Z. Barsukov[3], V.G. Khomenko[3], A.V. Tkachenko[3], and S.I. Chernukhin[4]

[1]*Institute for Sorption and Problems of Endoecology, 13, Gen. Naumov St., 03164 Kyiv, Ukraine;*
[2]*Institute of Inorganic Chemistry and Electrochemistry, 11, Mindeli St., 380086 Tbilisi, Georgia;*
[3]*Kiev National University of Technology and Design, 2, Nemirovich-Danchenko St., 01005 Kyiv, Ukraine;*
[4]*Joint Department of Electrochemical Power Engineering, 83A, Vernadsky Ave., 03680 Kyiv, Ukraine*

Abstract

Possible advantages of amorphous manganese dioxide as a cathode material for application in the electrochemical power sources are briefly outlined. A synthetic routine enabling one to produce amorphous manganese oxide from aqueous solutions is described. Preliminary galvanostatic tests reveal better performance of this material in coin-type cells with lithium anodes, if compared with conventional electrolytic manganese dioxide. Cycling studies demonstrate an ability of amorphous manganese oxide to reversibly form compounds with lithium.

1. INTRODUCTION

Crystalline chemical and electrochemical manganese dioxides are employed in billions of cells of Leclanché type which remain today probably the most popular and inexpensive power sources for wrist watches, cameras, flashlights, portable electronic products, etc. Worldwide annual output of power sources is estimated at 40 bln. pieces; more than 70% of them are those operating with MnO_2 chemistry as a cathode active material or a

[*] Corresponding author. E-mail: kir@i.kiev.ua

I.V. Barsukov et al. (eds.), New Carbon Based Materials for Electrochemical Energy Storage Systems, 473–479.

catalyst (zinc-manganese dioxide, lithium-manganese dioxide, zinc-air, etc.). Numerous commercial cells are produced on wide scale by many companies all over the world.

It is well known that manganese dioxides, being employed in the power source technologies are crystalline in nature. In the course of discharge of a power source with lithium anode, manganese oxide "absorbs" lithium cations and transforms into lithium manganese spinel oxide; when charging, spinel oxide "desorbs" lithium ions forming initial manganese oxide. Electrochemical activity of cathode materials made from crystalline manganese oxides is limited by diffusion of lithium cations inside (outside) the bulk of a manganese oxide (spinel) crystal.

The main drawback of lithium-ion batteries employing manganese oxides as cathode materials is their rapid capacity fade during cycling. Such a disadvantage could be avoided if cathode materials are obtained in their amorphous state (see, for instance, [1, 2]). This is easy to understand: in amorphous materials, the crystallites are of tens to hundreds angstrom size, hence diffusion complications do not exist or, at least, are significantly eliminated. This enables one to reach higher capacity values in power sources, to exclude capacity fade, and to improve cycling characteristics. On the other hand, high values of specific surface area of amorphous manganese oxides is a prerequisite of increasing catalytic activity of cathode materials in zinc-air cells, enabling one to reach higher capacity values. Another, and even better proven, electrochemical application, which explicitly exploits great specific area of amorphous manganese oxide, is its use in supercapacitors with non-acidic electrolytes [3].

As presented by this paper, our aim was to describe some preliminary results demonstrating that amorphous manganese-oxide-based materials synthesised by means of precipitation technique from aqueous solutions could serve as effective cathodes in lithium batteries.

2. TEMPLATE SYNTHESIS OF AMORPHOUS MANGANESE OXIDE

Manganese (IV) oxide enjoys numerous applications in modern technologies. The most widely known areas of its usage are sorption processes: one could recall that co-precipitation of contaminating cations with manganese oxide is still employed as part of the in-tank precipitation in processes of treatment of supernatant wastes at high concentration. Furthermore, co-precipitation data are usually used as benchmark results in studies of novel sorbents for strontium [4].

The legacy of Manganese oxides used in this work is as follows: initial chemistries were synthesized as part of the project directed towards optimization of amorphous manganese-oxide-based materials for waste water purification; it has started in 2003, and the works are now under way. In our preliminary studies we have managed to synthesize some mixed materials of MnO_2-TiO_2 system, prove their amorphous nature, estimate their properties and probe them as sorbents and catalysts [5]. Amorphous $xMnO_2$-$yTiO_2$ samples have surface areas of up to 400 m^2/g. It has been found that these samples are excellent sorbents for strontium and can be used for purification of drinkable and waste water contaminated by this hazardous pollutant. The distribution constants $\boldsymbol{K_d}$ of the samples towards strontium are ca. $1 \cdot 10^5$ mL/g; this value is comparable with characteristics of best sorbents for strontium [6]. Moreover, it is well known that manganese oxides are excellent sorbents for lithium and may be prospective for extraction of lithium cations from seawater [7]. In view of these facts, it became tempting to test if materials obtained in our studies could exhibit electrochemical activity in lithium cells.

Manganese (IV) oxide is known to be obtained by means of red-ox reactions. Several methods of precipitation of MnO_2 are mentioned in the literature sources: either reduction of manganese (VII) (permanganate) or manganese (VI) compounds, for example, by means of hydrochloric acid or hydrogen peroxide,

$$2MnO_4^- + 6HCl \rightarrow 2MnO_2 + 3Cl_2\uparrow + 2H_2O + 2OH^-, \tag{1}$$

$$2MnO_4^- + 3H_2O_2 + 2H_2O \rightarrow 2MnO_2 + 3O_2\uparrow + 4OH^-, \tag{2}$$

or oxidation of salts of manganese (II) or manganese (III), for example, using hydrogen peroxide

$$Mn^{2+} + H_2O_2 \rightarrow MnO_2 + 2H^+$$

Sometimes, reactions (1) and (3) are combined: permanganate serves as oxidant and a manganese-II salt as a reducing agent,

$$3Mn^{2+} + 2MnO_4^- + 2H_2O \rightarrow 5MnO_2 + 4H^+. \tag{4}$$

Main difference between the approaches used in [1-3] and in our work is as follows: in [1, 2], the authors operate with nonaqueous solvents: in order to obtain amorphous manganese oxide for battery applications, they reduce permanganate with lithium iodide and various alcohols in acetonitrile or propylene carbonate. In [3], the synthesis of amorphous manganese oxide for supercapacitor applications is made by mixing of $KMnO_4$ and $Mn(CH_3COO)_2$. Since in our studies we are largely interested in mixed MnO_2-TiO_2 oxides, the best synthetic approach employs reaction (3): oxidation of divalent manganese salts in aqueous solutions with hydrogen peroxide.

The discharge cycle of a lithium-ion battery with a manganese oxide cathode implies diffusion of lithium ions into the bulk of the cathode material forming lithium manganese spinel, and just structural limitations to the diffusion process can lead to the mentioned capacity fade during cycling. In order to facilitate the formation of spinel, it is desirable to create cavities of prescribed size in the structure of manganese oxide, so as to accommodate incoming lithium ions. To reach this goal, principles of the so-called template synthesis can be used. Tunnels or layers are arranged by using template ions (Li^+ in our case) to direct the cavity dimensions in embryos formed by means of co-precipitation. Depending on future applications, templates may be removed from the tunnel or inter-layer space by means of water (or acid) leaching.

As follows from the results obtained, numerous parameters influence the quality of precipitates, vs raw materials used (such as, manganese sulphate, nitrate, chloride); pH of initial and final solutions; the excess of the oxidant; the type of a neutralizing agent (lithium, sodium, potassium hydroxide); the washing method, and pH of leachates. More detailed description of the synthetic routine will be published elsewhere.

3. CHARACTERISATION OF MATERIALS OBTAINED

Air-dry samples have been studied using the methods of scanning electron microscopy, X-ray microanalysis, thermal analysis and X-ray diffraction. The latter two methods are crucial for conclusions regarding thermal pre-treatment of samples to be exploited in power sources.

It should be stressed that precipitates obtained do not follow the stoichiometric formula of MnO_2. Hydrous manganese oxides, depending on the conditions of synthesis, contain manganese in lower oxidation states, and their formula can be written as $MnO_{2-\delta}$. Actually, if the presence of water in the precipitate is taken into account, its formula should be written as $MnO_{n-z}(OH)_2 \cdot mH_2O$ ($z \leq n$). One can easily understand that, depending on z and n, substances of the variable degree of hydration and the number of hydroxyl groups could be obtained.

It is considered that hydrous manganese dioxides and their derivatives used as sorbents belong to the crystallographic group of α-MnO_2, where the atomic ratio O:Mn<2, and the presence of side ions is a prerequisite of the very existence of this type of the crystalline lattice. For example, in various synthetic samples, the O:Mn ratio can vary from 1.95 to 1.70. This means that the oxidation state of manganese in synthetic manganese oxides is seldom equal to 4.

As follows from the data of scanning electron microscopy, the mean size of particles is at around 2-5μm. The specific surface area of the amorphous material is of two orders of magnitude greater than that of crystalline samples. X-ray measurements and thermal analysis reveal that crystallization of amorphous manganese oxide does not occur up to 350°C. Even at this and the higher temperatures (up to 550°C), the intensity of reflexes from Mn_2O_3, if present, is much higher than from MnO_2. The existence of Mn_2O_3 can be easily detected by the increase of the mass of the sample due to an oxidation process taking place at T>350°C; however, we estimate an initial content of Mn_2O_3 as less than 3 %. Up to T≈600°C, MnO_2 remains amorphous or, at least, semi-amorphous, and does not crystallize. Upon further heating it loses oxygen, and transforms into Mn_2O_3. Finally, after 1000°C, all material undergoes transformation into Mn_3O_4. On the one hand, these data are significant as a proof of the amorphous nature of samples. On the other hand, they enable one to follow standard thermal treatment procedures established in manufacturing of power sources: for amorphous samples, thermal treatment at ca. 400°C removes all traces of Mn_2O_3.

4. ELECTROCHEMICAL TESTS

Preliminary electrochemical tests of materials obtained have been performed in two types of cells. Primary discharge measurements have been executed in standard 2325 coin-type cells (23 mm diameter and 2.5 mm height) with an electrolyte based on propylene carbonate – dimethoxyethane solution of $LiClO_4$. Cathode materials have been prepared from thermally treated amorphous manganese oxide in question (0.70±0.02g, 85wt%.) mixed with a conductive additive (10 % wt.) and a binder (5wt%). Lithium anodes of 0.45 mm thickness have been of slightly excess mass if compared to the stoichiometric amount, so as to ensure maximal possible capacity of a cell and full consumption of the cathode material.

For the sake of comparison, similar cells have been made containing electrochemical manganese dioxide produced by Chiaturi plant, Georgia. Just this material is employed in practically all commercial lithium power sources manufactured in the countries of the former Soviet Union.

Some data obtained at I=0.5 mA (fast discharge) are shown in Figure 1. It is well seen from this figure and Table 1 that cells with amorphous manganese oxide demonstrate higher capacity than those with electrochemical manganese oxide.

Promising results have also been obtained in cycling ability studies performed in model coin-type cells with this same amorphous cathode material. Moreover, it has been found that either thermally treated or even initial amorphous manganese oxide exhibit satisfactory results. Cyclic voltammetric studies have been carried out with the LP-30 electrolyte (Merck). Cathode materials have been prepared from amorphous manganese oxide in question (80wt%) mixed with a conductive additive (10wt%) and a binder (20wt%). Some results of these measurements are summarized in Table 2. They reveal that amorphous manganese oxide can be reversibly intercalated and deintercalated with lithium. The best values of the specific capacity obtained (ca. 40 mAh/g) are comparable with the standard capacity values of a commercial cathode material, lithiated manganese oxide ($LiMn_2O_4$), equal to 120-140 mAh/g.

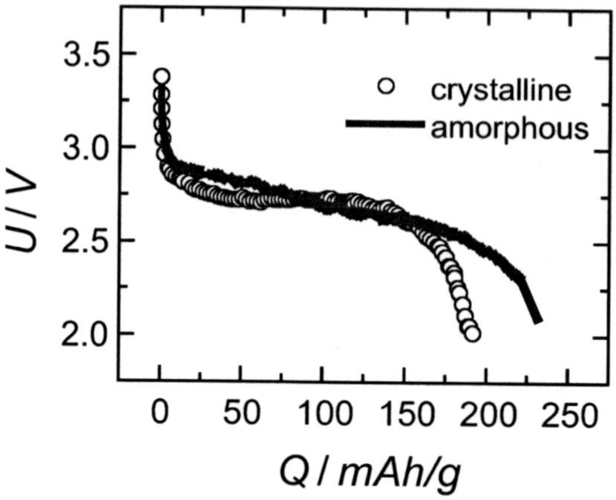

Figure 1. Discharge curves for 2325 coin cells with lithium metal anodes and electrolytic (crystalline) vs amorphous manganese oxide-based cathodes. For the latter sample, the best results are presented.

Table 1. Performance parameters of 2325 coin cells, as obtained from galvanostatic cycling regime (0.5 mA).

Cathode material	U_{mean}/V	Short circuit current / mA	Capacity / mAh	Specific capacity / mAh/g
Crystalline MnO_2	2.7	150	134	189
Amorphous MnO_2	2.6	120	175	247

Table 2. Cycling ability of amorphous manganese dioxide at various sweep scan rates.

Cathode material	Specific capacity / mAh/g					
	0.5 mV/s		0.2 mV/s		0.1 mV/s	
	Anodic	Cathodic	Anodic	Cathodic	Anodic	Cathodic
Before heat treatment	14.62	12.81	14.89	15.00	-	-
After heat treatment	-	-	42.92	40.43	42.30	39.56

5. CONCLUSION

Preliminary results presented in this paper seem to prove that amorphous manganese oxide can serve as a good cathode material in lithium battery applications. Preliminary galvanostatic tests reveal better performance of this material in model coin-type cells with lithium anodes, if compared with conventional electrochemical manganese oxide. Cycling studies demonstrate an ability of amorphous manganese oxide to reversibly intercalate and deintercalate lithium. Of course, much work is to be done in order to optimize the synthetic routine, so as to improve performance of the amorphous material, especially on cycling. On the other hand, it is clearly seen that simple aqueous chemistry routes are quite prospective for obtaining manganese oxide based cathode materials and can successfully compete with refined synthetic methods based on the use of non-aqueous solvents.

ACKNOWLEDGEMENTS

This research work has been funded by EC within the framework of the STCU Grant #GR85. S.A.K. is grateful to CRDF for financial support for visiting the NATO-CARWC. O.I. Pendelyuk acknowledges partial support of her stay at Argonne obtained through the U.S. Department of Energy Office of FreedomCAR and Vehicle Technologies, via the Lawrence Berkeley National Laboratory.

REFERENCES

1. Manthiram, A., Kim, J., Im, D., Amorphous Manganese Oxides for Lithium Ion Batteries, *ITE Letters*, **2**, 318-325 (2001).
2. Im, D., Manthiram, A., Amorphous Manganese Oxide Cathodes for Rechargeable Lithium Batteries, *Ceramic Trans.*, **127**, 205 (2002).
3. Lee, H. Y., Goodenough, J. B., Supercapacitor Behavior with KCl Electrolyte, *J. Solid State Chem.*, **104**, 220-223 (1999).
4. Hobbs, D. T., Blume, M. S., Tracker, H. L., *Screening Evaluation of Alternate Sorbents and Methods for Strontium and Actinide Removal from Alkaline Salt Solution*, WSRC-TR-2001-00072, 2001, 20 p.p. Available at http://sti.srs.gov/ fulltext/tr2001072/ tr2001072.html.
5. Lesnichaya, T. V., Visloguzova, N. M., Pendelyuk, O. I., Dzanashvili, D. I., Romanova, I. V., Farbun, I. A., Zazhigalov, V. A., Marsagishvili, T. A., Kirillov, S. A., Optimal design of nanodispersed amorphous manganese oxide as a sorbent towards strontium, *Trace Element Admixtures in Water*, Trans. of Int. Seminar, Kyiv, 2003, p.p. 35-36.
6. Behrens, E. A., Sylvester, P., Clearfield, A., Assessment of a Sodium Monotitanate and Pharmacosiderite-Type Ion Exchangers for Strontium and Cesium Removal from DOE Waste Simulants, *Environ. Sci. Technol.* **32**, 101-107 (1998).
7. Ooi, K., Miyai, Y., Katoh, S., Maeda, H., Abe, M., Topotactic Li^+ Insertion to λ-MnO_2 in the Aqueous Phase, *Langmuir*, **5**, 150-157 (1989).

INVESTIGATION OF CATHODIC MATERIALS BASED ON DIFFERENT TYPES OF MnO_2/CARBON

I. S. Makyeyeva [1*], N. D. Ivanova [2], G. V. Sokolsky [3]

[1]*Kiev National University of Technologies and Design, Ukraine*
[2]*Institute of General and Inorganic Chemistry of NAS, Kiev, Ukraine*
[3]*National Aviation University, Kiev, Ukraine*

Abstract

It is shown that electrolytic manganese dioxide, which has been obtained from fluorine-containing electrolytes differs from traditional types of MnO_2 obtained by state-of-the-art synthesis methods. This material is characterized by the increased amount of structural defects. It is shown that crystalline structure with a large number of defects has a higher catalytic and electrochemical activity.

The electrochemical testing in coin cells for systems of MnO_2 – Zn and MnO_2 – Li has been carried out. The comparison of characteristics of lithium elements of CR2325 configuration with known analogues demonstrated an increase by 50% in discharge capacity. The independent comparative testing of the experimental samples of our power sources of 1142 size and Toshiba's L 43 was carried out. It was shown that power sources with the manganese dioxide cathode obtained from fluorine-containing electrolytes have demonstrated 30% higher capacity in MnO_2 – Zn system. Such outstanding results are explained by apparent higher catalytic and electrochemical activity of subject MnO_2.

1. INTRODUCTION

Manganese dioxide has found wide use in different electrochemical power sources as cathode material. It is known that the structure and properties of manganese dioxide depend on the method of its synthesis.

* Corresponding author. E-mail: makeeva05@yahoo.com

I.V. Barsukov et al. (eds.), New Carbon Based Materials for Electrochemical Energy Storage Systems, 481–486.

Among five crystalline forms of manganese dioxide electrochemical γ-MnO_2 is known to be the most electrochemically active.

Manganese dioxide is the most extensively used electrode for different types of power sources (alkaline primary and rechargeable, lithium – and magnesium – based batteries). The production of power sources with manganese dioxide accounts for about 70 – 80 % of the world total output. In these system electrolytic γ – manganese dioxide is mainly used [1].

In [2] the authors suggested that naturally occurring and synthetic γ – MnO_2 phases used in batteries might contain lattice defects such as dislocation, chain defects and other irregular voids and that water, as well as impurity ions, may be accommodated in such lattice defects. They stated that such defects might account for the high water content and high electrochemical activity of γ – MnO_2.

In view of this situation we studied a number of samples of MnO_2 with different composition and various OH^- contents in order to estimate the correlation between the activity of MnO_2 and concentration of OH^- ions. This compound can be electrochemically deposited on the anode from various aqueous solutions, but electrolytes with sulfate and ammonia sulfate have found widest application [3]. It has been determined that the composition and structural parameters of the end-product are governed by the presence of fluoride ion in electrolyte.

2. EXPERIMENTAL

We investigated the following samples: № 1 – has been synthesized by electrochemical method from a manganese sulfate-containing electrolyte with additions of fluoride ion; № 2 – Battery grade Electrolytic Manganese Dioxide (EMD–2, Georgia); № 3 – Battery grade Electrolytic Manganese Dioxide γ–MnO_2 (Kerr – McGee Corp., USA); № 4 – chemical MnO_2 (Pridneprovsky Chemical plant, Ukraine); № 5 – commercial chemical β – MnO_2 (Ukraine).

Chemical analysis of the investigated samples has been carried out using standard methods [4]. The combined water content was determined by DTA and Infrared spectral studies in the range of 200 – 4000 cm^{-1}. X – ray diffraction analysis was carried out with Dron – 3 using MoK_α radiation.

The discharge test of manganese oxide compounds was performed at a constant external load at room temperature. Electrochemical behaviour was studied too. Manganese dioxide compounds were mixed thoroughly with battery-grade expanded graphite FormulaBT ABG1025 (Superior Graphite Co., USA) at the ratio 9:1. The purpose of graphite additions is conductivity enhancement. Conductive graphite has a mean particle size of 25μm. In order to stabilize the shape of the electrodes, 0,5wt% of a binder (Teflon suspension) had to be added (only for sample №3). The resulting active material was pressed at 10MPa into the test cells.

3. RESULTS AND DISCUSSION

It has been observed that MnO_2 samples were dramatically different from each other. They have different composition, structure and properties, some of which can be seen in Table 1.

Table 1. Characteristics of Manganese Dioxide Samples.

Sample №	Mn^{4+}, %	Mn^{3+}, %	Mn^{2+}, %	X – MnO_x	H_2O, %	OH^-, %	$k \cdot 10^5$, s^{-1}
1	50,8	5,4	0,28	1,95	9,9	2,2	21,7
2	58,0	1,8	0,05	1,97	4,9	2,1	10
3	89	-	-	-	5,3	2,0	9,4
4	61,0	1,7	0,75	1,98	3	1,3	18,1
5	62,1	1,74	-	1,99	0,4	0.2	3

The first sample has the greatest deviation from stoichiometry. Samples №1 - №4 have been found to be nonstoichiometric oxide – hydroxide type. They have a mixed conductivity – ionic (σ_i) and electronic (σ_e). The ionic one is due to the presence of OH^- - groups. Namely, they stabilize the defects of chemical nature in such compounds. These defects are determined by the presence of Mn^{4+} and Mn^{3+} in the same crystallographic position.

There is the correlation between water and OH^- – groups content, ionic conductivity and catalytic activity of compound, Figure 1, Table 1. The data shows that the concentration of OH^- - groups tends to increase in the order 1>2>3>4>5. In this order increases the catalytic activity too (value K, Table 1). This difference in the activity between samples seems to be related to the difference in the OH^- - group content and Mn^{3+}/Mn^{4+} concentration. This means that increasing of structure defects may lead to increasing of activity of compound. Additional structure distortion has been obtained in modified sample by insertion of small amount of ions of alkaline metals (sample № 4).

The observed higher rate of mass transfer in nonstoichiometric oxides makes it possible to operate at discharge current densities on the order of 2 – 5 mA/cm^2, which is several times greater than that used with the conventional manganese dioxide. Thus, the purposeful disordering of structure of a cathode material of manganese dioxide allows to utilize such power sources in extreme conditions – for element discharge at a low resistance (from 0,3 kOm), Figure 1, a, b. This result indicates that electric parameters of power sources are in agreement with data on the degree of deviation from stoichiometric composition.

The comparative testing of the experimental samples of power sources of LR2325 size (system MnO_2-Li) and 1142 size (system MnO_2-Zn), Tables 2, 3, support the above conclusion.

Table 2. Specific capacity, mAh/g, of LR2325 batteries based on various MnO_2 samples at different external loads (MnO_2 – Zn).

Load, KΩ	**Sample / Performance**			
	1	**2**	**3**	**5**
0,3	238	136	110	-
1,0	258	177	242	-
5,6	289	245	262	107

***Table 3. Technical Characteristics of LR2325-size** MnO_2-Li and 1142-size MnO_2-Zn cells based on MnO_2 synthesized electrochemically from manganese sulfate-containing electrolytes with additions of fluoride ion.*

Electrochemical System	**A.C. Voltage, V**	**R, Ω**	**Q, mA h/g**	**SE, Wh/kg**	**PD, W/kg**
MnO_2 – Li	3,5	5600	200	539	1,5
MnO_2 – Zn	1,5	5600	280	151	0,25

Table 4. Main performance characteristics of IGIC's 1142-size MnO_2-Zn cells based on MnO_2 synthesized electrochemically from manganese sulfate-containing electrolytes with additions of fluoride ion vs baseline LR43 Toshib'a cells (cell sizes are similar).

Cell size designation	**Diameter, mm**	**Height, mm**	**U_{nom}, V**	**U_{end}, V**	**I_{nom}, mA**	**C_{nom}, mAh**
LR43 (Toshiba)	11,6	4,2	1,5	1,2	0,12	70
1142 (IGIC)	11,6	4,2	1,5	0,9	5	≥ 100

a)

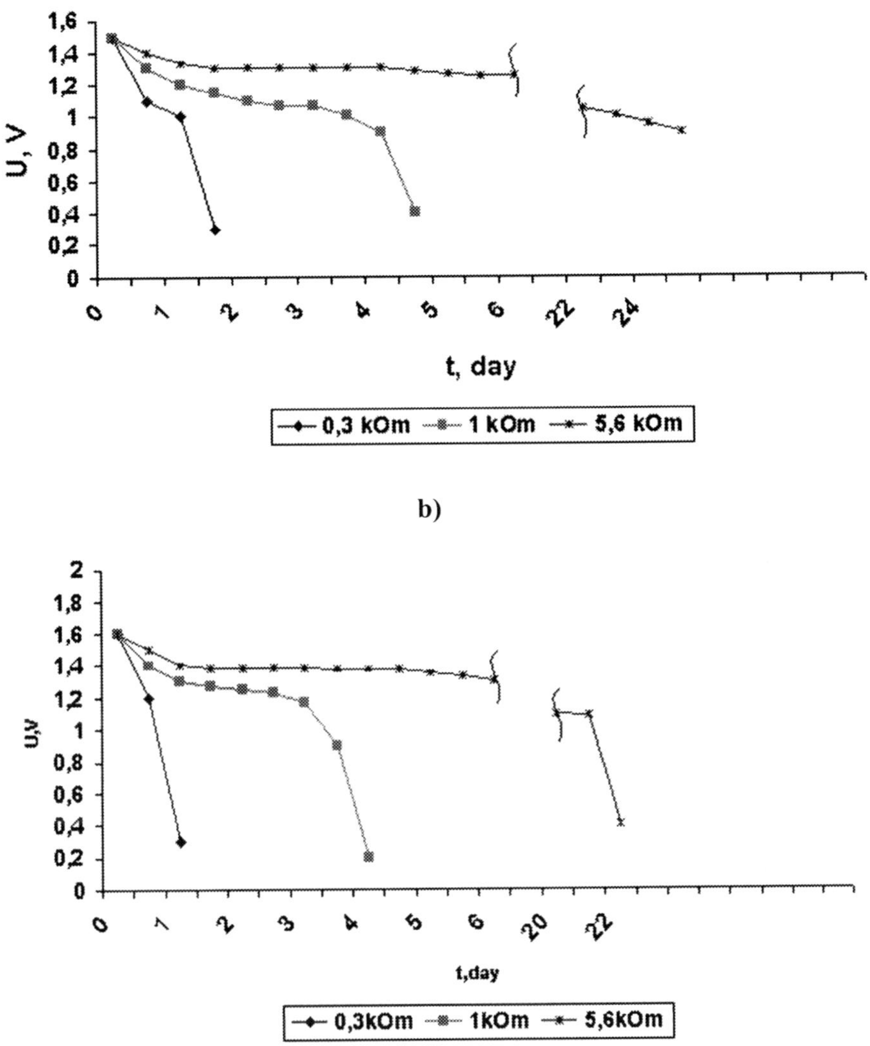

Figure 1. Discharge curves of MnO_2 – Zn power sources (IGIC 1142 cell size) at the low load (0,3 kΩ per cell); (a) - sample № 1; (b) - sample № 3.

The independent comparative testing of the experimental samples of the internally-made power sources of IGIC 1142 size and the Toshiba`s L43 (both size designations stand for similar cell sizes) were carried out. It was shown that power sources containing manganese dioxide cathode obtained from fluorine containing electrolyte (sample 1) have 30 % higher capacity in the MnO_2 – Zn system, Table 4.

4. CONCLUSIONS

The effect of structural defects in various types of MnO_2 on the electrochemical behavior has been investigated. It was shown that the basic types of defects in Mn3+ are in the same crystallographic position as in Mn^{4+}. Defects were seen to be are stabilized by the OH^- - groups. The catalytic and electrochemical activity of MnO_2 increases with increase in defect content.

A form of electrolytic manganese dioxide obtained by us from fluorine-containing electrolytes was seen to produce the best cathode active material for Zn/MnO_2 and Li/MnO_2 types of power sources.

According to the results presented, it appears that the higher the number of lattice defects in MnO_2 structure, the higher the discharge capacity will be.

ACKNOWLEDGEMENTS

The authors are grateful to Professor V. Barsukov (Kiev National University of Technologies and Design, Ukraine) and Dr. I. Barsukov (Superior Graphite Co., USA) for support of this work.

REFERENCES

1. G. Piao, M. Yoshio, H. Noguchi and A. Kozava. // J. Power Sources, 51 (1994) 391-395.
2. R. G. Burns, A. Kozava and M. Nagayama // Battery Materials Sump., Brussels, Belgium, 1983, 7-10.
3. K. Takahashi. // Electrochimica Acta, 26, 10 (1981) 1467-1476.
4. A. I. Vogel. // Quantitative Inorganic Analysis (1981).

INVESTIGATION OF THIN-FILM ELECTRODE MATERIALS AS CATHODIC ACTIVES FOR POWER SOURCES

Nataliya E. Vlasenko*, Nataliya D. Ivanova,
Evgeniy I. Boldyrev, Olga A. Stadnik

V.I. Vernadskyi Institute of General and Inorganic Chemistry
National Academy of Sciences of Ukraine
32/34 prospect Palladina, 03680 Kyiv-142, Ukraine

Abstract

The electrochemical behavior of thin-film oxide-hydroxide electrodes containing chromium, nickel and cobalt compounds was investigated. Experimental results have shown that such compounds can be successfully used as active cathodic materials in a number of emerging primary and secondary battery applications.

Keywords

oxides of chromium, nickel and cobalt; cathodic materials; thin films.

1. INTRODUCTION

Oxide compounds are widely used as cathodic materials in the power sources and electrochemical generators. Some literature data indicates that cathodic materials based on nonstoichiometric oxide compounds make it possible to increase the solid-phase reduction process. The kinetics of electrochemical reactions and consequently the current density are the higher, the greater the degree of deviation from stoichiometry, and the lager the number of the defects in the compounds' structure [1, 2].

In the rechargeable lithium-ion batteries, in which current-generating reactions are based on a "rocking chair" mechanism, the ions of lithium shift back and forth between the intercalation hosts of the cathode and anode. The

* Corresponding author. E-mail: vlasenko05@yahoo.com, nvlasenko@ukr.net

I.V. Barsukov et al. (eds.), New Carbon Based Materials for Electrochemical Energy Storage Systems, 487–493.

lithium-ion secondary battery mainly consists of a carbonaceous or lithium-alloy anode and a complex transition metal oxide as the cathode material. (Figure 1).

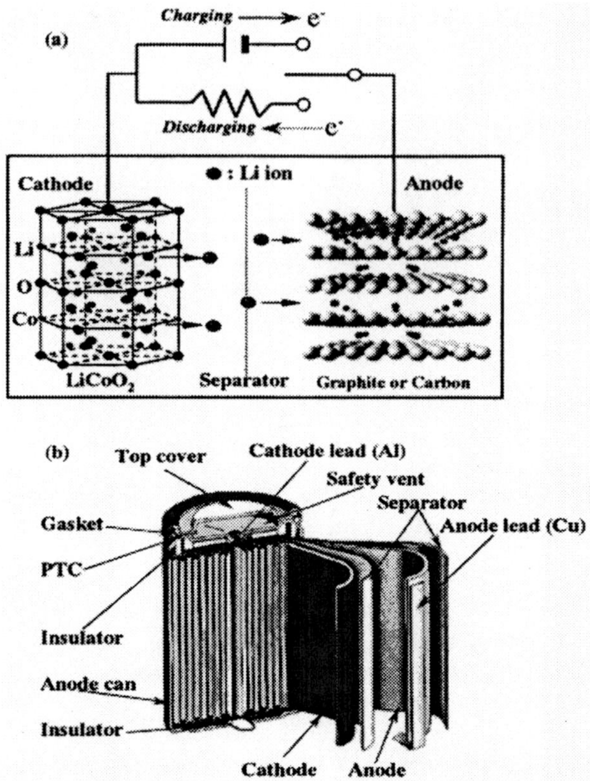

Figure 1. Recharging mechanism of batteries with the "rocking chair" concept (a) and structure of a rolled type battery (b), [1].

The use of thin-film cathodes for battery application usually results in a better performance due to shorter diffusion path of intercalated cation through solid matrix. The thin film electrodes are used in manufacturing of rolled type batteries and thin film cells.

The rolled type batteries are general-purpose batteries with high capacity and good characteristics. Thin film power sources have found a number of specific applications in the fields of portable accumulators with high energy density for application in electronic and integral circuits, sensors, monitors and for medicine [3].

Thin-film electrodes have promise in the development of flexible power sources (primary and secondary). It must be taken into account that change in cathodic material crystal lattice must not be over 20% as a result of intercalation of Li^+, Na^+ or H^+ [4]. The electrodes must be chemically active and have both electron and ion conductivity. In connection with these demands we have chosen oxide-hydroxide compounds of Cr, Ni, Co as the object of investigations.

This paper proposes utilization of such compounds as cathodic materials and identifying dependencies of charge-discharge processes on the composition of these compounds. We have studied the process of electrodeposition of thin films of nickel, chromium and cobalt oxides from aqueous solutions of salts of 3d-metals. The influence of electrolysis conditions and potentials on surface phase composition has been investigated. The original method of electrochemical synthesis [5] developed by us allows nonstechiometric oxide-hydroxide compounds of transition metals to be obtained. For film production we use cathodic and anodic electrodeposition on metal surface. For deposition of dense complex oxide films we have developed a number of original fluorine-containing electrolyte systems. The use of such complex electrolyte effects in deposition of 3d metals oxide thin films gives good electrochemical properties.

The mentioned method for synthesis of oxide-hydroxide compounds (Ni, Cr, Co) is more controllable and enables with production of electrode films definite amounts of components. Therefore it guarantees the reproducibility of their compositions and properties. Using the above method we were able to produce the following oxide compounds: zero valence metal and lowest oxidation state oxide-hydroxide compounds; in cathode process and oxide-hydroxide compounds; (in anode process the oxide compounds consist of highest oxidation state oxide-hydroxide compounds). Both type compounds possesses electronic and ionic conductivity.

2. EXPERIMENTAL

The film electrodeposition process was studied by means of linear sweep voltammetry. The rate of electrochemical reaction was determined from current density (current-potential curves). The film deposits were characterized by chemical analysis, IR - spectroscopy, XRD, TG, TGA and SEM methods.

The electrochemical activity of the products (Li intercalation) was studied in experimental coin cells of 2016 size. In the case of precipitated powder of Ni the electrode active masses were prepared by mixing the active material with 10 wt.% of graphite (conductive additive) and 5 wt.% of polyvinylidene fluoride (binder) stirred in 1-methyl-pyrrolidone solvent. Thin oxide films on steel lead were used as cathode in the case of dense and solid film covers. 1M $LiPF_6$ solution in mixture of ethylene carbonate (50 vol.%) and methyl carbonate was used as an electrolyte (Merck product LP30). The assembled cells were then tested using the method of linear sweep voltammetry (LSV) (potentiostat PI50) and/or galvanostatic charge-discharge (PC cotiolled automatic bench) methods. In the case of thin film of oxide-hydroxide of chromium and cobalt, the electrode active masses where used directly without conductive additives and binders. The electrolyte contained 1M $LiClO_4$ in propylene carbonate (PC).

3. RESULTS AND DISCUSSION

On the basis of obtained data of cyclic voltammograms for 3d metals oxides electrodeposition the optimal conditions (current density, potential, process time, electrolyte composition, temperature) for dense oxide films (Ni, Cr and Co) deposition on steel foil have been elaborated. Data relating to several best films are summarized in Table 1.

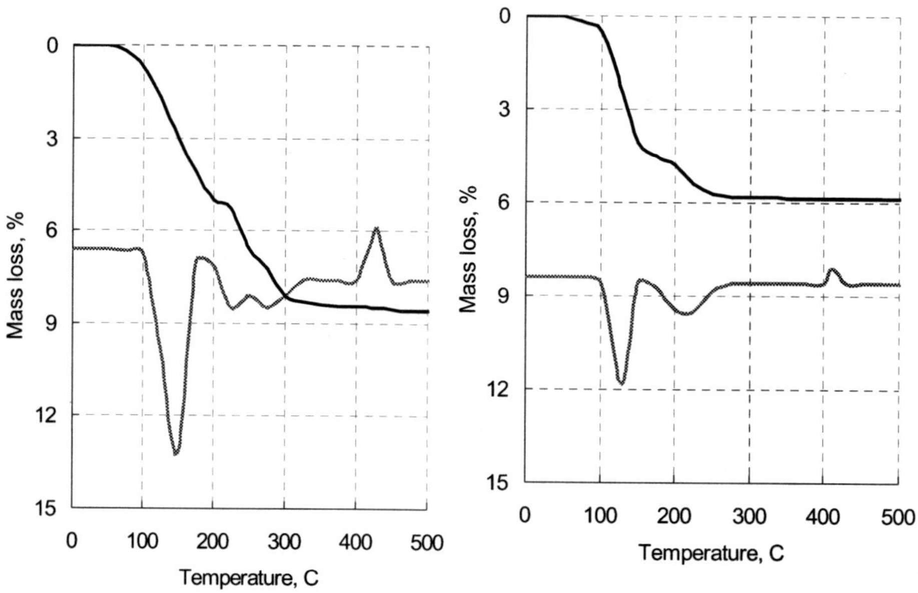

Figure 2. TG (1) and DTA (2) curves for samples 4 (a) and 6 (b).

According to TG and TGA measurements for all obtained samples (Figure 2) two steps of mass loss were observed. The first step in 80-190°C temperature range is an effect of bound water loss:

$$MeO_x(OH)_y \cdot zH_2O \rightarrow MeO_x(OH)_y + zH_2O\uparrow;$$

the second one is hydroxide water formation:

$$MeO_x(OH)_y \rightarrow MeO_x + yH_2O\uparrow.$$

As it can be seen from reaction mechanisms, all water can be removed at up to 300-350°C. The small exothermal area on DTA curve at 400-450°C can be attributed to recrystallization processes after H_2O and OH^- groups removal.

Table 1. Thin-film oxide chemical composition.

Sample #	Main components of electrolyte, g/l	Deposited metal oxide-hydroxide compounds	Metal content, %	OH^- - group content %	Bound water content %	O_2 cont. %
1	C_{CrO3}= 300, C_{HF}= 1.2.	Cr	70.0	6.0	5.0	19.0
2	C_{CrO3}= 300, C_{HF}= 0.8.	Cr	73.0	5.0	4.5	17.5
3	C_{CrO3}= 300, C_{HF}= 0.4.	Cr	82.0	4.5	4.0	9.5
4	$C_{NiSO4\cdot7H2O}$ =120, C_{HF}=10.	Ni	69.6	6.1	5.1	19.2
5	$C_{NiSO4\cdot7H2O}$=120, C_{HF}= 8.	Ni	72.3	5.2	4.6	17.9
6	$C_{NiSO4\cdot7H2O}$ =120, C_{HF}= 6.	Ni	78.4	4.6	4.2	12.8
7	$C_{CoSO4\cdot7H2O}$=250, C_{HF}= 6.	Co	71.0	5.8	5.2	18.0
8	$C_{CoSO4\cdot7H2O}$=250, C_{HF}= 4.	Co	73.5	5.5	5.0	16.0
9	$C_{CoSO4\cdot7H2O}$=250, C_{HF}= 2.	Co	80.4	4.4	3.8	11.7
10	$C_{CoSO4\cdot7H2O}$=10, C_{NH4F}=20.	Co	46	8.5	10.5	36
11	$C_{CoSO4\cdot7H2O}$=10, C_{NH4F}=25	Co	45	9.0	11	35
12	$C_{CoSO4\cdot7H2O}$=10, C_{NH4F}=30.	Co	44	9.5	12.0	34.5

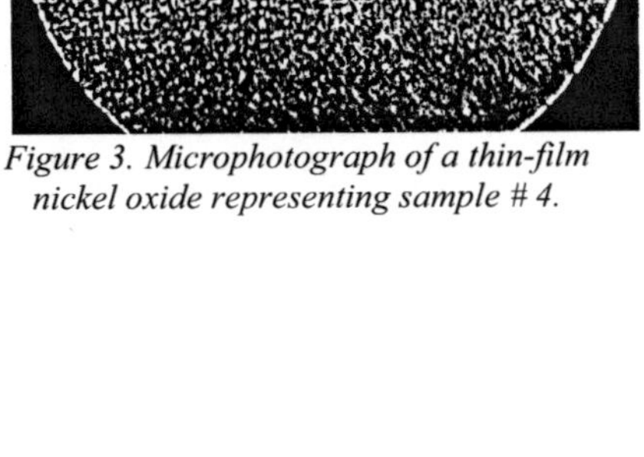

Figure 3. Microphotograph of a thin-film nickel oxide representing sample # 4.

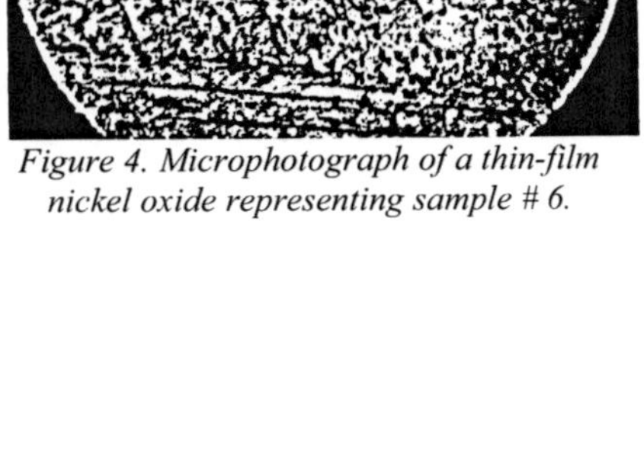

Figure 4. Microphotograph of a thin-film nickel oxide representing sample # 6.

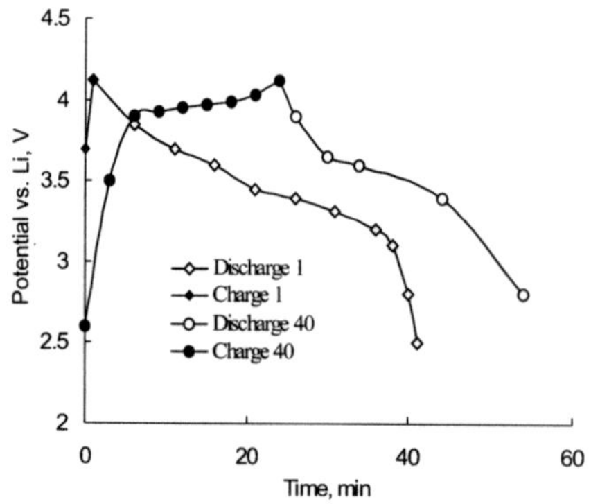

Figure 5. Specific capacities vs. cycle number for samples #4 and #5 in the 2,5-4,2V range.

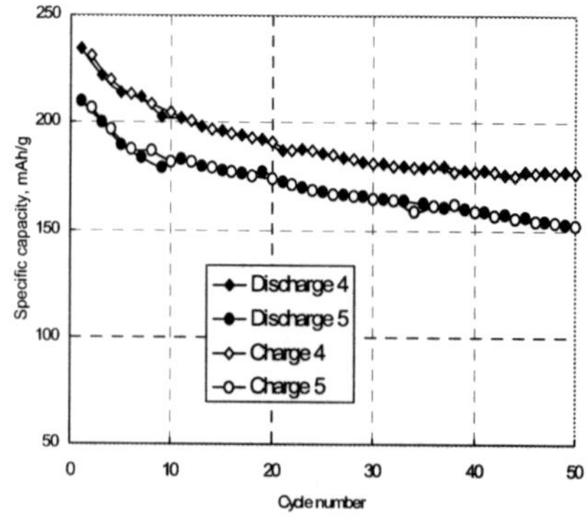

Figure 6. Charge discharge profile (1-st and 40-th cycles) for sample #1. Charge and discharge currents are 100 mA and 80 mA.

The mass losses for all samples are in a good agreement with values of water and hydroxide group content obtained by chemical analysis (Table 1). Thus, for electrochemical testing of Li- intercalation activity each sample was heated at 350°C to remove all types of bound water.

As it can be seen (Figures 3 and 4), film morphology is strongly dependent on deposition conditions. The film with lower oxygen content has larger crystallites and defects. The electrochemical activity for lithium intercalation for samples obtained from fluorine-containing electrolyte with higher F^- content is much higher. The electrochemical data are summarised in Figures 5 and 6, Table 2.

Table 2. Initial electrochemical capacity and cycle efficiency for different samples.

Sample #	Specific capacity mAh/g, first discharge	C_d/C_{ch}, second cycle
1	205	0.971
2	191	0.964
4	234	0.985
5	210	0.979
7	208	0.966
8	188	0.962

It is shown that the first step in the electrochemical reduction of cathodic material $Co_xO_y(OH)_z$ in power sources is controlled by kinetics. This process at the interface depends on the catalytic activity of compound, Figure 7 (a) the reduction of the bulk of cathodic materials is controlled by diffusion Figure 7 (b).

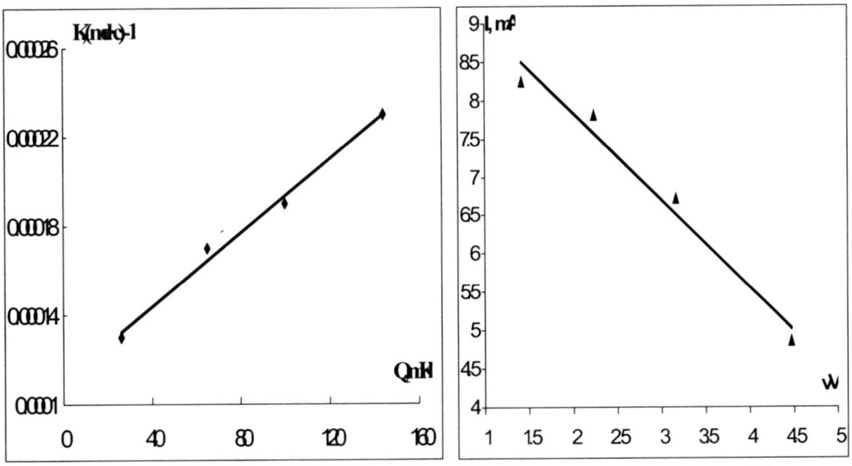

(a) *(b)*

Figure 7. Dependence of charge of the catalytic activity of cobalt oxide-hydroxide (a), I-√V dependence of cobalt oxide-hydroxide electroreduction (b).

4. CONCLUSIONS

We managed to obtain dense and solid thin films of 3d-metal oxides using the techniques of electrochemical deposition from aqueous fluorine-containing electrolytes. The films have been studied as a possible cathode material for secondary cells. The best samples show good cycle retention and acceptable specific capacity in the range of 180 mAh/g. They also feature a plateau of electrochemical potential at approximately 3,5 V, which is acceptable for present industrially produced electrochemical devices.

REFERENCES

1. J. O. Besenhard, M. Schwake, 4th Int. Meet. Li batteries. Progr. And Extended Abstr. May 24-27 (1988).
2. J. V. Goodenough 11th Int. Meet. Li batteries. Munchen (1992).
3. E. A. Nignikovskii Electrochemiya, 34, 7, 722 (1998).
4. I.A. Kedrinsky, V.E.Dmitrenko, I.I.Grydyanov. Li batteries. M.:Energoizdat 1992. 240p.
5. N.D.Ivanova, S.V.Ivanov. Russian Chemical Reviews 62(10) 907-918 (1993).

SYNTHESIS OF MIXED OXIDES USING POLYBASIC CARBOXYLIC HYDROXY- AND AMINO-ACID ROUTES: PROBLEMS AND PROSPECTS

S. A. Kirillov*, I. V. Romanova, and I. A. Farbun

Institute for Sorption and Problems of Endoecology,
13, Gen. Naumov St., 03164 Kyiv, Ukraine

Abstract

Polybasic carboxylic hydroxy and amino acid aided synthetic routes directed towards obtaining mixed inorganic materials, especially for battery and fuel cell applications, are overviewed. It has been shown that, in spite of enormous number of papers on the subject, significant efforts should be undertaken in order to understand the basic principles of these routes. Possible influence of the structure of reactants employed in the process (acids, poly hydroxy alcohols, metal salts) is put forward, and some directions of future work in the field are outlined.

1. INTRODUCTION

Transition-metal oxides and their mixtures are widely employed in numerous industrial applications, especially as cathode materials for batteries and fuel cells [1,2]. Practice poses certain well-known requirements to oxide materials, first of all, to uniformity of the size distribution of particles, to homogeneity of mixed oxides, etc. To meet these demands, two broad categories of methods are now in use, vs (i) mechanical methods and (ii) chemical methods.

In the first group of methods, small particles are produced from larger ones by mechanical treatment. Naturally, it is an uneasy task to obtain homogeneous powders with uniform size distribution by means of comminution, therefore, in modern applications, chemical methods prevail.

* Corresponding author. E-mail: kir@i.kiev.ua

I.V. Barsukov et al. (eds.), New Carbon Based Materials for Electrochemical Energy Storage Systems, 495–504.

The chemical methods can be divided into three types, in accord with three aggregate states of the matter: (i) gas phase reactions, (ii) liquid state reactions, and (iii) solid state reactions.

Gas phase reactions are not so common in producing solid oxides and will not be considered here. As far as liquid state reactions are concerned, these can be performed in aqueous and non-aqueous solvents and can be subdivided into (i) precipitation and co-precipitation, including the so-called template synthesis, (ii) evaporation, and (iii) sol-gel methods.

Precipitation is the most straightforward approach to the preparation of oxides, using hydrous oxides, carbonates, oxalates and other insoluble salts as precursors, with further thermal treatment (decomposition) of precursors. Co-precipitation is often utilised in order to obtain mixed oxides or to create tunnels or layers in the mother structure by inserting guest (template) ions. Depending on future applications, templates may be removed from the tunnel or inter-layer space by means of water (or acid) treatment. Precipitation (co-precipitation) can also be performed by means of complete evaporation of a solvent. Quite close to this method is the so-called freeze-drying, where the solution in the form of aerosol is sprayed into a coolant agent (e.g., liquid nitrogen). Another approach based on evaporation is spray drying where a solution is sprayed into a hot chamber.

In precipitation reactions, powder characteristics depend on the speed of the nucleation of particles and their growth due to the mass flow to the surfaces. In freeze drying and spray drying, powder characteristics primarily depend on the size of droplets, which in turn is determined by the parameters of a nozzle and characteristics of the flow of a carrier gas. Both these methods enable one to obtain powders with very high surface area.

Solid state reactions are also very common in producing oxide materials and are based on thermal treatment of solid oxides, hydroxides and metal salts (carbonates, oxalates, nitrates, sulphates, acetates, etc.) which decompose and react forming target products and evolving gaseous products. Solid-state chemistry states that, like in the case of precipitation, powder characteristics depend on the speed of the nucleation of particles and their growth; however, these processes in solids are much slower than in liquids.

There are numerous advantages and shortcomings of liquid state and solid-state reactions. Briefly speaking, solid-state routes are often unacceptable since it is notoriously difficult to achieve proper homogeneity of materials. Liquid state routes are much more flexible and enable one to find optimal precipitating agent and/or explicitly employ principles of chemical equilibria (hydrolysis, complex formation, etc.). On the other hand, some methods of liquid state chemistry, like freeze drying and spray drying, require the use of quite expensive equipment and therefore cannot be utilised on industrial scale.

2. SOL-GEL (PECHINI) METHODS

During past decades, a relatively new method attracted much of attention of experimentalists, namely, the sol-gel method widely implemented in numerous scientific and industrial applications. Actually, it caused a breakthrough in material science: it is worth mentioning that sol-gel routes are crucial for synthesis of nanomaterials, and one of methods of the 21st century; namely, commercial nanoporous glasses are readily available on the market just thanks to sol-gel technologies.

In the sol-gel process, an oxide network is obtained via inorganic polymerization reactions in solution. It offers many advantages if compared with conventional methods, first of all, a better control over morphology and texture of materials. Initially, it was considered that sol-gel routes are applicable to systems like silica, titania and zirconia, where hydrosols and hydrogels are formed due to the ability of hydrous oxides to set up a network where positively charged ions are linked by oxygen bridges. Later on, similar methods have been extended to almost all metal oxides: it has been found that a network could be created by hydrogen bonds formed by anions of polybasic carboxylic hydroxy and amino acids, and cations are presumably held in this network due to complex formation.

An idea to use polybasic hydroxy carboxylic acids in syntheses of oxides goes back to Pechini [3]. Evaporating solutions of metal salts in citric acid at presence of ethylene glycol he obtained a polymeric resin as a precursor of target oxides. Then this process was extensively used to manufacture various ceramic oxide powders in several publications [4-8].

One of the modern descriptions of the routine referred to as Pechini method reads as follows [5]. Citric acid forms chelates with the metal cations. These chelates undergo esterification, when heated with a poly hydroxy alcohol, such as ethylene glycol, at a temperature of 110-150°C to form a polymeric precursor resin. The cations are expected to be dispersed uniformly throughout the polymeric resin. Additional heating of the resin in air (at 400°C) results in the removal of organics and the formation of an amorphous product (sometimes called a char) with a controlled cation stoichiometry, little cation segregation, and high porosity. Sintering of the amorphous product at higher temperatures leads to the formation of oxide ceramics.

Other descriptions of Pechini-type syntheses presented by various authors are quite similar, and the only difference touches upon the definition of precursors. Some authors prefer to call them "polymeric resins" (see, for instance, [6, 8, 9]), although the terms "sol" and "gel" can be found with even higher probability [4,10-18].

Summarising, the following distinctive features of Pechini-type syntheses should be mentioned:

(1) Dissolution of metal salts in the aqueous solution of polybasic hydroxy (or amino) carboxylic acids (and poly hydroxy alcohols);
(2) Heating (evaporation) of the solution obtained and formation of a polymeric resin or a sol (gel) as a precursor;
(3) Decomposition of the precursor to obtain an oxide powder.

Single and mixed oxides synthesised using these procedures are widely probed as cathode materials, superionics, magnetic materials, catalysts, and other multipurpose materials [4-20].

In what follows, we try to shed the light on some mysterious features of Pechini-type methods, to overview major drawbacks in synthetic routines described in the past decade and put forward some problems to be solved in order to design advanced oxide materials in an optimal way.

3. UNSOLVED PROBLEMS OF PECHINI-TYPE ROUTES

3.1. Role of Polybasic Carboxylic Hydroxy (amino) Acids and Poly Hydroxy Alcohols

It should be emphasized that citric acid is not the only possible acid employed in Pechini-type syntheses. Other polybasic carboxylic hydroxy acids (malic, tartaric, hydroxyglutaric, etc.) and polybasic carboxylic acids (e.g., succinic) have been probed in Refs. [4, 13-16]. As far as amino acids are concerned, glycine seems to remain the only representative [13, 14]. However, the choice of each particular organic acid has never been justified, and no comparative studies are performed in order to find possible dependencies of the process (ability to form a sol, a gel, or a resin, easiness of thermal decomposition of precursors, etc.) on the steric factors, specifically, on the number of hydroxy and carboxylic groups in the molecule of an acid, as well as on the length of its carbon skeleton.

It is for long known that all aforementioned acids form stable complex (chelate) compounds with numerous, especially transition metal cations (see, e.g., Refs. [21, 22]). This fact is pointed out in practically all papers on the subject. However, the composition of complex compounds, their stability, the type of bonding between metal cations and carboxylic anions on various stages of the synthesis, and possible role of steric factors in these phenomena are discussed in a very small number of publications.

In [20], the composition of the citrate precursor of $CoFe_2O_4$ is proposed as $Co_3Fe_6O_4(C_6H_6O_7)_8 \cdot 6H_2O$, i.e., two protons are detached from each molecule of citric acid, and the complex compound could be classified as an acidic salt. Distinct signatures of complex formation are obtained by means of infrared (IR) spectroscopy and nuclear magnetic resonance (NMR) for citrate complexes of iron and yttrium, potential precursors of $YFeO_4$ and

$Y_3Fe_5O_{12}$ [7]. These complexes have been obtained at 110-120°C, isolated, and studied in the crystalline form and in an aqueous solution. IR spectra suggest that these are monodentate coordinated to the metal ions. NMR data are in accord with an assumption that hydroxy and carboxylate groups attached to the central carbon atom of the citric acid anion are coordinated to the metal cations forming a stable five-membered ring. However, these important data are obtained at conditions that do not correspond to the conditions of the synthesis of precursors.

From the point of view of complex formation, homogeneity of mixed oxides accessible by means of Pechini-type routines could be a result of the emergence of mixed complex compounds where two different cations are bonded to one and the same anion. Such a conjecture formulated in Ref. [7] yet has no definite proofs, and no attempts have been made in order to quantify the very role of complex formation in Pechini-type syntheses.

Furthermore, the role of a poly hydroxy alcohol, like ethylene glycol, seems ambiguous. As mentioned above, it is believed that the presence of ethylene glycol favours esterification of chelates. IR and NMR studies performed in [7] do not present solid proofs of such a belief. Synthetic routines with and without alcohols look the same, and the absence of alcohol seem not to influence the properties of precursors and final products. Some evidence exists enabling one to consider the esterification idea liable to more than one interpretation: it has been reported in [4] that the presence of ethylene glycol does not influence the morphology of oxides.

Accounting for these facts one should not be surprised coming across entirely different recommendations regarding the metal-acid-ethylene glycol ratio, which varies from 1-1.88-1.25 [5] or 1-4-1 [8] to 1-4.25-17.5 [7]. It is even sometimes noticed that a carboxylic acid and a poly hydroxy alcohol simply play the role of a "fuel" to burn a precursor and to form target oxides [16]. To shed some light on the aforementioned problems, the following steps could be of some help.

The influence of the structure of the acid could be clarified if extensive comparative studies of homologous acids would be performed in Pechini-type processes involving the same cations. From this point of view, the comparison of related acids [like malonic, $HOOC{-}CH_2{-}COOH$, succinic, $HOOC{-}CH_2{-}CH_2{-}COOH$, and glutaric, $HOOC{-}CH_2{-}CH_2{-}CH_2{-}COOH$), as well as acids and hydroxy acids [like succinic, $HOOC{-}CH_2{-}CH_2{-}COOH$, and hydroxysuccinic (malic), $HOOC{-}CH(OH){-}CH_2{-}COOH$] is required.

As far as the role of poly hydroxy alcohols is concerned, comparative studies of products obtained with and without their addition should be made. The use of homologous alcohols (like low cost substituted glycols employed as components of brake liquids, and glycerol) could be helpful if such role would be found significant.

In order to co clarify the role of complex formation, the new data on stability constants should be accumulated, being collected at strictly similar conditions. It should be also mentioned that any analysis of equilibrium in solutions involving anions of polybasic hydroxy carboxylic acids requires the data on the deprotonation constants of the acid in question. This information would be crucial for conclusions regarding the presence and stability of mixed complexes in the system. Valuable knowledge about the structure of complex compounds present in solutions (and in precursors as well, see later) may be gained by means of vibrational spectroscopy (IR and Raman spectra) and nuclear magnetic resonance.

3.2. Physical Nature of the Precursors

As it was mentioned above, there is no unanimous opinion concerning the nature of precursors obtained in Pechini-type syntheses, and these are called polymeric resins (glasses) or gels. Clearly, this contradiction has little to do with technical aspects of the synthesis. On the other hand, its clarification will be of great value for the physics of amorphous state.

In spite of the absence of clear physical evidence, there are little doubts that the early stage of formation of precursors involves creation of a network formed by hydrogen bonds between anions of polybasic carboxylic hydroxy and amino acids. The further direction of this process upon heating could be twofold: either formation of a sol and then a gel, or creation of a glass. To distinguish between these two possible directions, optical measurements, like light scattering, could be helpful. Being colloidal systems, sols and gels are optically inhomogeneous and exhibit significant light scattering (a Tyndall cone), whereas glasses are homogeneous and do not scatter the visible light.

Hydrogen bonded glasses are not so numerous [23], and glycerol is one of the best-known examples. Should the glassy nature of precursors be proven, the list of hydrogen bonded glasses would be considerably enlarged, and the studies of cooperative behaviour of glass forming systems (see, e.g., Refs. [24-28]) would get a new impetus.

3.3. Chemical Nature of Precursors, Their Behavior upon Heating, and Possible Role of Anions of the Metal Salt

One of the most remarkable attributes of Pechini-type routines is the behaviour of precursor resins (gels) upon heating. On the first stage of thermal decomposition of samples, being dried at approx. 120°C, they decompose forming very porous and voluminous sponge-cake-like materials [6]. Probably, just at this temperature, a highly porous intermediate is created, being a prerequisite of the exceptionally porous nature of target oxides. At higher temperatures (up to 400°C) this intermediate experiences further decomposition giving the amorphous oxide. Subsequent heating up to 1000°C is required to crystallise oxides in question and to reduce the content of carbon to ppm levels. On the other hand, the higher the temperature of calcination, the lower is the value of the surface area of the material obtained [5]. No definite influence on the cation of a metal salt taken as a raw material (nitrate [4-8,14], acetate [13-15], citrate [9,20]) on the properties of target oxides have been noticed.

It seems well established that decomposition of precursors takes place in two stages. Unfortunately, little efforts have been undertaken in order to study either the chemical nature of precursors and the decomposition process in greater detail. Say, it is found in Ref. [4] by means of thermal analysis that decomposition of citrate and tartrate precursors obtained from nitrates first involves the removal of water, and then the polymer itself decays. Gas analysis during decomposition shows the presence of H_2O, CO_2, and the products of the thermal degradation of the nitrate ion present in the raw material (N_2O, NO, NO_2). In Ref. [20], the citrate precursor of $CoFe_2O_4$, $Co_3Fe_6O_4(C_6H_6O_7)_8 \cdot 6H_2O$, first decomposes at 120°C forming an instable complex of acetonedicarboxylate acid and evolving H_2O and CO. On further heating to 220-230°C the citrate network breaks down giving the target oxides, CO_2, and acetone. Interestingly, at high heating rate (10°C min^{-1} and above), both processes can appear simultaneously.

All these data, though being quite incomplete, call for other valuable questions. Different carboxylic acids decompose in different ways, not obviously evolving simplest products, like H_2O, CO, and CO_2, hence it is clear that a carboxylic acid anion does not simply serve a fuel for burning. It is therefore tempting to find if an "optimal" carboxylic acid anion exists from the point of view of decomposition processes, and what is the most desirable metal-anion ratio. Further, participation of the anion introduced with the starting material is evident in the burning process. The question arises of what kind of anions favours performing the thermal decomposition in the most optimal way. These studies could be carried out by means of the mentioned methods (thermal analysis, gas chromatography).

4. CONCLUSIONS

In this paper we have tried to present a brief overview of state-of-the-art of the synthetic chemistry of oxide materials based on polybasic carboxylic hydroxy (amino) acid routes. It has been shown that, in spite of enormous number of papers on the subject (we have mentioned just a few latest references, significant efforts should be undertaken in order to understand the basic principles of this important method. Some gaps in our knowledge are mentioned, possible influence of the structure of reactants employed in the process (acids, poly hydroxy alcohols, metal salts) is put forward, and some directions of future work in the field are outlined.

Specifically,

(1) Extensive comparative studies of salts of homologous acids involving the same cations should be performed, aimed at clarification of the influence of the structure of acids on the properties of precursors and target oxides;
(2) Comparative studies of products obtained with and without addition of poly hydroxy alcohols should be made, so as to determine their role in the process;
(3) New data on stability constants accumulated at strictly similar conditions are required to explain the role complex formation;
(4) Valuable knowledge about the structure of complex compounds present in solutions and in precursors could be gained by means of vibrational spectroscopy (IR and Raman spectra) and NMR;
(5) Precursors obtained with various anions should be studied at comparable conditions, and the quality of target products should be compared in order to illuminate the role of anion in thermal decomposition processes.

It should be also mentioned that careful optical determination of the nature of precursors and solving the gel-glass dichotomy would be of great value to the physics of amorphous state.

Unravelling these problems one would be able to make well-grounded predictions and to justify optimal synthetic routes for the directed synthesis of advanced oxide materials.

ACKNOWLEDGEMENTS

This research has been funded, in part, by the EC, within the framework of the STCU Grant #GR85. S. A. Kirillov is grateful to CRDF Travel Grants Program for financial support of his visit to the NATO CARWC.

REFERENCES

1. *Materials for Lithium Ion Batteries*, Ed. by C. Julien and Z. Stoynov, NATO Science Series 3. High Technology, Vol. 85, Kluwer, Dordreht, 2000.
2. Kirillov, S. A., Tsiakaras, P. E., Romanova, I. V., Adsorption and oxidation of methanol and ethanol on the surface of metallic and ceramic catalysts, *J. Mol. Struct.*, **651-653**, 365-370 (2003)

3. Pechini, M., Method of preparing lead and alkaline earth titanates and niobates and coating method using the same form a capacitor, *U.S.Pat. # 3330697*, July 11, 1967.
4. Weidenkaff, A., Ebbinghaus, S. G., Lippert, T., $Ln_{1-x}A_xCoO_3$ (Ln=Er, La; A=Ca, Sr)/carbon nanotube composite materials applied for rechargeable Zn/Air batteries, *Chem. Mater.* **14,** 1797-1805 (2002).
5. Tas, A. C., Majewski, P. J., Aldinger, F., Chemical preparation of pure and strontium- and/or magnesium-doped lanthanum gallate powders, *J. Am. Ceram. Soc.* **83,** 2954-2960 (2000).
6. Alonso, J. A., Martinez-Lopez, M. J., Casais, M. T., Martinez, J. L., Demaseau, G., Largeteau, A., Garcia-Munoz, J. L., Munoz, A., Fernandez-Diaz, M. T., High-pressure preparation, crystal structure, magnetic properties, and phase transitions in $GdNiO_3$ and $DyNiO_3$ perovskites, *Chem. Mater.* **11,** 2463-2469 (1999).
7. Todorovsky, D. S., Dumanova, D. G., Todorovska, R. V., Getsova, M. M., Preparation and characterization of yttrium-iron citric acid complexes, *Croat. Chem. Acta* **75**, 155-164 (2002).
8. Kwon, Y.J., Kim, K.H., Lim, C.S., Shim, K.B., Characterization of ZnO nanopowders synthesized by the polymerized complex method via an organochemical route, *Journ. Ceram. Proc. Res.* **3**, 146-149 (2002).
9. Castro, R.H.R., Hidalgo, P., Mucillo, R., Gouvea, D., Microstructure and structure of $NiO\text{-}SnO_2$ and $Fe_2O_3\text{-}SnO_2$ systems, *Appl. Surf. Sci.* **214**, 172-177 (2003).
10. Pereira-Ramos, J.-P., Electrochemical properties of cathodic materials synthesized by low-temperature techniques, *J. Power Sources* **54**, 120-126 (1995).
11. Franger, S., Bach, S., Pereira-Ramos, J.-P, Baffier, N., Highly rechargeable $Li_xMnO_{2+\delta}$ oxides synthesized via low temperature techniques, *J. Power Sources* **97-98,** 344-348 (2001).
12. Franger, S., Bach, S., Farcy, J., Pereira-Ramos, J.-P, Baffier, N., Synthesis, structural and electrochemical characterization of the sol-gel birnessite $Mn_{1,84}\cdot 0,6H_2O$, *J. Power Sources* **109,** 344-348 (2001).
13. Julien, C., Letranchant, C., Rangan, S., Lemal, M., Ziolkiewicz, S., Castro-Garcia, S., El-Farh, L., Benkaddour, M., Layered $LiNi_{0,5}Co_{0,5}O_2$ cathode materials grown by soft-chemistry via various solution methods, *Mater. Sci. Eng.* **B76,** 145-165 (2000).
14. Julien, C., Letranchant, C., Lemal, M., Ziolkiewicz, S., Castro-Garcia, S., Layered $LiNi_{0,5}Co_{0,5}O_2$ compounds synthesized by a glycine-assisted combustion method for lithium batteries, *J. Mater. Sci.* **37,** 2367-2375 (2002).
15. Castro-Garcia, S., Julien, C., Senaris-Rodriguez, M. A., Structural and electrochemical properties of Li-Ni-Co oxides synthesized by wet chemistry via a succinic-acid-assisted technique, *Int. J. Inorg. Mat.* **3**, 323-329 (2001).
16. Prabahan, S. R. S., Saparil, N. B., Michael, S. S., Massot, M., Julien, C., Soft-chemistry syntethis of electrochemically-active spinel $LiMn_2O_4$ for Li-ion batteries, *Solid State Ionics* **112,** 25-34 (1998).
17. Chitra, S., Kalyani, P., Vebka, B., Mohan, T., Haro-Poniatowski, E., Gangadharan, R., Julien, C., Synthesis, characterization and electrochemical studies of $LiNiVO_4$ cathode material in rechargeable lithium batteries, *Mater. Chem. Phys.* **65,** 32-37 (2000).
18. Julien, C., Camacho-Lopez, M. A., Mohan, T., Chitra, S., Kalyani, P., Gopukumar, S., Combustion synthesis and characterization of substituted lithium cobalt oxides in lithium batteries, *Solid State Ionics* **135**, 241-248 (2000).
19. Prasad, S., Vijayalakshmi, A., Gajbhiye, N.S., Synthesis of ultrafine cobalt ferrite by thermal decomposition of citrate precursor, *J. Therm. Anal. Calorim.* **52,** 5-10 (1998).
20. Kostromina, N. A., Complex compounds of cerium, neodim, and samarium with glyconic acid, *J. Inorg. Chem.* (Russ). **5,** 95-101 (1960).
21. Kostromina, N. A., Comparative stability of glyconate complexes of rare earth metals, *Ukr. Khim. Zhurn.* **26,** 299-304 (1960).
22. *Frontiers in Material Science: Glasses and Amorphous Materials*, Special Issue, *Science* **267,** 1924–1953 (1995).

23. Kirillov, S. A., Spatial disorder and low-frequency Raman pattern of amorphous solid, with special reference to quasi-elastic scattering and its relation to Boson peak, *J. Mol. Struct.* **479,** 279-284 (1999).
24. Kirillov, S. A., Yannopoulos, S. N., Charge-current contribution to low-frequency Raman scattering from glass-forming ionic liquids, *Phys. Rev.* B, **61**, 11391-11399 (2000).
25. Kirillov, S. A., Yannopoulos, S. N., Vibrational dynamics as an indicator of short-time interactions in glass-forming liquids and their possible relation to cooperativity, *J. Chem. Phys.,* **117**, 1220-1230 (2002).
26. Kalampounias, A. G., Yannopoulos, S. N., Steffen, W., Kirillova, L. I., Kirillov, S. A., Short-time dynamics of glass-forming liquids: Phenyl salicylate (salol) in bulk liquid, dilute solution, and confining geometries, *J. Chem. Phys.,* **118**, 8340-8349 (2003).
27. Kirillov, S. A., Novel approaches in spectroscopy of interparticle interactions. Raman line profiles and dynamics in liquids and glasses, *J. Mol. Liq*. **110**, 99-103 (2004).

IMPROVED ELETROCHEMICAL PROPERTIES OF SURFACE-COATED $Li(Ni,Co,Mn)O_2$ CATHODE MATERIAL FOR Li SECONDARY BATTERIES

Sun-Ho Kang* and Khalil Amine

Chemical Engineering Division, Argonne National Laboratory, Argonne, IL 60439, USA

Abstract

The surface of $Li(Ni_{0.4}Co_{0.2}Mn_{0.4})O_2$ was coated with amorphous aluminum oxide using Al-isopropoxide. The effect of the surface coating on the cycling performance and high-temperature storage characteristics was investigated using coin-type cells. The surface-coated materials showed superior capacity retention at 55°C as well as at room temperature and limited impedance rise during high-temperature storage compared with the uncoated material. From the a.c. impedance measurement, it was found that the uncoated material exhibited very large increase in the cathode/electrolyte interfacial impedance during cycling whereas the surface-coated materials showed a limited increase in the interfacial impedance.

Keywords

Li-ion secondary batteries; $Li(Ni,Co,Mn)O_2$; surface coating; interfacial impedance.

1. INTRODUCTION

High-power lithium-ion batteries are promising alternatives to the nickel metal hydride batteries which are currently used for energy storage in hybrid electric vehicles (HEVs). Currently, $Li(Ni,Co)O_2$-based materials are the most widely studied cathode materials for the high-power lithium-ion batteries [1-4]. Although $Li(Ni,Co)O_2$-based materials meet the initial power requirement for the HEV application, however, it has been reported that they

* Corresponding author. E-mail: kangs@cmt.anl.gov

I.V. Barsukov et al. (eds.), New Carbon Based Materials for Electrochemical Energy Storage Systems, 505–512.

have major concerns related with calendar life, safety, and cost [1,4,5,6]. To address these concerns, we have been investigating the applicability of $Li(Ni_xCo_{1-2x}Mn_x)O_2$ ($1/3\leq x <1/2$) materials to the high-power applications due to their cost advantage and improved thermal safety compared with the $Li(Ni,Co)O_2$ materials [7-9]. In this article, we report the effect of surface modification on the electrochemical properties of the $Li(Ni_xCo_{1-2x}Mn_x)O_2$ cathode materials. Recently, a lot of research results have been reported on the improvement of electrochemical properties of cathode materials by modifying their surface with various metal oxides such as Al_2O_3, MgO, SiO_2, ZnO, and ZrO_2 [10-18]. In this work, we coated the surface of $Li(Ni_{0.4}Co_{0.2}Mn_{0.4})O_2$ particles with amorphous aluminum oxide using Al-isopropoxide and investigated its impact on the cycling performance, impedance, and high-temperature storage characteristics.

2. EXPERIMENTAL

2.1. Materials

$Li(Ni_{0.4}Co_{0.2}Mn_{0.2})O_2$ powder was prepared by the solid state reaction method using Li_2CO_3, $Ni(OH)_2$, $Co(OH)_2$, and $MnCO_3$ as starting materials. An appropriate amount of each starting powder was mixed using zirconia balls in acetone. The mixed powder was calcined at 700°C for 30 h and then at 1100°C for 5 h in air. The synthesized $Li(Ni_{0.4}Co_{0.2}Mn_{0.2})O_2$ powder was well dispersed in an ethanol solution of Al-isopropoxide (AIP, $[(CH_3)_2CHO]_3Al$), dried at 100°C, and then heat-treated at 300°C for 3hr in air. The amounts of AIP used were 0.5 and 1.0 wt%. Phase purity of the prepared materials was examined with a powder X-ray diffractometer using Cu Kα radiation.

2.2. Characterization

The electrochemical properties of the prepared materials were evaluated using coin-type cells. The positive electrode consisted of 80 wt% oxide powder, 10 wt% carbon, and 10 wt% polyvinylidene difluoride (PVDF) binder on aluminum foil. The negative electrode was either metallic lithium or carbon on copper foil. The electrolyte was 1 M $LiPF_6$ in a 1:1 mixture of ethylene carbonate (EC)/diethyl carbonate (DEC). The coin cells were galvanostatically cycled at 2.8~4.3 V. The a.c. impedance of the coin-type cells was measured in the frequency range of 10mHz-100KHz using an impedance analyzer (BAS-ZAHNER IM6).

The high-temperature storage characteristic was evaluated as follows. After initial formation cycles, the C/Li($Ni_{0.4}Co_{0.2}Mn_{0.2}$)O_2 cells were charged to 3.8V and stored at 50°C; every couple of days, the cells were cooled to room temperature and the impedance of the cells was measured. By repeating the above steps, the change of the cell impedance with the 50°C-storage time was obtained.

3. RESULTS AND DISCUSSION

3.1. Crystal Structure

Figure 1 shows the powder X-ray diffraction (XRD) pattern of the as-prepared Li($Ni_{0.4}Co_{0.2}Mn_{0.4}$)O_2 material. All of the peaks could be indexed based on the α-$NaFeO_2$ structure ($R\bar{3}m$). The lattice parameters in hexagonal setting obtained by the least square method were a=2.86$\underline{8}$Å and c=14.2$\underline{5}$Å. Since no second-phase diffraction peaks were observed from the surface-coated materials and it is unlikely that the Al ions were incorporated into the lattice at the low heat-treatment temperature (300°C), it is considered that the particle surface was coated with amorphous aluminum oxide.

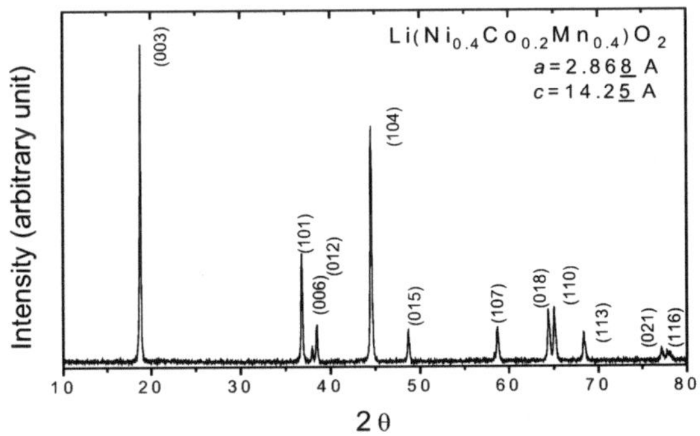

Figure 1. The power XRD pattern of the Li($Ni_{0.4}Co_{0.2}Mn_{0.4}$)O_2 material prepared at 1100°C for 5hr in air.

3.2. Cycling Performance

Cycling performances of the uncoated and coated cathode materials cycled versus Li at room temperature and 55°C are given in Figs. 2(a) and 2(b), respectively. It can be seen that the coated materials exhibit cycling performance superior to the uncoated one. At room temperature, the uncoated material delivered only 27% of the initial capacity at the 100th cycle whereas the materials coated with 0.5 and 1.0 wt% AIP retained 71 and 75% of their initial capacities, respectively, at the 100th cycle. We investigated the XRD pattern of each cathode after 100 cycles; neither impurity peaks nor changes in peak breadth and $I_{(003)}/I_{(104)}$ ratio were observed in all of the cycled cathodes. This observation indicates that the significant capacity fading of the uncoated material was not due to the degradation of the sample bulk such as phase segregation, amorphization and cation mixing during cycling.

3.3. High-temperature storage characteristic

In Figure 3 are given the variations of the area specific impedance (ASI) of the $C/Li(Ni_{0.4}Co_{0.2}Mn_{0.2})O_2$ cells as a function of 50°C-storage time. The ASI was determined by $A \cdot \Delta V/I$, where A is the cross-sectional area, ΔV the voltage variation during current interruption for 30 s at each state of charge (SOC), and I the current applied during the galvanostatic cycling. The data shown in Figure 3 are the ASI values at 60% SOC. Initially, all of the materials showed similar ASI values (35-40 Ω-cm^2). However, the uncoated material showed very large increase in the ASI; the ASI after storage at 50°C for 90 days increased to 160Ω-cm^2. On the contrary, the coated materials exhibited limited increase in the ASI with storage time compared with the uncoated one.

3.4. Impedance

As can be seen in Figure 3, the surface coating improved the high-temperature storage characteristic so that the surface-coated materials possessed enhanced calendar life. The variation of the ASI during room temperature cycling was also found to be significantly suppressed by the surface coating. The ASI is the overall impedance from various impedance sources in a cell that are related with cathode(anode) /electrolyte interfacial overpotential, electrolyte resistance, and Li-diffusion kinetics in the oxides. To distinguish the major source responsible for the very large impedance rise of the cell with uncoated material, we measured the a.c. impedances of the coin-type cells. Figure 4 shows the variations of the a.c. impedance spectra during cycling. The impedances were measured at the open cell voltage of 3.7V at each cycling number, which corresponds to ~60% SOC [6].

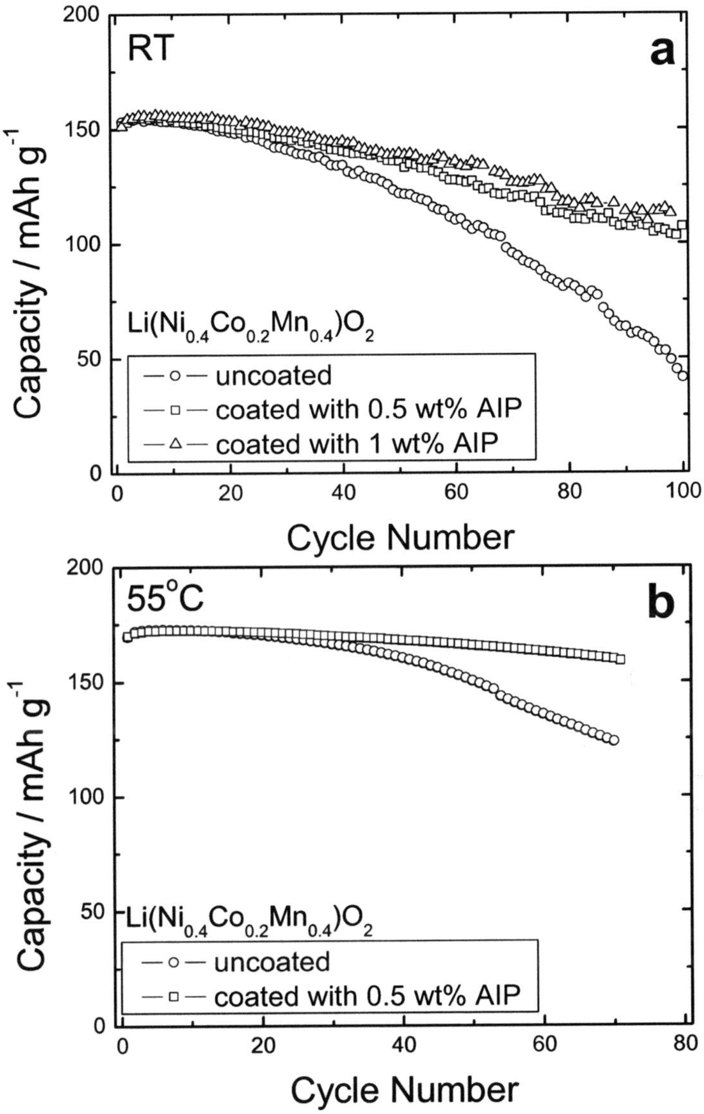

Figure 2. Change of discharge capacity with cycle number of $Li/Li(Ni_{0.4}Co_{0.2}Mn_{0.4})O_2$ cells in the voltage range of 2.8-4.3V at 0.2mA/cm^2. (a) at room temperature; (b) at 55°C.

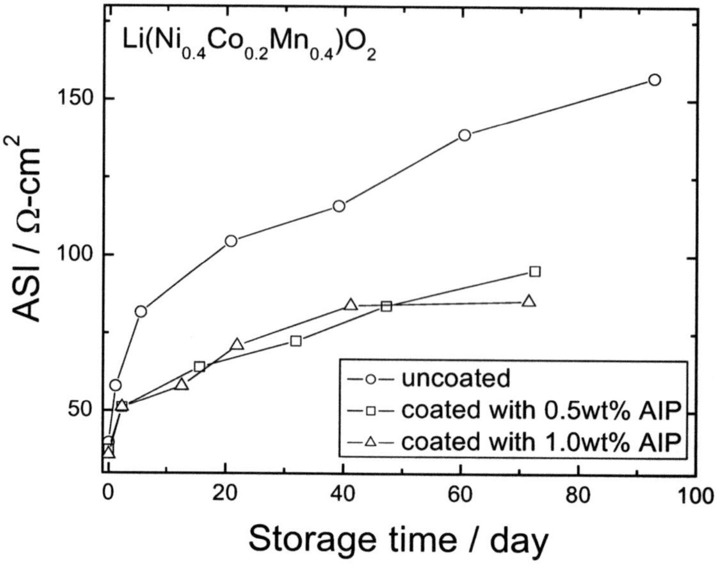

Figure 3. Changes of the ASI at 60% SOC with storage time at 50°C.

It can be seen in Figure 4(a) that the uncoated material showed significant increase in diameter of the semicircle at medium frequency, which is generally attributed to Li-ion transport and charge-transfer kinetics at the cathode/electrolyte interface Consequently, the very large increase of the ASI during the 50°C-storage and galvanostatic cycling of the uncoated material was caused mainly by the impedance rise at the oxide particle surface/electrolyte interface.

The experimental results in Figures 2-4 clearly show that the surface phase suppresses the degradation of the cathode/electrolyte interface, resulting in improved capacity retention and limited impedance rise. At this stage, the detailed mechanism for the role of the surface phase is not clear. One possible scenario is the role of the surface aluminum oxide as an acid scavenger. When water molecules are present as an impurity in the electrolyte, they react with $LiPF_6$ salt to produce HF. The generated HF attacks the cathode surface, which results in destruction of the SEI layer and dissolution of transition metal ions into the electrolyte. We speculate that the aluminum oxide protects the surface from the acid attack by reacting with the acid to form another compound such as AlF. A similar mechanism was also proposed by Zheng *et al.* [14] and Sun *et al.* [15] for SiO_2- and ZnO-coated spinel, respectively.

4. SUMMARY

The surface of $Li(Ni_{0.4}Co_{0.2}Mn_{0.4})O_2$ was coated with amorphous aluminum oxide using Al-isopropoxide. The as-prepared and coated materials exhibited almost no differences in initial capacity and impedance. However, during the extended cycling, the uncoated material exhibited significant capacity fading.

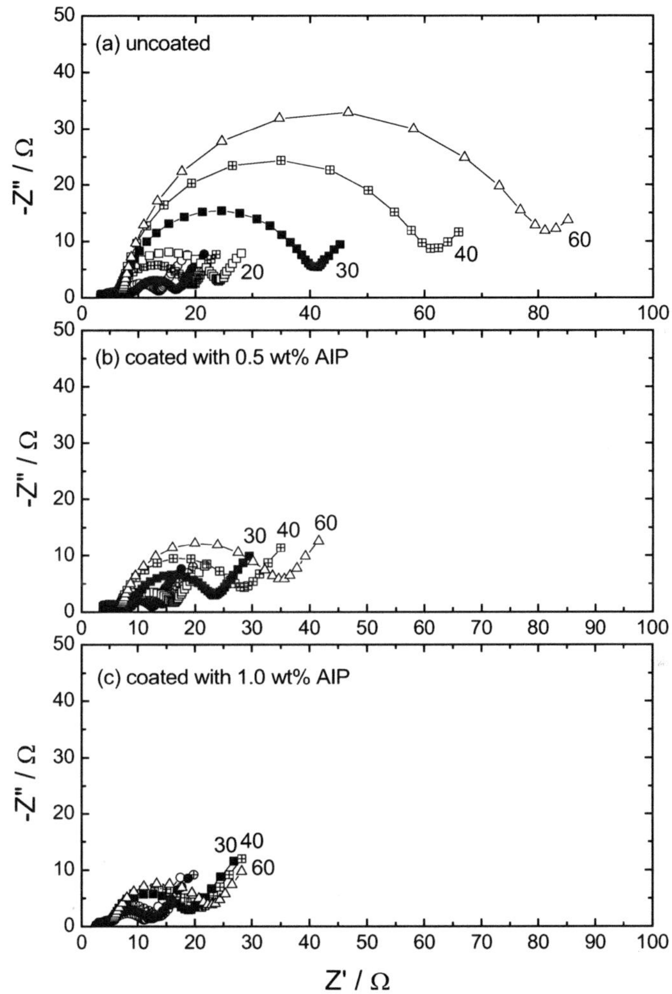

Figure 4. Variations of the a.c. impedance with cycling measured with $C/Li(Ni_{0.4}Co_{0.2}Mn_{0.4})O_2$ cells at the open cell voltage of 3.7V.

On the other hand, the coated cathodes showed superior cycling performance. The surface-coated materials also exhibited better high-temperature storage characteristics than the uncoated material. From the variations of the a.c. impedance with cycling, we observed that the cell with the uncoated cathode showed a significant increase in cathode/electrolyte interfacial impedance during cycling. In other words, the poor performance of the uncoated material was caused by the significant degradation of the cathode surface in contact with the electrolyte, and the surface-coated phase stabilized the cathode/electrolyte interface. We speculate that the surface aluminum oxide reacts with HF present in the electrolyte so that the cathode surface is protected from the acid attack. Further studies are needed to explore the specific role of the surface phase in improving the electrochemical properties of the cathode material. In summary, the surface coating of cathode materials is a simple and effective way to improve cycle life as well as calendar life of the Li-ion cells.

ACKNOWLEDGMENTS

This paper has been created by the University of Chicago as Operator of Argonne National Laboratory under contract No. W-31-109-Eng-38 with the U. S. Department of Energy. This work was supported by the U.S. Department of Energy, FreedomCAR and Vehicle Technologies Program, under contract No. W-31-109-ENG-38. The authors also wish to thank Dr. S. –Y. Yoon (Quallion LLC, Saugus, CA) and Dr. H. –J. Kwon (Samsung SDI, South Korea) for helpful discussions.

REFERENCES

1. K. Amine et al., J. Power Sources 2001; 97: 684.
2. M. Balasubramanian, X. Sun, X. Q. Yang, J. McBreen, J. Power Sources 2001; 92: 1.
3. X. Zhang et al., J. Electrochem. Soc. 2001; 148: A463.
4. D. P. Abraham et al., Electrochem. Commun. 2002; 4: 620.
5. A. M. Andersson et al., J. Electrochem. Soc. 2002; 149: A1358.
6. D. P. Abraham et al., J. Power Sources 2003; 119: 511.
7. S. –H. Kang et al., J. Power Sources 2002; 112: 41.
8. S. –H. Kang, K. Amine, J. Power Sources 2003; 119: 150.
9. I. Belharouak, Y. –K. Sun, J. Liu, K. Amine, J. Power Sources 2003; 123: 247.
10. J. Cho, Y. J. Kim, B. Park, Chem. Mater. 2000; 12: 3788.
11. L. Liu, Z. Wang, H. Li, L. Chen, X. Huang, Solid State Ionics 2002; 152: 341.
12. H. –J. Kwon, D. G. Park, Electrochem. Solid-State Lett. 2000; 3: 128.
13. Z. Wang et al., J. Electrochem. Soc. 2002; 149: A466.
14. Z. Zheng et al., Solid State Ionics 2002; 148: 317.
15. Y. –K. Sun, K. –J. Hong, J. Prakash, K. Amine, Electrochem. Commun. 2002; 4: 344.
16. Z. Chen, J. R. Dahn, Electrochem. Solid-State Lett. 2002; 5: A213.
17. Y. J. Kim, J. Cho, T. –J. Kim, B. Park, J. Electrochem. Soc. 2003; 150: A1723.
18. M. M. Thackeray et al., Electrochem. Commun. 2003; 5: 752.

INDEX

A

B

C

D

E

N

O

P

Q

R

S

T

U

V

W

X

Z

Printed in the United States
54151LVS00001B/34-48